W9-BCC-276

Mathematical Software

ACM MONOGRAPH SERIES

*Published under the auspices of the Association for
Computing Machinery Inc.*

Editor ROBERT L. ASHENHURST *The University of Chicago*

A. FINERMAN (Ed.)   University Education in Computing Science, 1968
A. GINZBURG   Algebraic Theory of Automata, 1968
E. F. CODD   Cellular Automata, 1968
G. ERNST AND A. NEWELL   GPS: A Case Study in Generality and
   Problem Solving, 1969
M. A. GAVRILOV AND A. D. ZAKREVSKII (Eds.)   LYaPAS: A Programming
   Language for Logic and Coding Algorithms, 1969
THEODOR D. STERLING, EDGAR A. BERING, JR., SEYMOUR V. POLLACK,
   AND HERBERT VAUGHAN, JR. (Eds.)   Visual Prosthesis:
   The Interdisciplinary Dialogue, 1971
JOHN R. RICE (Ed.)   Mathematical Software, 1971

*Previously published and available from The Macmillan Company,
New York City*
G. SEBESTYN   Decision Making Processes in Pattern Recognition, 1963
M. YOVITS (Ed.)   Large Capacity Memory Techniques for Computing Sys-
   tems, 1962
V. KRYLOV   Approximate Calculation of Integrals (Translated by A. H.
   Stroud), 1962

MATHEMATICAL SOFTWARE

Edited by *JOHN R. RICE*

DEPARTMENTS OF MATHEMATICS
AND COMPUTER SCIENCE
PURDUE UNIVERSITY
LAFAYETTE, INDIANA

*Based on the proceedings of the
Mathematical Software Symposium
held at Purdue University, Lafayette,
Indiana, April 1–3, 1970*

 1971

ACADEMIC PRESS New York and London

ACADEMIC PRESS, INC.
111 Fifth Avenue, New York, New York 10003

United Kingdom Edition published by
ACADEMIC PRESS, INC. (LONDON) LTD.
Berkeley Square House, London W1X 6BA

LIBRARY OF CONGRESS CATALOG CARD NUMBER: 78-169082

AMS (MOS) 1970 Subject Classification: 68-02, 65-02, 08-02,
68-03, 68A10

PRINTED IN THE UNITED STATES OF AMERICA

CONTENTS

PART ONE   PROLOGUE

Chapter 1.   Historical Notes

John R. Rice

Chapter 2.   The Distribution and Sources of Mathematical Software

John R. Rice

Chapter 3.   The Challenge for Mathematical Software

John R. Rice

Chapter 4. **Discussion of the Papers**

       John R. Rice

PART TWO PROCEEDINGS OF THE SYMPOSIUM

Chapter 5. **The Papers**

    5.1 A USER'S EXPERIENCE WITH SOPHISTICATED LEAST-SQUARES SOFTWARE IN THE DISCOVERY OF THE LUNAR MASS CONCENTRATIONS (MASCONS)

       P. M. Muller

    5.2 USER MODIFIABLE SOFTWARE

       Robert C. Bushnell

    5.3 NUMBER REPRESENTATION AND SIGNIFICANCE MONITORING

       R. L. Ashenhurst

PART THREE SELECTED MATHEMATICAL SOFTWARE

Chapter 6. Self-Contained Power Routines

N. W. Clark, W. J. Cody, and H. Kuki

Chapter 7. CADRE: An Algorithm for Numerical Quadrature

Carl de Boor

LIST OF CONTRIBUTORS

Numbers in parentheses indicate the pages on which the authors' contributions begin.

R. L. ASHENHURST (67), The University of Chicago, Chicago, Illinois

D. BARTON (369), Cambridge University, Cambridge, England

E. L. BATTISTE* (121), IBM Corporation, Houston, Texas

R. BAYER† (275), Boeing Scientific Research Laboratories, Seattle, Washington

CARL DE BOOR (201, 417), Purdue University, Lafayette, Indiana

K. M. BROWN‡ (391), Cornell University, Ithaca, New York

ROBERT C. BUSHNELL (59), Ohio State University, Columbus, Ohio

N. W. CLARK (399), Argonne National Laboratory, Argonne, Illinois

W. J. CODY (171, 399), Argonne National Laboratory, Argonne, Illinois

J. E. DENNIS§ (391), Cornell University, Ithaca, New York

A. W. DICKINSON (141), Monsanto Company, St. Louis, Missouri

C. B. DUNHAM (105), University of Western Ontario, London, Ontario

P. A. FOX (477), Newark College of Engineering, Newark, New Jersey

C. W. GEAR¶ (211), Stanford University, Stanford, California

M. GOLDSTEIN (93), Courant Institute of Mathematical Sciences, New York, New York

---

* Present address: International Mathematical Statistical Libraries, Inc., Houston, Texas
† Present address: Purdue University, Lafayette, Indiana
‡ Present address: Yale University, New Haven, Connecticut
§ Present address: Yale University, New Haven, Connecticut
¶ Present address: University of Illinois, Urbana, Illinois

V. P. HERBERT (141), Monsanto Company, St. Louis, Missouri

S. HOFFBERG (93), Vassar College, Poughkeepsie, New York

O. G. JOHNSON (357), International Mathematical and Statistical Libraries, Inc., Houston, Texas

D. K. KAHANER (229), Los Alamos Scientific Laboratory, University of California, Los Alamos, New Mexico

H. KUKI (187, 399), The University of Chicago, Chicago, Illinois

C. L. LAWSON (347), Jet Propulsion Laboratory, Pasadena, California

T. G. LEWIS (331), Washington State University, Pullman, Washington

P. M. MULLER (51), Jet Propulsion Laboratory, Pasadena, California

A. C. R. NEWBERY* (153), The Boeing Company, Renton, Washington

B. N. PARLETT (357), University of California, Berkeley, California

A. C. PAULS (141), Monsanto Company, St. Louis, Missouri

W. H. PAYNE (331), Washington State University, Pullman, Washington

JOHN R. RICE (3, 13, 27, 43, 451), Purdue University, Lafayette, Indiana

E. M. ROSEN (141), Monsanto Company, St. Louis, Missouri

J. E. SAMMET (295), IBM Corporation, Cambridge, Massachusetts

L. R. SYMES† (261), Purdue University, Lafayette, Indiana

H. C. THACHER, Jr. (113), University of Notre Dame, Notre Dame, Indiana

J. F. TRAUB‡ (131), Bell Telephone Laboratories, Inc., Murray Hill, New Jersey

I. M. WILLERS (369), Cambridge University, Cambridge, England

R. V. M. ZAHAR (369), Cambridge University, Cambridge, England

* Present address: University of Kentucky, Lexington, Kentucky
† Present address: University of Saskatchewan, Canada
‡ Present address: Carnegie-Mellon University, Pittsburgh, Pennsylvania

PREFACE

Mathematical software is the set of algorithms in the area of mathematics. Its exact scope is not well defined; for example, we might include any algorithms that result from the creation of a mathematical model of nature and an analysis of that model. The scope defined by this book is more narrow and includes only topics of a definite mathematical nature or origin. Even so, the scope is much broader than the pure mathematician's view of mathematics, for it includes things such as programming languages and computer systems for mathematics. The scope is also broader than traditional numerical analysis for it includes clearly mathematical areas such as symbolic analysis, statistics, and linear programming.

The Mathematical Software Symposium was held at Purdue University on April 1–3, 1970. Its primary objective was to focus attention on this area and to provide a starting point and meeting place for a more systematic development. The editor believes that mathematical software will develop into an important subdiscipline of the mathematics–computer science area and that it will make a significant impact on both theoretical developments and practical applications.

This book has three distinct parts, the second (and major) part contains the proceedings of the symposium. The first part serves as a prologue and establishes a context for the symposium and its papers. Chapter 1 gives some historical background in the form of a chronological list of events that trace the development of computing in general and mathematical software in particular. Chapter 2 discusses the present situation with emphasis on where and how mathematical software is being created and how it is being disseminated to the eventual consumers. A number of important shortcomings are identified. The third chapter considers the future of mathematical software, but not in the sense of making predictions. Two challenges face mathematical software. One of these is the technical challenge of creating quality mathematical software; various facets of this challenge are presented. The scientific public is still unaware of this challenge and regards this process as mechanical, even if tedious and time consuming. The problem is primarily due to lack of a means of discrimination, and one might make an analogy with short story writing in that they are unable to distinguish between the efforts of a college freshman and of O'Henry, Chekhov, Faulkner, or Hemingway. It is hard to make a case that the O'Henrys of mathematical software have yet appeared, but there are multitudes of college freshmen and a few people who are much, much better. The second challenge is sociological in that mathematical software is not yet a viable scientific community or entity. Some possible reasons and remedies for this are discussed.

A capsule summary and discussion of the twenty-three technical papers is given in Chapter 4 and is not repeated here. Suffice it to say that they cover a broad range of topics and serve well to outline the present state and scope of mathematical software.

The third part of the book contains four chapters, each of which presents and discusses a significant piece of mathematical software. These items were selected by the editor as examples of programs that are well documented and tested and whose properties have been well analyzed. They also are basic components of a library of mathematical software and have widespread applications. It is too early to hope for anything close to perfection in mathematical software, but these programs have made an attempt. They can at least serve as bench marks for future better and more effective software.

ACKNOWLEDGMENTS

I thank the symposium's organizing committee, Robert Ashenhurst, Charles Lawson, Stuart Lynn, and Joseph Traub, for their aid and advice. The Association of Computing Machinery and SIGNUM officially sponsored the symposium and their aid in publicity, etc., is much appreciated. Many of the myriad of arrangement details were handled by the conferences division of Purdue University and I thank them for this service and Purdue University for providing facilities. A most crucial element, money, was provided by the Mathematics Branch of the Office of Naval Research through contract NOOO 14–67–A–0226–0011, Project NRO44–382. I thank them very much for listening to my proposal and providing an essential element for the success of this symposium. Finally, I thank Academic Press for their assistance and consideration in publishing this book.

Mathematical Software

Part One

PROLOGUE

The first three chapters of this book form a prologue to the Proceedings of the Symposium on Mathematical Software held at Purdue University in April 1970. The texts of the papers presented at this symposium are given in the second part of this book. The third part of this book presents a small selection of mathematical software that appears interesting and significant to the editor.

CHAPTER 1

HISTORICAL NOTES

John R. Rice

PURDUE UNIVERSITY

LAFAYETTE, INDIANA

I. Introduction

One might make a temporal classification of the three chapters in this prologue: namely, the past, the present, and the future. Unfortunately, the memories of the past are somewhat dim, the appreciation of the present is distorted by prejudices and a lack of perspective, and only the foolhardy claim to make meaningful statements about the future. Thus, these three chapters are the views and thoughts about mathematical software of only one person, the editor of this book, and the reader must take into account that there are many divergent views and thoughts which are current.

This chapter simply presents a chronological list of events which seem to the editor to help place the present into the context of the past. Some of the events noted are obviously very significant in the development of mathematical software and computer science in general; some are irrelevant to mathematical software; and others are only incidental curiosities. It is almost certain that some important events have been overlooked or forgotten by the editor, but hopefully enough of them have been recorded to give the reader some feeling for how mathematical software got to where it is today.

II. Chronological Record

1943

We begin with the initial publication of the journal *Mathematical Tables and Aids to Computation* (*MTAC*). The first issues were primarily concerned with tables of mathematical functions. One can view these tables as the "precomputer" analogs of mathematical software, but there was nothing therein that one would view today as mathematical software. These were small sections entitled "Mechanical Aids to Computation" which contained mostly reviews of books.

This journal was to evolve into the primary publication medium for computer science and then to evolve further so as to exclude almost all of computer science not associated with numerical analysis or number theory (and considerable segments of these areas are not currently represented there, e.g., mathematical software).

The first paper in this journal to present formulas for computing was by Lehmer (1943), although this topic appears in the "Queries–Replies" section earlier (p. 99 of the first volume). An error in these formulas is pointed out on p. 309.

1944

A "computer"-oriented article by Pitteget and Clemence (1944) discusses variable spacing of tabulated functions so as to minimize the number of cards needed (to store the tables) to give a specified accuracy with a particular interpolation scheme.

A routine for $\sin x$ is written for the Mark I. It is an external routine on paper tape.

1945

A compilation of tables of functions known to exist on punched cards is given in *MTAC* on pp. 433–436 by W. J. E.

1946

The multiplication of matrices on an unnamed "general purpose" IBM machine is described by Alt (1946). Two 9×9 matrices with 6-digit entries could be multiplied in 9 min—after setup time. This did not include checking, etc.

A general description of ENIAC is given by Goldstine and Goldstine (1946). Of particular interest is the fact that this machine had three "function table units" which can hold up to 104 entries with 2 sign digits and 12 decimal digits. The ENIAC could also punch out intermediate results which could then be placed in these table units (auxiliary storage?). Other miscellaneous facts include the following: add time is 1/5000 sec, ENIAC had 18,000 tubes, 1500 relays, 20 accumulators, one multiplier, and one combination division–square-root unit.

A paper by L. J. C. (pp. 149–159 in *MTAC*) mentions that a "master deck" is available for a particular IBM machine which evaluates a triple Fourier sum from coefficients given on cards.

The review by D. A. L. of a manual for a particular Harvard computer includes the following quote: "As it is, coding appears to be an almost insurmountable barrier between the machine and the mere mathematician with a problem to solve."

The section entitled "Automatic Computing Machinery" is introduced on p. 354 of *MTAC*.

1947

The Eastern Association for Computing Machinery has its first meeting in September at Columbia University. By the end of 1947 it has 246 members and has dropped the word Eastern from its name.

The first general symposium on automatic computing machinery is held at Harvard University in January.

1948

By this time (Vol. 3 of *MTAC*) basic numerical analysis for digital computing is appearing regularly.

Alt (1948) describes a Bell Telephone Laboratories computer which has "routine tapes" and "table tapes." These are five routine tapes which are physically in a loop for repetitive use. The table tapes could move forward or backward and the contents are read into an associated "table register." Part II of this paper notes the existence of permanent (hardware) function tables for $\sin x$, $\cos x$, $\tan^{-1} x$, $\log x$, and 10^x.

1949

A "semialgorithm" for plotting is given by King (1949) in a cryptic English jargon. It concerns using an IBM 405 with the data on punched cards.

The first actual program segment or algorithm to appear in *MTAC* is on pp. 427–431 of Vol. 3. It is a program for EDVAC (in machine language) to convert 10 decimal digit integers to binary. It is presented and explained much in the spirit of mathematical software.

The first computations by a stored program computer are made in May 1949 by the EDSAC at Cambridge University.

1950

The following quote appears in Huskey (1950): "All the more frequently used subroutines will be stored on the magnetic drum, thus being comparatively easily accessible." These routines are invoked by name (as in

macros) and included things like floating point arithmetic, vector operations, integration formulas, and iteration formulas.

1951

The first UNIVAC computer is delivered.

The first joint computer conference is held in Philadelphia. There are about
900 attendees at the meeting sponsored by the American Institute of Electrical
Engineering, the Institute for Radio Engineers, and the Association of
Computing Machinery.

The predecessor to Computers and Automation appears. The name is
changed the next year.

The book by Wilkes *et al.* (1951) includes a thorough discussion of the
subroutine library development and use for the EDSAC.

1952

Brooker *et al.* (1952) describe an error in the floating point square-root
routine for the EDSAC. The situation is analyzed in detail and it is shown
that, for random arguments, the probability of an incorrect result is
$1/256 = 0.0039$.

1953

A bibliography of coding procedures is reviewed on p. 42 of Vol. 7 of
MTAC. The National Bureau of Standards announces the formation of a
central library of notes and reports which will include, for example, coding
manuals, computing routines, etc. The contents of one computing center's
library is summarized and it includes routines for linear equations, systems
of first-order differential equations, quadrature, post-mortem dump, eigenvalues and eigenvectors of symmetric matrices, address searches, and square
root.

The first Defense Calculator (IBM 701) is delivered. It has a 2048-word
cathode-ray-tube memory backed up by magnetic tapes and drum storage.
IBM also announces its IBM 650 in the small computer field. It has a 2000
10-digit word drum memory and the $1 + 1$ addressing system appropriate
for minimum access coding.

1954

The first complete program to be published in a journal (as far as the
editor knows) appears in LaFara (1954). A flow chart and 60 lines of IBM 605
code are given for a method to compute inverse trigonometric functions.

A set of 20 test matrices are given by Lotkin (1954). They are expressly for testing matrix routines.

The *Journal of the Association for Computing Machinery* appears.

1955

The book by Hastings (1955) appears which gives a small collection of formulas for the efficient computation of some of the elementary functions.

The NORC, perhaps the first "supercomputer," is delivered to the US Naval Weapons Laboratory. It is designed to perform 15,000 3-address operations per second. Floating point addition and multiplication take 15 and 31 μsec, respectively. Only one NORC was built.

1956

The first IBM 704's are delivered. It has a core memory (at least 4096 36-bit words) and a 12-μsec cycle time. It has three index registers and floating point hardware.

The first vacuum tube computer is built in Japan. It has 256 words of memory and an add time of 1.1 msec.

1957

The first version of Fortran is issued for the IBM 704.

The British Computer Society is formed.

1958

The *Communications of the Association for Computing Machinery* appears. The original statement includes "descriptions of computer programs" as appropriate material to publish. The techniques department invites: "Digital computer routines ... of general interest to a large class of users and normally independent of a particular computer." The first issue contains a short 15-statement IBM 650 program.

The language IAL (International Algorithmic Language) is defined. It later becomes Algol, then Algol 58, then Algol 60, then Algol 60 revised; there is also an Algol 68. The word Algol currently refers to Algol 60 revised and the fate of Algol 68 is not yet known.

1959

The second generation starts—the first transistorized computers appear. The IBM 7090 is first delivered late in 1959 and is shortly followed by the CDC 1604, the Philco 2000, and another "supercomputer," the Remington Rand Univac LARC. These are delivered in 1960. The 7090 has a basic cycle time of 2.18 μsec compared to 12 μsec for the IBM 704 and 709 and the number of cycles required is reduced for many instructions.

The journal *Numerische Mathematik* appears. The first issues contain typical numerical analysis papers and a report on Algol.

A bibliography of nuclear reactor codes is published in *Comm. Assoc. Comput. Mach.* About 200 programs are included, and some of them are in fact, general purpose mathematical software. The information given for each code is (1) author, (2) status, (3) problem statement and program description, (4) estimated running time, (5) references, and (6) comments and limitations.

Volume 2 of *Comm. Assoc. Comp. Mach.* contains a program in IAL for decimal to binary conversion. Perhaps it is a translation of the 1949 EDVAC algorithm.

Aegerter (1959) gives a class of text matrices.

A compilation of statistical programs known to exist for the IBM 650 is given by Hamblen (1959). About 90 programs are listed (along with a brief description) from various computing centers (mostly universities).

1960

The "Handbook for Automatic Computation" is announced by *Numerische Mathematik*. The purpose "is to provide a collection of tested algorithms for mathematical computations," and it is stated that "it is evident that such a collection could have no general utility unless written in some common program language."

The first book by Ralston and Wilf (1960) appears. Although there are no complete algorithms (or programs), there are many detailed flow charts aimed at guiding someone through the process of constructing an algorithm. There is considerable pertinent discussion of things like memory requirements and estimates of running times.

1961

The journal *BIT* (*Nordisk Tidskrift for Informations-Behanling*) appears. The first issue has a small Algol program. A medium-size Algol program appears in the second volume and the first substantial algorithm appears in the fourth volume.

The "supercomputer" Stretch is delivered by IBM to the Los Alamos Laboratories of the Atomic Energy Commission. It did not meet its design specifications and only seven were built.

The journal *MTAC* changes its name to *Mathematics of Computation*. It has stopped publishing material primarily related to hardware or software for computers.

The first computer is installed in Rhodesia to process personnel records for the civil service.

Almost 30 large companies now submit their Social Security reports on a single reel of magnetic tape, once each quarter. These replace typewritten reports that often exceeded 1000 pages per company per quarter.

1962

Brenner (1962) gives a set of test matrices.

Comm. Assoc. Comput. Mach. gives an index of the algorithms published in it during 1960–1961. There are 73 algorithms noted.

1963

The book by Lyusternik *et al.* (1963) appears which presents material for computing the elementary mathematical functions. In content and spirit it stands midway between the book by Hart *et al.* (1968) and that by Abramowitz and Stegun (1964). That is to say, midway between direct consideration of how to write elementary function routines and the classical formulas of the theory of elementary and special functions.

1964

The third generation of computers is announced by IBM in its 360 line. It is not clear just what constitutes a third-generation computer (i.e., is it the use of integrated circuits, of multiprogramming, of time sharing?); however, almost all computers on the market after 1965 are declared to be third (even fourth!) generation computers by their manufacturers.

The first "supercomputer" from Control Data Corporation, the CDC 6600, is delivered to Lawrence Radiation Laboratory at Livermore. The central processor executes about 3,000,000 operations per second. The internal organization is, in fact, a collection of over 20 specialized processors (10 of these have independent memories and serve input–output functions).

The first system to provide general symbolic formula manipulation appears. It is Formac, an extension of Fortran, made available on the IBM 7090/7094.

1965

There are five universities (Columbia, M.I.T., Princeton, Purdue, and Yale) which offer Bachelor, Master, and Ph.D. degrees in computer science. There are over 1100 computers installed or on order at universities (about 50% have a computer). There are over 35,000 computers in the United States.

1966

The SICNUM (later renamed SIGNUM) Newsletter appears. It announces the existence of a subcommittee interested in the certification of subroutines. The membership of SICNUM passes 1000.

1967

The ILLIAC IV's design concept of many (256?) central processors in parallel is announced and vigorously debated. Parallel processing (when it becomes available) opens up new possibilities (some think they are very limited) for the construction of algorithms for mathematical software.

The book by Forsythe and Moler (1967) gives a detailed study of algorithms for solving systems of linear equations along with a number of algorithms (in various languages) which are at the state of the art.

The book of Davis and Rabinowitz (1967) contains a provocative chapter on the automatic evaluation (numerical) of integrals. This is, I believe, the first systematic discussion of automatic numerical analysis to appear in a text [earlier discussions do appear in the literature, and a more general and detailed discussion is given by Rice (1968)].

The second volume of Ralston and Wilf (1967) appears which now contains a number of Fortran programs as well as extensive flow charts and numerical analysis background for mathematical software.

On its twentieth anniversary the Association for Computing Machinery has over 15,000 members (and about 4000 student members).

The *Australian Computer Journal* appears.

There were over 35 time-sharing systems in operation by the summer of 1967.

Lawrence Radiation Laboratory at Livermore installs a photodigital storage device with a capacity of over one-trillion bits and a retrieval time of a few seconds. This is equivalent to about a billion cards or a stack of magnetic tapes 1000 ft high.

The total value of computers delivered in 1967 is 5.9 billion dollars.

1968

The proceedings of a symposium on systems for applied mathematics appears [Klerer and Reinfelds (1968)]. This book presents a snapshot of many varied views and efforts on producing better systems, languages, hardware, and algorithms for mathematical software.

The book by Hart *et al.* (1968) is a direct attempt to provide the background, the specific techniques, and sets of coefficients needed to produce good subroutines for the elementary functions and a few special functions.

A rather long Fortran program appears in the *Computer Journal* (pp. 112–114).

The first Fortran program appears in *Comm. Assoc. Comp. Mach.* (it is number 332) on p. 436 in June 1968.

The journal *Applied Statistics* initiates an algorithms section with Algol, Fortran, and PL/1 as admissible languages.

Mass. Inst. of Tech. releases ICES, an Integrated Civil Engineering System.

The Central University of Venezuela initiates a masters degree program in computer science.

The first of three books [DeTar (1968)] appears entitled "Computer Programs for Chemistry." They contain Fortran programs presented much in the same spirit as the algorithms in the third part of this book.

1969

Control Data Corporation announces the 7600 which is to be able to execute 36 million operations per second, has two core memories of 650,000 and 5,000,000 characters, and has a disk storage of 800,000,000 characters.

The paper by Cannon (1969) gives a survey of computational attempts in group theory and lists over 75 references.

REFERENCES

Abramowitz, M., and Stegun, I. (1964). Handbook of Functions, *Nat. Bur. Standards Appl. Math. Ser.* **55** U.S. Government Printing Office.

Aegerter, M. J. (1959). Construction of a set of test matrices, *Comm. Assoc. Comp. Mach.* **2,** 10–12.

Alt, F. L. (1946). Multiplication of matrices, *MTAC* **2,** 12–13.

Alt, F. L. (1948). A Bell Telephone Laboratories computing machine, *MTAC* **3,** part 1, 1–13, part 2, 69–84.

Brenner, J. L. (1962). A set of matrices for testing computer programs, *Comm. Assoc. Comp. Mach.* **5,** 443–444.

Brooker, R. A., Gill, S., and Wheeler, D. J. (1952). The adventures of a blunder, *MTAC* **6,** 112–113.

Cannon, J. J. (1969). Computers in group theory: a survey, *Comm. Assoc. Comp. Mach.* **12,** 3–12.

Davis, P. J., and Rabinowitz, P. (1967). "Numerical Integration." Ginn (Blaisdell), Boston, Massachusetts.

DeTar, D. F. (1968). "Computer Programs for Chemistry," Vol. 1. Benjamin, New York (Vol. 2, 1969; Vol. 3, 1969).

Forsythe, G., and Moler, C. B. (1967). "Computer Solution of Linear Algebraic Systems." Prentice Hall, Englewood Cliffs, New Jersey.

Goldstine, H. H., and Goldstine, A. (1946). The electronic numerical integrator and computer (ENIAC), *MTAC* **2,** 97–110.

Hamblen, J. W. (1959). Statistical programs for the IBM 650, *Comm. Assoc. Comp. Mach.* **2,** Part 1, 13–18, Part 2, 32–37.

Hart, J. F. *et al.* (1968). "Computer Approximations." Wiley, New York.

Hastings, C. B. (1955). "Approximations for Digital Computers." Princeton Univ. Press, Princeton, New Jersey.

Huskey, H. D. (1950). Characteristics of the Institute for Numerical Analysis computer, *MTAC* **4**, 103–108.

Klerer, M., and Reinfelds, J. (1968). "Interactive Systems for Experimental Applied Mathematics." Academic Press, New York.

King, G. W. (1949). A method of plotting on standard IBM equipment, *MTAC* **3**, 352–355.

LaFara, R. L. (1954). A method for calculating inverse trigonometric functions, *MTAC* **8**, 132.

Lehmer, D. H. (1943). Note on the computation of the Bessel function $I_n(x)$, *MTAC* **1**, 133–135.

Lotkin, M. (1955). A set of test matrices, *MATC* **9**, 153.

Lyusternik, L. A. *et al.* (1963). "Handbook for Computing Elementary Functions." Pergamon Press, Oxford (English translation 1965).

Pitteget, P., and Clemence, G. M. (1944). Optimum-interval punched card tables, *MATC* **1**, p. 173.

Ralston, A., and Wilf, H. S. (1960). "Mathematical Methods for Digital Computers." Wiley, New York.

Ralston, A., and Wilf, H. S. (1967). "Mathematical Methods for Digital Computers," Vol. 2. Wiley, New York.

Rice, J. R. (1968). On the construction of polyalgorithms for automatic numerical analysis, *in* "Interactive Systems for Experimental and Applied Mathematics" (M. Klerer and J. Reinfelds, eds.), pp. 301–313. Academic Press, New York.

Wilkes, M. V., Wheeler, D. J., and Gill, S. (1951). "The Preparation of Programs for an Electronic Digital Computer." Addison–Wesley, Reading, Massachusetts.

CHAPTER 2

THE DISTRIBUTION AND SOURCES OF MATHEMATICAL SOFTWARE

John R. Rice

PURDUE UNIVERSITY

LAFAYETTE, INDIANA

I. Introduction

This chapter is aimed at describing part of the context of the symposium on mathematical software. It contains an assessment of the current status of mathematical software distribution and sources. The distribution considered is *local distribution*; i.e., how the programs get into the user's hands once they are at the local computing center. The section on sources then considers the places from which a computing center might obtain programs to distribute locally. This might be termed *global distribution*.

Each of these topics is subdivided into more specific mechanisms and the strengths and weaknesses of these mechanisms are outlined. It is seen that there are problems with almost every aspect of the operation. The final section contains a summary of the difficulties and classifies them into three main categories: (1) dissemination, (2) motivation and organization of talent, and (3) evaluation and reporting.

The complete range of mathematical software is included in this discussion and many sections are relevant only to certain subsets. Three typical examples from the range of mathematical software are (a) the sin–cos routines in the language package, (b) a linear equation solver in the subprogram library,

13

and (c) an integrated set of programs for mathematical optimization (linear and nonlinear programming) problems.

II. Local Distribution Methods

This section is concerned with the mechanisms used to get a program into the user's hands once that program is at his local computing facility. The primary mechanism is the subroutine library, which is discussed next. A second mechanism (which is essentially untested at this time) is the self-contained system whose language includes a large amount of mathematical software internally.

A. COMPUTER CENTER SUBPROGRAM LIBRARIES

The three papers by J. F. Traub, A. C. R. Newbery, and A. W. Dickinson *et al.* in these proceedings describe the mechanics and development of good subroutine libraries in considerable detail. The mechanics may be summarized by saying that there are two basic parts: the programs (which may be in the form of card decks, tape files, or disk files) and documentation. The documentation may be in separate write-ups or in comments attached to the programs. There is also some sort of index or directory of the library so that a user can locate a possibly helpful program.

1. *Strengths of Libraries*

The advantages of subroutine libraries are very evident, and libraries have been an integral part of computing facilities from almost the beginning. The two most obvious advantages are: (1) *one saves writing a program;* (2) *one should get a better program than one would write himself for a single application.* These two advantages are sufficient to justify the existence of program libraries in every computing facility.

2. *Weaknesses of Libraries*

In spite of these great advantages, there is a considerable body of opinion that program libraries have fallen far short of their potential. It has been repeatedly observed that ordinary users write their own programs rather than use the program library. Many reasons have been advanced to explain this situation, including such extremes as "ignorant users," "worthless and unreliable libraries," and "incompetent administration of libraries." Some users simply want to write their own programs, no matter what.

It seems that the most likely reason for the shortcomings in the effectiveness of libraries is a lack of appreciation of the motivations and habits of users. A typical and reasonable attitude for the computing facility to take is as follows:

> We have gone to a lot of trouble to collect these programs, check them out, and document them. We have made the library readily available. All the user has to do is become familiar with our index, read the documentation on how to use the program, and he is ready to go. If he still has trouble with this, we provide regular consultants to help him. Surely this is little enough to ask of the user, especially in view of all the work we've done.

The flaw in this apparently reasonable position stems from the fact that the primary objective of the computing facility is to provide *computing service*. If any part of that service is substantially short of its potential, then that part of the service is not satisfactory, no matter how "reasonable" the operation might appear.

Several specific points that cause users to neglect program libraries are discussed below.

a. *Location of Program Documentation*   The documentation of the programs in the libraries is normally available from someone in the computing center. There are a number of trivial difficulties that can arise in obtaining these write-ups; e.g., the computing center is closed, they are out of write-ups, the computing center is inconvenient to reach. Each of these trivialities contributes to the "nuisance level" of using the program library.

It is frequently less trivial to obtain copies of the exact source language of subprograms in the library. It is not uncommon for source programs to be completely missing or to be significantly out of date. This prevents the knowledgeable user from modifying these programs either to correct them or to adapt them to his specific needs.

b. *Understanding the Program Write-Ups*   The documentation is normally (and, in many peoples' opinion, best) written by the person who wrote the program. Documentation is viewed by most programmers as a dull, thankless task and many program write-ups reflect this view. Equally pertinent is the fact that the program originator is too familiar with the program and omits stating certain "obvious" facts. The typical user often has difficulty because he does not know these facts. Further, the quest for generality leads to a write-up that explains how to do many things that the user does not want to do and this generally makes it all the more difficult to find out how to do one specific task. All of these problems (and similar ones) contribute further to the nuisance level of using the program library.

c. *Program Reliability*   A chronic complaint of subroutine libraries is that the programs are unreliable. It is undoubtedly true that the quality of programs in the typical subroutine library falls very short of what is desirable or possible. It is also undoubtedly true that the bulk of the unreliability is due to ignorance, incompetence, and carelessness of the user. However, the user is not likely to continue using an unreliable library even if he is the sole source of the unreliability. Furthermore, he does not admit to very much ignorance, incompetence, and carelessness and attributes the unreliability to other causes (e.g., poor write-ups, lack of built-in safeguards, etc.).

In summary, there is considerable evidence that the typical computer center library is not yet near its desired usefulness to the computing community. The reasons proposed above for the neglect of these libraries are mostly of the trivial variety and there might be a more fundamental reason. Whatever the reasons might be, it is appropriate for those responsible for libraries to seriously review their aims and operations toward increasing the use and usefulness of these libraries.

Even if libraries are near perfection, it is likely that they will fall short of the objective of making every user's mathematical software have high quality. The difficulty is not peculiar to subprogram libraries, but applies to all libraries (and other stores of knowledge). It has been estimated that 60–90% of all research and development work is a duplication of previous work, and it is easy to believe that this applies to mathematical software with perhaps an even higher percentage than 90. This indicates that libraries in general are not very effective, and it is unlikely that subprogram libraries will significantly surpass the effectiveness of libraries in general. One might hope—and certainly it should be tried—that information retrieval methods can be particularly effective for subprogram libraries.

B. Extended Systems of Mathematical Software

Several papers (R. Bayer, C. W. de Boor, C. W. Gear, J. E. Sammet, and L. R. Symes) in these proceedings discuss some aspects of special systems for mathematical software at the level of programs found in a typical library. Furthermore, several papers (R. L. Ashenhurst, C. B. Dunham, M. Goldstein and S. Hoffberg, and H. C. Thacher) discuss special arithmetic and two more (W. J. Cody and H. Kuki) discuss software for higher-level language built-in libraries. All of these efforts attempt to make certain selected mathematical software automatically available to the user as an integral part of the language he is using. There is a tendency to divide the extended systems into two categories: numerical and symbolic (or nonnumerical). This division is not really pertinent to a general discussion of the attributes of such systems.

In the discussion that follows we *assume* that the extended system has become the ordinary programming language for the user. Thus the user has learned language XXX, which includes not only arithmetic and elementary functions but also an extensive set of additional mathematical software. Further the language incorporates the usage of this software in a natural way and the user is unaware of any essential difference in computing $x + 3.1$ or $\sin(x + 3.1)$ or $\int_2^3 (x + 3.1)\, dx$. He simply writes them down as part of the language.

1. Strengths of Extended Systems

The primary attraction of extended systems is the possibility to avoid all (or at least most) of the essential weaknesses of computer center libraries (at least from the user's point of view). The essential weaknesses are the communication and documentation problems. This does not avoid the unnecessary (but currently very real) weakness of poor and ineffective software. It is very possible to have a beautiful language, efficient compiler, etc., which incorporates terrible mathematical software.

Simple, yet very important, examples of this occur in the built-in function libraries of a higher-level language [e.g., $\cos(x)$, $\log(x)$, $\mathrm{abs}(x)$] and in the special arithmetic variables in many languages (e.g., double precision, complex). The typical user does not know (or care) that $\mathrm{abs}(x)$ is not a software package (since it is usually implemented by one machine language instruction) and that $\cos(x)$ is a software package of considerable size. What the user knows and cares about is that these procedures are available to him in a completely natural and obvious way.

Not only does this approach avoid the weaknesses of libraries but it retains the strong points of libraries. Furthermore, this approach has the potential to resolve the very obnoxious dissemination problem. Thus if an extended system has quality software embedded in it, then the user obtains access to all this software as soon as he has access of the system.

2. Weaknesses of Extended Systems

A special system and language with added capabilities is necessarily more complex than Fortran and similar languages. This means that the system and language required additional learning on the user's part. The actual amount of additional learning can vary by a great deal, depending on the scope of added capabilities and the effort put into making the language natural.

This is the primary and perhaps only disadvantage of extended systems in the eyes of the user. There are, however, additional real difficulties that are not inevitable, but which exist for all current systems of this nature. Foremost among these is the almost complete lack of transportability. Thus even though these systems hold out the promise of solving the dissemination problem for quality mathematical software, this potential depends on the transportability of these systems.

The reasons for the lack of transportability are numerous and lead to long discussions in otherwise unrelated areas of computing. Some of these reasons are mentioned briefly in the following.

a. *Machine Dependence*   Significant portions of these systems are (and often must be) written in machine language.

b. *Operating System Dependence*   Many of these systems interact extensively with the operating system and thus depend on its capabilities and idiosyncrasies.

c. *System Configuration Dependence*   Considerable auxiliary storage is often required and the extended system often is written for one specific configuration of main and auxiliary memory. Furthermore, some of the systems are on-line and depend on the specific consoles used and the specific interface operation.

d. *Large Size*   The size of these systems is large (say compared to Fortran or Algol) for two reasons. First, they have considerable additional capability. Second, they have not gone through enough rewrites to reduce the size to a minimal amount. Many computing centers with time sharing and/or multiprogrammed operations are reluctant to have a large system extensively used (unless it was written by the local people, in which case its merits are well enough recognized to allow it to be put into general use).

e. *Lack of Manufacturer's Commitment*   Fortran and Cobol are universally available because the computer manufacturers see to it that their machines have these languages available. Such a commitment has not been made to any extended system of higher-level mathematical software (no doubt such a commitment is debatable on economic grounds at this time). Some software, primarily for linear programming, has been distributed and maintained by the manufacturers but with a secondary priority (e.g., LP90 by IBM, OPTIMA by CDC).

In summary, the weakness of the extended system approach to the local distribution of mathematical software is that such systems are generally unavailable. Contributing to this is the fact that most systems of this type are still first generation and have the problems that come with this status.

The potential of extended mathematical systems to give the user access to mathematical software is clearly higher than that of even a first-rate computer center subroutine library. The crucial questions are when will reliable, efficient, and natural systems become available, and how will it be possible to disseminate them (and the presumably quality mathematical software they contain) throughout the computing community.

III. Assessment of General Sources

We have seen that there are two ways to get mathematical software into the user's hands once it arrives at the local computing facility. This section discusses briefly the sources that a computing facility might use to obtain programs for its library, etc. A few remarks are made about the strong and weak points of each of the general sources mentioned.

A. MANUFACTURERS

The amount of mathematical software supplied by manufacturers varies from a considerable quantity (e.g., IBM) to a very minimal quantity (only the elementary Fortran functions).

1. *Strengths*

There are three strong points for this software: (1) It is sometimes abundant. (2) It probably has very few compatibility problems with the machine and operating system. (3) It probably has standardized documentation. Note that standardized and poor documentation is much better than unstandardized and poor documentation.

2. *Weaknesses*

The main weakness is in the quality. It varies from very poor (even in some elementary Fortran functions) to adequate. It contains little software with close to the state of the art quality. Furthermore, the software is not likely to have been thoroughly (or even minimally) tested. While this level of quality may seem (and be) unsatisfactory, one must keep in mind that the manufacturer's software is at least as good as the average contents of a computer library and sometimes constitutes the best programs in the library.

B. USER GROUPS

Groups of users of various machines realized long ago that they had a high level of duplication of effort and they formed groups which (among other things) exchanged mathematical software.

1. *Strengths*

The strong points here are the same as those for manufacturers (abundance, good compatibility, standard documentation) except that the level of standardization in the documentation tends to be lower.

2. *Weakness*

Again the main weakness is quality. The variation in the quality is higher. Some excellent and some terrible programs have been disseminated by these groups. The average quality is probably somewhat lower than mathematical software obtained from manufacturers. A secondary weakness is that many supposedly general purpose programs contain small, but annoying, quirks due to the specific application of the originating user. Some programs appear to run correctly only for this application. There is an additional difficulty (or perhaps nuisance) in locating a program available through these groups.

C. LOCAL COMPUTING CENTER

Most computing centers have some effort (sometimes informal) to supplement the set of programs available from external sources. The motivation is either special needs of the local center or lack of access to and/or confidence in the external sources.

1. *Strengths*

The primary strength of this software is that there is someone on-site who really knows what the program does. Thus when it doesn't work or is not directly applicable, there is someone who can quickly modify it.

2. *Weakness*

Again the main weakness is in quality. It is highly variable and not easily ascertainable. The programs are usually not well tested and often the people who write them are not the best people in the computing center.

D. INFORMAL DISSEMINATION

A fair number of programs for mathematical software are distributed in an informal way to people who have somehow heard of them. The programs so distributed tend to have unusual capabilities or quality, otherwise there would be no reason for them to be distributed in this manner. However, this is only the source of an occasional program and can only supplement the development of a program library.

E. SERVICE BUREAUS

A fair number of software service bureaus have been organized recently to provide software support for a fee. The influence they will have on mathematical software is unknown at this time. Various factors make mathematical software less profitable for these bureaus. In general, these bureaus hope to have better software than is available from manufacturers and to provide maintenance and limited tailor-fitting to a user's needs. If a service bureau decides to enter the mathematical software market, they have the potential to provide better software than most computing centers now possess. Further, they may be able to supply advice on how to effectively use this software and such advice and knowledge is in very short supply at many computing centers.

F. SELF-WRITTEN BY THE USER

This is probably still the biggest source of mathematical software that is not provided as part of a system (e.g., the Fortran elementary functions). Some users simply refuse to use other people's programs.

1. Strengths

The overriding strength of this software is its familiarity and convenience to the user. He knows it well, he does not depend on external help or services, and he can modify it rapidly when needed.

2. Weaknesses

The average quality of such programs is probably the lowest of all sources of mathematical software. Hopefully, the user is aware enough to recognize when he is obtaining erroneous results and he continues to modify his program until it works for his specific application. The cost-effectiveness of writing one's own basic mathematical software is low, but many users prefer this to the inconvenience of locating and understanding someone else's program. This preference is reinforced by a few instances where the program so located could never be made to work correctly.

G. FORMAL PUBLICATION—ALGORITHMS

There are a number of sources of programs in the literature (e.g., *Numerische Mathematik*, *Comm. Assoc. Comp. Mach.*, *BIT*) and in some books. These programs are widely adapted for use in computing center libraries.

1. Strengths

The primary strength is assurance of quality and significance. In theory at least, these programs have been carefully written and tested and then evaluated by an independent expert. Some of the early items were neither of

significance nor of quality, but the editors of the journals now attempt to
maintain high standards. These programs tend to be well documented and to
contain discussions or references for the mathematical developments needed.
The level of quality of the programs that appear in books is more variable.

2. *Weaknesses*

These algorithms almost invariably must be adapted in one or more of the
following ways before they can be used.

a. *Character Set Change*   This is a trivial but often time-consuming nuisance,
especially with Algol.

b. *Language Translation*   Frequently the best (and sometimes the only)
way to obtain a working program is to translate it (e.g., from Algol into
Fortran).

c. *Machine and System Adaptation*   Even though many of these programs
are explicitly designed to ease this adaptation, it usually requires some
analysis of the program.

d. *Input–Output*   It is often missing or must be heavily modified.

e. *User Conveniences and Safeguards*   Such programs rarely have the
conveniences and safeguards needed in computer program library routines.
For example, input arguments are often very rigid, few checks are made for
erroneous input, or intermediate results and diagnostics are rare.

In spite of all of these problems, these algorithms are a very useful source of
mathematical software and the resulting quality is generally high.

As an indication of the quality of programs that one might obtain from
books, all the programs in a well-known book of this type were examined
and the results are summarized in Table I (some programs appear more than
once). It is plausible that this is typical of the quality of the programs in
such books.

H. Textbook Methods

A large number of programs are based on a direct adaptation of numerical
methods described in standard textbooks. These programs probably form the
bulk of the programs in computer center libraries. There are two major sources
of difficulty with such programs.

First, there are significant points of mathematical software that are not

2. DISTRIBUTION AND SOURCES   23

considered in the current texts, particularly in introductory texts. For example, realistic appraisals of the efficiency or condition of a computation cannot be made in an introductory text. This difficulty is aggravated because many of the more advanced texts also ignore these points.

Second, the transformation of a two-page discussion into a computer

TABLE I

SUMMARY OF AN ESTIMATION OF THE QUALITY OF PROGRAMS
IN A TYPICAL BOOK

Remark	Number of programs
Machine dependent	2
Incomplete	2
Errors noted upon casual reading	2
Documentation of program	
Poor	5
Fair	2
Good	1
Programming quality	
Poor	2
Fair	5
Good	1

program is not at all well defined and the result depends to a large extent on the ability of the programmer. There are widespread instances of different implementations of the same "method" which have a great variation in their performance capabilities. Examples of the aspects of a program that are generally undefined by a textbook description are:

(a) program organization and modularity,
(b) computational strategy and techniques,
(c) error detection and associated diagnostics,
(d) efficiency of storage allocation,
(e) input–output conventions,
(f) documentation.

Decisions about these aspects can make the difference between excellent and terrible software.

A final unfortunate point about texts is that some have programs in them which apply to some mathematical problem but which are primarily designed to be illustrative examples about a programming language. These programs have a tendency to be used no matter how inefficient or ineffective they might be.

IV. Summary

The preceding two sections indicate that there is some difficulty in almost every aspect of the local and global distribution of mathematical software. These problems are put into three categories as follows:

A. DISSEMINATION

There is a large amount of quality mathematical software in spite of the low percentage of mathematical software that is of good quality. The conclusion from this observation is that the primary problem is the dissemination problem. The computing community is simply not able to identify and distribute those quality programs that already exist. It is easy to conclude further that the current mode of operation will make slow, but probably steady, improvement in this area.

B. MOTIVATION AND ORGANIZATION OF TALENT

The development of good mathematical software requires three basic ingredients:

(1) understanding of the relative effectiveness and value of the possible computational procedures,
(2) programming talent,
(3) awareness of the eventual user's needs and views.

The first ingredient normally is found only in professional numerical analysts who have considerable computational experience. The second ingredient is harder to identify and is much scarcer than one might expect. The third ingredient is the easiest to obtain but also the most neglected.

It is worthwhile to point out the pitfalls in a common arrangement for mathematical software development; that is, to form a team of a professional expert, say a numerical analyst (college professor type), and a programming assistant (graduate student type). The expert has considerable knowledge of the theory and the assistant has considerable experience in programming. This arrangement often leads to mediocre software because the expert has little direct computational experience and most programming assistants are not talented programmers. The result can be that some critical problems (e.g., storage allocation, information representation, program organization) are not recognized, let alone solved.

The solution is, of course, either to find one person with the first two ingredients or to form a team of two people, a talented programmer and a computationally oriented expert. It seems highly unlikely that, in fact, one can be a computationally oriented expert without being a talented programmer.

C. EVALUATION AND REPORTING

It is necessary that objective means be established to evaluate mathematical software if there is any hope of eliminating poor software from libraries. Furthermore the results of these evaluations must be reported in a way that makes them widely available. This chapter contains a great deal of critical evaluation of mathematical software, but it is 90% subjective and based on a very haphazard sampling and on hearsay. It is possible that mathematical software is much better (or even much worse) than indicated here. It is essential that an objective and scientific approach be made to this question.

CHAPTER 3

THE CHALLENGE
FOR MATHEMATICAL SOFTWARE

John R. Rice

PURDUE UNIVERSITY

LAFAYETTE, INDIANA

I. Introduction

This chapter attempts to outline the problems and difficulties currently faced in the area of mathematical software. These problems cover the broad range from a lack of fundamental knowledge and scientific know-how to organizational problems and low professional status. The presentation is divided into three parts. The first concerns algorithm construction and is primarily a discussion of technical problems with a few remarks about non-technical issues. The second concerns the evaluation of algorithms and the discussion is entirely about technical points. There is a real lack of theoretical foundation for some aspects of the construction of algorithms, but the evaluation of algorithms is almost entirely virgin territory. The questions range from the foundations of mathematics to measurements of consumer preferences. The final part concerns dissemination and this is the most complex of the challenges. A number of different organizations are involved, each with different capabilities and objectives. Steps should be taken, so that it becomes in the interests of each of these organizations (or people), to do what needs to be done. It is not clear what these steps are. The final section presents two recommendations (the establishment of a journal and of a coordinating center or focal point for mathematical software) which are believed to be feasible and which would significantly contribute to the advancement of knowledge and usefulness in mathematical software.

II. Algorithm Construction

It has already been noted that there is a shortage of good mathematical software. Part of the reason for this is that the creation of such programs has not been identified as a high-status activity which requires expert and professional talent. Indeed many people have not identified it as a separate, independent activity at all. In this section we briefly outline the objectives of algorithm construction and then discuss some of the problems that arise. The objectives are, of course, a list of all the desirable characteristics one can imagine, so the bulk of the discussion concerns the difficulty. Some of the views expressed here are elaborated upon in Rice (1968).

A. Technical Objectives

We list nine desirable characteristics (not necessarily in order of preference) of mathematical software:

reliability	flexibility
simplicity of use	modularity (modifiability)
reasonable diagnostics	common sense
good documentation	efficiency
transportability	

There are direct conflicts between some of these objectives. The essence of mathematical software construction is to achieve a reasonable balance while optimizing with respect to two or three of these objectives. One does not really realize what the trade-offs and conflicts are until one has a prototype algorithm constructed. Fortunately, experience shows that many conflicts can be resolved by using some ingenuity rather than the most obvious approach. However, even this process is often in conflict with modularity and, especially, user modifiability.

B. Technical Difficulties

The most obvious and a crucial difficulty is the selection of an appropriate mathematical procedure upon which to base the program. This difficulty is so well recognized that it is not discussed further here. We cannot present a discussion of the *real* difficulty in algorithm construction because it is a highly creative mental process similar to problem solving, theorem proving, and creative writing. None of these processes are understood at this time. There are some aspects of algorithm construction that might not be evident and some of these difficulties are listed and discussed below. However, the heart of the technical difficulties in mathematical software is how to construct the algorithm.

1. *Verification that Objectives Are Met*

This is really an aspect of the evaluation of mathematical software which is discussed in the next section. However, this question must be thoroughly considered even in the construction of a prototype program. It is highly non-trivial and requires considerable effort even to make tentative evaluations of the performance of an algorithm.

2. *Error Control*

This touches closely upon the choice of mathematical procedure, but two aspects of this question are somewhat independent. First, there are many instances where minor variations in the implemented version of the basic mathematical procedure result in dramatically different error control characteristics. Thus once a basic approach is chosen one often needs to consider the exact organization and sequencing of the computation in order to achieve reasonable error control. Second, error control is sometimes implemented through a posteriori accuracy checks. These checks may be independent of the computation of the proposed solution of the problem. Some problems lend themselves well to this approach and it can greatly increase the realiability with a small additional computational effort. More substantial computations are required to obtain a posteriori bounds in other problems.

3. *Transportability*

This aspect of mathematical software is repeatedly emphasized throughout this book, and it is very unfortunate when really good, generally applicable programs are developed which cannot be transferred to another computing facility. There are examples of quality programs that depend on the particular equipment configuration and/or the special facilities of the local operating system. On the other hand, complete transportability is sometimes impossible to achieve. There are programs whose performance depends in an unavoidable way on, say, machine word length. When the word length is changed, certain modifications must be made. One must keep the question of trans-portability in mind throughout the construction of an algorithm.

Programs that are destined for exclusive use on a particular computer (or series of computers) can, of course, be machine dependent and written in machine or assembly language. Such programs are normally ingredients in a larger system such as a language processor, linear equation solver, etc. These programs have the potential to be the best software of all because not only are transportability constraints weakened but so are modularity and modi-fiability constraints. In other words one has more latitude to exploit the particular situation envisaged.

Programs aimed at really general distribution (e.g., to computer center libraries) face severe problems. They must be in a higher-level language that is widely known and used. They must run on a wide variety of language processors with a minimum of effort and change. Furthermore, the modifications that are essential must be clearly identified in the documentation and the program should be organized in a way that makes the modification easy. Sometimes certain computations can be made in machine language with a very significant improvement in performance. In these cases the requirements on this small piece of code must be exactly specified and a substitute section of higher-level language code provided.

It is not at all easy to identify a universal subset of the three languages one can consider (these are Algol, Fortran, and, to a lesser extent, PL/1). Almost all Fortran compilers now include features which are nonstandard but nice. Even identical features are usually included in the language with different rules, syntax, etc., by various compilers. While one is grumbling about the restrictions of a language subset one might be consoled by the following. One sometimes obtains more efficient object code when using a larger number of simpler higher-level language statements. In most cases one can accomplish the same thing without any significant loss of object code efficiency. One sometimes obtains more understable algorithms in this way. Unfortunately, one equally often obtains less understandable algorithms.

4. *Documentation and Readability*

It is easy to say that good documentation and readability are needed and it is easy to believe that achieving this requires a certain amount of effort. The amount of thought required to do a good job is more than most people expect, though a little reflection shows that this must be the case. For we want to communicate to another person in a clear concise way what the algorithm does. The ability to communicate clearly (never mind concisely) is not that common and comes only after considerable effort. The difficulty is further compounded by the fact that the author of the algorithm understands it very well and has a hard time visualizing what information is really needed. Needless to say, this objective is not going to be met by giving the program to some junior assistant who neither understands it well nor is proficient at exposition.

5. *Software Interface and Input–Output*

A small but extremely crucial step in the construction of an algorithm is the specification of the interface between the algorithm and the user. There are several facilities available (e.g., arguments, global variables, printing, card reading) and the primary criterion in the selection of facilities is convenience and flexibility for the user. Most users dislike long argument lists for subprograms. Further, these lists tend to contain some variables whose meaning is unclear and others which are irrelevant to his needs. The possible approaches to this problem are not discussed here, but one must keep in mind that it is not sufficient merely to establish this interface; it should be done well.

Printed output is particularly troublesome because almost every subprogram is used in situations where absolutely no messages are to be printed. This has led many to say that subprograms in a library should be input–output free. This simplifies the subprogram at the expense of inconveniencing the user. In general, he should be able to obtain a reasonable summary of the results of the subprogram without writing his own code. Furthermore, certain facts should be brought to his attention unless he specifically suppresses them. Thus a square-root routine should print a diagnostic for a negative argument and a polynomial root finding subroutine should do the same for a negative polynomial degree. The most obvious solution is to provide these subprograms with a print switch to control the output.

6. *Computer Resource Allocation*

The advent of more sophisticated operating systems has introduced some really difficult problems. The easiest to visualize is the trade-off between computation time and central memory use. There are frequent instances where significantly faster programs result by using significantly larger amounts of memory. In the past, when only one program at a time was in main memory, this trade-off was frequently easily resolved because there was an excess of memory available. In multiprogramming operating systems, this trade-off is very real even for small programs, and there seem to be no firm guidelines to use. Moreover, what is best today might not be best tomorrow and what is best at one place might not be best at another. Nevertheless, a choice must be made about how to organize the computation and, as a consequence, about resource allocation.

This problem will undoubtedly get much worse in the future when the standard situation is likely to be computers with multiprogramming, multiprocessing, some parallel capabilities, and a hierarchy of memories. While the primary responsibility at this point must be assumed by the operating system, the writer of a program will still be able to affect the allocation of resources by the way he writes the program.

C. NONTECHNICAL DIFFICULTIES

The difficulties discussed here are not those that arise in constructing an algorithm, but rather those that prevent one from attempting it. Thus this discussion is aimed more at the reasons for the shortage of mathematical software than at considerations which arise during the construction. One might say that we are giving the reasons why people do not attempt to create quality mathematical software.

1. Low-Status Activity

The bulk of the scientific community assigns a very low status to any form of computer programming. This implies that anyone who has or wants to achieve a high professional standing is wasting his time by creating mathematical software. The low status assigned to programming stems from the belief that "anybody" can learn to do it in a few weeks. This is correct but completely irrelevant. One can learn how to write poems in a few weeks. The result is terrible poetry. One can learn how to play golf in a few weeks; the result is terrible golf. One can learn how to prove theorems in a few weeks; the result is worthless theorems. The result of a few weeks' training in programming is that one is able to write terrible programs.

2. Inadequate Publication Media

The publications of results are the milestones by which the scientific community currently measures professional competence. At this time there are very few journals which publish mathematical software and even these restrict in various ways the algorithms that they publish. There are several reasons why journals are reluctant to publish algorithms. Probably the foremost among these is the belief that programs are really beneath the "dignity" of a distinguished journal. A second reason, which is very real to the editors in charge, is that there are no accepted standards for algorithms and no accepted ways to see if a particular algorithm meets these standards. A third reason, which is very real to professional societies, is that the type-setting, etc., of algorithms is often done in a way that maximizes the expense and difficulties of publication. The net result of this is a real—and perhaps justified—reluctance of professional journals to publish algorithms.

The effect of the shortage of publication media on the individual who might create mathematical software is very direct. He must either avoid this activity or disguise the product of this activity as some more familiar (and acceptable) type of research. The people most affected by this situation are the most competent people. They realize that they have some prospects of achieving a high professional standing and they are thus most reluctant to engage in creating quality mathematical software.

D. SUMMARY

The aim of this section is to establish the fact that the construction of quality algorithms requires the highest level of scientific competence and knowledge. Unfortunately, quite the opposite view is widely held in scientific circles. Scientists tend to be quite conservative in scientific matters and one can expect this opposing view to last for many years. Nevertheless, more and more people will give the matter serious thought or attempt to create good algorithms. They will then appreciate the intricacies and difficulties that are involved.

III. Evaluation—Charting the Unknown

The point has been made several times that quality software is needed and that means to objectively measure quality and performance must be developed. Several aspects of the evaluation of algorithms are considered here. Some of these are primarily of theoretical interest (e.g., proofs of algorithms) and others concern practical approaches (e.g., the experimental investigation of performance). Perhaps the most important of these aspects is the determination of the domain of applicability of an algorithm. This aspect deserves much more attention and a primary defect in current mathematical software is that this domain is unknown. The characteristic weakness of low-quality software is that its domain of applicability does not include many problems on which it is naturally used. Furthermore, such software rarely makes any effort to verify that a problem posed to it is within its domain of applicability.

A. DETERMINATION OF THE PROBLEM DOMAIN

There are two different approaches one might use here. The first is to determine the domain of problems to which a particular algorithm is to be applied. Once this is done one would then construct an algorithm which is effective for these problems. The second approach is to describe a large class of interesting problems for which a particular algorithm is effective. Neither one of these approaches is easy or likely to be feasible in the near future.

The difficulty with the first approach is rather obvious. It requires a large effort in order to identify the ultimate consumers of mathematical software and to collect meaningful information from them about the problems they have. Such a data collection effort should not be restricted to just one special group of users, even though this limited information would be of considerable use. The difficulty is further compounded because most people who would develop quality mathematical software have little contact with consumers of it. There is a real danger that software will be developed that is especially effective for a large class of problems that no one wants to solve. Some hard information in this area is extremely desirable and it would be a valuable contribution.

The difficulty with the second approach is less obvious but much deeper. In a nutshell, there is no way to describe most problem domains of interest. What one really wants to do is to classify problems in some systematic way. This means that one must have well-defined and well-understood terms to make this classification and, further, these terms must be able to delineate the classes of interest. Such terms do not now exist in most problem areas. This difficulty is illustrated for three typical, but somewhat different, problem areas.

1. Example 1—Sorting

Consider a list of items that are to be ordered according to some criterion. Three attributes that might enter into a classification are

(a) the length of the list,
(b) the size of the items in the list,
(c) the difficulty of the comparisons required.

These attributes seem to be simple and one could reasonably hope to make quantitative statements about the effectiveness of an algorithm in terms of these attributes.

2. Example 2—Solving Linear Equations

Consider the solution of the standard problem of n linear equations in n unknowns. There is a much larger list of attributes that are relevant to classifying these problems. Typical ones are

(a) the size of n,   (e) diagonal dominance,
(b) sparseness,   (f) conditioning,
(c) symmetry,   (g) positive definiteness,
(d) bandedness,   (h) scaling.

It is much more difficult to make quantitative statements about the effectiveness of an algorithm in terms of these attributes. This is true in spite of the fact that this is perhaps the best understood of all mathematical procedures of the standard numerical analysis type. Even if the domain of applicability is known in terms of these attributes, it is not likely to be of much use to the typical user. He will not understand the meaning of some attributes (e.g., condition number, positive definiteness) and will be unable to measure some of them (e.g., condition number) even roughly. It is typical that classification schemes involve attributes that cannot be reasonably measured for most specific problems. By "reasonably measured" we mean measured with an amount of effort significantly less than that of solving the problem.

3. Example 3—Numerical Integration

Consider the problem of accurately estimating the value of

$$\int_a^b f(x)\, dx$$

The primary attributes pertain either to the accuracy required or to the nature of the function $f(x)$. Mathematics contains a vast array of attributes for functions; e.g.,

(a) k-times differentiable,	(g) integrable,
(b) analytic,	(h) singularities of type xxx,
(c) entire,	(i) meromorphic,
(d) convexity,	(j) piecewise analytic,
(e) total variation,	(k) Lipshitz continuity,
(f) continuity,	(l) positive.

Other less classical attributes include conditioning and the level of round-off contamination.

The classical mathematical attributes are almost all completely irrelevant to the computational problem at hand. Furthermore, most of them cannot be measured or verified outside a small (but admittedly interesting) class of problems.

The development of means to classify problems is one of the most challenging areas in mathematical software. It will require a combination of experimental judgment (in order to achieve relevance) and mathematical analysis (in order to achieve understanding) of the highest caliber.

B. Experimental Performance Evaluation

The "analytic" determination of the domain of applicability of algorithms is very inviting and important. It is not, however, going to provide the quantity of useful results needed in the immediate future because the necessary foundations are still missing. This means that experimentation will be the primary approach to algorithm evaluation in the immediate future.

The approach is deceptively simple. One merely generates a representative set of test problems and evaluates the performance of algorithms for this set of problems. The delicate yet crucial point is the appropriate choice of the set of test problems and the criteria to measure the performance. There do not seem to be general guidelines for these choices, but rather each problem area and algorithm requires special analysis. In fact, a thorough understanding of computation, of the problem area, and of the types of algorithms is required.

There is a natural tendency for one to describe a domain by describing its boundary. This means that one tends to choose test problems which delineate the limits of domain of applicability of an algorithm. Performances of algorithms are then compared near the boundary of their usefulness. This overemphasis on the boundary may lead one to overlook the most critical area for performance evaluation, namely the run-of-the-mill routine problems. Perhaps the best algorithms are those that are extremely efficient for these problems but nevertheless are able to handle difficult problems near the boundary of the problem area domain.

There is an annoying complication to these experiments: they must be run on a particular computer, with a particular compiler, and with a particular operating system. The peculiar characteristics of these otherwise extraneous things can materially affect the experimental results. It requires very careful study to foresee or detect some of these effects, and yet it increases the work considerably to make parallel evaluations in a variety of environments.

C. Numerical Analysis and Algorithms

The mathematical background for the majority of the algorithms in mathematical software lies in the field of numerical analysis. A few comments are in order on the relationship between actual programs and theoretical numerical analysis. These comments also apply to most nonnumerical algorithms as well.

The basic technique of theory is to construct an abstract model of the problem and associated procedures and then rigorously analyze various aspects, including algorithms. There are two fundamental difficulties that prevent one from obtaining in this way conclusions about specific algorithms. The first of these has already been mentioned; namely, mathematics does not have the terms and framework needed to formulate good abstract models. Many abstract models which are formulated and analyzed are even irrelevant to actual computational problems. The second difficulty is that theoreticians are unable to carry through the analysis of a complex model and thus they continually simplify the model in order to make analysis possible. Some significant features (sometimes the most significant ones) are inevitably lost in this simplification process.

The net result is that theory only provides guides about the computational properties of algorithms, not specific information about their effectiveness or efficiency. An understanding of the theoretical background is a necessary, not sufficient, ingredient to an understanding of mathematical software.

D. Proofs of Algorithms

The idea of constructing proofs of algorithms has a number of attractions. First, it is a challenging problem area. Second, it is mathematical in nature and will appeal to those who wish to maintain mathematical respectability. Third, it will probably lead to worthwhile insight into the nature of specific algorithms and of algorithms in general. These attractions are sufficient to assure the success and development of this area.

Let us consider briefly what these proofs might be. The serious study of algorithms now resides in a branch of mathematics that is closely related to foundations. Proofs in this area of mathematics tend to be very rigorous with detailed attention to special cases and with well-oiled mechanisms. One might say that proofs are viewed somewhat in the sense of Russell and Whitehead. Attempts are already underway to prove programs, and these attempts seem to involve a good deal of exhaustive examinations (computer aided) of various possibilities.

I believe that this approach will not be fruitful for mathematical software; that is to say, few, if any, nontrivial results will be obtained in this way. The basis for this belief is that the algorithms and problems are too complex to yield (in a reasonable time) to techniques that heavily involve enumeration. The guide for proofs of algorithms in mathematical software should be taken from modern analysis rather than from the foundations of mathematics. In spite of the appearances, there is no pretext of maintaining complete rigor in the proofs of analysis. The arguments labeled proofs are nothing more or less than arguments which are accepted by the community of analysts as convincing. It should be possible to develop proofs for the algorithms of mathematical software which meet these specifications. The proofs that currently occur in numerical analysis are not close to this objective (because they convincingly support conclusions with little relevance to algorithms). For some first efforts along these lines see Hull (1970).

IV. Dissemination—Some Alternatives

The current alternatives for the dissemination of mathematical software have been examined in some detail in Chapter 2. None of the mechanisms described there has reached perfection and we can expect a steady, albeit slow, rise in the quality of software due to the increased effectiveness of these established mechanisms.

On the other hand, it is reasonable to assume that these mechanisms will always provide the user with software tools that are considerably less than the state of the art. Direct evidence of this can be seen in the software for the elementary functions. Many current compilers incorporate software that would have been substandard ten years ago. This is in spite of the fact that there are literally dozens of people who know how to produce quality routines for the elementary functions. There is now a considerable literature on the subject, including at least one book which allows even a competent novice to create routines significantly better than those now supplied by some manufacturers.

It appears that some new approach to dissemination must be made if it is to become significantly more effective. Several new approaches are presented below and discussed in some detail.

A. Professional Society Support

The professional societies hold a key position in almost any attempt to take organized action on a scientific matter. They form the common meeting ground for the experts in the field and their membership includes almost all of the leaders in both the technical and organizational aspects of a scientific area. The professional society most relevant to mathematical software is the Association for Computing Machinery. Its members are involved at every step in the creation, evaluation, and dissemination of mathematical software. We note below the activities currently supported by this society that are relevant to mathematical software.

The Association of Computing Machinery supports the publication of algorithms in its communications. The algorithm section is probably not intended to provide the quantity of publication required to support the dissemination of mathematical software and its scope is larger than mathematical software. The membership and goals of the Association of Computing Machinery are so broad that it is unlikely that it would (or should) fully support such a narrow area in one of its general journals.

Within the Association for Computing Machinery there are several special interest groups, SIGMAP, SIGSAM, and, especially, SIGNUM, which are concerned with mathematical software. SIGNUM sponsored the symposium on mathematical software and officers of the other groups cooperated on various points during its organization. Further, SIGNUM has had for several years a special committee on the evaluation and certification of subroutines. Recall that these special interest groups are rapidly evolving into professional societies complete with journals, newsletters, dues, etc.

These activities do not form a solid support for mathematical software efforts and they should probably be described as sporadic. The primary deficiency (e.g., SIGNUM's committee on subroutine certification) has been the lack of a person (or persons) with both technical and organizational ability who will stake his professional reputation on his ability to organize and accomplish something significant in the area of mathematical software. Until such people appear, one can expect the professional society efforts to continue to be sporadic.

It should be emphasized that the professional societies have an essential role to play in the development of mathematical software as an area of scientific inquiry. It is these societies which can make high professional status available to workers in this area. They can do this by sponsoring special meetings in the area, by putting papers in this area in their programs at general meetings, and, especially, by providing journals in which new ideas and results may be communicated throughout the professional community. These activities that increase the status of work in the mathematical software area have as a side effect the provision of means of dissemination.

However, their main value to the dissemination effort is to provide status so that top-quality people can work in the area without penalty.

B. REGULAR GOVERNMENT AGENCY SUPPORT

The federal government has many efforts which provide general support for the scientific and technical communities. It would certainly be appropriate for an agency of the government to take on some of the responsibility of providing quality mathematical software.

The existence of a single, respected organization which would provide quality software would have a dramatic effect on dissemination. Both manufacturers and independent computing centers would quickly adopt these programs and the time lag between the creation and consumer use of mathematical software would be greatly reduced.

A logical candidate for this activity would be the National Bureau of Standards. It has the rudiments (i.e., established activities in applied mathematics, numerical analysis, and computer science) of an organization to undertake this, but a little analysis shows that a substantial additional effort would be required. Disadvantages of the National Bureau of Standards are that it has, at times, been particularly susceptible to budget cutting and it has experienced strong buffeting by the winds generated from political pressures of divergent economic interests.

C. GOVERNMENT DICTATION

A simple, but perhaps idealistic, approach would be for the federal government to dictate that quality mathematical software be provided by the manufacturer. Such a dictum might be made in several ways, but the simplest is to require that this software be provided (or be available for purchase) on all computers leased or bought by the federal government. The federal government is the one consumer of computers which can make such a dictum stick and, incidently, the consumer which stands to gain the most from it.

It is unlikely that this approach will be used by itself. It would place a significant burden upon the manufacturers of computers and some of them are probably unable to carry it. Even if they were, it would create considerable duplication of effort, though less than exists at present.

This approach is very feasible and logical if used in conjunction with other approaches. Thus if a government, semigovernment, or professional organization exists which provides a wide range of quality software, then it would be reasonable that all manufacturers obtain this software and incorporate it into their systems. Such an organization would very likely be supported by federal funds, and it is only natural that the federal government ask to receive the benefits that accrue from its expenditures.

D. New Organizational Support

We have already noted that a regular government agency might take on the responsibility of providing quality mathematical software. The same functions could be carried out by other organizations as well. A small new organization might be formed within an existing research and development organization. The funding would undoubtedly still be primarily federal. Organizations which meet the basic requirements in this area include a number of universities, some Atomic Energy Commission laboratories (e.g., Argonne, Brookhaven, Oak Ridge), the Jet Propulsion Laboratory, and the Rand Corporation. There has been some serious discussion of a federally supported "Institute for Computer Science" analogous to the atomic energy laboratories. The support of research and development into mathematical software would be a natural function of such an institute. It is conceivable, but not likely, that a professional society could create an organization to take on this task.

V. Two Recommendations

In this chapter I have attempted to outline the problems and prospects of mathematical software for the future. Many, if not most, of the shortcomings will be overcome only by the efforts of individual workers in the field. There are, however, two steps that can be taken which would serve a useful purpose and, perhaps more important, which would encourage talented people to take up work in the area of mathematical software.

A. A Journal of Mathematical Software

This is not the place to analyze the publication policies of the scientific community, but it is rather clear that lack of a suitable publication media has seriously restricted the activity in mathematical software. One cannot launch this journal as easily as the typical scientific journal for several reasons. Among these are the following:

(1) Satisfactory editorial policies are not yet well defined or widely accepted and whatever policy is adopted will be difficult to enforce.
(2) Competent referees are scarce.
(3) The effort required to referee an algorithm is substantially larger than that required for an ordinary scientific paper.
(4) The reliability of current typesetting procedures for computer languages is too low and the cost is too high.

These difficulties are very real but not insurmountable.

B. A CENTER OR FOCAL POINT FOR MATHEMATICAL SOFTWARE

A journal would serve as somewhat of a focal point for those doing research in mathematical software, but there is also a real need for one organization which has access to the knowledge and know-how in this area. Such a center would be extremely valuable to those outside the field, i.e., to the consumers of mathematical software. For example, computer manufacturers could be expected to establish liaison with such a center and to incorporate up-to-date mathematical software into its systems and software support packages.

These two recommendations are aimed at creating an environment to attract talented people into this area. The creation of mathematical software is a very challenging and stimulating intellectual activity and will inevitably attract some talented people. However, the current environment is such that it is not in their own self-interest to work in this area and the environment must be changed.

REFERENCES

Hull, T. E. (1970). The effectiveness of numerical methods for ordinary differential equations, *SIAM Stud. Numer. Analysis* **2**, 114–121.

Rice, J. R. (1968). On the construction of polyalgorithms for automatic numerical analysis, *in* "Interactive Systems for Experimental Applied Mathematics" (M. Klerer and J. Reinfelds, eds.), pp. 301–313. Academic Press, New York.

CHAPTER 4

DISCUSSION OF THE PAPERS

John R. Rice

PURDUE UNIVERSITY

LAFAYETTE, INDIANA

A brief discussion is given of each of the twenty-two papers presented at the Symposium on Mathematical Software. The remarks are not intended to abstract or summarize the main results, but rather are intended to present the "flavor" of each paper and perhaps to whet the reader's appetite for more information. The papers are grouped in a certain way, but many touch on a variety of areas so the grouping is for convenience rather than a definitive classification.

I. The User's Voice

The first two papers present the point of view of people outside the professional community of mathematical software. Both of the authors are sophisticated in computing, but they are consumers, not producers, of software. What they want to have in mathematical software is not at all identical with what the creators of mathematical software normally want to supply.

In the first paper, Muller gives a fascinating account of how the lunar mass concentrations were discovered and emphasizes the important role that *high-quality* mathematical software played. He states that he is against the "black box" approach to library programs and that he finds most library programs are out of date, from both the mathematical and the systems point of view. He strongly supports the development of modular, well-documented programs which the user can modify to suit his own needs.

The second paper by Bushnell chooses "user modifiability" as its theme and presents a strong case for such programs. His arguments are based on his background and experience in linear programming applications to business and economics.

II. Arithmetic

Ashenhurst's paper is the first of four papers on arithmetic, and he presents somewhat of a treatise on the monitoring of significance in computation and its relation to number representation. Detailed definitions essential to a serious study are given, and the relationship between errors of various kinds, significance analysis, and significance monitoring are analyzed in detail. He concludes with a discussion of the relation between these questions and overall programming systems (e.g., can significance monitoring be successful in the environment of user modified software?).

The paper by Goldstein and Hoffberg explains how a particular significance monitoring system has been incorporated into the Fortran compiler for the CDC 6600. The idea is to replace double-precision variables by "significance" variables, and it is shown how this can be implemented with the hardware available.

The papers of Dunham and Thacher both make a strong case for the usefulness of a variety of special arithmetics. The question is: How can each computing center provide this variety of special arithmetics? A number of alternative implementations are presented and discussed, but there seems to be no clear "best" choice. In view of the trade-offs between implementation effort, ease of use, efficiency, storage requirements, etc., we can expect this controversy to continue for some time.

III. Libraries

There are six papers about libraries, and the first of these by Battiste sets the scene with a "white paper" on production in quantity of mathematical software. This paper is essential reading for those chronic complainers about libraries or those who claim (hope?) that the Utopian library is just around the corner. The author has had extensive experience with IBM, particularly in connection with SSP/360, and he points out some very real and difficult problems that are not so obvious.

The next paper by Traub relates to the efforts at Bell Labs to produce a very-high-quality library of mathematical software. The work is carried on in almost ideal conditions: highest priority goes to quality and thoroughness; there are no pressing deadlines; there is good access to knowledgeable people (both inside and outside Bell Labs). The fact that this project produces one or two programs per year is indicative of the level of effort required to produce quality software.

The next two papers, by Newbery and by Dickinson, Herbert, Pauls, and Rosen, describe how two large industrial computing centers chose to build their subroutine library. Their approach had to differ from that in the previous paper and they differ considerably from one another.

The last two papers in this group consider the library that most users never think about, but merely use. This is the library of elementary functions which is usually imbedded in a compiler. Cody presents the results of a rather thorough survey of the quality of these libraries for the Fortran compilers of the major manufacturers. Only IBM can be proud (or even unashamed) of the results reported. Most other libraries commit faux pas that were well advertised a decade ago.

The quality of the IBM elementary function library is due in considerable part to Kuki, and his paper gives a penetrating analysis of the construction of quality routines. While quality routines are sometimes faster, shorter, and more accurate than simple-minded routines, one must inevitably face the trade-offs and decisions of value inherent in mathematical software.

IV. The Automation of Numerical Analysis

The first paper of de Boor considers the mathematical procedure, integration, which has received the most study from the point of view of automatic numerical analysis. After pointing out the pointlessness of observing that the problem is impossible, he discusses a meaningful framework in which to consider the problem. Various approaches, tried and untried, are analyzed along with some of the basic questions that any such routine must face. The thoughtful reader can ponder both the difficulties and promises of automating integration.

The paper by Gear builds on the long experience in the automatic solution of ordinary differential equations. He outlines past attempts and the lessons learned from them and then goes on to describe his current project in this area. This system goes a very long way toward meeting the user on his own ground and it incorporates some novel and provocative ideas in system organizations, man–machine interface, and numerical mathematics.

V. Comparative Evaluation

Eleven algorithms for numerical quadrature are compared in the paper of Kahaner. The criteria of comparison are discussed and detailed results given of the tests. Some algorithms are clearly poor, but not one of them stands out as best in all situations. One may compare the results reported in Chap. 7 for the same tests made on another algorithm, CADRE.

VI. Systems for Mathematical Software

These are two papers which discuss aspects of complete systems which incorporate higher-level mathematical software. One, by Symes, considers the analysis and evaluation of general expressions in the Numerical Analysis Problem Solving System (NAPSS). These expressions involve more complex entities and structures than occur in the Algol–Fortran type languages. The other paper, by Bayer, describes a self-contained system for mathematical programming (of the nonlinear optimization problem type, not writing programs for mathematics). He outlines the language features, operators, and associated data structures that are required and gives some indications of how the language processor works.

VII. Nonnumerical Software

Sammet gives an authoritative survey of the past and present in non-numerical mathematical software with the emphasis on recent activities. There is an extensive bibliography and a provocative discussion of the ideal "scientific assistant" which incorporates (efficiently and naturally) all of the good things of mathematical software.

VIII. Mathematical Procedures

The last five papers consider various mathematical procedures in detail. The first of these by Payne presents some versatile algorithms for generating random variables from a specified distribution. In the background is the stimulating suggestion that random numbers are best generated via micro-programming or even by (pardon the bad word) hardware.

Lawson gives very convincing evidence that a singular value analysis of a matrix can be extremely useful in a variety of problems related to linear algebra. It is particularly valuable when poorly conditioned matrices are involved and where standard methods tend to generate garbage.

The paper of Johnson and Parlett contains an instructive analysis of the solution of an eigenvalue problem. They identify a sequence of intermediate problems between the original mathematical problem and the computation actually made. The interrelationships of these problems are studied and it is seen that to achieve an overall accuracy requirement requires higher accuracy on some of the intermediate problems than one might initially guess.

Barton, Willers, and Zahar give an analysis and comparative evaluation of the Taylor series method for ordinary differential equations. This paper is intriguing on two counts: First, it shows that this old, but usually maligned, method can, in fact, be much more effective than methods currently in fashion. Second, and more significantly, this paper indicates that a combination of numerical *and* symbolic analysis might lead to great improvements in some areas of mathematical software.

The final paper of Brown and Dennis gives a new algorithm for solving nonlinear least-squares problems. Tests made indicate that this new algorithm is quite efficient compared to some other methods.

Part Two

PROCEEDINGS OF THE SYMPOSIUM

This part is the heart of the book. It contains the complete text of the papers presented at the symposium. The papers are preceded by a brief discussion, which is intended to indicate the flavor of the papers.

CHAPTER 5

THE PAPERS

5.1 A User's Experience with Sophisticated Least-Squares Software in the Discovery of the Lunar Mass Concentrations (Mascons)

P. M. Muller

JET PROPULSION LABORATORY
PASADENA, CALIFORNIA

I. Nature of the Data Reduction

The data was reduced in the manner detailed in Muller and Sjogren (1968). It involved two steps, one using the raw doppler observations themselves, the other utilizing residuals (observed doppler minus that calculated by the orbit determination program).

The Jet Propulsion Laboratory Deep Space Network uses large radio telescopes to track our unmanned US spacecraft with a coherent two-way doppler system. On the ground an atomic oscillator controls a 2.3-GHz transmission to the spacecraft which in turn coherently multiplies to a slightly different frequency which is retransmitted. This spacecraft signal is received at the same ground antenna, multiplied back by the reciprocal of the spacecraft multiplier, and is then beat against the master oscillator. The difference signal thereby obtained is the well-known doppler tone introduced by the relative antenna–spacecraft velocity. These measurements are made by continuously counting (and resolving) the difference frequency at discrete times, usually every minute. These measurements are in hertz, and have the incredible precision of approximately 0.001 Hz under good conditions. At 65 mm/sec/Hz, the ability to measure spacecraft velocity, in 1 min, is approximately 0.05–0.1 mm/sec.

In making a circuit of the moon, with the spacecraft visible on the front side only, we can see this small change on top of a total velocity ± 2 km/sec. Clearly this data is an extremely precise tool, and very small physical effects on the spacecraft can show up.

The orbit determination program is designed to model the entire solar system with sufficient precision to reach the level of data noise, and on most missions either it has done so successfully or it has been possible to eliminate whatever model deficiency was causing an observed disagreement between the program computed doppler and that which we observed. From the beginning of the Lunar Orbiter program, on the other hand, it was apparent that the doppler residuals (observed minus calculated) were about three orders of magnitude higher than the random noise, and were systematic. This was clear evidence that either (1) the model (Orbit Determination Program) was incomplete or in error, or (2) the spacecraft was returning spurious signals not related to an important physical effect. It was immediately suspected, but not proven for nearly two years, that the cause of these residuals ("periwiggles" as they were called) was local variations in the lunar gravitational field (which had not been suspected or modeled in the Orbit Program).

To eliminate program errors as a spurious source of error, we attempted to fit the raw doppler (having an excursion over one orbit of at least 2×10^4 Hz) with a polynomial of some kind, down to the data noise. If the same residuals were to come out of such a fit, then the Orbit Program would be cleared of blame. We tried Chebychev polynomials and others, but none were able to even begin to reveal 1-Hz signatures out of the kilohertz "parabola" formed by the raw doppler.

Drs. Peter Dyer and Steven McReynolds of Jet Propulsion Laboratory (JPL) were able to devise a Fourier-like system using terms for the earth's rotation and spacecraft rate about the moon, with harmonics and phase coefficients, which did adequately fit this data (a first breakthrough in this type of analysis). They employed a least-squares solver devised by Drs. Charles Lawson and Richard Hanson of JPL. This new system was the result of several years of (1) research into the nature of least-squares solutions and the attendant problems of computer implementation, and (2) development of programs to implement the newly discovered systems. Without this sensitive and stable least-squares system, the analysis of our problem would have been impossible. With this new, basically applicable tool, we were home free, and the residuals from the two independent programs agreed. We had proven the effect was really in the data.

The next step was to prove that it was not some spurious and meaningless spacecraft problem. This could be accomplished by showing that the same as well as distinct spacecraft produced the same or similar velocity residuals along the same or similar lunar tracks (when they overflew the same lunar paths). The same spacecraft, on subsequent orbits, inevitably flies almost identical tracks because of the moon's slow axial rotation rate. We were also able to find distinct spacecraft orbits, months apart, which happened to coincide. The residuals in all cases agreed, both from the Orbit Program and from the polynomial fits.

Knowing that the residuals represented velocity variations induced by the lunar gravity field meant that, if we could fit these raw doppler residuals with some sort of polynomial, we could simultaneously smooth the data and differentiate it, producing acceleration data, the desired measure of gravity. Because of the extremely high frequency variations present in the observations, other more sophisticated methods for obtaining the bulk of gravity information had not succeeded. The proposed fitting of the residuals (instead of the raw doppler as was usually attempted) could reasonably be expected to work, providing a suitable fitter could be found. The nature of the residuals is illustrated for two typical examples in Fig. 1. The results obtained by patched cubic fits (discussed below) are also shown.

Once again we tried standard techniques from Chebychev to Fourier to algebraic, all without success. We now needed to fit the 1–2-Hz variations seen in Fig. 1 down to a noise level of about 0.01 Hz at least. Dr. Lawson and his staff came to our rescue once again. They had just finished a system of spline-fit programs using the linear least-squares package already discussed above.

I do not want to go into detail about splines, which I respect as a highly involved and subtle branch of numerical analysis in its own right. It should be noted, for those who may be unfamiliar, that splines are made up of ordinary polynomials of low order (in our case, cubics) which are patched together at break points with some constraint (such as first-derivative continuous). It was precisely this type of system we solved for each of 80 consecutive orbits of data (about 90 min of data each) in order to achieve the sample results of Fig. 1, which are self-explanatory.

We have since used an improved spline fitter which uses cubic polynomials constrained to second-derivative continuous. This resulted in greater ease in selection of break points, and freedom from the nodes which can be seen in the accelerations (derivative) plotted in Fig. 1. These 80 consecutive orbits were plotted, acceleration point by acceleration point, on a lunar map. The result, when contoured, appears in Fig. 2, and constitutes discovery of the mascons, or mass concentrations, revealed by the *higher* gravity over the larger lunar ringed sea *basins*. (Higher gravity over a basin must imply greater density and mass there.)

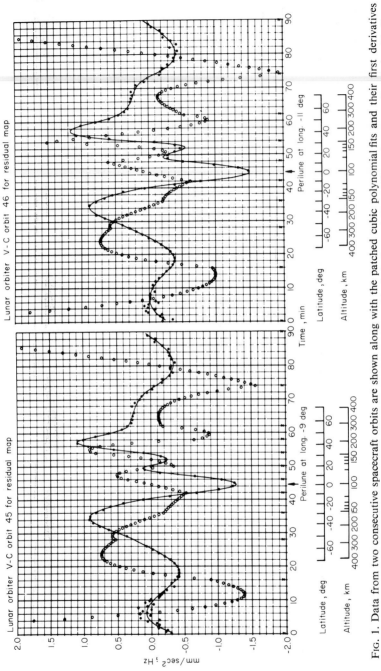

FIG. 1. Data from two consecutive spacecraft orbits are shown along with the patched cubic polynomial fits and their first derivatives (which indicate accelerations or gravitational forces). *, Raw doppler residuals in hertz, 1 Hz = 65 mm/sec; ——, Patched cubic fit to residuals; ○, Normalized accelerations, in millimeters per square second; ×, Patched cubic break points.

In retrospect, and this is a basic point of this paper, the whole analysis was elegantly simple and easy to apply, particularly compared to the more sophisticated methods employed both before and since to obtain slightly more precise and complete results. The reason it was successful and the conclusions we draw from this experience which apply to computer program development are noted in the next section.

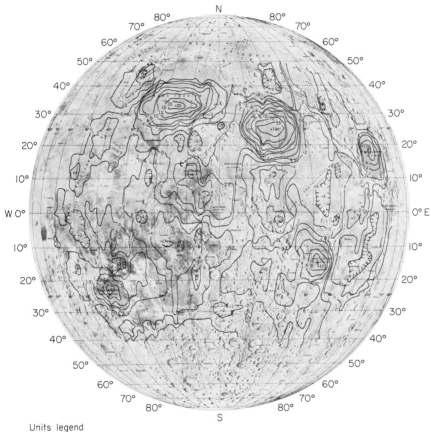

Units legend
Above × 0.1 = mm/sec^2
Above × 10.0 = milligals

FIG. 2. The contours of gravitational force deduced from the analysis of the residuals in the raw doppler data.

II. Implication for Program Development and Distribution

The author has developed large programs himself (the MOP System, General Technical Document 900–20, Jet Propulsion Laboratory, and distributed by COSMIC) and has used many developed by others. In addition he has programmed most tools used in day-to-day analysis. Four basic empirical observations have resulted from this experience.

First: All attempts to employ main programs, designed for specific purposes (such as regression analysis or matrix packages) while searched for and found, were never successfully applied to any problem I have faced. The reason is simple to state; none of the programs were either general or specific enough to do the job at hand with reasonable efficiency. Some were too general, and therefore difficult if not impossible to apply. Some were too specific, and would not handle the problem at hand. All suffered from the disadvantage that it was cost- and time-ineffective to convert them into subroutines suitable for call. This meant that each stage of the problem had to be attacked sequentially and separately, and data being passed from tape to tape to card, etc.

Second: Most available programs were years old, employed arbitrary or obsolete mathematical methods, and therefore failed to solve our state-of-the-art problems, even if the system could be brought to bear on the problem at hand.

The programs we used in the mascon analysis were in subroutine form, with little documentation, but they solved a basic problem in a new and much improved fashion. They were modular, loaded with comment cards, and could be readily brought into simple programs written to handle the data and I/O. While my type of work may seem to be of a kind shared by only a few engineers and scientists, and many program developers have made this comment in defense of "blackbox one-application" programs currently available, I believe this is, in fact, not so. *Most* engineers coming out of school today are competent programmers, and most older engineers have learned. For example, of some 50 engineers in my section when I came to JPL six years ago, only one wrote programs. I was the second. Now, I would estimate there are not more than 20% who do not write programs. In this connection, it should be noted that I refer to Fortran coding of moderate efficiency. For most engineering work done by one or two individuals, highly efficient programs are not necessary, and the added complexity and cost of same is unwarranted.

Third: Most engineers do write programs and the rest should learn. The program developers should give more attention to writing what could be called "user modifiable programs." It was interesting to me that the only other "user" at this meeting, Dr. R. C. Bushnell, gave a paper with this exact title. That modifiable programs are needed is a conclusion of his paper and my talk. Such programs, we agree, should be modular subroutines accomplishing specific tasks. They would be 1–100 instructions in length, cite a reference for the mathematical theory, and be loaded with comment cards giving the role of each variable and step. No further documentation should be needed other than an abstract somewhere, permitting contact with the program by a potential user.

Fourth: Each user program would be a one-of-a-kind, user modifiable subroutine. We would not be so much concerned with 10% faster and 10% shorter as with ten times better; better in the sense that the program does an old or new job in an improved way. This is a plea for more research and development in numerical mathematics such as that exemplified in our work by Dr. Lawson and his colleagues. Without their previous research and development, the mascon discovery would have been delayed years, and cost much more to accomplish.

III. Summary of Conclusions

(A) I have found little use for one-purpose or multipurpose blackbox main programs designed for stand-alone use.

(B) Most available programs are highly duplicative and employ years-old techniques.

(C) Most engineers program, and are eager and willing to use well-constructed, user modifiable subroutine blocks.

(D) Finally, let's meet the need for support programs by placing heavier emphasis on modular, user modifiable subroutines, backed up with theoretical research and clever programming and design. For documentation place comment cards in the program, and publish a brief abstract emphasizing the mathematics and not the programming.

ACKNOWLEDGMENT

Many successful efforts I have participated in have been made possible by top-quality programming done by others. I owe a great debt to those who work in that field, and it was with great pleasure and respect that I was able to join with some of the best program developers in this country at the Mathematical Software meeting. I hope this modest effort on my part serves to help clear, rather than muddy, the turbulent and difficult waters of program development, and to help answer the pertinent questions why, how, and for whom.

REFERENCE

Muller, P. M., and Sjogren, W. L. (1968). Mascons: Lunar mass concentrations, *Science* **161**, 680–684.

5.2 User Modifiable Software

Robert C. Bushnell

OHIO STATE UNIVERSITY

COLUMBUS, OHIO

I. The Argument for Easy-To-Modify Software

A. WHO ARE USERS?

The purpose of this paper is to argue for more easily modified software and to indicate a method whereby this may be achieved. The first question that must be posed in such a quest is: "Who is the user?" It is my feeling as a user of software that the typical user for whom the software fabricator believes he is writing is a rather dull fellow who surely has not ever before written a line of coding. While this description may continue to fit some users, in a day where almost every college graduate has had at least a little of some programming language, it is becoming less appropriate. Increasingly, users are, I believe, people variously knowledgeable about programming, but not people who in any sense pretend to be programmers. To them a computer is something that is to do work they want done. They are aware of the wide latitude programming permits and software already written represents work they do not have to do. The best software from their point of view is that which is easiest to use.

But how should we identify ease-of-use? One attitude which I believe to be characteristic of software designers is to attempt to anticipate *all* the options that *all* users might wish to have and to provide these together with a system whereby digits or keyword parameters in control cards select the options. Fancy versions of such systems operate on the compilation level, including those options specified by the user. Such programs are fine and I should be the last to fault such efforts, but it is fairly obvious that one can always posit a user who would like something just a little bit different. How often have I heard my colleagues say, "but it won't give me that data," or "it won't accept that kind of input," or "that coefficient isn't calculated by this program." For anyone with any programming experience at all the normal reaction to such statements is to insist that one really ought to be able to modify the program to make it do what is desired. Unfortunately this is never very easy. Even if the programmer has written without using such labels as "Amy," "Betty," "Carol," "Doris," and avoided such enlightening comments such as "this will zap it" it is a rare program that is sufficiently well documented that a user can find the information that he wishes to use and the point in the program where he should insert his commands.

B. EASE-OF-USE

A referee on this paper has suggested that "ease-of-use" is (1) almost certainly going to be in conflict with ease-of-modification, and (2) is almost certain to decrease efficiency severely in frequent situations. He has also asked: "Isn't there something a software programmer can do to give you the advantages you seek without requiring the user to modify the programs?" Since his comments are likely to be characteristic of most of you at this symposium let me speak to his points directly. There is a difference in philosophy between us. I, of course, concur with him that ease-of-use is a *bonum*; however, ease-of-use to me means getting what I want.

I think the best way to answer is to assert that I do *not* favor ease-of-use if ease-of-use means additional layers of software. It is sometimes asserted, as we used to joke in the offices of the computer manufacturer for whom I once worked, that the perfect program was one which, when given a vague description of the customer's problem, would deliver *exactly* the output required. Now I have no doubt, such is my faith in programmers and analysts, that someday such a program will exist. However I have the feeling it would be a rather large program with an exceedingly complicated logical structure. Being myself an economist, who is, by *definition*, a man interested in control systems which work with a minimum of tampering, I wonder if a little local option would not work as well as one big federated program. It is for this alternative approach that I am arguing. If I have a supermarket of compatible but independent items which I can combine rather facilely into a program which meets my particular needs I have an ease-of-use which is not incompatible with user modifiability.

I can appreciate the power of keyword parameters and the desire of at least some programmers to have their users communicate their needs to the program in natural language. I am at the same time rather awed by the software needed to implement this goal. Is it not possible, I ask, to go another direction as well? After all, can a program really be made so open-ended as to include options that the specialized applied researcher, let alone the software programmer, has not even yet conceived? Such open-endedness requires, I believe, ease-of-modification. Complexity of structure *conflicts* with ease-of-modification. For example: Most linear programming codes include a sensitivity analysis with respect to right-hand sides. This interesting exercise determines the domain within which each so-called right-hand-side parameter can vary individually, under the assumption that all others do not, without causing any element of the solution vector to become negative. This procedure has a simple variant. It may be described as the answer to this question: "What is the maximum percentage by which all right-hand sides could simultaneously vary without causing any element of the solution vector to become negative?" To implement this calculation under a system designed for ease-of-modification is relatively simple: I write it. In an easy-use system, since it is, perchance, an operation desired by only one half-mad economics professor in Columbus, it isn't even an option. Do I then discount the possibility of obtaining the advantages I seek without requiring the user to modify the program? Yes, I do, for the advantage I seek as a user is a method whereby I may implement procedures which which I am experimenting. In such a case efficiency of execution is not the principal problem.

C. MODIFIABILITY DOES NOT CONTRADICT EFFICIENCY

What about the criticism that to allow the user to modify the program will ruin the efficiency of the job. To this I can only reply that users *will* appropriate existing programs and they *will* modify them as best as they can. If the person who is modifying is inexperienced he will make less efficient modifications than he who is crafty. With programs that are hard-to-modify, the user will pay dearly to make his modification. Let us not forget that the relevant costs are not solely in terms of machine time but in machine time plus programming expense. When programs are not formulated for ease-of-modification, a user may make his modifications on a pragmatic basis. If he is not sure of the use of all storage at all stages of the computation he may alter the contents of a storage cell and cause erratic program behavior. In debugging he will typically find the cause of the problem and make alterations to solve *that* problem. As problems arise they will be solved on an *ad hoc* basis. The program that results will be patched, hard to follow, and inefficient in running. Why else are there so many regression programs and so many linear programming codes if not that successive groups of researchers, finding that existing programs do not meet their special needs, commission new programs? Any experienced programmer would far rather start a new system from scratch instead of fighting his way to an incomplete understanding of another programmer's production. I cannot help but believe that even a professional programmer would rather start from a system in which most of the parts were satisfactory to him but which had the added feature of being easily modifiable by him in those parts he would like to change.

D. THE NEED FOR EASY-TO-MODIFY SOFTWARE

It may be interesting to you for me to describe how I became interested in this subject. In my work as an economist I am interested in so-called general equilibrium systems. In such systems, multiple decision-making units, all pursuing their own goals, interact. By their own individual decisions they collectively alter the environment faced by other decision-making units. The tool with which I approach such problems is linear programming. Typically one has a rather large number of such decision units involved in a problem, each a proxy for a class of real-world decision units. The question at root is usually how a technological innovation or a change of taste or a legislative prohibition will change the demands and alter the prices for decision units not directly involved with the change. To solve such problems one solves a large decomposed linear program and then observes the effects of changed activities, altered parameters, or added constraints on each of the decision-making units. The easiest way to solve such problems is to solve severally the decision problems of each of the decision units and then also the decision problem posed by their interactions one with another. In the course of these computations I often wish to plot certain key variables or otherwise especially handle a single unit. More important for our thinking here is that I don't know ahead of time what I want to do and may think of things as I go along. Therefore I want a central body of coding to work on any number of problems presented to it within their own data contexts and with their own named arrays, and I want the ability to modify the process and record data. In short I want a modifiable program.

My first thought of how to obtain this capability was to modify an existing system. I received a grant for the purpose and hired an excellent programmer. After struggling for two months to understand documentation such as "this will fake them out," we came reluctantly to the conclusion that modifying someone else's system was costly enough so as to justify cutting our losses and starting completely over on our own. We called our effort OSMOSIS for the Ohio State Mathematical Optimization System of Integrated Sub-programs. Why not? As we have developed the system we have found flaws and I have come to the conclusion that writing an easily modifiable system is in itself an art. Since I am not a professional programmer, I am not forwarding my system as a prototype to anyone, but in formulating it I did learn what I want in the way of a modifiable system.

II. Writing Easy-To-Modify Software

A. DESIRABLE PROPERTIES

The basic premise from which this discussion starts is that every large software system should be built up of subprograms, each of which performs a well-defined function. Further it is my opinion that each such subprogram should be replaceable in the sense that a user subprogram should be able to replace the supplied subprogram. From the user's point of view there are real advantages to this premise. Most obviously one may have extensive data already punched or on media in a form not readily acceptable by a software package. Or, one may want output in a different format or want different output from a program than contemplated by the software designer. Or, one may want to perform additional calculations not included by the software designer. Or, one may really wish this procedure to be but a subprogram in a larger scheme.

Providing software systems which meet the need for user modifiability is not easy. If not written in some language like Fortran such subprograms must at least share some common calling conventions. Then, some rather obvious documentation, flow charts, comment lines between coding lines, etc., is required. Also, some method must be devised to permit easy intercommunication among subprograms so that user programs easily mesh with supplied subprograms. A system for avoiding a particular required set of variables or a particular setup of "common" should be implemented so the software system may be employed simultaneously on more than one set of data. Finally the set of building-block matrix subprograms utilized throughout the system should be available to the user so he may construct alternate or additional routines without duplicating routines for rather basic manipulations.

Let us consider what is required for a program to be modifiable: First, the use of every data cell must be clearly identified. Second, there must be some manner whereby every decision and control parameter is revealed. Third, the logical structure of the program must be transparent to the user. Last, so that the user may himself write additions or alterations using routines already in the program, the function of each subprogram must be clearly identified.

B. An Example: The OSMOSIS System

Other methods of resolving the problems posed by ease-of-modification may exist, but, as an example, OSMOSIS handles these problems by means of the following conventions:

(1) All data are in one singly dimensioned array (we may identify this *data array* as *DA*).
(2) All parameters not data are stored in one singly dimensioned array (we may identify this *constant utility* array as *KU*).
(3) Each subprogram has only one return point.

The use of a common data array does not mean that the routines must regard all data as in a single array nor lack mnemonics for designating parameters. Rather every "master procedure" (what would be called a main program in a "stand-alone" package) defines a configuration of storage called an "environment" and establishes a common communication standard called a "*KU* list." These concepts are best explicated by example.

The *environment* consists of a positional and dimensional definition of a list of matrices and vectors to be used in solving the problem. If writing arrays out on paper, one might conceive of the format in Fig. 1 for handling a

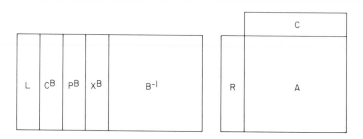

FIG. 1. A sample environment for a revised simplex computation. B^{-1} is the inverse of the current basis, X^B is the current solution, P^B is the current dual variable values, C^B is the ordered objective function coefficients of in basis activities, L is the list of basic activities, R is the original right-hand-side vector, A is the matrix of original activities, C is the objective function vector.

revised simplex problem. One of the functions of the master procedure is to calculate the base address or cell element of *DA* in which the (1, 1) element of each array will be located. For the example given, if KBM is the number of rows in the inverse and KAM the number of rows in the problem statement, the statements in Fig. 2 would be used to establish the environment.

$$KU(KLBASE) = 1$$
$$KU(KCBBAS) = KU(KLBASE) + KBM$$
$$KU(KPBBAS) = KU(KCBBAS) + KBM$$
$$KU(KXBBAS) = KU(KPBBAS) + KBM$$
$$KU(KBBASE) = KU(KXBBAS) + KBM$$
$$KU(KRBASE) = KU(KBBASE) + KBM * KBM + 1$$
$$KU(KCBASE) = KU(KRBASE) + KAM - 1$$
$$KU(KABASE) = KU(KCBASE) + 1$$

FIG. 2. Fortran statements required to establish the sample environment.

The *KU list* is a reservation of cells in the *KU* array. For example, the statements given in Fig. 2 for establishing an environment would have to have been preceded by a "*KU* list" which at least contained the statements given in Fig. 3 (or an equivalent DATA statement): Names in the *KU* list

$$KLBASE = 1$$
$$KCBBAS = 2$$
$$KPBBAS = 3$$
$$KXBBAS = 4$$
$$KBBASE = 5$$
$$KRBASE = 6$$
$$KCBASE = 7$$
$$KABASE = 8$$

FIG. 3. A sample "*KU* list."

will be passed as parameters to subprograms which will use these names as subscripts of *KU* to obtain the desired parameters. This practice permits parameter references to subprograms to be written as mnemonics, even though all parameters are in one *KU* array.

Finally let us consider the effect of the single return rule. To accomplish its task the master procedure calls other subprograms. These are of two varieties: (1) *Atoms* are subprograms that are independent of an environment. An example is matrix multiplication. Except for the fact that their parameters are expressed as subscripts of a *KU* array these programs are similar to routines such as make up the matrix section of the IBM Scientific Subroutine Package. (2) *Procedures* are environmental dependent. An example is the selection of an entering activity in a linear programming problem. Since no atom or procedure may return to any but the next step, the master procedure displays a complete flow chart for the procedure.

From a user's point of view these conventions are very convenient. (1) The master procedure spells out exactly and explicitly the environment; (2) the description and calling parameters of every atom and procedure define quite clearly what each subprogram accomplishes and what a substitute routine has to accomplish; (3) the master procedure itself displays a complete flow chart for the procedure.

C. APPLICATIONS OF OSMOSIS

In concluding I would like to allude to some particular instances in which the modifiable structure of OSMOSIS has been exploited. As one example, a colleague of mine has a transportation problem with additional constraints. When set up as a linear programming problem, there are a few constraints of the conventional sort but a great number of columns in which each vector has only two unit elements, all other elements in the column being zero. With a revised simplex procedure which stores the original activities on disk, only the few nonzero coefficients need be stored. The regular revised simplex procedure for this environment was modified to fit his circumstances merely by writing a special entering variable routine to "price out" the potential entering activities. As another example, a second colleague utilizes recursive programming, a scheme whereby successive tableaux are modified both by the outcome of the previous tableaux and also by the inclusion of exogenously determined parameters. In his application the linear programming algorithm is but a subprogram of a larger program. Finally, I myself have found it very instructive to plot demand and supply curves derived from activity arrays. This involves essentially parametric variation of an element of the objective function. Many software packages will perform this operation but none of which I know except OSMOSIS will permit me to store the numbers from successive iterations in my own array for use with the printer plotting routine available at my computer center.

D. Other Applications

Finally, although the system I have developed is within the area of mathematical programming, I cannot help but believe that the principle has applicability in other areas. In the linear regression area, for example, I would think that as researchers wish to add other test statistics to their programs or incorporate least-squares procedures within larger programs and procedures a modular approach would be very useful. I can't help recalling remarks, perhaps now outdated, that the reason there were no software systems to handle standard industrial jobs such as payroll was that each application differed in just enough respects that writing a routine to handle all the idiosyncrasies of a particular firm was easiest done by starting from scratch. Would not a library of state tax routines or a standard payroll package which was modifiable be a very profitable item for a software marketer? As a user I find myself continually wanting to use a program package just a little differently than the designer intended. With OSMOSIS I am able to do so. Why not with everything?

ACKNOWLEDGMENTS

The author wishes to record his appreciation to Mr. Paul Rappoport who aided materially with the exposition and to Mr. Lawrence J. Shoemaker and Mrs. Patricia Weed, programmers on the OSMOSIS project. Thanks are also due to the former College of Commerce and Administration for a grant which facilitated programming and to the Instruction and Research Computer Center of the Ohio State University which made free computer time available for the project.

5.3 Number Representation and Significance Monitoring*

R. L. Ashenhurst

THE UNIVERSITY OF CHICAGO
CHICAGO, ILLINOIS

In the quarter century since the first large-scale digital computer started functioning, questions concerning the representation of numbers for such computers have received continuing attention. Factors affecting the choice of a number system may be roughly grouped into those bearing on engineering efficiency (e.g., using binary rather than decimal), programming convenience (e.g., using floating point rather than fixed), detection and correction of machine malfunction (e.g., introducing extra digits to permit a redundancy check), and assessment of computational error.

The presence of computational error is, of course, inevitable in computation confined to finite resources. Although it would seem that the sacrifice of efficacy in error assessment for the enhancement of performance with respect to objectives in the other three categories must represent a trade-off of questionable utility, this nevertheless seems to have been the tendency in computer system design. And while (at least some) numerical analysts may criticize computer designers for failing to take the needs of error assessment into account, they themselves have for the most part failed to incorporate considerations of number representation into the theory and practice of numerical analysis in an effective way. The purpose of this paper is to investigate some of the issues involved in these questions, in a context appropriate to the use of mathematical software. This latter consideration dictates that computational error effects be not merely understood and analyzable, but be treated explicitly in a manner which permits them to be followed through a chain of references to a collection of numerical routines. The realities of such "significance monitoring" are seldom treated adequately in works on numerical methods, even those having a primary emphasis on error analysis. The present point of view suggests that through the medium of number representation, in the broadest sense, the needs of significance monitoring in the application of mathematical software may be served.

* Work reported here has been supported by the US Atomic Energy Commission.

I. Number Representation

If x is any real number, then x has an infinity of exponent-coefficient representations (e, f) with respect to a given base b, one for each integer e. That is, $x = b^e f$ for any $e = 0, \pm 1, \pm 2, \ldots$ and an appropriate real number $f = b^{-e} x$.

Although the form of (e, f) is suggestive of a "floating point computer number," the intention here is to encompass a more general class of representation. Thus it is assumed that f may be any real number, finitely representable or not, and e may take on any integral value without restriction. Hence (e, f) is a representation of x in the mathematical rather than the symbolic sense. The advantage of this point of view is that some, although by no means all, of the features of a symbolic representation may be conveniently incorporated in the representation by placing mathematical, rather than formal, restrictions on e and f, and this facilitates a general discussion of error considerations.

For example, the customary normalized floating point symbolic representation is modeled by imposing a restriction $e_L \leq e \leq e_U$ on exponent and requiring that for $f \neq 0$ coefficient f obey $1/b \leq |f| < 1$. This not only restricts the range of nonzero representable real numbers to $b^{e_L - 1} \leq |x| < b^{e_U}$, but also defines a unique representation for each number x within this range. The restriction on exponent is intended to reflect a symbolic representation of e in terms of a fixed number of digits, while the restriction on coefficient is intended to reflect a representation of $|f|$ as a string of digits following an (implied) radix point, the first digit of which is not 0. Questions as to precisely what system is used (sign-magnitude, complement, etc.), or even the digital number base (b or some other), are essentially irrelevant to the type of argument under consideration here. At most these imply minor changes in the restrictions, such as allowing $f = -1$ in the case of a complement representation. By contrast, the question of the finiteness of the digit string for f is a crucial one for error analysis, and it is just here that the mathematical approach is found wanting, since the class of finitely representable numbers f is not at all straightforward to handle. The fact that rounding a coefficient f to n (radix-b) places introduces a perturbation of order b^{-n}, and by virtue of normalization this is the relative as well as the absolute order of magnitude, is used as the basis for a mathematical approach to analysis of rounding error. Here again, success is achieved only by taking what is essentially a symbolic problem (finite digit string) and translating it into mathematical terms (bound of b^{-n}). The effect of using increased precision at some stage in a calculation can be taken into account by changing the perturbation to b^{-n-p}, where, for example, $p = n$ for double precision, and so forth.

The exponent-coefficient representation can easily encompass the notion of "fixed point computer number," as traditionally defined, by simply requiring that $e = 0$, so that $x = f$, and removing the restriction that $f \geqq 1/b$, so that $-1 < f < 1$. From the point of view of symbolic representation, e then need not be explicitly included at all, and it is the usual custom to assign the digit places thus saved to extend the precision of f by p, where p is the number of places assigned to exponent in floating point. This introduces an incompatibility between floating point and fixed point representations which can lead to confusion, although it has the advantage that the ingenious subroutine designer can convert to fixed point at crucial stages of a calculation where just a little extra precision is needed, and then convert back to floating point once more.

In the early days of computers, floating point facilities were not available, and arithmetic units worked directly on fixed point symbolic representations only. Since normally one would not expect to encounter only numbers less than unity in magnitude, the programmer had to assign "scale factors" to numbers represented in the computer, and keep track of them himself. Although much of the early literature explained this in terms of recasting a problem so that each number x was replaced by a scaled number $b^{-d}x$ in the required range, a much more straightforward way to look at the situation is in terms of "virtual exponents" attributed to fixed point numbers. Thus two operands are conceived of as (e_1, f_1) and (e_2, f_2), but only f_1 and f_2 are symbolically represented in the computer. The computer arithmetic unit computes their product as $f_0 = f_1 f_2$, and the programmer must know that this is to be assigned virtual exponent $e_0 = e_1 + e_2$. In this system "numeric shifting," which obtains $b^{-d}f$ from an operand f, is to be thought of as an "adjustment" operation leaving a number unchanged, but altering its virtual exponent from e to $e + d$ (since (e, f) and $(e + d, 2^{-d}f)$ represent the same number). By use of this operation a computed product at exponent $e_1 + e_2$ can be readjusted to any desired exponent, be it e_1, e_2, or some other. The way most arithmetic units are designed allows the product $f_1 f_2$ to be developed in double precision, and shifted as a unit, so the rounding error associated with the final adjusted product to be stored is still of order b^{-n}. These techniques can also be applied (only a little less straightforwardly) to handle addition/subtraction and division of numbers with virtual exponents.

If fixed point calculation is conceived of in this fashion (which all too often it is not), floating point then appears to be mainly a device to relieve the programmer of the burden of keeping track of exponents, by representing them explicitly (at the accompanying sacrifice of precision in the fixed point part f). The adjustment to normalized form is then justified on the basis that one would like to have numbers uniquely represented, and besides, this seems to be the way to preserve as many "significant digits" as possible (since rounding errors are then of relative as well as absolute magnitude b^{-n} to f). Unfortunately, the formal nature of the normalization criterion (which can require adjustment of results of addition/subtraction operation by arbitrary shift amounts) makes significance very difficult to assess. This gives rise to a situation where results, even if optimally adjusted (which assumption is indeed only intuitively justified in any case), may be quite inaccurate, with no indication being given of the fact. Particular trouble in this regard may be expected in the region around zero, since the number $x = 0$ has no normalized representation (it being necessarily an exception to the $|f| \geq 1/b$ restriction), and the calculation of a series of convergents to zero, all of which must be represented normalized, leads in certain cases to massive loss of significance. One might argue that since the numbers are small the loss of accuracy does not matter, but the validity of this depends very much on what is to be done with the numbers subsequently.

The foregoing discussion leads naturally to the idea of an unnormalized floating point representation, where numbers are carried as (e, f) with the exponent e explicitly represented, but with no requirement that the fixed point part f be normalized. Although floating point arithmetic units frequently permit operations which leave results unnormalized, and are even defined for unnormalized operands, these operations are usually not systematically defined in a way to permit their exploitation for purposes of significance monitoring.

In order to fulfill this latter objective, an unnormalized arithmetic system must be flexible enough to permit adjustment to be determined by a combination of automatically applied rules and program-determined options in a straightforward way. The University of Chicago Maniac III computer is designed to achieve this in hardware (Ashenhurst, 1962). Based on a single unnormalized exponent-coefficient number format, several varieties of arithmetic manipulation are made possible through the inclusion of operations which employ "specified point," "normalized," and "significance" adjustment rules for results. The unnormalized format permits smooth transitions to zero, since there exist a multiplicity of "relative zeros" with coefficient 0 and arbitrary exponent. There is also an incidental advantage in the avoidance of "exponent underflow," by never allowing a result to be adjusted so that its exponent exceeds the lower limit e_L.

Unnormalized arithmetic operations have also been made available by modification of a conventional arithmetic unit (Goldstein, 1963), or by design of interpretive software (Blandford and Metropolis, 1968). The same considerations of flexibility beyond the basic arithmetic operations also apply to these cases. At a higher level, there is the question of providing for unnormalized representation in standard programming languages. The "precision" attribute of a variable in the PL/I language, for example, is stated[1] to specify the number of digits to be carried in the representation of that variable. No provision is made, however, for letting this attribute to be determined dynamically by the process. What is wanted is an implicit attribute of a number, say "effective precision," which is determined along with assignment of values to a variable, and can also be altered arbitrarily by appropriate program statements. This, coupled with the ability to manipulate unnormalized numbers in assembly language when subroutine efficiency is at a premium, would make it possible to use the unnormalized representation as a measure of significance in mathematical software.

The spirit of computing with unnormalized arithmetic is that the non-uniqueness of representation gives an extra dimension to the process, which can be exploited for purposes of significance monitoring. This will be elaborated on in later sections; suffice it to say here that there is no implication that the use of significance adjustment rules automatically takes care of all error propagation problems, without further attention by the programmer.

Unnormalized number format is only one way of incorporating a potential significance indicator into a representation. Other possibilities include the explicit inclusion of additional information in a representation. One proposal is to carry along an "index of significance" i, and so represent a number in exponent-coefficient-index form (e, f, i) (Gray and Harrison, 1959). In this scheme f is normalized, and i is an integer in the range $0 \leq i \leq n$, giving an estimate of the digit place at which uncertainty is to be attributed to f. Determination of i then corresponds to adjustment rules for the unnormalized representation, and the system should be provided with comparable flexibility to determine indices by combination of automatically applied rules and program determined options. Similar considerations to those already mentioned also govern the use of a significance index with standard programming languages.

Another proposal for significance monitoring which has received considerable attention is the use of interval arithmetic (Moore, 1965). Here an "interval number" is represented by a pair of real numbers (x_1, x_2) giving lower and upper endpoints. Arithmetic is then defined abstractly in such a way that, if the corresponding real arithmetic operations are performed on any numbers encompassed by the interval operands, the real result will be encompassed by the interval result. This can be implemented by letting endpoints be represented by a pair of floating point numbers (e_1, f_1), (e_2, f_2), and executing interval arithmetic interpretively. By paying careful attention to rounding it can be insured that the "encompassing" criterion is rigorously adhered to even with finite representations. For this reason, many have argued that interval arithmetic is the most promising tool for significance

---

[1] In the IBM Systems Reference Library publication PL/I: Language Specifications.

monitoring, because it permits rigorous statements about error bounds to be made. Unfortunately, as might be expected, the price paid for this certainty is that intervals are often generated which are so large as to be useless for error assessment. To be successfully applied to significance monitoring, an interval arithmetic system must be equipped with operations comparable to those cited above for unnormalized and index systems, which permit knowledge obtained analytically to be brought to bear by the programmer. This is indeed suggested in the description of some of the interval methods which have been developed (Hansen, 1965).

Some possible representational variants of interval arithmetic suggest themselves. One could consider an implementation in terms of a double-coefficient (unnormalized) floating point number (e, f_1, f_2), implying that both endpoints be adjusted to the same exponent e. This would be somewhat more compact, and also simpler to implement in hardware. Another possibility is to use the so-called "centered form" of interval representation (Chartres, 1966), where an exponent-coefficient-increment form (e, f, h), with h a nonnegative integer, is used to represent the interval with end points $(e, f - 2^{-n}h)$, $(e, f + 2^{-n}h)$ (here n is the number of digit places of f, as usual). The reason for defining h as an integer is to suggest a scheme whereby h is carried with fewer digit places than f, and thus may be thought of as another variety of "significance index." Indeed, if h is of order 2^m in magnitude, then the unnormalized number $(e + m, 2^{-m}f)$ can be regarded as approximately representing the centered interval number (e, f, h). This leads to the notion of unnormalized arithmetic as approximate interval arithmetic, which permits concepts relative to each to be applied to the other (Ashenhurst, 1969).

It may appear that interval representation is naturally incorporated into standard programming languages, being realizable as a type "interval" comparable to types "real," "complex," and so forth. Given suitable conversion conventions, arithmetic expressions involving interval numbers could then be evaluated in interval arithmetic to give an interval result. For effective use of this expedient, however, auxiliary operations corresponding to adjustment in the case of unnormalized representation would have to be made available, for the reasons of flexibility cited earlier and explored in more detail subsequently.

Throughout the foregoing discussion it has been assumed that the reader generally understands how the representations cited can be thought of as providing an indication of "significance," but the necessity of giving a more precise definition of the latter concept has purposely been avoided. This is because the concept itself is a subtle one, and seems to require an exposition independent of a discussion of number representation, as is attempted in the next two sections.

II. Error Classification

It is customary to classify sources of computational error as follows: discrepancies due to uncertainties in input data, discrepancies due to the use of approximation formulas, and discrepancies due to the necessity of rounding or otherwise truncating symbolic representations of numbers obtained as computed results. This classification is often cited informally in the first chapter of a numerical analysis text, and all too often neglected in the sequel. In particular, the question of how the errors arising from these "sources" interact is not paid sufficient attention.

The formalization of this concept was presented at a very early stage in the development of computer numerical analysis (Householder, 1954). It is instructive to resurrect this early model for the insight it gives into contemporary techniques of error analysis and significance monitoring.

Consider the conceptually simplest of all computations, the development of a value $y = F(x)$ of a function F of a single argument x. Suppose the function F to be fixed and known (e.g., square root, exponential, etc.), but the argument x to be (possibly) different from its representation at hand, which may be designated x^*. A component of *inherent error* $F(x) - F(x^*)$ is thus defined. This is conceived of as the error attributable to uncertainty in the input value, and as such is not correctable by any amount of computational sophistication in the evaluation procedure, hence the name. Now, the said procedure generally introduces further approximation, first by replacing the function F by some arithmetically specifiable function Φ, so that the procedure actually is programmed to produce $\Phi(x^*)$ by a sequence of arithmetic operations, and then by further approximation of Φ by a function Φ^* such that $\Phi^*(x^*)$ is produced from x^* by rounding or otherwise truncating intermediate results to a finite number of digit places. Thus is defined the *analytic error*, as $F(x^*) - \Phi(x^*)$, and the *generated error*, as $\Phi(x^*) - \Phi^*(x^*)$. The advantage of framing the definitions in just this way is that, if $y^* = \Phi^*(x^*)$ is taken as the computed approximation to $y = F(x)$, the total error $y - y^*$ is the sum of the three components:

$$y - y^* = F(x) - \Phi^*(x^*)$$
$$= [F(x) - F(x^*)] + [F(x^*) - \Phi(x^*)] + [\Phi(x^*) - \Phi^*(x^*)]$$

Note that this additivity is not quite trivial; if, for example, one had decided (as a theoretical numerical analyst might well do) to define the analytic error as $F(x) - \Phi(x)$, then to get the same additive effect one would have to redefine the inherent error as $\Phi(x) - \Phi(x^*)$, which renders it inappropriately dependent on the approximation function Φ. Furthermore, an analogous redefinition of generated error is not meaningful at all, since the generated effects can only be defined with respect to the function Φ which is actually evaluated.

Now, in terms of the above model, it is customary to approach the assessment of error for each of the three components separately. Note that inherent error is defined in terms of a single function evaluated at two different arguments, while analytic and generated error are each defined in terms of two functions evaluated at a single argument. This leads to a distinction in the style of error analysis. For inherent error, one seeks a formula for a bound on $| F(x) - F(x^*) |$ as a function of a bound on $| x - x^* |$. Assuming that the function F is differentiable in an interval containing both x and x^*, this is at once expressible in terms of the usual differential approximation of elementary calculus,

$$F(x) - F(x^*) \approx \mathbb{D}F(x)\,(x - x^*)$$

where $\mathbb{D}F$ is the derivative of F. Thus the magnitude of $| \mathbb{D}F(x) |$, or its approximation $| \mathbb{D}F(x^*) |$, can be used to assess the magnitude of error $| F(x) - F(x^*) |$ in function value induced by an uncertainty of magnitude $| x - x^* |$ in argument value. This may be used as the basis for an automatic error assessment feature of a subroutine for calculating F, in those cases where the derivative $\mathbb{D}F$ is a simply evaluable function (as, for example, when F is expressible as a polynomial), or is related in some simple way to the function itself (as in the case of elementary functions such as square root, exponential, etc.).

By contrast, the type of analysis appropriate to the assessment of analytic error is that which expresses a bound on $| F(x^*) - \Phi(x^*) |$ for x^* known or assumed in some given interval, which is a standard problem in approximation theory. Much of the traditional and continuing effort in theoretical numerical analysis is addressed to this type of problem, and very powerful error estimation methods exist. It is also frequently the case that the function Φ is not selected in advance but is one of a class of closer and closer approximants, values of which can be computed by recursion. This is the basis of iterative methods which evaluate successively better approximating functions Φ, until one is reached which is found by some test to be sufficiently close to F for the given argument x^*. It is frequently suggested that the tolerance for goodness of approximation be selected partially on the basis of an assessment of inherent error, since it seems fruitless to compute a value $\Phi(x^*)$ such that $| F(x^*) - \Phi(x^*) |$ is much smaller than $| F(x) - F(x^*) |$. For many functions, however, known methods for computing Φ are so efficient as to make the saving involved in a conditional computation based on inherent error not worth the extra decision process entailed.

Superficially it would seem that the problem of assessing generated error would be similar to that for analytic error; that is, it would consist of ascertaining a bound $| \Phi(x^*) - \Phi^*(x^*) |$, for x^* known. But unfortunately the function Φ^*, depending as it does on the order in which arithmetic operations are performed and extraneous characteristics of the digital representations used, is not in general analytically tractable. The great advance in this area has been the development of the concept of backward analysis, which would in the present case consist of determining the existence of a value $\bar{x}^*$ such that, rigorously, $\Phi^*(x^*) = \Phi(\bar{x}^*)$, and obtaining a bound on $|x^* - \bar{x}^*|$. This is the spirit of modern round-off error analysis (Wilkinson, 1960), which has been extended to a variety of more complicated problems depending on a multiplicity of input values, and is usually framed in terms of relative rather than absolute error.

This transformation of generated error into inherent error calls attention to another point on which confusion often arises, the case where the argument x is "known" exactly, but is not representable as a computer number (e.g., a case in point is $x = 0.2$ on a computer which works in binary). The best representable approximation to x is then some value x^* differing from it by something of the order of a single rounding error, and instead of thinking of this as inherent error it is preferable to consider it as a generated error introduced at the zeroth stage of the computation. The reason this presents any problem at all is that it is so easy to overlook, the error ordinarily being generated in the input value conversion (from decimal to binary) instead of in the subroutine evaluating F. Although it may be thought that cases where a pure known input number is subjected to a single functional transformation and nothing more are out of the ordinary, this is precisely the situation which is likely to obtain when tests of a function subroutine are being constructed, and this sometimes leads to misleading reporting of the results of such tests.

The important point for the present discussion, one which is brought out very clearly for this simple case by the three-component error model, is that the backward analysis, if successful, merely converts the generated error component into a form where it looks like inherent error; i.e.,

$$\Phi(x^*) - \Phi^*(x^*) = \Phi(x^*) - \Phi(\bar{x}^*)$$

The assessment of the magnitude of $|\Phi(x^*) - \Phi(\bar{x}^*)|$ then depends on a knowledge of $\mathbb{D}\Phi(x^*)$, along with information about the magnitude of $|x^* - \bar{x}^*|$. Having noted this, one may look back at the analytic error component and see that a corresponding assessment could often be made for it, when it can be shown that $\Phi(\bar{x}^*) = F(x^*)$ for some $\bar{x}^*$ close to x^*. This expedient is usually not needed, however, since analytic error can be controlled separately and made negligible.

Thus in an idealized sense, at least, the effective assessment of total error can be said to depend on the effective assessment of inherent error. But here is where the problem is often left in the specification of numeric procedures. That is, it is usually stated that the assessment of inherent error is "up to the programmer," and is not to be taken into account in the procedure specifications (i.e., the procedure is designed to produce as good an approximation as possible to $F(x^*)$ for the x^* given it, nothing more). Furthermore, any definitive statement about generated error, if framed with reference to backward analysis, only begs the question of inherent error assessment again. Only analytic error is satisfactorily taken care of, by designing the procedure so that the magnitude of this component is less than the effect due, say, to initial rounding.

Although the task of assessing inherent error may not seem so formidable for the single function evaluation case discussed, consider that in general a computation consists of many stages, and that the total error at each stage acts as inherent error for the next. Thus the final result must ordinarily be considered as a multiargument function derived by composition from a long series of multiargument functions which are so derived in turn. It is this fact which makes an external analysis virtually unfeasible, and argues strongly for designers of mathematical software to provide some automatic method of inherent error assessment, in a form that it can be passed along from stage to stage. To put it simply, if one wishes to consider an evaluation $z = G(F(x))$, one would like to have some indication of the total error in the evaluation of $F(x)$ produced along with $y = F(x)$ and passed along with y as inherent error in the evaluation of $z = G(y)$. This is the sense of the term "significance monitoring" as it is used here. In the following section these error considerations are related to the conventional meaning of "significance," in a way that shows the relevance of the earlier discussion of number representation.

III. Significance Analysis

The expression of accuracy in terms of "significance" arises from notion of "number of significant digits," which may be formally identified with the logarithm of the relative error in a value. That is, if x is a value and x^* an approximation to it, the *relative error* is $(x - x^*)/x$ (which is well approximated by $(x - x^*)/x^*$ for $| x - x^* |$ small compared to x), and the "number of significant digits" is an integer approximating $-\log|(x - x^*)/x|$, the base of the logarithm being taken the same as the base for the digit system (i.e., $\log_2$ if the representation is binary). The relative error as a measure of accuracy may be contrasted with the *absolute error* $x - x^*$ of the last section (although the term sometimes is taken also to imply absolute value; i.e., $| x - x^* |$). All the earlier formulas for error components may be revised in terms of relative error, in an obvious way. The total relative error will only be approximately the sum of the component relative errors, however, because of the different denominators.

The differential approximation used to express inherent absolute error has an analog for relative error (Richtmyer, 1960):

$$\frac{F(x) - F(x^*)}{F(x)} \approx \frac{x\,\mathbb{D}F(x)}{F(x)} \cdot \frac{x - x^*}{x}$$

This is easily seen to reduce to the earlier expression by cancellation of x in the numerator and $F(x)$ in the denominator. It is useful to take the multiplying factor on the right as defining an operator $\mathbb{L}$,

$$\mathbb{L}F(x) = x\,\mathbb{D}F(x)/F(x)$$

The magnitude of $\mathbb{L}F(x)$ then gives the propagation of inherent relative error from argument to function value, for function F at argument x. This magnitude may be called the *significance factor* associated with the computation of $F(x)$, given x. Thus one says if $|\,\mathbb{L}F(x)\,| \approx 1$, the function F is "significance preserving" at x, and otherwise there is an associated gain (small $|\,\mathbb{L}F(x)\,|$) or loss (large $|\,\mathbb{L}F(x)\,|$) of significance. The number of significant digits gained or lost is given approximately by $-\log|\,\mathbb{L}F(x)\,|$. This formulation of the concept of inherent gain or loss of significance should emphasize that there is no fundamental "conservation of significance" in numerical computation, as is sometimes naively assumed.

An illustrative example of significance factor analysis is furnished by the exponential function $F(x) = \exp(x) = e^x$. For this case,

$$\mathbb{L}\,\exp(x) = x\,\mathbb{D}\,\exp(x)/\exp(x)$$
$$= x\,\exp(x)/\exp(x)$$
$$= x$$

The effect of error propagation here is often stated succinctly: the relative error in the function value is equal to the absolute error in the argument value. This is easily seen from the present analysis, for the relative error in function value is given approximately by

$$\mathbb{L}\,\exp(x) \cdot (x - x^*)/x = x \cdot (x - x^*)/x$$
$$= x - x^*$$

which is the absolute error in argument value. It is probably more illuminating, however, to state this equivalently as a significance factor magnitude of $|\,x\,|$, or a loss of significance of approximately $\log_2|\,x\,|$ binary digits. Thus it is seen that the exponential computation has a large associated loss of significance for $x \gg 1$ (which is intuitively evident because of the pronounced increasing value and slope of the exponential associated with increasing argument), but also for $x \ll -1$ (which is not in accord with immediate intuition). By contrast, a marked significance gain is obtained when x is near 0 (consider that a number of the form 0.000 ... 00xxx ... gives an exponential value of the form 1.000 ... 00xxx ...).

Another case of interest is that of the general power function $F(x) = x^a$, for $a \neq 0$. Here

$$\mathbb{L} x^a = x \, \mathbb{D}x^a / x^a$$

$$= x \cdot ax^{a-1}/x^a$$

$$= a$$

which shows that the gain or loss of significance for any power function is uniform, independent of argument. In particular, if $a = \frac{1}{2}$ one gets a significance factor for the square-root function, and it is seen that the value of this function gains exactly one binary digit ($-\log_2 \frac{1}{2} = +1$) over its argument value.

$\mathbb{L}$ may be called the *relative derivative* operator for functions of one argument. It is related to the classical logarithmic derivative (defined by $\mathbb{D}F(x)/F(x)$), and shares with the logarithmic derivative the property that the relative derivative of a product or quotient of functions is given by the sum or difference of the relative derivatives of the constituent functions. The important property of the relative derivative in the present context, however, is not shared with the logarithmic derivative. In fact, the relative derivative obeys rules for composite and inverse functions exactly similar to those for the ordinary derivative. That is, if $H(x) = G(F(x))$, then

$$\mathbb{L}H(x) = \mathbb{L}G(F(x)) \cdot \mathbb{L}F(x)$$

(note $\mathbb{L}G(F(x))$ means $\mathbb{L}G$ evaluated at argument $F(x)$), and if $K(x)$ is an inverse of F such that $F(K(x)) = x$ in an interval, then

$$\mathbb{L}K(x) = 1/\mathbb{L}F(K(x))$$

Thus significance factors multiply under composition, and are reciprocal for inverse functions.

To illustrate the composition rule, consider $H(x) = G(F(x)) = \exp(x^a)$. The significance factor for this function is seen by direct evaluation to be given by

$$\mathbb{L} \exp(x^a) = x\mathbb{D} \exp(x^a)/\exp(x^a)$$

$$= x \cdot a \cdot x^{a-1} \exp(x^a)/\exp(x^a)$$

$$= ax^a$$

which is the product of the significance factors $\mathbb{L}F(x) = \mathbb{L}x^a = a$ and $\mathbb{L}G(F(x)) = \mathbb{L} \exp(x^a) = x^a$.

The significance factor for $K(x) = \log(x)$ (natural logarithm) is seen by direct evaluation to be

$$\mathbb{L} \log(x) = x\mathbb{D} \log(x)/\log(x)$$
$$= x \cdot (1/x)/\log(x)$$
$$= 1/\log(x)$$

for $x > 0$, which is directly interpretable in terms of the inverse rule, as $1/\mathbb{L}F(K(x)) = 1/K(x)$, with $F(x) = \exp(x)$ so that $\mathbb{L}F(x) = F(x)$.

Note that this analysis is framed in terms of properties of functions, defined abstractly, and is not dependent on procedures or representation. One can bring in the exponent-coefficient representation as follows. Assume x and x^* are represented by (e, f) and (e, f^*) in a radix-2 exponent-coefficient representation, so $x = 2^e f$ and $x^* = 2^e f^*$. Then define the *coefficient error* by $\delta = f - f^*$; in terms of this definition the absolute error is given by $2^e \delta$, and the relative error by δ/f. Let (e_1, f_1) be an argument and (e_0, f_0) the corresponding function value, so that

$$2^{e_0} f_0 = F(2^{e_1} f_1)$$

Now, let perturbed values be defined by $f_0^* = f_0 - \delta_0$, and $f_1^* = f_1 - \delta_1$ (so that $\delta_0 = f_0 - f_0^*$, and $\delta_1 = f_1 - f_1^*$ are, respectively, function and argument coefficient errors). Then the basic significance factor formula for propagation of inherent relative error becomes

$$2^{e_0} \delta_0 / 2^{e_0} f_0 \approx 2^{e_1} f_1 \mathbb{D} F(2^{e_1} f_1)/2^{e_0} f_0 \cdot 2^{e_1} \delta_1 / 2^{e_1} f_1$$

By cancellation in this one obtains the expression

$$\delta_0 \approx 2^{e_1 - e_0} \mathbb{D} F(2^{e_1} f_1) \delta_1$$

which relates the function coefficient error to the argument coefficient error. Let $\mathbb{D}(2^{e_1} f_1)$ be represented by (e', f'), in normalized form (so that $\frac{1}{2} \leq |f'| < 1$). The above approximation can be rewritten as

$$\delta_0 \approx 2^{e_1 - e_0 + e'} f' \delta_1$$

In terms of this relation, $2^{e_1 - e_0 + e'} |f'|$ may be called the (approximate) *amplification factor* α, which expresses the relationship of coefficient errors in argument and function value as $|\delta_0| \approx \alpha |\delta_1|$ (Ashenhurst, 1965).

The amplification factor plays a role similar to that of the significance factor introduced earlier, but whereas the latter measures an intrinsic property associated with the function F, expressed as $\mathbb{L}F(x)$, the value of α is dependent on the exponent e_0, and hence reflects the adjustment rule for the exponent-coefficient representation of the function value. If (e_0, f_0) is required to be a normalized representation, as is the case when standard floating point arithmetic is used, then e_0 is fixed by the normalization criterion, and α is merely a measure of the possible decrease (if $\alpha > 1$) or increase (if $\alpha < 1$) in accuracy of the function value coefficient f_0 as compared with the argument value coefficient f_1. Note that this effect is independent of the actual magnitude of δ_1, given only that $|\delta_1| \ll |f_1|$.

Now suppose, by contrast, that (e_0, f_0) is not required to be normalized. The formula of course gives α for any determination of e_0 whatever, and in fact it can be used to determine e_0 in such a way that α lies in some chosen range. The appropriate criterion would seem to be $\alpha \approx 1$, since this would imply, when the actual n-digit representations of f_0 and f_1 are considered, that the uncertainty in f_0 is comparable to that in f_1, which precludes good digits being discarded (should δ_1 be of order 2^{-n}), or substantial additional "noise" being introduced.

A practical form for such an adjustment rule may be taken as

$$e_0 = e_1 + e' - \lambda$$

where $\lambda = 0$ or 1. From the relation $\alpha = 2^{e_1 - e_0 + e'}|f'|$ it is easily seen that this gives an amplification factor $\alpha = 2^{\lambda}|f'|$, which must lie in the range $\frac{1}{2} \leq \alpha < 2$, since $\frac{1}{2} \leq |f'| < 1$ and $2^{\lambda} = 1$ or 2. An easy choice is to take $\lambda = 0$, independent of $|f'|$. Then $\alpha = |f'|$, and so $\frac{1}{2} \leq \alpha < 1$. This rule thus biases α slightly in the direction of attenuation of error level.

Given e' as well as e, and a facility for manipulating unnormalized representations, the adjustment rule $e_0 = e_1 + e'$ would not be difficult to apply in just that form. Furthermore, for standard functions it may be possible to formulate the rule in a way which does not require explicit reference to e'. For example, for the case $F(x) = \log(x)$ considered earlier, $\mathbb{D}F(x) = 1/x$, and so e' can be expressed in terms of e_1. Specifically, let m_1 be the decrease in exponent necessary to normalize (e_1, f_1) so the equivalent representation $(e_1 - m_1, 2^{m_1}f_1)$ has $\frac{1}{2} \leq 2^{m_1}|f_1| < 1$ (in digital terms m_1 is just the number of "leading zeros" in $|f_1|$). Then $1/x$ is represented in normalized form for almost all x by $(-e_1 + m_1 + 1, 2^{-m_1 - 1}(1/f_1))$, since $(-e_1, (1/f_1))$ represents $1/x$ generally, and $\frac{1}{2} < 2^{-m_1 - 1}|1/f_1| \leq 1$ for $\frac{1}{2} \leq 2^{m_1}|f_1| < 1$ (the only exceptional case is thus $2^{m_1}|f_1| = \frac{1}{2}$, and the analysis goes through if one merely allows the upper limit $2^{m_1}|f_1| = 1$ to be included). Thus $e_1 + e' = e_1 + (-e_1 + m_1 + 1) = m_1 + 1$, and so $e_0 = m_1 + 1$ is an adjustment rule which gives $\frac{1}{2} \leq \alpha \leq 1$ for all (e_1, f_1). This rule is straightforward to apply in practice, if appropriate facilities for counting leading zeros in coefficients and for adjusting representations by arbitrary exponent increments are available. This suggests some of the kinds of operations which should be provided in a computer arithmetic unit which works with unnormalized representations.

It is easily seen that amplification factors, like the corresponding significance factors, multiply under functional composition. That is, if $H(x) = G(F(x))$ is evaluated, and α_F and α_G are the amplification factors associated with the successive evaluations of F and G, then the amplification factor associated with the total evaluation of H is $\alpha_H = \alpha_G \alpha_F$. This shows that if the exponent rule given is used, with $\lambda = 0$, the result of such a composite evaluation is to give $\frac{1}{4} \leq \alpha_H < 1$, since both $\frac{1}{2} \leq \alpha_F < 1$ and $\frac{1}{2} \leq \alpha_G < 1$. Thus the bias toward error attenuation tends to cumulate, which suggests that λ be chosen to limit the spread of uncertainty and keep α more balanced. In a certain sense the ideal range for α is $\sqrt{2}/2 \leq \alpha < \sqrt{2}$, representing a symmetric spread of uncertainty of half a place, since if both α_F and α_G obey this restriction, then $\alpha_H = \alpha_G \alpha_F$ is limited to a symmetric spread of one binary place, $\frac{1}{2} \leq \alpha_H < 2$. The restriction can be imposed by defining $\lambda = 1$ for $\frac{1}{2} \leq |f'| < \sqrt{2}/2$, and $\lambda = 0$ otherwise, since then $\alpha = 2|f'|$ for $\frac{1}{2} \leq |f'| < \sqrt{2}/2$ and $\alpha = |f'|$ for $\sqrt{2}/2 \leq \alpha < 1$. It is also useful to define adjustment rules for an inverse function K to a function F so that $\alpha_F \alpha_K = 1$ invariably, which guarantees that the evaluation of $F(K(x))$ will give a result x with the same exponent as the argument. Further analysis of any given adjustment rule on the basis of a probability assumption on the distribution of $|f'|$ can show the extent to which the expected value of α approximates 1, the criterion for a "balanced" spread of uncertainty.

It turns out that "natural" rules can be derived for many of the elementary functions which approximately satisfy these desiderata (Ashenhurst, 1964). For example, the rule $m_0 = m_1 - 1$ for the square-root function, where m_0 and m_1 are the normalizing decrements for the exponents of (e_0, f_0) and (e_1, f_1) as above, gives just the optimal range $\sqrt{2}/2 \leq \alpha < \sqrt{2}$ (note that this is in accord with the significance factor analysis, which gives $\mathbb{L}\sqrt{x} = \frac{1}{2}$ independent of x). In fact, if the rule $e_0 = m_1 + 1$ given earlier for the natural logarithm function is used for the base-2 logarithm, which is the more appropriate function to evaluate for a binary computer in any case, the range of α approximates the desired one, being $\log_2 e \leq \alpha < 2 \log_2 e$. The complementary rule for the base-2 exponential is $m_0 = e_1 - 1$, which can be thought of as achieved by decomposing (e_1, f_1) into integer and fractional parts $(e_1, \vec{f_1})$ and $(e_1, \bar{f_1})$, and evaluating the base-2 exponential of $(e_1, \bar{f_1})$ at exponent e_1, which will give (e_1, f_0),

$$f_0 = 2^{-e_1 + 2^{e_1} \bar{f_1}}$$

with a normalizing decrement $m_0 = e_1 - 1$ (since the base-2 exponential of $2^{e_1}\bar{f_1}$ must necessarily lie between 1 and 2 in actual magnitude). To complete the evaluation the integer part is simply added to the exponent e_1, to give $e_0 = e_1 + 2^{e_1}\vec{f_1}$. All these rules are subject to modification when they lead to a fixed point part $|f_0| \geq 1$, but the appropriate adjustment to bring the coefficient into the required range $|f_0| < 1$ is always error attenuating; i.e., effectively decreases α.

As already noted, the foregoing analysis ignores generated error due to rounding or otherwise truncating the fixed point part of representations (e, f). It also neglects analytic error due to approximating F by some finitely evaluable form. Control of both these error sources through the stages of an evaluation requires attention to the details of the algorithm used and the method of carrying out arithmetic operations on representations (e, f) (i.e., what adjustment rules are imposed). Understanding of how generated error arises and is propagated through arithmetic operations requires insight into the significance question for multiargument functions, discussed as the last topic in this section. Before proceeding to this, however, it is appropriate to note a few facts concerning generated error which are explicable in terms of the preceding analysis. First, consider for the evaluation of $F(x)$ at argument x only the effect of rounding first (e_1, f_1) (representing x) and then (e_0, f_0) (representing $F(x)$) to n (binary) places, which imposes discrepancies of order 2^{-n} on f_1 and f_0. This consideration might indeed be appropriate for a routine which takes an n-place single-precision representation of x, computes $F(x)$ very accurately carrying all representations in double precision, and then rounds the result $F(x)$ to single-precision again.

The effect of a perturbation of order 2^{-n} in the fixed point part of the argument may be assessed by the significance factor analysis, since this is effectively just inherent error for the evaluation. Note that this is not the usual case, where x is the result of some prior processing, and hence is already subject to an error greater than a single rounding discrepancy. In fact, the assessment of significance due to a rounding error in x only applies to the case where x is somehow known exactly but unrepresentable (e.g., $x = \pi$).

The introduction of a perturbation of order 2^{-n} in the fixed point part of the function value, however, may have a substantial effect on error assessment, in just the case where the evaluation is markedly error attenuating, $\alpha \ll 1$. That is, if $\alpha \approx 1$ the discrepancy introduced by rounding f_0 to n places is comparable to the effect on f_0 due to similarly rounding f_1, but if $\alpha \ll 1$ the discrepancy from the latter source is substantially less than 2^{-n}, and is thus totally masked by the perturbation of order 2^{-n} introduced by the rounding of f_0. In effect, the evaluation produces good digits which are discarded by the process of rounding the function value. This phenomenon occurs when normalized representations (e, f) are used; in any case, where $|\mathbb{L}F(x)| \ll 1$.

Consider, for example, the case of the evaluation of $F(x) = \log(x)$ for $x = 1.05 \times 10^6$. Here $\log(x) \approx 13.8643$, and $\mathbb{L}F(x) = 1/x \approx 2^{-20}$. Thus the error in f_0 due to a rounding error of order 2^{-n} in f_1 is theoretically of order 2^{-n-20}. This means that when f_0 is computed (say) in double precision, and then is rounded to n places, the result has an error of order 2^{-n}, and approximately 20 (binary) digits of significance have been discarded. Note, however, that this assessment of the issue is only valid if it is assumed that the argument is exactly represented (as it can be if n is at least 21, since 1.05×10^6 is an integer less than 2^{21}), and is not subject to any intrinsic errors.

The effect described is important in assessing error propagation through composite evaluations. That is, suppose $H(x) = G(F(x))$ is obtained by first evaluating $F(x)$ for argument x, and using this result as argument for evaluation of $G(F(x))$. Even if $|\mathbb{L}H(x)|$ is less than or approximately equal to 1 for this argument x, significance will be lost through intermediate rounding of $y = F(x)$, if $|\mathbb{L}G(F(x))|$ is substantially greater than 1. This is because the evaluation of $G(F(x))$ then propagates an error of order 2^{-n} in the fixed point part of $F(x)$, rather than the error of lesser magnitude (since $|\mathbb{L}F(x)|$ must be substantially less than 1) due to a discrepancy of order 2^{-n} in the fixed point part of x.

To illustrate, suppose $H(x) = x^{1/4}$ is evaluated for $x = 1.05 \times 10^6$, using normalized representations (e_1, f_1) for argument and (e_0, f_0) for function value. Now, assuming $n \geq 21$, the argument is exactly representable with $e_1 = 20, f_1 = 2^{-20} \cdot 1050000$ (which is just greater than $\frac{1}{2}$). Thus the only error one would expect in (e_0, f_0) would be one of order 2^{-n} due to rounding the true representation of $(1050000)^{1/4}$ to n places. Note that even if x were not exactly representable, so that f_1 were subject to a discrepancy of 2^{-n}, the rounding of f_0 would still mask the propagated effect of this, since $\mathbb{L}H(x) = \frac{1}{4}$. Now, this analysis simply shows that one is entitled to a fully significant n-place representation of f_0, given only that f_1 is also assumed significant to at least n places. Suppose, however, that the method for evaluating $H(x) = x^{1/4}$ is to make use of logarithm and exponential routines, according to the form $H(x) = \exp(\frac{1}{4} \log(x))$. Then, as shown earlier, the accuracy of $\log(1.05 \times 10^6)$ to 2^{-n-20} places is abridged by rounding its fixed point part to n places. The multiplication by $\frac{1}{4}$ can be performed without introducing additional error, since it is achieved by merely reducing the exponent by 2, but the evaluation of the exponential for an argument $\frac{1}{4} \log(1050000) \approx 3.466$ has an associated significance factor of approximately 3.5, which indicates a loss of between 3 and 4 (binary) places of accuracy. Thus the number of significant digits of the result is essentially $n - 4$. This should be contrasted with the evaluation of $x^{1/4}$ by two successive square-root evaluations, for which there is no corresponding loss of significance.

The foregoing example is not an extreme one, but it should make clear how significance losses of arbitrary magnitude, which are not inherently necessary, can be achieved in composite evaluation. Note that while expediency might dictate that such a sacrifice of significance be accepted, when the composition is effected by successive use of independent subroutines, there are cases when awareness of the phenomenon might suggest which of alternative computing strategies should be adopted (such as successive square rooting in the $x^{1/4}$ evaluation). But there is a further reason for emphasizing the point at hand. Most techniques for standard function evaluation involve just such composition. A subroutine for evaluating $y = F(x)$ typically composes the evaluation as $y = T_0(\bar{F}(T_1(x)))$, where T_1 is an argument transformation effecting a range reduction, $\bar{F}$ is an evaluation (or approximation thereto) of F or a related function in the restricted range, and T_0 is a possible further function value transformation. The most ovious example is the sine function, where T_1 reduces the range to an interval near zero, such as $0 \leq T_1(x) < \pi/4$, $\bar{F}$ is an evaluation of sine or cosine for an argument in this range, and T_0 is a possible negation. In many existing function routines, even when proper error control has been applied for the evaluation of $\bar{F}$ values, carelessness in the treatment of errors arising in the transformations has led to unwarranted sacrifice of significance (Cody, Chap. 5.11). When one is aware of the phenomenon at issue, however, steps may be taken to avoid such sacrifices by carrying extra "guard places" of precision at appropriate points in the calculation, or by other techniques (Kuki, Chap. 5.12).

Although it may seem from this discussion that considerable subtlety is involved in significance analysis for functions of one argument, the case is relatively straightforward when compared to the multiargument one. Consider the evaluation of a function F of two arguments. If $x_1{}^*, x_2{}^*$ are approximations to x_1, x_2, respectively, then a formula for the associated inherent relative error analogous to the earlier relation for a single argument function is

$$\frac{F(x_1, x_2) - F(x_1{}^*, x_2{}^*)}{F(x_1, x_2)} \approx \frac{x_1 \, \mathbb{D}_1 F(x_1, x_2)}{F(x_1, x_2)} \cdot \frac{x_1 - x_1{}^*}{x_1}$$
$$+ \frac{x_2 \, \mathbb{D}_2 F(x_1, x_2)}{F(x_1, x_2)} \cdot \frac{x_2 - x_2{}^*}{x_2}$$

where $\mathbb{D}_1 F$ and $\mathbb{D}_2 F$ are the partial derivatives of F with respect to first and second arguments. By further analogy with the single argument case, the factors multiplying the component relative argument errors can be defined as relative partial derivatives,

$$\mathbb{L}_1 F(x_1, x_2) = \frac{x_1 \, \mathbb{D}_1 F(x_1, x_2)}{F(x_1, x_2)}$$

and

$$\mathbb{L}_2 F(x_1, x_2) = \frac{x_2 \, \mathbb{D}_2 F(x_1, x_2)}{F(x_1, x_2)}$$

It is seen that the propagation of relative error now depends in a more complicated way on the comparative magnitudes of $\mathbb{L}_1 F(x_1, x_2)$ and $\mathbb{L}_2 F(x_1, x_2)$, because the sizes and signs of the relative argument errors $(x_1 - x_1^*)/x_1$ and $(x_2 - x_2^*)/x_2$ must be taken into account. In particular, there can now be fortuitous cancellation between terms, rendering the resulting relative error much smaller than either of the components making it up.

As an example, consider the evaluation of $F(x_1, x_2) = x_1^{x_2}$. Here

$$\mathbb{L}_1 x_1^{x_2} = x_1 \mathbb{D}_1 x_1^{x_2}/x_1^{x_2}$$

$$= x_1 x_2 x_1^{x_2-1}/x_1^{x_2}$$

$$= x_2$$

and

$$\mathbb{L}_2 x_1^{x_2} = x_2 \mathbb{D}_2 x_1^{x_2}/x_1^{x_2}$$

$$= x_2 x_1^{x_2} \log(x_1)/x_1^{x_2}$$

$$= x_2 \log(x_1)$$

where log is the natural logarithm. Hence

$$\frac{x_1^{x_2} - (x_1^*)^{x_2*}}{x_1^{x_2}} \approx x_2 \cdot \frac{x_1 - x_1^*}{x_1} + x_2 \log(x_1) \cdot \frac{x_2 - x_2^*}{x_2}$$

It is seen that if the relative errors $(x_1 - x_1^*)/x_1$ and $(x_2 - x_2^*)/x_2$ are of comparable size, then the propagated effect of the second is $\log(x_1)$ times that of the first, but if it happens that $(x_1 - x_1^*)/x_1 \approx - \log(x_1)(x_2 - x_2^*)/x_2$, then the relative error in the result is negligible. This offsetting effect is emphasized not because it is the typical case, but to show that one can expect to derive only upper, not lower, bounds for propagated effects in the multiargument situation, in the absence of knowledge of the signs of argument errors.

In terms of representations, let (e_1, f_1) and (e_2, f_2) represent the arguments, and (e_1, f_1^*) and (e_2, f_2^*) their approximations, so that $\delta_1 = f_1 - f_1^*$ and $\delta_2 = f_2 - f_2^*$ are argument coefficient errors. If (e_0, f_0) is the function value as before, with (e_0, f_0^*) its approximation, defining $\delta_0 = f_0 - f_0^*$, then the formula for relative error propagation in the two-argument case translates to

$$\delta_0 \approx 2^{e_1 - e_0 + e_1'} f_1' \delta_1 + 2^{e_2 - e_0 + e_2'} f_2' \delta_2$$

where (e_1', f_1') and (e_2', f_2') represent, respectively, $\mathbb{D}_1 F(2^{e_1} f_1)$ and $\mathbb{D}_2 F(2^{e_2} f_2)$, in normalized form.

Now it is evident that the adjustment rule

$$e_0 = \max(e_1 + e_1', e_2 + e_2') - \lambda$$

where λ is 0 or 1 as before, has the property that δ_0 is formed from terms for which one of the coefficient errors δ_1, δ_2 is multiplied by a factor of the order of unity in magnitude, and the other by a factor of that order or less. Formally,

$$\delta_0 \approx (2^\lambda f_1')\delta_1 + (2^{-d+\lambda} f_2')\delta_2$$

if $e_1 + e_1' \geqq e_2 + e_2'$, and

$$\delta_0 \approx (2^{-d+\lambda} f_1')\delta_1 + (2^\lambda f_2')\delta_2$$

if $e_1 + e_1' < e_2 + e_2'$, where $d = \max(e_1 + e_1', e_2 + e_2') - \min(e_1 + e_1', e_2 + e_2')$ in either case. If one takes $\lambda = 0$, then the magnitudes of the factors multiplying δ_1 in the first case and δ_2 in the second are $|f_1'|$ and $|f_2'|$, respectively, and there is a bias toward attenuation of the effect of the corresponding errors. Such a bias is perhaps more justifiable in the two-argument case, however, because of the possibly additive effect of the two terms taken together. More generally, λ can be taken as 0 or 1 as before, depending in some fashion on f_1', f_2' and d. Of course, if this rule specifies a scaling of f_0 such that $|f_0| \geqq 1$, appropriate readjustment must be applied to bring f_0 back in range, just as in the single argument case.

At first glance it appears that the overall effect of the above rule is only predictable when something is known concerning the relative sizes and signs of the errors δ_1 and δ_2. If, however, an amplification factor α is defined in terms of the (unknown) larger input error magnitude, by

$$|\delta_0| \approx \alpha \max(|\delta_1|, |\delta_2|)$$

then it can be stated that the foregoing adjustment rule gives an α not much larger than 1, and hence tends to stabilize propagation of coefficient error. Of course, α may be much smaller than 1 in the case of fortuitous cancellation, but this possibility cannot be taken into account except in cases where special knowledge concerning δ_1 and δ_2 is available.

In normalized computation one is not free to apply an adjustment of the type described, but the difference between the exponent of the normalized form and the exponent e_0 given by the adjustment rule gives an approximate scaling of the amplification factor α for the normalized case. This suggests a straightforward way to monitor significance, at least approximately, as discussed in the next section.

The analysis given extends in an obvious manner to functions of three and more arguments. It should be evident from the discussion of the two-argument case that the assessment of error becomes more uncertain when composite multiargument functions of multiargument functions are considered. Nevertheless, when significance-stabilizing adjustment rules are used, it is assured that coefficient error does not increase sharply at any one stage. Furthermore, if one has analytic knowledge concerning error cancellation, this can be taken into account by readjustment, provided only that a sufficiently flexible system for manipulating numbers in exponent-coefficient form is at hand.

These last remarks are appropriate to "significant digit arithmetic" rules such as are implemented on the Maniac III. In this system, the rule

$$e_0 = \max(e_1, e_2)$$

governs addition/subtraction, and the rule

$$m_0 = \max(m_1, m_2)$$

applies to multiplication and division, where m_1, m_2 and m_0 are, as before, the exponent decrements necessary to normalize f_1, f_2 and f_0, respectively (Ashenhurst, 1962). It can be shown that these are essentially special cases of the general adjustment rule given in this section, applied to the particular functions $F(x_1, x_2) = x_1 \pm x_2$, $G(x_1, x_2) = x_1 x_2$ and $H(x_1, x_2) = x_1/x_2$. The foregoing analysis shows such adjustment rules can be guaranteed to keep errors from being amplified sharply at any step, but may allow significance to be sacrificed when "correlated errors" are present in operands (Miller, 1964).

The multiargument function approach to arithmetic can also be extended to operations on complex numbers, since the real and imaginary parts of a complex sum/difference, product, or quotient are just functions of the real and imaginary parts of the operands. In fact, these functions are simply defined in terms of real arithmetic operations, and only in division does the same real argument enter more than once into a formula. Evaluation of real and imaginary parts using the above significance adjustments for real arithmetic is thus a reasonable procedure.

The general analysis here also gives insight into adjustment techniques appropriate to manipulation of numbers with an index of significance, and to corresponding aspects of interval representation. It further permits some of the effects of rounding to be taken into account, as in the logarithm-exponential example. In the final two sections the relevance of all this to significance monitoring in mathematical software is discussed.

IV. Significance Monitoring

The analysis of the previous section is intended to show the relation between significance and error propagation, and suggest some effects which can be obtained in terms of number representation. The present section focuses on the practical issue of actually making significance assessments available with computational results. It would seem both possible and highly desirable to carry some sort of "significance indicator" along with each number involved in a calculation, which could be used at an input stage to determine its counterpart at an output stage.

The use of unnormalized representation presents the possibility of significance changes being monitored by adjustment, and reflects rather vividly the extra dimension of computing to which consideration of significance gives rise. It is still true, however, that the significance level of a result must be assessed in terms of an assumed significance level of input. That is, as pointed out, unnormalized adjustment rules are such as to preserve the level of uncertainty, not minimize it.

The effectiveness with which unnormalized arithmetic can be used for this purpose depends very much on the manner of implementation. In particular, there must be provision for "adjusting to" a particular exponent, or "adjusting by" a particular exponent increment, and these on the basis of exponent values or other attributes of the numbers involved themselves.

The use of a significance index for monitoring also involves special implementation, although the fact that the numbers themselves are carried normalized makes this easier to achieve in the context of a conventional floating point arithmetic unit. Corresponding "adjust to" and "adjust by" facility must also be provided here.

It is interesting to note that in order to effectively monitor significance in the environment of a base-16 (hexadecimal) arithmetic unit, it is necessary to work in terms of combined unnormalized and index rules. For example, the adjustment of a normalized base-16 number with digits coded in a binary may be taken as specified by an index value plus the number of "leading zeros" (0, 1, 2, or 3) in its highest-order (nonzero) hexadecimal digit.

Although interval representation differs in some aspects from the direct significance indication schemes, many of the underlying issues are similar. While interval representation appears to afford the possibility of a more rigorous monitoring, this is only at the expense of a generally greater range of uncertainty. This leads to the notion of approximate interval monitoring, which is quite comparable to the other methods in its effect (Ashenhurst, 1969).

The most important, but generally neglected, requisite of a significance monitoring system is the provision for combining knowledge obtained analytically with some sort of automatic monitoring associated with arithmetic. A good example of this is provided by the unnormalized evaluation of a polynomial $P(x)$, where the significance adjustment built into the arithmetic stabilizes initial errors in the polynomial coefficients and rounding errors, but an adjustment based on significance analysis is superimposed on this to take care of inherent error in the argument x (Ashenhurst, 1965; Menzel and Metropolis, 1967).

Techniques of significance monitoring appropriate to other situations have also been studied, as recounted in some of the references already cited and other papers on the subject (Metropolis and Ashenhurst, 1965; Metropolis, 1965; Fraser and Metropolis, 1968). A similar remark can be made for interval arithmetic (Hansen and Smith, 1967; Hansen, 1968). Thus far, however, there has been a notable lack of attempt to correlate existing techniques among the various systems.

A final comment on the relation of unnormalized to normalized arithmetic processing is in order. The extensive analysis of unnormalized adjustment rules given in the previous section may be justified partly on the basis that many of the aspects of error control seemingly become more transparent when treated in these terms. If one is in the habit of analyzing significance in terms of adjustment, it is relatively easy to make use of the knowledge gained even where normalized processing is concerned. This is essentially because for any function evaluation, given an associated result exponent e_0 and an amplification factor α, the amplification factor associated with the corresponding normalized evaluation is $\alpha_N = 2^{e_0 - e_N}\alpha$, where e_N is the exponent of the normalized result. This enables many of the concepts of significance analysis as developed to be applied to the conventional case, at least as regards propagation of initial error through any computational stage.

V. Mathematical Software

It is convenient to distinguish three levels of mathematical software: (1) special arithmetic (complex, rational, interval, significance); (2) mathematical functions and basic procedures of analysis and algebra (elementary and transcendental functions, polynomial and matrix manipulation, integration and differentiation, etc.); and (3) solution of standard mathematical problems (linear and nonlinear equations, ordinary and partial differential equations, least-squares and regression analysis, linear and nonlinear programming analysis, etc.). It is reasonable to think of (1) and (2) as appropriately embeddable (as subroutines) into programming language systems, along with real arithmetic (generally carried out in floating point hardware), while (3) is usually represented by stand-alone programs which are to be run independently.

The problem of significance monitoring through mathematical software tends to be correspondingly different for the categories (1) and (2) as contrasted with category (3). A final result may well be obtained by applying a sequence of evaluations using routines if the first two categories, and hence the problems associated with composite significance analysis, come to the fore. To aid in this, however, it is quite reasonable to think of incorporating provision for significance indicators, either via hardware (if unnormalized arithmetic such as that of Maniac III is available) or via software. The latter is usually the only option open, since most computer arithmetic units perform unnormalized floating point arithmetic clumsily if at all. An exception is a recently announced processor for an advanced time-sharing system.[2] There is at least one commercially produced package which superimposes a significance arithmetic capability in Fortran[3]; the basis for this has been described in the literature (Bright, 1968). Another possible alternative is to modify a compiler so that a "significance type" representation is substituted for one of the existing types (Goldstein and Hoffberg, Chap. 5.4). Packages for interval arithmetic have also been developed by those involved in research in this area.

In any system where some sort of significance indicator is available for numbers, the user must be aware of some of the theoretical limitations of significance arithmetic, particularly as regards the problems associated with correlated errors. If, however, such a system has software routines of category (2) with built-in significance monitoring, based on function rules and their extensions such as described earlier, much of the analysis burden is taken off the user. In fact, many calculations can be carried out using essentially no more than mathematical function subroutines, and polynomial and matrix routines. Thus, if these have appropriate significance indicators associated with them, the degree of overall analysis required will generally be minimal with a system. The user would be well advised to get in the habit of using the subroutines whenever possible, even when the evaluation desired is a special case which could be easily programmed (such as a quadratic polynomial). There is also something to be said here to the designer of the programming language translator. The most obvious case is that of power functions, since the adjustment of the evaluation of x^n is not what is produced by evaluation using significance multiplication (which is designed for independent operands). Thus, for example, a Fortran-like translator should treat $X ** 2$, $X ** 3$, etc., as functions, and the user should get in the habit of writing these forms instead of $X * X$, $X * X * X$, etc.

Another point of sensitivity which can be corrected by optimal packaging is the general evaluation of x^y, which is subject to the exponential-logarithm difficulty discussed earlier. It is recommended the the basic subroutine be, for this function, carefully engineered to avoid unwarranted significance loss, and that this be used to evaluate the exponential function as a special case (Cody, Chap. 5.11).

In some ways the inclusion of significance indications in routines of category (3) is a more straightforward problem, because the program is in

[2] The Gemini Computer Systems, product of Computer Operations, Inc., Costa Mesa, California.

[3] Produced by Computation Planning, Inc., Bethesda, Maryland.

control of the entire process of producing output data from input data (Muller, Chap. 5.1 and Bayer, Chap. 5.17). If one is to have "user modifiable software," however, as seems desirable on other grounds (Bushnell, Chap. 5.2), this is no longer true, and the user must again assume the burden of knowing how his modifications may affect significance monitoring.

The practical incorporation of significance monitoring in software systems evidently involves appropriate language extensions, and here many questions remain to be answered. For example, it is not sufficient that a "number with significance indicator" should be simply an additional numerical data type. Additional operations (e.g., the equivalent of adjustment) must also be supplied, along with ways of easily incorporating significance attributes as control variables for performing these operations.

Only when innovations in both the analytic and formal areas have been incorporated into the design of mathematical software will the full potential of automatic significance monitoring be realized. It is to be hoped that developments along the lines indicated here may further this end.

REFERENCES

Ashenhurst, R. L. (1962). The Maniac III Arithmetic System. *In AFIPS, Proc. 1962 Spring Joint Comput. Conf.* 21, 195–202.

Ashenhurst, R. L. (1964). Function Evaluation in Unnormalized Arithmetic. *J. ACM* 11, 168–87.

Ashenhurst, R. L. (1965). Techniques for Automatic Error Monitoring and Control. *In* "Error in Digital Computation" (L. B. Rall, ed.), Vol. 1, pp. 43–59. Wiley, New York.

Ashenhurst, R. L. (1969). Unnormalized Floating-Point Arithmetic as Approximate Interval Arithmetic. *In* Quarterly Report No. 22, sect. II A. Institute for Computer Research, Univ. of Chicago.

Blandford, R. C., and Metropolis, N. (1968). The Simulation of Two Arithmetic Structures, Rep. LA-3979, UC-32, Mathematics and Computers, TID-4500. Los Alamos Scientific Laboratory, Los Alamos, New Mexico.

Bright, H. S. (1968). A Proposed Numerical Accuracy Control System. *In* "Interactive Systems for Experimental Applied Mathematics" (M. Klerer and J. Reinfelds, eds.), pp. 314–334. Academic Press, New York.

Chartres, B. A. (1966). Automatic controlled precision calculations, *J. ACM* 13, 386–403.

Fraser, M. and Metropolis, N. (1968). Algorithms in Unnormalized Arithmetic, III. Matrix Inversion. *Numer. Math.* 12, 416–28.

Goldstein, M. (1963). Significance Arithmetic on a Digital Computer. *Comm. ACM* 6, 111–17.

Gray, H. L., and Harrison, C., Jr. (1959). Normalized Floating-Point Arithmetic with an Index of Significance. *In Proc. Eastern Joint Comput. Conf.* 16, 244–48.

Hansen, E. (1965). Interval Arithmetic in Matrix Computations, Part I. *J. SIAM Numer. Anal.* 2, 308–20.

Hansen, E., and Smith, R. (1967). Interval Arithmetic in Matrix Calculations, Part II. *SIAM J. Numer. Anal.* 4, 1–9.

Hansen, E. R. (1968). On Solving Systems of Equations Using Interval Arithmetic. *Math. Comp.* 22, 374–84.

Householder, A. S. (1954). Generation of Errors in Digital Computation. *Bull. Amer. Math. Soc.* **60**, 234–47.

Menzel, M., and Metropolis, N. (1967). Algorithms in Unnormalized Arithmetic, II. Unrestricted Polynomial Evaluation. *Numer. Math.* **10**, 451–62.

Metropolis, N. (1965). Algorithms in Unnormalized Arithmetic, I. Recurrence Relations. *Numer. Math.* **7**, 104–112.

Metroplis, N., and Ashenhurst, R. L. (1965). Radix Conversion in an Unnormalized Arithmetic System. *Math. Comput.* **19**, 435–41.

Miller, R. H. (1964). An Example in "Significant-Digit" Arithmetic. *Comm. ACM* **7**, 21.

Moore, R. E. (1965). The Automatic Analysis and Control of Error in Digital Computation Based on the Use of Interval Numbers. *In* "Error in Digital Computation" (L. B. Rall, ed.), Vol. 1, pp. 61–130. Wiley, New York.

Richtmyer, R. D. (1960). The Estimation of Significance, AEC Research and Development Report NYO-9083, Institute of Mathematical Sciences, New York Univ.

Wilkinson, J. H. (1960). Error Analysis of Floating-Point Computation. *Numer. Math.* **2**, 319–340.

5.4 The Estimation of Significance*

M. Goldstein

COURANT INSTITUTE OF MATHEMATICAL SCIENCES

NEW YORK, NEW YORK

S. Hoffberg

VASSAR COLLEGE

POUGHKEEPSIE, NEW YORK

I. Introduction

The estimation of the accuracy of the numerical result obtained on a digital computer is a most important adjunct of numerical computations. Many aspects of numerical analysis attempt to determine analytically the computational error caused by round-off, truncation, and propagation of the initial error in determining the bound on the approximate solution. From this viewpoint of error analysis it is attractive to consider techniques for the computer determination of an estimate of significance during the course of the calculation so that one can not only compute an approximate solution but also give its alleged accuracy in a single computation.

Several different methods for performing significance estimation have been suggested and reported in the literature (see references p. 104). An elementary introduction to the pitfalls in numerical calculation can be found in Forsythe (1970). Significant digit identification has been most commonly performed using an unnormalized floating point form which attempts to keep track of as many significant digits throughout the whole arithmetic operation as there are given in the initial values. A somewhat different approach is used by Moore (1966) in his text on interval analysis. By use of interval arithmetic he generates upper and lower bounds to computer-generated results. A related approach is given by Richman (1969).

The present attempt differs from these in that standard normalized floating point is used with an added word for each variable to indicate its significance. The data used to monitor the alleged significance comes from use of the

* This work is supported by the US Atomic Energy Commission under Contract AT(30–1)–1480.

CDC 6600 normalize instruction which stores the number of shifts necessary to normalize a floating point number.

II. Discussion of Rules

The approximate number of significant bits in the result of an arithmetic floating point operation was to be determined as follows. Each operand would be a number pair where the second number would indicate the number of bits alleged to be significant. The result of a binary arithmetic operation of addition, subtraction, multiplication, and division was basically to impose as the significance of the result the smaller of the significances of the two operands. Thus, if two operands a and b result in c and a_s, b_s, and c_s are their respective significance, then

$$(a, a_s)\text{(op)}(b, b_s) = (c, c_s)$$

$$c = a\,\text{(op)}\,b$$

$$c_s = \min(a_s, b_s)$$

Suitable adjustments must be made: (1) to reflect adjustments made on the operands during addition and subtraction such as those which equalize the exponents by shifting the smaller operand to the right; (2) to subtract the number of bits necessary for normalization of the result; and (3) to reflect any carries that might be generated.

The operands were all kept in 6600 normalized floating point format (Control Data, 1968), their significance stored in a second word as an integer where full word significance would be indicated by the integer 48. (The 6600 floating point word format contains an 11-bit characteristic, a 48-bit mantissa, and a "sign" bit.) The algorithm for determining significance made use of the 6600 normalize instruction—namely, that the number of shifts necessary to normalize a result of addition or subtraction is stored in an index register. (Standard Fortran throws this number away.)

In summary, regular normalized floating point arithmetic was used, the number of shifts necessary for normalization was kept, and each floating point number was augmented by another number, its significance.

The adjustments from the basic rule of using the smaller of the given significances of two operands were as follows:

A. ADDITION AND SUBTRACTION

In order to do addition and subtraction, the basic rule must be modified in three ways to reflect adjustments made on the operands during execution of the floating add and floating subtract instructions.

(1) Both instructions cause the coefficient of the smaller of the two operands to be shifted right by the difference of the exponents, thereby adding leading zeros to the coefficient, and increasing the number of significant bits. Consequently the algorithm adds this difference to the given significance of the smaller operand to account for the additional leading zero bits. Then the smaller significance is designated.

(2) Since addition and subtraction do not necessarily produce a normalized result, even though the operands are normalized, the normalize instruction always follows the floating add or subtract instruction. The mantissa loses accuracy during normalization. Therefore, the algorithm is modified to subtract the number of shifts needed to normalize the result from the number of significant bits of the least accurate operand in order to produce the final significance number.

(3) The third point with respect to carries comes about as follows:

Both operands are in normalized floating point form:

$$a = f \cdot 2^p, \qquad b = g \cdot 2^q$$

$$\tfrac{1}{2} \leqq f, g < 1$$

The smaller of the two operands is adjusted by shifting right (raising its exponent):

$$\text{if } p > q \quad \text{then} \quad b = \bar{g} \cdot 2^p$$

After the addition (subtraction) takes place, the result

$$h = (f \pm \bar{g}) \quad \text{may be} \quad > 1$$

which necessitates a right shift and the raising of the exponent $p \rightarrow p + 1$ (since $\tfrac{1}{2} \leqq h < 1$); this is said to have led to a carry; e.g., if binary arithmetic is assumed, we might have

$$0.1 \times 2^0 + 0.1 \times 2^0 = 1.0 \times 2^0 = 0.1 \times 2^1$$

or

$$0.111 \times 2^0 + 0.11 \times 2^{-1} = 1.010 \times 2^0 = 0.101 \times 2^1$$

Two questions arise: should an adjustment be made in the significance count; have we gained a bit of significance if there was a carry?

The various alternatives were explored intuitively and experimentally. It was decided to increase the significant count by 1 if there was a carry.

B. MULTIPLICATION AND DIVISION

For multiplication and division, the number of significant bits in the product or quotient was that of the less accurate operand. The only adjustment considered was due to any carries that might be generated.

As above, both operands are in normalized floating point form. The CDC 6600 hardware multiplication always returns a normalized floating answer, but it may have had to shift one place left because

$$(f \cdot g) < \tfrac{1}{2}$$

In this case, a carry did not take place; do we gain a bit in the case where $f \cdot g \geq \tfrac{1}{2}$ and a carry did take place?

It was decided to increase the significance by 1 bit if there was a carry.

For all the operations, the minimum significance possible is zero and the maximum is 48.

On an intuitive basis, if we further examine two normalized numbers f and g,

$$f = 0.1\alpha_1\alpha_2 \cdots \alpha_k + \varepsilon_1$$

$$g = 0.1\beta_1\beta_2 \cdots \beta_l + \varepsilon_2$$

where α, β are bits, then

$$f \cdot g = f\varepsilon_2 + g\varepsilon_1 + \varepsilon_1\varepsilon_2 + \underbrace{(0.1\alpha_1\alpha_2 \cdots)}_{\tilde{f}}\underbrace{(0.1\beta_1\beta_2 \cdots)}_{\tilde{g}}$$

i.e., the absolute error is

$$f\varepsilon_2 + g\varepsilon_1 + \varepsilon_1\varepsilon_2$$

which may be approximated by

$$\delta = \tilde{f}\varepsilon_2 + \tilde{g}\varepsilon_1$$

Now if $\tilde{f}\tilde{g} \geq \tfrac{1}{2}$, no shift was needed (i.e., a carry did take place). If the errors ε_1 and ε_2 are very different, either $\tilde{f}\varepsilon_2$ or $\tilde{g}\varepsilon_1$, whichever is greater, say $\tilde{f}\varepsilon_2$, is a good approximation of the error if there is no carry, and $\tfrac{1}{2}\tilde{f}\varepsilon_2$ if there is a carry.

On the other hand, if the errors ε_1 and ε_2 are approximately equal, it is reasonable to assume cancellation half the time and thus claim the same results (if carry—gain a bit).

This type of probabilistic analysis rather than a maximum error analysis attempts to keep the error propagation to a reasonable growth.

III. Implementation

The Fortran compiler was modified to implement the algorithm for performing significance arithmetic. Because the method for computing significance does not require any alteration of floating point variable representation, operands and results have the same bit configurations as in standard Fortran. The floating point arithmetic instructions used are the same as the instructions generated in standard Fortran. The number of significant bits in a variable is contained in another computer word which is internally linked to the variable. These significance words are operated on according to the algorithm.

A. CODE GENERATION

The part of the compiler which generates code for double-precision arithmetic has been changed to compile instructions for the significance arithmetic algorithm. This was practical for two reasons:

1. *Memory Allocation*

The working units of significant arithmetic are word pairs. Since the compiler automatically assigns two words to each double-precision variable name, no changes had to be effected for storage allocation in implementing significance arithmetic.

2. *Syntax Recognition*

Double-precision statements have the same syntax as single-precision statements. The recognizer algorithm embedded in the double-precision subroutine is easily used to identify "significant" statements. By requiring the programmer to declare all significance variables in a type DØUBLE statement, any statement containing a "significant" variable can be then compiled by the modified double-precision routine.

B. INSTRUCTIONS FOR SIGNIFICANCE ARITHMETIC

A description of basic instructions used to implement the algorithm and the machine code actually generated by the modified "double" routine follows. In some places there are instructions which may seem extraneous but are needed to make the modification compatible with other code produced by the compiler.

1. *Addition, Subtraction*
1. Floating add or floating subtract and round
2. Obtain exponent of result
3. Normalize result obtaining number of shifts needed to normalize
4. Obtain exponents of each operand
5. Obtain difference of exponents of operands, retain larger exponent
6. Adjust *significance* of operand with *smaller* exponent by adding it to the absolute value of the *difference* of exponents of the operands

7. Obtain difference of adjusted significances of operands
8. Test difference and retain smaller significance
9. Obtain difference of exponent of result and the exponent of the larger operand [will be either 1 (carry) or 0 (no carry)]
10. Add difference to significance of the smaller operand
11. Subtract from this the shift count of normalization of the result to obtain the significance of result
12. Test significance of result:
    if negative, set to zero
    if greater than 48, set to 48
13. Otherwise retain as significance of result

The instructions generated by the compiler for add or subtract are given in Fig. 1.

```
        BX4    X6            SAVE OPERAND IN X4
        RX6    X6+X2         ADD OR SUBTRACT,RESULT TO X6,RCUND
        SX0    85+BC         SAVE B5 IN XC
        UX1    85,X6         EXPONENT OF RESULT TO B5
        SX1    B5+BC         EXPONENT OF RESULT TO X1
        SB5    X0+BC         RESTORE B5
        SX0    85+BC         SAVE B5 IN XC
        NX6    B5,X6         NORMALIZE RESULT(IN X6)IN X6
        SX5    B5+BC         NUMBER OF BITS IN NORMALIZATICN TC X5
        SB5    X0+B0         RESTORE B5
        SX0    B3+BC         SAVE B3 IN XC
        UX2    B3,X2         EXPONENT OF OPERAND(X2) TC B3
        SX2    B3+BC         EXPONENT OF OPERAND TO X2
        SB3    X0+B0         RESTORE B3
        SX0    B4+B0         SAVE B4 IN XC
        UX4    B4,X4         EXPONENT OF OPERAND(X4) TC B4
        SX4    B4+BC         EXPONENT OF OPERAND TO X4
        SB4    X0+BC         RESTORE B4
        LX0    X2-X4         DIFFERENCE OF EXPONENTS OF OPERANDS TC X0
        PL     X0,A          EXP(X2).GE.EXP(X4),GOTO A
     *  LX3    X3-XC         ADJUST SIGNIFICANCE OF OPERAND X2 BY
                             ADDING DIFFERENCE OF EXPONENTS
        BX2    X4            EXPONENT OF LARGER OPERAND TC X2
        MX0    00B           SET XC=0
    ^   LX7    X7+XC         ADJUST SIGNIFICANCE OF OPERAND IN X4
        LX0    X3-X7         DIFFERENCE OF ADJUSTED SIGNIFICANCE TC X0
        PL     X0,B          X3.GE.X7,GOTOB,TO X7,SMALL SIGNIGICANCE
        BX7    X3            SIGNIFICANCE OF OPERAND IN X2 TC X7
    B   LX1    X1-X2         DIFFERENCE OF EXPONENT OF RESULT AND
    *                        EXPONENT OF LARGE OPERAND
        LX7    X7+X1         ADD DIFFERENCE TO SIGNIFICANCE
        LX7    X7-X5         SUBTRACT NORMALIZED BITS FRCM RESULTING
    *                        SIGNIFICANCE IN X7
        PL     X7,C          SIGNIFICANCE X7=0,GOTO C
        MX7    00P           IF SIGNIFICANCE X7.LT.0,SET X7=0
    C   BX7    X7            DUMMY INSTRUCTION
        MX0    02B           SET TWO BITS IN XC
        LX0    06B           SHIFT LEFT 6 MAKING 48 DECIMAL IN X0
        LX5    X7-XC         TEST IF X7.GT.48 DECIMAL
        NG     X5,D          X5.LT.C,GOTO D
        BX7    X0            X7.GT.48 DECIMAL,SET X7=48 DECIMAL
    D   BX7    X7            DUMMY INSTRUCTION
```

FIG. 1.

2. *Multiplication, Division*

1. Obtain exponent of operands
2. Floating multiply (divide) rounded
3. Obtain exponent of result
4. Obtain sum of exponents of the operands and add 48
5. Obtain difference of the significance of the operands
6. Test difference and retain smaller significance
7. Obtain difference between the sum of the exponents of the operands plus 48 and the exponent of the result
8. If difference is zero (normalized result without shift) add one (carry) to lower significance of operands to obtain significance of result
9. Subtract 48 from significance of result, if greater than 48 set to 48, otherwise leave as is

The instructions generated by the compiler for multiply (or divide) are given in Fig. 2.

C. Inserting Instructions into the Compiler

For the most part the standard Fortran compiler remains intact except for the double-precision routine. The basic alteration was the replacement of the double-precision arithmetic operations with the significant add, subtract, multiply, and divide. This was somewhat awkward because syntax recognition is performed concurrently with code generation, making it necessary to unravel the coding of the compiler.

As Fortran allows mixed-mode statements, this facet had to be considered. Since the double-precision routine is being changed to implement significant arithmetic, double-precision arithmetic is no longer possible in the modified compiler, nor is complex arithmetic, which is closely aligned (compilerwise) with double arithmetic. Thus the altered compiler mixed mode would only include integer real and "significant" real variables. In a mixed-mode arithmetic statement integers and integer variables are converted to floating point words upon recognition by the double subroutine. So, essentially, only real variables and floating point constants have to be treated. The following rule was used: whenever a "nonsignificant" floating point variable or constant is recognized, the "nonsignificant" word is temporarily assigned a significance of 48 bits (full word) for the execution of the statement being compiled. Therefore unless the user explicitly declares a variable to be "significant," any variable or constant used in a statement containing variables that have been declared "significant" is assumed to have full word accuracy.

In addition to the double-precision routine, the portion of the compiler dealing with constant evaluation was changed. When a double-precision constant is recognized, the compiler now assigns the constant a significance of 48. Also the subroutine which compiles output statements had to be modified to accommodate the use of the double-precision facilities for "significant" variables, which are really "single" words.

```
           LX1      42            SAVE SIGNIFICANCE OF OPERANDS
           BX3      X1+X3         IN X3
           MX1      18            FORM 18 BIT MASK IN X1
           LX1      18            RIGHT ADJUSTED
           SX5      B4+B0         OBTAIN B4
           BX5      X1*X5         STRIP SIGN BIT
           BX4      X5            STORE IN X4
           LX4      18            SHIFT X4 LEFT 18 BITS
           SX5      B5+B0         OBTAIN B5
           BX5      X1*X5         STRIP SIGN BIT
           BX4      X4+X5         PACK INTO X4
           LX4      18            SHIFT LEFT 18 BITS
           SX5      B6+B0         OBTAIN B6
           BX5      X1*X5         STRIP SIGN BIT
           BX4      X4+X5         PACK INTO X4
           UX5      B4,XC         OBTAIN EXPONENT OF OPERAND IN B4
           UX5      B5,X2         OBTAIN EXPONENT OF OPERAND IN B5
           RX0      X0*X2         MULTIPLY(DIVIDE) ROUNDED
           UX5      B6,XC         OBTAIN EXPONENT OF RESULT IN B6
           MX5      2             FORM 48 DECIMAL IN
           LX5      6             X5
           SB5      B4+B5         ADD EXPONENTS OF OPERANDS
           SB5      X5+B5         ADD 48 TO SUM OF OPERAND EXPONENTS
           LX3      18            UNPACK SIGNIFICANCES
           BX1      X1*X3         OF OPERANDS INTO X1
           AX3      18            AND X3 RESPECTIVELY
           IX5      X1-X3         DIFFERENCES OF SIGNIFICANCES
           PL,X5    AA            X1.GT.X3,GOTOAA
           SX3      X1+B0         X3.GT.X1,STORE X1 INTO X3
AA         SX1      X3+B0         LOWER SIGNIFICANCE IN X1
           SX5      B5-B6         OBTAIN DIFFERENCE BETWEEN(SUM OF(EXPONENTS
*                                 + 48)AND EXPONENT OF RESULT )
           NZ,X5    BB            IF DIFFERENCE IS ZERO(CARRY)
           MX5      1             ADD 1 TO LOWER SIGNIFICANCE OF
           LX5      1             OPERAND TO FORM
           IX1      X1+X5         SIGNIFICANCE OF RESULT
BB         MX5      2             48 DECIMAL INTO LOWER BITS OF
           LX5      6             X5
           SB4      X5+BC         STORE 48 DECIMAL INTO B4
           IX5      X1-X5         SUBTRACT 48 DECIMAL FROM SIGN OF RESULT
           NG,X5    CC            IF LESS THAN 48,LEAVE AS IS
           SX1      B4            IF SIG GREATER THAN 48,SET TO 48
CC         SB6      X4            RESTORE B6
           LX4      18
           SB5      X4            RESTORE B5
           LX4      18
           SB4      X4            RESTORE B4
```

Fig. 2.

IV. Elementary Functions

Consider a floating point number which is unnormalized; then we can represent the fractional part x in a 48-bit word with $lz(x)$ leading zeros as

$$x = f \cdot 2^{lz(x)}$$

and δx, the error of x, as

$$\delta x \approx 2^{lz(x) - 48}$$

and

$$| \, x/\delta x \, | = | \, f \cdot 2^{48} \, |$$

or

$$\log_2 | \, x/\delta x \, | = \log_2 | \, f \, | + 48$$

Now normalize f to f_1 by shifting left $lz(x)$ places (the number of leading zeros in f); then

$$f = f_1 \cdot 2^{-lz(x)} \quad \text{with} \quad \tfrac{1}{2} \le f_1 < 1$$

Now

$$\log_2 | \, f \, | = \log_2 | \, f_1 \, | - lz(x)$$

$$\log_2 | \, x/\delta x \, | - 48 = \log_2 | \, f_1 \, | - lz(x)$$

or

$$\log_2 | \, x/\delta x \, | = \log_2 | \, f_1 \, | + 48 - lz(x)$$

Let $\text{dig}(x) = 48 - lz(x)$ denote the number of significant bits of x. Since $\tfrac{1}{2} \le f_1$ and $\log_2 | \, f_1 \, | < 0$ and ≥ -1, we have

$$\text{dig}(x) > \log_2 | \, x/\delta x \, | \ge \text{dig}(x) - 1$$

If F is now a function of x, then

$$\text{dig}(F(x)) \sim \log_2 \left| \frac{F(x)}{\delta F(x)} \right| = \log_2 \left| \frac{F(x)}{-F(x) + [F(x) + \delta x \, F(x)]} \right|$$

$$= \log_2 \left| \frac{F(x)}{\delta x \, F(x)} \right|$$

or

$$\text{dig} \, F(x) - \text{dig}(x) \sim \log_2 \left| \frac{F(x)}{\delta x \, F'(x)} \right| - \log_2 \left| \frac{x}{\delta x} \right|$$

$$\sim \log_2 \left| \frac{F(x)}{x \, F'(x)} \right|$$

As has been discussed previously [Goldstein (1963)], if functions of one variable such as the elementary functions are to be incorporated in a significance arithmetic calculation, the above approximation may be used to compute the number of significant bits of the function given a certain number of significant bits in the argument. In particular:

(a) For $F(x) = \sqrt{X}$,

$$\text{dig}(F) - \text{dig}(X) \sim \log_2 | \, 2 \, | = 1$$

(b) For $F(x) = e^x$,

$$\text{dig}(F) - \text{dig}(X) \sim \log_2 |\, 1/x \,| = -\log_2 |\, x \,|$$

(c) For $F(x) = \ln X$,

$$\text{dig}(F) - \text{dig}(X) \sim \log_2 |\, \ln x \,|$$

(d) For $F(X) = \sin X$,

$$\text{dig}(F) - \text{dig}(X) \sim \log_2 \left| \frac{\sin x}{x \cos x} \right|$$

V. Numerical Experiments

A number of numerical experiments were conducted, including those previously reported (Goldstein, 1963). We report here three of these experiments.

A. Series Evaluation (Table I)

Let $X = \pi/4 + (\pi/2)n$ and evaluate the series

$$\text{SERIES} = X - X^3/3! + X^5/5! - X^7/7! + \cdots$$

using standard floating point arithmetic until the one-word sum is unchanged. Because of "catastrophic cancellation" the result will only approach

TABLE I

$$\text{SERIES} = X - X^3/3! + X^5/5! - X^7/7! + \ldots$$

$$X = \pi/4 + (\pi/2)n$$

n	SERIES
0	7.071 067 811 865 36 E − 01 (48)
1	7.071 067 811 865 28 E − 01 (46)
2	−7.071 067 811 864 68 E − 01 (44)
3	−7.071 067 811 866 49 E − 01 (42)
4	7.071 067 811 859 56 E − 01 (40)
5	7.071 067 811 845 71 E − 01 (38)
6	−7.071 067 811 951 65 E − 01 (36)
7	−7.071 067 811 908 09 E − 01 (34)

the true value $\pm\frac{1}{2}\sqrt{2}$. If we now compare this result with a multiprecision evaluation we find errors in the computed value in the digit underlined. Our significance calculation gives in parentheses following the result the number of *bits* alleged to be accurate. As is shown in Table I, this closely approximates the expected result.

B. RUNGE-KUTTA (Table II)

In solving the ordinary differential equation

$$y' = \cos x$$

with initial conditions $x = 0$ and $y = 0$ by Runge-Kutta we obtain the results shown in Table II. The digit underlined is in error when compared to a multiprecision evaluation and the number in parentheses is our evaluation of the significance (given as the number of bits alleged to be significant between 48-all significance and zero-none).

TABLE II

$y' = \cos x$

x			y						
0.000 E − 01 (48)			0.						(48)
1.250 E − 01 (48)			1.246	747	439	589	52 E − 01		(48)
⋮			⋮						
2.875 E + 00 (48)			2.634	460	157	063	75 E − 01		(47)
3.000 E + 00 (48)			1.411	200	200	282	94 E − 01		(47)
3.125 E + 00 (48)			1.659	189	363	648	57 E − 02		(43)
3.250 E + 00 (48)			−1.081	951	437	062	18 E − 01		(48)
⋮			⋮						

C. MATRIX INVERSION AND DETERMINANT EVALUATION (Table III)

The results show the number of bits alleged to be significant in the inversion of the Hilbert matrix and in the evaluation of its determinant as a function of the order n. Again the underlined digit indicates the number of digits that are believed accurate by a multiprecision calculation and the number in parentheses our alleged significance with which it is to be compared.

TABLE III

$$H = h_{ij} = 1/(i + j - 1)$$

n	Det H	Max$\lvert H^{-1} \rvert$
2	8.333 333 333 333 2$\underline{2}$ E − 02 (46)	1.200 000 000 000 0$\underline{2}$ E + 01 (46)
3	4.629 629 629 627 $\underline{8}$2 E − 04 (42)	1.920 000 000 000 $\underline{7}$6 E + 02 (42)
4	1.653 439 153 43$\underline{3}$ 18 E − 07 (39)	6.480 000 000 02$\underline{3}$ 31 E + 03 (38)
5	3.749 295 132 1$\underline{1}$6 02 E − 12 (35)	1.792 000 000 1$\underline{8}$8 01 E + 05 (34)
6	5.367 299 87$\underline{0}$ 569 24 E − 18 (29)	4.410 000 013 $\underline{8}$07 68 E + 06 (30)
7	4.835 802 $\underline{8}$13 810 29 E − 25 (24)	1.334 024 $\underline{9}$47 495 55 E + 08 (24)

REFERENCES

Ashenhurst, R. L. (1965). Experimental Investigation of Unnormalized Arithmetic, *In* "Error in Digital Computation" (L. B. Ralls, ed.). Wiley, New York.

Ashenhurst, R. L. (1964). Functional Evaluation in Unnormalized Arithmetic, *J. ACM* **11**, 168–187.

Ashenhurst, R. L. and Metropolis, N. (1965). Error Estimation in Computer Calculation. *Amer. Math. Monthly* **72**, No. 2, part 11, 47–58.

Ashenhurst, R. L., and Metropolis, N. (1959). Unnormalized Floating Point Arithmetic, *J. ACM* **6**, 415–428.

Control Data Corp. (1968). 6400/6500/6600 Computer Systems Reference Manual. Publ. No. 60100000.

Forsythe, G. E. (1970). Pitfalls in Computation, or Why a Math Book Isn't Enough. Stanford Univ., Computer Science Dept., Tech. Rep. No. CS 147, January.

Goldstein, M. (1963). Significant Arithmetic on a Digital Computer, *Comm. ACM* **6**, 111–117.

Gray, H. L., and Metropolis, N. (1959). Unnormalized Floating Point Arithmetic with an Index of Significance. *PJCC* **16**, 244–248.

Lewis, G. (1960). Two methods using power series for solving analytic initial value problems. AEC Res. and Develop. Rep. NYO-2881.

Metropolis, N., and Ashenhurst, R. L. (1958), Significant Digit Computer Arithmetic. *IRE Trans.* **EC-7**, No. 4, 265–267.

Moore, R. E. (1966). "Internal Analysis." Prentice-Hall, Englewood Cliffs, New Jersey.

Richman, P. L. (1969). Variable-Precision Interval Arithmetic and Error Control and the Midpoint Phenomenon, Bell Telephone Laboratories, Tech. Memo.

Richtmyer, R. D. (1957). Detached-shock calculation by power series, I. AEC Research and Development Report NYO-7973. [Also see Richtmyer, R. D. (1960). Power series solution by machine of a nonlinear problem in two-dimensional fluid flow. *Ann. N.Y. Acad. Sci.* **86**, 828–843.]

Richtmyer, R. D. (1960). The Estimation of Significance, AEC Res. and Devel. Rep. NYO-9083.

Wadey, W. G. (1960). Floating-Point Arithmetic, *J. ACM* **7**, 129–139.

Yasui, T., and Winje, G. (1969). Significant Digit Arithmetic on ILLIAC IV, File No. 789, Feb. 18.

5.5 Nonstandard Arithmetic

C. B. Dunham

UNIVERSITY OF WESTERN ONTARIO

LONDON, ONTARIO

I. Reliability

To be anything but a toy, the arithmetic must be absolutely reliable and the subroutine library highly reliable. It might be thought that reliability could be guaranteed by random tests, but this turns out to be wishful thinking. An example of an erroneous certification of a function routine based on random tests appeared recently (Turner, 1969). The author has personal experience of infrequent bugs. An IBM 7090 FAP package was adapted to an IBM 7040 by using macros for missing instructions. The program ran well for four months before an error was found (caused by use of a macro) and the case turned out to be a very odd one. The author checked a Digital PDP-10 subroutine to see what method it used and found out that for exactly one fraction in the 2^{27} possible ones the answer was garbage. The only way to ensure reliability is for the coder to completely document his method and for the certifier to check the method and code and then devise test cases which are likely to cause trouble.

II. Subroutine Library

Without a subroutine library, a nonstandard arithmetic package is useful for only a few specialized purposes. The minimum requirement for general use is that there be provision for defining and outputting nonstandard numbers. In some cases routines to convert standard numbers (for example, floating point numbers) to and from nonstandard numbers are adequate. In many cases some means of accessing mathematical constants would be preferable to defining or reading the constants: in Fortran, block data would be convenient for this. Next most important are the basic mathematical functions (integer part, sine, cosine, log, exponential, arctangent, and square root). This still leaves the bulk of numerical analysis routines, such as linear equation solvers, zero finders, and integrators to be written. When one realizes that the subroutine libraries for standard arithmetic on many machines leave much to be desired in quantity and quality, one may shudder at the difficulty of writing a good library in an intrinsically more difficult arithmetic. This difficulty will be minimized if one can write in an algorithmic language or convert standard programs (see Sect. IV).

III. Efficiency in Execution

The principal feature of programmed arithmetics is that they are slow in relation to hardware arithmetic, and the more powerful the arithmetic the slower it is. In the case of multiple precision, for example, execution time goes up linearly with the precision for addition and with the square of the precision for multiplication. In the case of multiple-precision functions things are even worse, for in addition to slower arithmetic we have the penalty of more arithmetic being required (for example, if a series is used we need more terms). The classic example the author encountered was the calculation of the Bessel functions P_0, Q_0 used in the asymptotic for J_0, Y_0. To compute these in 40 decimals required a 40-decimal sine and cosine plus two long series in 80-decimal arithmetic (needed to control cancellation). A function evaluation took around $\frac{1}{10}$ min on the IBM 7040 (cycle time: 8 μsec).

The only reasonable conclusion is that if a nonstandard arithmetic package is to be used frequently and if it takes a long time to execute compared with machine arithmetic, the arithmetic and other frequently used subroutines should be efficiently coded.

It might be desirable to include subroutines to speed up special computations. For example, a routine to add to base-2 exponents enables us to halve a number without a multiplication. A routine to divide a multiple-precision number by an integer executes much faster than a multiple-precision divide and enables one to speed up a Taylor series evaluation or a binary to decimal conversion.

Attention might also be paid to efficient use of storage, as nonstandard numbers may occupy several words and as programs for nonstandard arithmetic generate substantially more instructions than standard programs. In addition, the basic arithmetic and input–output routines occupy a substantial amount of space.

IV. Ease of Use

Standard arithmetic can be done in an algorithmic language. All too frequently, nonstandard arithmetic must be done in something that resembles or is assembly language. For example, in all three packages used at Western, if one wanted to do the Fortran statement

$$Z = SIN(X) + Y * A(3)$$

in nonstandard arithmetic, one wrote something of the form

CALL PROQ(Y,A(1,3),TEMP)

CALL QSIN(X,TEMP2)

CALL SUMQ(TEMP2,TEMP,Z)

There is nothing intrinsically difficult about such coding, but a long program in an algorithmic language tends to become much larger and much more tedious to write in this form. What was a minor part of one line in an algorithmic language tends to become a line in itself and a perfectly readable program in an algorithmic language becomes an unreadable string of three address codes. When one realizes the length that present programs in algorithmic languages can attain, the drudgery of writing the equivalent in pseudo-assembly code becomes overwhelming. The ideal would be for the non-standard variables to be another data type in a standard algorithmic language, so that a standard program in an algorithmic language can be converted to nonstandard arithmetic by changing a type statement.

V. Implementation of Nonstandard Arithmetic

It seems reasonable to favor an implementation which can be done piece by piece, which is cheap, and which requires a minimum of special software. We presently describe such an implementation. First, however, we give an example where this strategy was not followed. In the early 1960's it was decided at the University of Western Ontario to write a full Algol compiler for the IBM 7040 which would incorporate variable-precision data. One group worked on the arithmetic. Another group worked on the compiler and developed elaborate linkages for execution of Algol. It was finally found that Algol had so many bugs without even considering variable precision that variable precision was dropped. As the variable precision depended on special linkages which never materialized in Algol, it could only be used in assembly language. It was therefore junked. There are several morals to this story. First, don't pick the most complex language to add nonstandard arithmetic to. Secondly, don't depend on compiler writers. Thirdly, develop the package so it can be adapted to other systems as easily as possible.

By taking advantage of special hardware–software characteristics one may be able to get very cheap implementations. For example, on the IBM

7090 and Digital PDP-10, double precision is done by software. To make any double arithmetic (for example, interval arithmetic) or single-word arithmetic available in Fortran, one merely writes arithmetic routines with the same names as the double-precision routines. One can then replace double-precision functions by one's own functions.

The cheapest general way of implementing nonstandard arithmetic is to use subroutines to do all operations and to make the subroutines callable from an algorithmic language. A three-argument subroutine seems to be most convenient for arithmetic operations. For example, to subtract B from A to get C, we could write

CALL DIFQ(A,B,C)

Other examples are given in Sect. IV. The principal advantage of this form is convenience for human use, making it possible to generate code rapidly without having a special compiler. If it gets tedious coding in this form, it is not too difficult to add a translator from algorithmic language to this form (see the next section). One alternative is use of subroutines with pseudo-accumulators instead of three addresses. This may result in faster execution but is less convenient for human use. A second alternative is use of sub-routines callable from assembly language only. These calls may take less space and execute faster. They are useful only by the professional systems programmer and make any complex piece of code almost impossible to write. Both of these alternatives make a full compiler almost mandatory.

VI. Use of Precompiler

We have already seen in Sect. IV that extensive coding using subroutines for operations is inconvenient. Coding can be made slightly more convenient by permitting writing nonstandard arithmetic expressions in the form

var. = var. op. var.

and writing a translator to convert these to three-argument subroutine calls in algorithmic language. It seems worthwhile to go further and write a translator (which we will call a precompiler) to convert algorithmic state-ments involving nonstandard arithmetic to a series of subroutine calls in standard algorithmic language, thus converting a program in an extended algorithmic language to a program in standard algorithmic language. The precompiler can be written with the names of the subroutines and the number of words occupied by nonstandard data as variables, so that one precompiler can handle all types of nonstandard arithmetic, as long as only one is used per subprogram. The precompiler approach has many advantages. We need only concentrate on translation of nonstandard statements and thus the compiler does as much of the work as possible. The code generated by the precompiler is simple, making it easy to debug the precompiler. The precompiler can be largely written in an algorithmic language. The precompiler can be started accepting only simple statements and the ability to handle more complex statements added: there are good reasons (cost, reliability) for *not* having the precompiler accept very complex statements. The user can do part of the translation by hand if he wants to do something tricky or the precompiler does not work. If there are changes to the operating system no changes need be made in the precompiler (two changes that could be unpleasant for a true compiler are a change in calling sequences or a change in register saving conventions). The user does not have to wait for the precompiler to be finished before starting coding.

Let us consider writing a precompiler for the Fortran language. As arithmetic statements make up most of a long job, a precompiler that handles only arithmetic statements would be a worthy beginning. There seems to be no difficulty to converting arithmetic expressions into Polish notation and then using each operator, operand, and operand triplet to generate a line of code. Nonstandard arrays must be converted to standard arrays. For example, a reference to A(2, 3) becomes a reference to A(1,2,3) and, if a number occupies five words, the statement

DIMENSION A(3,3)

becomes

DIMENSION A(5,3,3)

Facilities for converting type, DIMENSION, COMMON, and EQUIVALENCE statements should be added at the same time. Next might be added facilities for interpretation of nonstandard constants in expressions and DATA statements. Most IF statements will probably not involve tests of nonstandard quantities, and so these are easily handled. The most general way of implementing relational expressions involving nonstandard quantities is to do a subtraction, a conversion to a standard number, and a standard test against zero. The user could do this himself and save us the trouble of implementing this. In many cases tests can be made more efficiently. For this and other purposes we might have a special symbol preceding a statement to tell the precompiler that the statement has already been precompiled by hand. Input–output statements can be handled in several simple ways. To make things simple, mixed standard and nonstandard expressions would probably not be allowed; if one wants such expressions, an appropriate conversion subroutine could be used in the expression. For example, if X is standard and Y is nonstandard, X * Y could be written NONST(X) * Y, where NONST is a conversion routine.

VII. Type Other

One alternative to a precompiler is to use a compiler with a type "other." One way of doing this is to have in the algorithmic language a type called "other." The compiler then generates a standard call for each operation. The user then writes routines compatible with the standard call and tells the compiler the name of his routines. The compiler then compiles type other statements as calls to these routines. A more general approach is to have a compiler which generates whatever call is given to it to be used for a given operator. This facility is available on some Control Data machines. For an example of the use of type other see Frankowski and Zimmerman (1968). If no type other is available, one has to write a special compiler: as type other will handle *all* cases, there is little advantage in writing a compiler just for one type of nonstandard arithmetic.

It would be nice if type other were available on all machines, but in fact it is on only one manufacturer's machines. Why is this? Other manufacturers would give several reasons. One claim might be that such compilers would be more costly to write and maintain. A second claim might be that such features would degrade the performance of compilers for standard jobs, making them somewhat bulkier, slower, and less reliable. Another reason might be that type other would appeal only to a minority of users. The answer to such claims is that there already exist infrequently used types [complex, double complex, extended precision (on IBM 360/85, 195)] and that one way of handling them would be to put *all* under type other. Compilers could be therefore simplified by use of type other.

VIII. Conclusion

We consider what features a nonstandard arithmetic package should have to be useful for large jobs and how the package should be implemented. The two key factors in usefulness turn out to be reliability and an adequate subroutine library, with execution efficiency and ease of use being almost as important. It is recommended that nonstandard arithmetic be implemented in the form of three-argument subroutines callable from an algorithmic language and a precompiler be written to translate nonstandard algorithmic statements.

It is hardly an exaggeration to say that when arithmetic routines are written only 5% of the work in getting a useful package has been done. The arithmetic routines by themselves are as useful as a computer with an assembler but no compiler or library. This accounts for infrequent use of nonstandard arithmetic, since nonstandard arithmetic is difficult to use unless someone pays the cost of developing a complete package. Neither users nor manufacturers want to pay the cost. If a manufacturer offers non-standard arithmetic, the users demand better software. This is very expensive, so the safe thing is to not offer any nonstandard arithmetic.

REFERENCES

Frankowski, K., and Zimmerman, C. (1968). Algebraic manipulation on computers for scientists and engineers, *In* "Interactive Systems for Experimental Applied Mathematics" (M. Klerer and J. Reinfelds, eds.). Academic Press, New York, review in (1969). *Comput. Rev.* **10**, Rev. 17297, 394.

Turner, L. R. (1969). Difficulties in SIN/COS routine, *SICNUM Newsletter* **4**, No. 3, 13.

5.6 Making Special Arithmetics Available

H. C. Thacher, Jr.

UNIVERSITY OF NOTRE DAME

NOTRE DAME, INDIANA

Special arithmetics are not used for many tasks for which they would be appropriate because they are not easily available to the user who needs them. A particular reason for this lack of availability is that, for the most part, the codes were written in assembly language, and with fragmentary documentation. Thus their use on any computer other than that for which they were originally written requires several man-months of tedious recoding and testing. In many cases this labor, which often means that the job is not done at all, would be avoided if a suitable machine-independent (or machine-adaptable) version of the program were published, or otherwise made widely available. It is the purpose of this contribution to consider some of the applications which such special arithmetic packages have, or might have, to discuss experience with one such package, and to attempt to state guide lines for the development of new ones. We will begin by citing some of the types of special arithmetic which may be of interest, and some of the uses to which they should be put.

Probably the most important single special arithmetic is multiple precision, extended range floating point, i.e., a floating point arithmetic which carries at least 10% more significant digits in the fraction than the most precise floating point arithmetic in normal use on the particular computer. It is even more advantageous if the precision is under the control of the programmer. Although not as essential, it is often convenient for such a package to provide an extended exponent range, to minimize overflow and underflow problems.

The uses of multiprecision arithmetic are manifold. It is essential (unless a second, more precise computer is available) for testing double-precision function libraries. An indication of the need for ready access to such arithmetic is the number of libraries for which only the single-precision functions have been tested at all. Another application in which multiprecision arithmetic is needed is in the generation of accurate constants to assist in the preparation of full-accuracy double-precision routines. In many cases, these will be the coefficients in expansions of functions: in particular, as power series, series of Chebyshev polynomials, and continued fractions. Accurate zeros for functions are needed to preserve relative accuracy in constructing function routines, and are often of interest in their own right. The development of special Gauss-type quadrature formulas is another application requiring extra precision. In some cases, only one or two words of extra precision will be needed, but the availability of arbitrarily high precision will often permit the use of mildly unstable methods, such as the qd algorithm, and, by comparing results at various levels of precision, allow an easy estimate of their reliability.

Computations with integers and rational numbers avoid the difficulty of round-off at the expense of rapid growth in the length of the numbers. The drawback to the use of integers is the rapid exhaustion of memory if all integers are allowed to occupy the same amount of space. A variable-length integer and rational arithmetic package is the answer to such problems, and would have many of the same applications as multiprecision floating point arithmetic, particularly in the generation of expansion coefficients of various types. Another important application of this type of arithmetic may be in the area of linear algebra, where algorithms such as integer preserving Gauss elimination (Bareiss, 1966, 1968) would permit the construction of exact inverses for test matrices. Some users of this algorithm have found it desirable to take advantage of the indefinite-length rational facility embedded in FORMAC.

A third class of arithmetic, the use of which has been restricted because of the limited availability of programs, is that of error-estimating arithmetics: interval arithmetic, nonnormalized arithmetic, and precision-index arithmetic. Except where hardware implementations are available, the consensus seems to be that these arithmetics are too expensive for routine use. There seems little doubt, however, that they are of considerable value in analyzing special situations, such as the conditioning of polynomial and rational forms.

In addition to the particular uses of the specialized arithmetics which have been mentioned, one should mention their importance in formal and informal instruction. Both students and more mature scientists often learn best by experience, and experience with special arithmetic requires the availability of suitable implementations.

The applications of special arithmetics mentioned above share an important characteristic which is significant for implementation. Although of considerable importance, they are all relatively low-volume problems. The coefficients of an expansion or of a quadrature formula may be used millions of times, but only need to be evaluated once. A function subprogram need only be certified once or twice, no matter how heavily it is used. Thus machine economy is of secondary importance compared to economy of programming effort. Furthermore, many of the programs run with special arithmetic are relatively simple, so that the major programming effort is in the preparation and testing of the special arithmetic programs, and not in their use. Unfortunately, this effort, particularly if it is to be done in assembly language, is often so large as to result in postponing the task requiring it until someone else has so pressing a need that he is forced to produce the required arithmetic programs. The package produced then circulates widely among the users of the particular computer.

By the time they become obsolete, some of the more popular computers have been provided with a variety of special arithmetic packages. Users of less popular, or more modern, computers are less fortunate. The provision of assembly-coded routines for the more important special arithmetics on all computers for which they may be needed would be a prodigal waste of programming effort. A more reasonable approach, and one which is strongly recommended, is to write such packages in a higher-level language, and preferably in one which is available, and maintained at most computer installations. It would be particularly desirable if the programs were well enough documented to publish in the open literature, instead of merely circulating informally or through user's groups. The use of a higher-level language introduces certain inconveniences, such as the inability to shift and mask in standard Fortran IV. The major benefits of the higher-level language, however, lie in the wider number of users who can benefit from the effort.

The feasibility of this approach has been demonstrated by the published Algol procedures of Gibb (1961) for range arithmetic, and of Hill (1968) for arbitrary-precision floating point arithmetic. Unfortunately, these interesting procedures do not seem to have been exploited as widely as they deserve—Gibb's procedures have not even been certified. It seems probable that this neglect is due to the language barrier. Few United States' installations have Algol translators which are as readily available or as well maintained as their Fortran counterparts. Even the task of transcribing from Algol publication language to the appropriate hardware representation is often inconvenient. Moreover, even though the special arithmetic procedures require none of the special capabilities of Algol, and are thus readily transcribable into Fortran, or any other comparable language, this minimal effort seems to have been sufficient to inhibit wide use, or even experimentation.

For widespread adoption at this time in the United States, it thus appears pragmatically that machine-independent special arithmetic procedures should be written in Fortran and, if possible, in American Standard Basic Fortran. Where nonstandard basic Fortran, or even assembly coding, would improve efficiency significantly, it may well be supplied by comments, but the primary aim should be to present a program which will run correctly with minimum modification on as many computers as possible.

Adoption of a standard, and popular language for expressing special arithmetic procedures only partially attains the goal of machine independence. The differences in machine arithmetic must also be accounted for. Because of these differences, and also differences in procedures at different centers, the user will probably have to supply a number of global parameters defining the arithmetic of his computer, the unit numbers of his input and output devices, and also the details of the special arithmetic being implemented. Unfortunately, machine floating point arithmetics are highly variable, and extremely difficult to parameterize. In addition to the number base, and the allocation of parts of the word to exponent, fraction, and sign, details of the rounding or truncating process, and its interaction with normalization, the representation of negative numbers and so on must be known to insure transferability. Integer arithmetic at the Fortran level is, by contrast, far less variable, and may be characterized for most purposes by a single parameter, the magnitude of the largest representable integer. Hence it is likely that programs for special arithmetic which rely primarily upon Fortran integer arithmetic will require minimum modification for transfer among machines, and thus will be most widely adopted.

Special arithmetic programs written in Fortran have obvious limitations. Fortran is not really designed for working at this level of detail, and many intrinsically simple procedures, such as masking and shifting, require tedious circumlocutions and correspondingly inefficient object code. The exclusion of floating point operations, which is desirable to minimize the effect of machine differences, sacrifices the inherent speed advantages of hardware floating point operations over interpretive floating point. When these effects are added to the nonoptimality of the code generated by many compilers, and to the relatively long algorithms used to implement special arithmetics, slow programs are almost inevitable. For example, to compute $\log 7$ using a Fortran version of Hill's long floating point algorithms required 3.2 sec in $35 + S$ arithmetic and 5.9 sec using $45 + S$ arithmetic, compared to 0.15 msec in 8S single-precision arithmetic on the Univac 1107.

A second disadvantage is that programming with packages of this type is rather clumsy. Standard Fortran makes no better provision for implementing special arithmetics than providing a set of subroutines, each, perhaps, equivalent to a single, three-address, pseudo machine instruction. This forces the user to code his application program at essentially the assembly code level, a somewhat tedious task. Fortunately, most applications for which this approach is feasible require relatively short amounts of coding, so that this characteristic is an inconvenience rather than a serious disadvantage. It could, of course, be eliminated if precompilers were more universally available, or if a suitable extendable Fortran gained widespread acceptance.

A third source of minor inconvenience in using Fortran-programmed special arithmetics is input–output. Although this is an inherent problem with many forms of special arithmetic—no formatting will make a 200-digit long-precision number fit on a single line, or be comprehensible at a single glance— standard Fortran input–output facilities are not as flexible as might be desired. For example, it is not possible to print leading zeros in integer output. An additional nuisance is the variation in unit numbers among various installations, which suggests including these with the machine-dependent parameters describing the arithmetic. My own solution to the formatting problem has been to accept inelegant output for preliminary work, and, for final results, to output the data on cards or tape and convert them to the desired form by a short postprocessing program. The nonstandard Fortran ENCODE–DECODE facility is also helpful for this purpose.

For some important applications, such as testing of double-precision sub-routines, or the provision of extra precision at critical points in a large computation, input–output is replaced by conversion between standard and special number representations. If moderate accuracy is sufficient, this is not too difficult, but the extent to which it can be made machine in-dependent if the last bit or two are significant is doubtful.

As an indication of possible applications for Fortran-programmed special arithmetic, and as the basis for these comments, a brief review is appropriate of work which has been done at the University of Notre Dame using a Fortran version of Hill's long arithmetic algorithms.

Since this account represents personal experiences, rather than firmly established, independently verifiable facts, a personal anecdotal style has been adopted.

So far as I know, assembly language subroutines for n-precision floating point arithmetic have never been available for the Univac 1107, although I have tried to promote their construction for the last four years. The Hill package was transcribed from Algol to Fortran and partially tested during the Christmas vacation of 1968. The first application, in January 1969, was the transcription of a qd algorithm for converting power series to the correspond-ing continued fractions, and the use of this program to find, among others, the elements of the expansion of the exponential integral $E_1(z)$ in a continued fraction about the origin. In addition to revealing certain errors in the Fortran transcription—Algol "calls by value" require special treatment when converted to Fortran—the tests using up to $60+$ significant digit arithmetic revealed that some of the high-order elements given by Cody and Thacher (1968), which were computed in 40S arithmetic, and published to

25S, are in error in as many as the last 7 digits printed. Errors of this magnitude in the high-order elements do not, however, affect the 25D accuracy of function values computed from them. The same program was used to determine the elements of continued fraction expansions of the integral sine and cosine about the origin, and of the functions appearing in the asymptotic forms of these functions for large x.

Over the mid-year examination period of 1969 a long square-root routine and a long logarithm routine were programmed and tested. Although not highly polished, the square-root routine obtains 25 + S values in about 0.35 sec, and 45 + S values in about 0.65 sec. Times for the logarithm were of the order of 10 sec for 45 + S, and 20 sec for 55 + S. This routine used a rather crude argument reduction scheme, followed by evaluation of a power series, so that the times were not only relatively long but also showed considerable variability.

The next task was the computation of zeros of the integral sine and cosine. For this purpose, a simple Regula Falsi code was written, and used with the Maclaurin series. With these routines, Mr. John O'Sullivan and I have computed the first 5 zeros of $Ci(x)$ and of $si(x)$ with accuracies ranging from 45D for the first to 30D for the fifth.

During the spring, it became apparent that the algorithm being used at Argonne for computing the Riemann zeta function was losing accuracy by cancellation near $s = 1$. The Gram power series expansion about this point promised a solution to the difficulty, but the coefficients, which are complicated limits, had only been evaluated to 16D, compared to the 25D needed for a double-precision routine on the CDC 3600. It was therefore decided to recompute these coefficients to higher precision and, at the same time, to rearrange the series to produce a series of Chebyshev polynomials for the appropriate intervals. So far as can be determined from independent checks and internal consistency, the results are good to 40D or more.

Returning to the integral sine and cosine, it became necessary to construct long arithmetic sine and cosine subroutines and, most recently, to compute Chebyshev expansions for the asymptotic forms of the integral sine and cosine. This work is still in progress, but indications are that 25S master routines for producing best rational approximations on the 3600 will be attained.

Thus, in the course of less than a year, in addition to teaching and other faculty tasks, it was possible not only to implement a usable long-precision floating point package, and to use it to compute accurate values needed for other calculations, but also to construct the beginning of a library of elementary function routines and basic procedures. Since these subroutines were developed as the need for particular functions arose, the task has been less tedious than constructing a complete library at once. The latter approach would, however, be necessary for users with less sophistication in the computation of functions.

This account also indicates a major advantage of this set of routines. It makes the user essentially machine independent so that computing can be done at the location which is convenient at the particular time rather than where the computer and library suit the job to be done. Moreover, the programs which have been written are, with minor exceptions, completely portable, and can be made available to other workers, regardless of their facilities.

REFERENCES

Bareiss, E. H. (1966). Multistep Integer-Preserving Gaussian Elimination, Argonne National Laboratory Rep., ANL-72/3.

Bareiss, E. H. (1968). Sylvester's Identity and Multistep Integer-Preserving Gaussian Elimination, *Math. Comput.* **22,** 565–578.

Cody, W. J., and Thacher, H. C. (1968). Rational Chebyshev Approximations for the Exponential Integral, $E_1(x)$, *Math. Comput.* **22,** 641–650.

Gibb. A. (1961). Algorithm 61. Procedures for Range Arithmetic, *Comm. Assoc. Comput. Math.* **4,** 319–320.

Hill, I. D. (1968). Algorithm 34. Procedures for the Basic Arithmetical Operations in Multiple-Length Working, *Comput. J.* **11,** 232–235.

5.7 The Production of Mathematical Software for a Mass Audience

E. L. Battiste*

IBM CORPORATION

HOUSTON, TEXAS

I. Introduction

IBM's decision to enter the software field formally was initially considered at a 1961 conference. One hundred programmers, applications people, and the officers of the corporation attended. Before the conference, many attendees had felt that a positive decision was obviously called for. After serving on the committee investigating applications software, and after listening to reports from systems committees and language committees, it became apparent that decisions on formal entry were not to be reached easily. The critical question was "how far past the line of hardware development into the realm of user responsibilities should the corporation venture"?

It was generally recognized, in 1961, that stable hardware could be mass produced. There were few examples available which implied that the corporation had entered a field where its responsibilities to its customers could not be met. Whether or not this could be achieved in software development was certainly not clear. It seemed that the production of software had entered its most chaotic stage to date, and that the experience gained starting in 1955 was not being used to solve software problems.

This discussion is intended as a statement concerning the attitudes of a software development manager toward the production of mathematical software. In making this statement, the problems which arise in that production will be discussed; implicitly, their discussion should point out why totally acceptable mathematical software is not as yet available, give a view of the scope of the total job, and address the concern of a manufacturer about this area of endeavor. In the following, library will be termed mathematical software.

---

* Present address: International Mathematical Statistical Libraries, Inc., Houston, Texas.

II. Discussion Assumptions

The development of a library is considered, in this discussion, in the following light. Bases for element definition are available, in the knowledge within the corporation, in consultants, and in libraries of the type presented in *Numerische Mathematik* or *Communications of the Association for Computing Machinery*. Given these bases, a library, running as distributed, in a high-level language on a class of machines, is to be constructed.

The assumptions on which this discussion is based are as follows:

(a) Everyone in the audience is interested in the development of mathematical software, and some have been involved in this development. However, in general, this audience has not been deeply involved in development for a mass audience peopled mainly with nonmathematicians.

(b) Software development has been and is chaotic, must be controlled, and is not being controlled. We know that potential producers are faced with a number of heretofore unsolved problems, and have finite-sized funds available for production.

III. Problems in Mathematical Software Production

A. MARKETING

Unfortunately, there are few nonprofit organizations concerned with mathematical software development for world consumption. Even though a slow, incremental attack on library development should take place, economic considerations require that the first subset of a library be marketable, and have a reasonable set of abilities. Thus, before interested users are given the opportunity to impact package design, ground rules have been set so as to have lasting effect. Most of the problems discussed below arise in the production of the first subset.

The investment in such an initial production is not small, and is extremely portable to other machines and languages. Fortunately, computer manufacturers have not allowed this fact to restrain their efforts. Such restraint of effort as may seem evident is due to other factors, to be described.

Direct production costs in 1970 for a group concerned with this development, including programmers and applied mathematicians, is in the range of 20,000 to 30,000 dollars per man. These costs include working facilities and computing costs. They do not include the cost of testing by independent groups, distribution expenses, and overhead costs, like insurance.

Considering the above "manufacturing" situation, other important issues may take a subservient position. Decisions on the following problems are basic to the production of a library.

B. General Problems

(1) How is a long-lived library produced and maintained in a rapidly changing environment?

(a) Should code be produced so as to be easily modifiable for specific user needs? Or, if a library is to be centrally maintained, is it almost mandatory that users accept the ability as given so that their maintenance facility is manageable?

(b) Are there standards of excellence which are generally accepted, on coding? Who can make decisions such as "accuracy versus efficiency versus space" for a mass audience? Who can define whether or not a code is good? After production of a code, and before release, should it be "honed," by expert programmers? Is it realized that such a procedure may retard or halt release? Can a class of "honers" be bred? Can the independent test groups accept this task, in groups of 100 codes? Can they be restrained from redesign which impact sections of the library, unbeknownst to implementers of the other sections?

Can general decisions be made which define the level of diagnostic aid which should be handled within a library element? Can a decision be made, in this regard, which will not raise the issue of "unnecessary overhead"?

(c) Is there an adequate statement, easily and inexpensively available, on algorithmic state of the art?

(d) What are the correct categories for breakdown of a mathematical computer library? What mesh should be used to cover each category? How does the mesh change across machines?

(e) How can the requirement for cross-referencing of information among implementers of a library be handled? What control over impact of a convention change is feasible? How do separate groups control coding, and how is between-group variance minimized (all within a production schedule time frame)?

Since no development group has acceptable answers to these questions, a method must be available for production which will allow answers to be formed over a time span during which libraries are available. Many answers cannot be obtained before production and release.

Even nonbasic questions arise which inhibit production, release, and usage of libraries. It is difficult to differentiate between important and unimportant considerations in a field which is new and with products which are applicable to almost all environments. For example, who should dictate the names of the elements of the library? Recently, a computer manufacturer, having being deluged by letters from customers concerned with the recurrence of names—cross software recurrence—created a well-paid position, replete with computation (sorting and matching) abilities, with the title "Keeper of the Names." No one who understood the concept would accept the job.

(2) Should (can) a library be produced which is acceptable to a mass audience?

Current libraries in mathematics are used by the following types of installations:

(a) Small computer users, who make up the largest audience, and who are mainly nonmathematicians.

(b) Medium-scale computer users.

(c) High-performance, or large-scale, scientific facilities. This is by far the smallest set, in numbers, and in dollars spent.

The large-scale user group is not as interested in the availability of mathematical software products as are the other groups. Often, talent resides with the high-performance group which is capable of producing excellent, tailored codes and which may be loath to use any product produced by any one else especially if the product is designed for a large audience.

There is some concern about giving "black boxes" to any user set, since users are inclined to accept results without internal testing, and since the availability of such tools may cause users to omit "in depth" studies of the problems of interest. Our industry has been presenting such "black boxes" to the industrial and scientific communities for 15 years now, and it is difficult to imagine a 1970 user who does not know that the word beware goes with the code. We are now at the stage of at least being able to release codes which give results as documentation infers, but description of these codes is not good. Since such codes often relate to pure mathematics on one hand, and very dirty engineering on the other, since the implementer is the best candidate for documenting, and since he generally knows too much about the code to document it concisely for the general reader, documentation preparation is a real problem.

C. COLLABORATION

If the audience for a library is to be world-wide, then each major geographical set with an expected audience could be expected to participate in the production of the library. Such participation will allow costs to be spread, and more knowledge will be available for package design. However, if the number of individual participants in library development exceeds one, agreement becomes a severe production constraint. Production schedules are necessary; one generally wishes to produce a package before the need for it vanishes.

Agreement on conventions of the following types becomes more difficult as the number of participants increases.

1. *Basic Intent*

This would seem to be an idea which could be framed in a statement which is acceptable world-wide. Yet, the average degree of sophistication of the user and developer is not constant across countries.

2. *Efficiency–Accuracy–Space Conventions*

Vague statements can be made in this regard, and they will be interpreted vaguely. Finally, their interpretation is left to the individual implementer.

3. *"Standards" Conventions* (*Error Detection, Precision, Naming*) *Which Allow Production of an Interactive Package by Multigroups*

In February, 1969, my group agreed on a standard convention to be applied in the detection of errors. We felt that utter simplicity was of overriding importance. A participating group disagreed. We took the resulting conventions to a user group, who felt that neither was correct, and that the most important element of such a convention was its ability to warn a user if the user ignored documented warnings, and if the user chose not to question error flags in his main program. After the three groups each framed a convention, a compromise somewhat unsatisfactory to each individual group was implemented. This "agreement" took nine months, and no code could be finalized until agreement was reached.

After conventions are agreed to and production has taken place, and when the individual elements of a library are brought together for dissemination, it may be discovered that various contributing groups did not follow the rules, and others interpreted them incorrectly. The consistency and even the usage of the resultant package may be seriously impaired, and quick revisions to intricate codes are often necessary in order to meet production schedules.

D. PERSONNEL—TESTING

The task of preparing robust mathematical codes for a mass audience is formidable mainly because of the range of abilities required of the personnel involved with the task. For example, it is necessary that an analyst who is to implement a set of eigenanalysis abilities be conversant with the literature, which implies a degree of formal training. He must have been previously faced with scientific computing problems, and know, for example, how to use the theory of error analysis in implementing on a particular machine. He must see the total problem set which spans the area he is working in, and understand where it might overlap with other library abilities. "In-depth" knowledge of the language being used is necessary, whether he programs or directs programmers. The design of the program set which spans the area should imply that usage is easy. The test set must be designed to encompass the error possibilities which can be made, and to span input possibilities in the ranges considered.

After the set of codes has been implemented and tested internally, the analyst must be prepared to justify his detailed decisions about specific lines of code, and his general decisions about design. If these cannot be justified, perhaps a considerable time after implementation when the program set is being tested as a product, then release may be postponed until modifications are made. To meet release dates, the analyst must be prepared to modify quickly his codes and designs, realizing that his original decisions, thought out during the process of construction, were probably correct.

The mathematician–analyst who is capable, and who works in the software field today, is faced with the *total* task of engineering the product, from justification through design through being involved with the detailed engineering task of writing, justifying, and changing thousands of lines of code. This task continues well beyond release of the product.

If one were to devise reasonable methods for testing and checking a library of codes, independent testing would certainly be applied. One of the difficulties encountered in such testing is that the testers must be individuals who are competent in the area being tested and in scientific computing. They are not inclined to accept a design or code and not perturb it. This perturbation comes very late in the process of development, after much effort has been expended by the developers. The developers have supposedly concluded their task.

The test group is also required to accept a large number of codes spanning a broad discipline. Applying more than a check for consistency and readability may be a waste of effort; at this point, some reliance must be placed on the implementers. Yet, talent must be available to test, and time may not allow application of that talent. The effect of such a mass effort cannot possibly be certification of codes.

Thus, the internal tests required of individual implementers cannot be relaxed because there is independent testing. In addition to these, the library developers must implement a test mechanism which attempts to catch errors caused by perturbation and vague interpretation of conventions. This general test must detect departures from normal procedures of development, made because programmers were not communicative, were not experienced, or were not interested. Such checking and communication mechanisms often cause so much overhead that their implementation replaces the final goal of producing a running element for a library.

If a library is to be developed and maintained, should programs that test the individual elements be maintained? It may be that these programs, and their correction, updating, and maintenance, will create more difficulties than does the library. At this time, it seems better to trust individual implementers to read conventions, to know mathematics, to understand corporate needs, and finally to produce tested codes than to establish comprehensive checking mechanisms. Such mechanisms, simple in concept, grow in complication and may burst forth as goals.

All of the above problems are finally faced by the individual implementer. They make one of the most important requirements for good library program production almost unachievable. A continuity of qualified and experienced personnel, on the job long enough to see that long-range plans are carried out, may not be available. Because of this, below the level of mathematician–analyst, many software firms have a planned program for personnel transfer. Two or three years of producing and reproducing software is as long as can be tolerated by most human beings.

E. CODE WARRANTY

When software is to be sold, developing groups are concerned with the warranty attached to their product. The meaning of a warranty, in this new area of automation, is vague. When one writes a program to run in a particular language on a rather large group of machines under a versatile operating system in the environment of 10,000 different operating shops, it may be impossible for the developing group to apply robustness to the procedure which implies correct results in all situations. A lengthy look at conventions may aid in the solution of this problem, but, since the developing group cannot imagine the union of environmental conditions in which the codes will be utilized, warranties are dangerous.

IV. Environmental Conditions Affecting Mathematical Software Production

From mid-1965 through mid-1968, one-thousand four-hundred 1130 computers and three-thousand eight-hundred 360 computers (models 30 and above) were installed in user organizations in the United States. The task of installing these and other computers faced by computer manufacturers, and the software task faced by users, was massive. Manufacturers did try to aid early users with fundamental software abilities. In hindsight, software jobs which were produced in 1965 and 1966 resulted in products which had to be maintained, but which were not worthy of maintenance.

Perhaps the only term which can be applied to the current state of affairs is that it developed because of lack of experience. However, "lack of experience" may not be relevant in a situation where decisions are to be made on problems which have not been previously examined and which have not been clearly defined or solved by the entire computer-using community over a fifteen-year period. Software production has always been carried out without the standard research and development phases. This is true because if time is taken for research and development the ground rules and need for the product will change before production can be completed. Today, in the library situation, we have some experience for establishing bases. The need will not vanish; research and development is being applied. The product is basic to computer usage and its utility will last.

The computer manufacturer can take blame, however. The need for a solid library base should have been understood and promoted years ago.

V. Production of Mathematical Software

(1) Why should a computer manufacturer build a library?

(a) Community requirements—the transient and chaotic nature of the problem of communicating with computers requires that as much of the total problem be standardized as is possible. In the late 1950's, it was expected that the elementary functions would be presented along with high-level languages. Today, it should be expected that proven mathematical techniques will be available to users, as machines change.

(b) Mappings to new computers and new languages; if the basic library is well constructed and documented the costs of building it for new computers are minimal.

(2) What should be provided?

(a) Robust input–output free procedures written in high-level languages. These procedures must run efficiently as distributed, on a class of computers, and produce accurate results.

(b) Concise documentation, describing usage requirements, references to algorithms, perturbations to algorithms, and error-detecting abilities.

(c) A maintenance mechanism which allows upgrading on a continuous basis without undue user concern or expense.

(d) A user response mechanism which allows consideration of input from the only community which has answers.

(3) What are the major ingredients required?

(a) After initial work is released, upgradings should be prepared on a continuous basis.

(b) Consultants should be made available to the small development group to view the total library, aid in the direction of its growth, and review user responses.

(c) Each major high-level language should be considered for development, but the basis library should be written in the most versatile high-level language available.

(d) Heavy weight should be given to the needs and desires of high-performance computer users and applied mathematicians. Currently, the time lag before the effects of good production are felt is longer than the life of a particular software product. However, the life of a well-constructed and well-maintained library is so long that the benefits of strong bases will reward the manufacturer.

Many of the problems mentioned in this discussion could be avoided if the above ingredients were available to development groups. However, there is another requirement which, if met, would allow the circumvention of much of "collaboration, personnel, and testing" problem set. At this time, the mechanisms which will provide this ingredient are being considered by the community. If successful, a current state-of-the-art basis will be available to all manufacturers for development of mathematical software on specific machines. This talent, available to the community, could never be matched within one organization, and the absence of such a basis has slowed library development.

Consideration of the above mechanisms should include the following topics.

(a) Techniques in use today for accepting articles for journal publication should be applied.

(b) The question of "machine-dependent" documents and codes should be given lengthy consideration. Many early library attempts failed, not because the basic ideas were incorrect but because attempts were made to provide "in-depth" information which became too unwieldy and clerical in nature to control.

(c) Dissemination control and publications must consider the changing state of the art. Aid to the user community which allows easy maintenance of revised publications is of utmost importance.

(d) Funding, although initially necessary, must become implicit, and independent of changes in the economy.

(e) Foundations of the project must be considered with as much time and ability as is available. Too many restarts of such a basic idea will destroy confidence in abilities to provide solutions to this problem.

VI. User Attitudes

During this dynamic growth period, users should understand that most of the decisions made in the initial production of a library are temporary, justifiable, and incorrect. They are modified if input from users is available. Growth cannot be sacrificed to perfection.

It should be clear that no clear-cut answers to problems are available to developers, and that only incremental improvements can lead to perfection in this situation. Experience provides answers.

It should be clear that library production for a mass audience is a conservative venture in that impact of new ideas and new conventions can seldom be foreseen. Any code produced for a library can be tailored and made more efficient as soon as it is introduced. A smaller audience, or a fixed environment, implies that a better code can be built for the specific user.

VII. Summary

Production of mathematical software requires expense and effort to the extent that a large set of potential buyers is necessary. Difficulties which have inhibited good development in the past can be circumvented if it is realized that the construction of a library is not a discrete process, ending with the death of a class of machines. Simple construction rules, set today, and modified by experience, will allow us to bypass the recurring problem of mathematical software availability as hardware evolves. Manufacturers and users must consider that this conservative approach is necessary, in order that stability enters the field.

5.8   High Quality Portable Numerical Mathematics Software

BELL TELEPHONE LABORATORIES, INC.

MURRAY HILL, NEW JERSEY

I. Introduction

Numerical mathematical software has elements common to both scientific publication and engineering practice. Although it may be viewed as a medium for communicating the latest results, it must also work in the real world.

Yet we see software of a quality which we would never countenance in scientific publication or in engineering practice. Software is still being written based on algorithms obsolete for a decade. The same or similar software is written in scores of installations with little exchange between the installations. We have software without adequate documentation and, even when documented, it doesn't meet specifications. There are also examples of excellent software, painstakenly built, which cannot be moved from the environment for which it was written. I must emphasize that there are some important exceptions to the pattern of mediocre software. Due to the efforts of some dedicated men a few outstanding program libraries have been constructed.

In this paper I want to discuss two efforts to build high-quality portable software. These are the Bell Laboratories Numerical Mathematics Program Library One and ZERBND, a package of routines for calculating all the zeros of a real or complex polynomial. (ZERBND is not included in Library One because it has not been subject to comparative testing against all other serious candidates at Bell Laboratories.)

After discussing these two efforts I shall focus on two matters which deserve special consideration: portability and testing.

II. The Bell Laboratories Numerical Mathematics Program Library One

Library One is a high-quality portable library. A discussion of how it fits into our picture of the total program library may be found in Gentleman

---

* Present address: Carnegie-Mellon University, Pittsburgh, Pennsylvania.

and Traub (1968). The work on Library One was initiated in 1968. Plans for
its construction may be found in Gentleman and Traub (1968). We outline
these plans below.

Major problem areas in numerical mathematics are selected. Routines
which solve these problems are collected. These routines are tested and the
tests are analyzed. A "best" program is selected and certified. Interlanguage
interfaces are inserted which enable programmers in other languages to
access the program. Calls are designed for simplicity of use. Finally a report
documenting the use of the program and reporting the criteria by which it
was chosen is published.

For the remainder of this section I will discuss some of the general
philosophy of Library One. In the following section I will describe the current
status of Library One and describe some of what we have learned by building
the library.

First of all, it is our environment which makes Library One possible.
Since the manufacturer supplies software, there is no pressure to produce a
large quantity of software by a certain deadline. Instead we can concentrate
on doing a careful job on a limited number of areas.

Earlier I observed an analogy between software and scientific publication,
which I want to draw on again here. At the center of the scientific publication
system stands the impartial referee. Similarly a key element in the Bell
Laboratories project is an impartial referee who compares the candidate
routines.

A number of rather large-scale library projects have been attempted with
volunteer help and have resulted in libraries of uneven quality. An important
element of the Bell Laboratories project is that all work is done in one
group and hence uniform documentation standards can be imposed.

A typical computing community has users of a variety of languages. In
Library One there is only one program for each problem area. Users writing
in other languages access the program through user interfaces. For example,
although the nonsymmetric eigenvalue program was written in Algol, a
Fortran user can access it in Fortran. He needn't know that the eigenvalue
program is an Algol program.

III. Status of Library One

Since the initiation of the construction of Library One in 1968, the following has been accomplished. Software for the absorption of candidate routines
and for interlanguage calls has been written (Gentleman and Traub, 1968).
Three areas have been investigated and reports published. They are eigenvalues and eigenvectors of nonsymmetric matrices (Businger, 1968), systems

of ordinary differential equations (Crane and Fox, 1969, 1969a), and solution of linear algebraic equations (Businger, 1970). Businger is currently working on eigenvalues and eigenvectors of symmetric matrices.

This is certainly slow progress. There are two reasons for this. The testing, certification, and documentation of a high-quality routine is a long and difficult job. Each of the routines accepted in the Bell Laboratories library has required over six man-months of work. Furthermore this work is being done by independent workers, who are also conducting other lines of research. Therefore the areas which are investigated depend on the research interests of the members of the group.

Our experience with the library differs in several important respects from what we expected. We had hoped that a significant portion of our testing would be blackbox testing. By this we mean deciding on the quality of a routine only by its output. We found, however, that the investigator must look at the programming details of a routine.

We had anticipated that the investigator would confine himself to the comparison of existing routines. Instead we found that he finally writes his own program. That is, rather than using or slightly modifying one of the programs he has tested, he incorporates the best features of existing programs and adds his own ideas to his own program. The objection could be raised that we now have just one more iterate. This is, of course, true. But if the investigator has done his work well, then we expect that no or few additional iterations will be required.

Users care very much about the speed of their program. If one program is significantly faster than a second program, then, other things being equal, the faster program will be used. Given present compilers, Fortran generally produces faster codes. Therefore, contrary to our original plans, we anticipate that the source program will be written in Fortran. The nonsymmetric eigenvalue program, which was the first routine that we certified, was written in Algol. We expect, in the future, to replace this by a Fortran program.

To gain speed, segments of the routine must be written in machine language. For example, in the routine for the solution of linear algebraic systems (Businger, 1970) fast inner loops written in machine code make the program two and one half times as fast as a program written entirely in Fortran. These machine language modules should be clearly isolated and marked in the documentation. We return to this point in the section on portability.

The Library One project has led to new information about the state of the art in algorithms and raised interesting research questions.

The comparative testing revealed that, for the solution of the initial value problem for ordinary differential equations, the Bulirsch–Stoer rational extrapolation algorithm seemed to have some definite advantages over predictor-corrector methods. The same conclusion was reached independently by Clark (1968) in an Argonne National Laboratories study.

Most packages for the solution of ordinary differential equations use an adaptive rule. That is, the step size h is adjusted based on results monitored during the integration. In the course of the investigation into ordinary differential equations, the following observation was made.

The rational extrapolation routine was used to integrate the system

$$y_1' = y_1^2 y_2, \qquad y_2' = -1/y_1$$
$$y_1(0) = 1, \qquad y_2(0) = 1$$

from $x = 0$ to $x = 4$. The relation between the initial step size h and the total number of function evaluations is given in Fig. 1. This shows that too small or too large an initial step size may cost many additional function

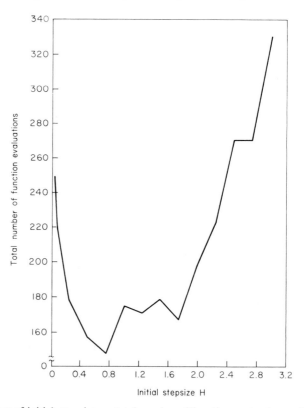

FIG. 1. Effect of initial step size on total number of function evaluations using program EXTRAP RAT on test problem at $x = 4$ for $\varepsilon = 0.0001$ [taken from Crane and Fox (1969)].

evaluations. In this example we see that too small an initial step size is especially expensive.

Although this example is of a particular routine on a particular problem, there are indications that this phenomenon may be rather general and deserves further investigations. In any case the problem of optimal initial step size is an important one which requires further research.

New theoretical results have been stimulated by the library work. For example, although complete pivoting (Wilkinson, 1963) is known to be "safer" than partial pivoting, it is far more expensive. In MIDAS, Businger (1970) uses partial pivoting. The magnitudes of intermediate results are estimated by a new technique (Businger, 1971) in case of alarming growth the program switches to complete pivoting. Thus, except when complete pivoting is required, the calculation proceeds essentially at the speed of partial pivoting.

At Bell Laboratories this technique is affectionately known as P^8: Peter's Perfect Pseudo Partial Pivot Picker, Patent Pending.

IV. ZERBND

ZERBND is a Fortran package, consisting of three independent programs, which calculates all zeros of complex or real polynomials and provides the option of calculating a posteriori error bounds. The bounds include the effects of round-off error.

The zero finder is based on a globally convergent algorithm (Traub, 1966; Jenkins and Traub, 1970, 1970a). The programs have an interesting history. The programs were first written by Jenkins in Algol W (Wirth and Hoare, 1966), a dialect of Algol 60. They were translated by Jenkins into Algol 60, which is more widely available. K. Paciorek of Argonne National Laboratories translated the package into IBM Fortran. Finally, M. T. Dolan of Bell Laboratories translated the package into a portable subset of ASA Fortran. We will have something more to say on this portable subset in Sect. V.

Although ZERBND has been thoroughly tested and documented (Jenkins and Traub, 1970b), it is not part of the Bell Laboratories Library One. The reason is that it has not been subject to comparative testing against all other serious candidates at Bell Laboratories.

V. Portability

The importance of portability cannot be overemphasized. To paraphrase Santayana: The man who doesn't write portable software is condemned to rewrite it.

By portable software we mean that only small easily identifiable changes are necessary to transfer the software to a new environment. This is not the same as machine independence. Portability is achieved by writing in a higher-level widely available language. Brown (1969) gives a perceptive discussion of software portability.

As we noted earlier, it may be necessary to write segments of a routine in machine language. Such machine language modules should be clearly isolated and marked in the documentation. When the routine is sent to another facility, this module should be replaced by a higher-level language module. If the second facility wants optimum speed it must replace the module by a machine-dependent segment. Note that this kind of replacement comes within our definition of portable.

We can get closer to machine independence by writing in "intersection Fortran." This is a subset of ASA Fortran which is the intersection of Fortrans on several different machines.

ZERBND was written in such a subset of Fortran. It was compiled and run, without any change, on CDC, GE, and IBM machines. The fact that it is a subset of ASA Fortran was verified by running it through a compiler (Hall, 1969) which checks this.

ZERBND gives guaranteed a posteriori bounds (including the effects of round-off error) on the zeros. This is certainly machine dependent. The effect of the machine dependence is minimized by the following technique. Machine dependence is isolated in the following five parameters.

BASE   Base of floating point arithmetic
PRECIS   Number of base digits in the mantissa of a double precision word
EXPHI   The maximum exponent
EXPLO   The minimum exponent
ROUND   TRUE if double-precision arithmetic is rounded

When ZERBND is installed at a new facility, these five parameters are specified. This is done once and the object programs can then be stored in the program library.

VI. Testing

Testing is the central problem involved in the selection of high-quality routines. Various lists of tests for important problem areas have been published. The SIGNUM Subroutine Certification Group is collecting tests to be used for certification. Crane and Fox (1969a) published some twenty examples which test various aspects of ordinary differential equation solvers.

Jenkins (1970b) performed extensive testing of ZERBND. I will summarize a number of these tests to illustrate some of the principles of testing. The results may be found in Jenkins (1969). The tests may be divided into those for difficult polynomials and random polynomials.

A. DIFFICULT POLYNOMIALS

A particular polynomial may be difficult for a routine based on one kind of algorithm and easy for a routine based on a different algorithm. Those tests should be applied which are as stringent as possible for the particular routine being tested.

Example A1
$$P(z) = 10^{40}(z^{10} + z^9 + z^8 + z^7 + z^6) + 10^{-50}$$
This example tests whether the zero finder works for polynomials with widely varying coefficients.

Example A2 The 21st-degree complex polynomial with zeros at
$$-1 + 2i, i, i, 2i, 3i, 3i, 1, 1, 1, 1 + i, 1 + 2i, 1 + 3i, 1 + 4i, 1 + 4i,$$
$$1 + 4i, 2 + i, 2 + i, 2 + 2i, 2 + 3i, 2 + 3i, 3 + 2i$$
This difficult problem may be found in Durand (1960). It has a number of multiple zeros.

Example A3 Polynomial of degree 15 with zeros at $1, ..., 15$. This polynomial (Wilkinson, 1963) has some very poorly conditioned zeros, and this example tests whether the a posteriori error bounds give a reasonably tight bound on the actual error.

Example A4 The polynomial of degree 60 constructed by taking the 60th roots of unity and multiplying the zeros in the left-half complex plane by 0.9 and then calculating the coefficients of the new polynomial. This polynomial is difficult for any algorithm which finds the zeros one at a time and with increasing magnitude because the original well-conditioned polynomial is deflated to an ill-conditioned polynomial with zeros only in the right-half plane. A discussion may be found in Jenkins and Traub (1970a).

B. RANDOMLY GENERATED POLYNOMIALS

Totally different classes are generated, depending on what is randomized. For example, polynomials with uniformly distributed coefficients (Example B1) will tend to have zeros distributed around the unit circle. They are well-conditioned and are quite easy for ZERBND. On the other hand, we give a class of randomly generated polynomial (Example B3) which is ill-conditioned.

138 J. F. TRAUB

Example B1 The coefficients a_j are uniformly distributed random numbers with $|a_j| \leq 1000$. These polynomials have zeros which tend to lie around the unit circle.

Example B2 For each coefficient the mantissa is chosen uniformly in $(-1, 1)$ while the exponent is set equal to 10_j, where j is an integer picked randomly in $[-20, 20]$. This generates polynomials with widely varying coefficients.

Example B3 For polynomials of degree n, k zeros are chosen uniformly on $(0.8, 1)$ and $n - k$ zeros from $(-1, 1)$, $k = 2, 3, \ldots, n$. Polynomials with these zeros are formed and solved. As k increases, the zeros in the cluster become ill conditioned.

REFERENCES

Brown, W. S. (1969). Software Portability. To appear in *Proc. NATO Conf. Techn. Software Eng. Rome.*

Businger, P. A. (1968). NSEVB—Eigenvalues and Eigenvectors of Nonsymmetric Matrices. Numerical Mathematics Computer Programs, Vol. 1, Issue 1. Bell Telephone Laboratories.

Businger, P. A. (1970). MIDAS—Solution of Linear Algebraic Equations, Numerical Mathematics Computer Programs, Vol. 3, Issue 1. Bell Telephone Laboratories.

Businger, P. A. (1971). Monitoring the Numerical Stability of Gaussian Elimination. *Numer. Math.* (to appear).

Clark, N. W. (1968). A Study of Some Numerical Methods for the Integration of Systems of First-Order Ordinary Differential Equations. Rep. ANL-1728, Argonne National Laboratory.

Crane, P. C., and Fox, P. A. (1969). DESUB—Integration of a First Order System of Ordinary Differential Equations. Numerical Mathematics Computer Programs, Vol. 2, Issue 1. Bell Telephone Laboratories.

Crane, P. C., and Fox, P. A. (1969a). A Comparative Study of Computer Programs for Integrating Differential Equations. Numerical Mathematics Computer Programs, Vol. 2, Issue 2. Bell Telephone Laboratories.

Durand, E. (1960). "Solutions Numeriques des Equations Algebriques," Vol. 1, p. 264. Masson, Paris.

Gentleman, W. M., and Traub, J. F. (1968). The Bell Laboratories Numerical Mathematics Program Library Project. *Proc. ACM Nat. Conf.*, pp. 485–490.

Hall, A. D. (1969). A "Portable" Fortran IV Subset. Unpublished Bell Telephone Laboratories Rep.

Jenkins, M. A. (1969). Three-Stage Variable-Shift Iterations for the Solution of Polynomial Equations with a Posteriori Error Bounds for the Zeros. Stanford Dissertation available as Computer Science Report 138.

Jenkins, M. A., and Traub, J. F. (1970). A Three-Stage Variable-Shift Iteration for Polynomial Zeros and Its Relation to Generalized Rayleigh Iteration. *Numer. Math.* **14**, 252–263.

Jenkins, M. A., and Traub, J. F. (1970a). A Three-Stage Algorithm for Real Polynomials Using Quadratic Iteration, *SIAM J. Numer. Anal.* **7**, 545–566.

Jenkins, M. A., and Traub, J. F. (1970b). ZERBND—A Package for Polynomial Zeros and Bounds. Unpublished Bell Telephone Laboratories Report.

Traub, J. F. (1966). A Class of Globally Convergent Iteration Functions for the Solution of Polynomial Equations, *Math. Comput.* **20,** 113–138.

Wilkinson, J. H. (1963). "Rounding Errors in Algebraic Processes." Prentice-Hall, Englewood Cliffs, New Jersey.

Wirth, N., and Hoare, C. A. R. (1966). A Contribution to the Development of ALGOL, *Comm. ACM* **9,** 413–431.

5.9 The Development and Maintenance of a Technical Subprogram Library

A. W. Dickinson, V. P. Herbert, A. C. Pauls, and E. M. Rosen

MONSANTO COMPANY

ST. LOUIS, MISSOURI

I. Introduction

The growth of technical computations in the Monsanto Company has been characterized by three major factors: First, the number of people doing technical computations has steadily increased. Second, these computations have been done on a variety of computers in diverse locations. Third, a large central computer has been made available to people in diverse locations through remote terminals.

The development and distribution of highly reliable, well-documented mathematical software has become an economic necessity for the following reasons:

(1) There was a great deal of duplicate effort in the development of mathematical software.

(2) Well-documented mathematical software was not easily available from outside the company.

(3) If the software were available it was often not in a form which could be easily used by a wide variety of technical programmers.

(4) The software which was available was not sufficiently reliable.

Because of these problems many potential users of computers had been discouraged and many possibly valuable computer applications had not been initiated.

As a result efforts were made beginning in 1967 to develop a technical subprogram library which would overcome some of these difficulties.

Many of the first routines put into the library had existed for a number of years but were poorly documented. Their existence was often discovered only through the informal grapevine and no one person or group was responsible for their maintenance. A committee of the authors was formed to oversee the effort to correct this situation.

II. Coding Standards

The principal objectives to be satisfied were:

(1) that the routines should fill a definite need,

(2) the object code should be reasonably efficient,

(3) the routines should be easy to use, and

(4) the library and its documentation should be easy to maintain.

In those cases where objective (2) conflicted with (3) or (4), more weight was given to ease of use and maintenance than to efficiency of object code.

In pursuit of the above objectives, the following standards were suggested as guides to writing of the software:

(1) All subprograms should be written in a high-level language. Initially this was Fortran, but the scope has recently been expanded to include PL/I. The use of Fortran allows a transfer of the source decks between different machines with a minimum of conversion problems. It also minimizes the impact of a new release of the operating system!

(2) If a subprogram requires a functional evaluation (as in equation solving, optimization, integration, etc.), it should return to the calling program to do this rather than calling another subprogram. Returning to the calling program to do a function evaluation slightly complicates the coding but pays rich dividends when the need to overlay is encountered, the function to be evaluated is very complex, or the routine is used repeatedly. In addition, it enables the user to follow the logic of his routine more easily and provides him with the ability to test convergence and print selected variables on option.

(3) No input or output should be done by a subprogram. Error or message conditions will set an error code variable. The use of an error code allows the user to determine the action to be taken on an error return rather than cluttering up his output with unwanted printing or one with an inconvenient format.

(4) All subprograms should be serially reusable. To be serially reusable the routine must store in a retention vector all information used or developed during one call and needed on subsequent calls. This is especially important if the routine is reentered in a number of different loops; e.g., with overlay or nested loops.

(5) Input parameters transferred to a subprogram must never be altered by the subprogram.

(6) All arguments (both input and output) should be transferred through the calling sequence. This simplifies the documentation, use, and maintenance of the routine at the expense of a small loss in execution-time efficiency.

(7) Liberal use should be made of default options. Variable step-sizes, convergence criteria, etc., will be set to specified values if not defined by the user. This policy facilitates the application of the routines in those cases where the default options are adequate, and provides the user with the capability of overriding the defaults when this is necessary.

(8) The routines and their documentation should be as nearly machine independent as it is possible to make them. Algorithms which depend upon the word size, floating point range, or other special feature of a particular machine or class of machines will be avoided if possible.

(9) In the event that an iterative procedure fails to converge, the routine should return the best answer found prior to the time that the failure occurred. Examples of areas in which this has proved helpful include equation solving, optimization, and nonlinear estimation. This standard may require that the user reevaluate related functions which depend on the unknown parameters when the best answer is not the last one tried.

(10) All convergence criteria should be relative rather than absolute. This, to a large extent, frees the user from having to worry about the magnitude of the values of his variables.

(11) All data-dependent storage vectors required by the routine must be contained in the calling program. Variable dimensions will be used for all multiply subscripted arrays. In this way, arbitrary limits on the number of variables or number of equations handled by the routine can be avoided.

III. Documentation Format

The following document is issued with each new program. It is intended to be a stand-alone document. Though a certain amount of flexibility is allowed, by and large this format has proved to be adequate for a broad range of programs. A sample of the documentation (without the listing) is given in Sect. XI.

1. DESCRIPTION

1.1. *Scope*

A short statement is made indicating what the program does in non-mathematical terms. For example, it may be stated that an equation-solving program "supplies the trial and error values of an independent variable for the trial and error solution of a one-dimensional function."

1.2. *Model*

Here the model which the program uses is stated generally in mathematical terms. It states the "what" rather than the "how." The details of the algorithm, "the how," are reserved for the Appendix of the documentation.

1.3. *Limitations*

Here all known limitations of the routine are listed. For example, it may be stated that in polynomial fitting, "the number of distinct points must be greater than the maximum-order polynomial specified."

2. SUBPROGRAM USAGE

2.1. *General*

Here general directions are given for the use of the subprogram. A check list is given for the programmer to follow which includes allocating the required storage, setting required bounds, specifying starting values, etc. In addition, any other general information to aid the programmer in the use of the subprogram is given.

2.2. *Declaratives*

A list of the needed DIMENSION statements is given and the required storage for each array is indicated.

2.3. *Argument List*

The calling sequence together with the definition of each variable is given. The role each argument plays (whether an input or output or both) is also carefully detailed.

2.4. *Error Returns*

Generally these will be described in the argument list. However, any special emphasis can be given here. Error returns are given for any invalid arguments on entry, round-off error detections, nonconvergence, etc.

3. EXAMPLE OF USE

This is a complete sample of how the routine is called from a calling program. The example illustrates at least the major capabilities of the subprogram. A listing of the calling program as well as its output is given in Sect. XI.

4. REFERENCES

Any literature sources or internal documents are cited. The major contributions to the routines are listed here as "private communications."

APPENDIX A—ALGORITHM AND AUXILIARY STORAGE

A.1. *Algorithm*

Here a detailed description of the way in which the model is implemented is given. If the algorithm is lengthy or complex then a separate internal report is issued and referenced. If the algorithm is an implementation of a literature procedure, only an outline of the procedure is given, together with any changes made.

A.2. *Auxiliary Storage*

Here the use of any scratch storage is indicated.

APPENDIX B—PROGRAM LISTING

This is a complete source listing of the subprogram. Source decks are reformatted by a program which aligns "equal" signs, sets indentations, assigns sequence numbers, etc. The listing is always well commented according to a standard format.

APPENDIX C—SAMPLE CALLING PROGRAM

Here a listing of the sample calling program and output is given. A major effort is made to make the example simple and easy to follow. Thus, the number of variables is usually limited to two or three and the output is restricted to one page or less. The sample problem is almost always different from the problems used to test the routine, which are frequently numerous and may be large or complex.

IV. Review Procedures

Once the program has been tested by the author and the documentation has been prepared it is turned over to the Manager of the Technical Subprogram Library. The Manager checks the program and documentation for any obvious errors. He then makes the program and documentation available to at least five reviewers. The reviewers are requested to comment on the documentation. They may carry out their own testing of the program in the process of doing this. It is very important that the Manager select reviewers who are actual or potential users of the subprogram so that he can feel confident that the subprogram is properly user oriented and reliable. The participation of the reviewers is very necessary since the author rarely recognizes all of the users' needs or anticipates all of the problems that may be encountered in the use of the routine.

After all the reviews are received the Manager confers with the author on needed changes in the documentation and/or program. The documentation is revised and again distributed to the review panel. Generally three reviews are needed to turn out a subprogram of acceptable quality. This may extend over several months time.

V. Maintenance Procedures

Once the documentation has received acceptance from the reviewers it is distributed as part of the library in a loose-leaf notebook to the technical programming community. Persons who have difficulty with the routine or documentation are instructed to contact the Manager, who makes a note of the difficulty or error. After approximately a six-month period, corrections to the documentation are made and distributed on a page-by-page basis. In order to build confidence in the reliability of the library, errors are always rapidly corrected, usually within hours after a reported difficulty. Major changes, however, must go through the entire review procedure.

VI. Multiple Precision in Fortran

On the IBM 360 the standard Fortran versions of the library subprograms are maintained in single precision. However, double-precision versions are also made available on a separate library. The calling sequences for the double-precision routines are the same as those for the corresponding single-precision routines with the exception that double-precision arguments are required. In order to simplify library maintenance, the single-precision versions are always written in such a way that they can be converted to double precision by only adding an IMPLICIT statement and arithmetic function statements, such as $ABS(X) = DABS(X)$.

VII. Support and Maintenance Requirements

Our experience has indicated that the choice and coding of the appropriate algorithms, while important, is only a small part of the process of library development and management. We believe that ease of use of the routines and documentation and the ease of maintenance of the library are major factors which, in the past, have often been overlooked. Proper attention to these factors requires a well-planned approach and a long-term commitment from management of the necessary resources and facilities. It is our belief that the following requirements must be satisfied:

(1)   The routines must be well tested, preferably by many different users over a period of several months or more.

(2)   The library should be supported by an active group which is committed to its maintenance and has access to the facilities for doing so.

(3)   The documentation must be issued in a standard format which is capable of being updated easily.

(4)   Updates should be issued on a regular basis, preferably at least twice a year.

(5)   The documentation should be subject to a critical review by a group of qualified actual or potential users of the routines.

VIII. Access Procedures

The potential user in the technical community determines the existence of a subprogram in either of two ways. The first is by means of a computer program index which has a Keyword-In-Context-Index prepared from title cards for each subprogram. Included in the title is the name of the subprogram and where it can be located; e.g., in the Technical Subprogram Manual. The second means of locating the subprogram is a direct scan of the Technical Subprogram Manual itself. Included in the manual is an Introduction, a Subprogram Index arranged alphabetically by name of the routine, and a Category Index arranged according to the function of the subprogram. Finally the Program Descriptions are arranged alphabetically by name.

The routines themselves are all stored on direct-access disk storage. Both single- and double-precision routines can be directly accessed by the programmer with simple Fortran CALL statements.

IX. Summary and Conclusions

The effort to develop and maintain a well-documented Technical Subprogram Library of high reliability has met with wide acceptance in the technical computing community. It is difficult to measure the actual economic savings, but if usage is a satisfactory measure of success then the effort has been quite successful. The library has not met every need but appears to fill in many of the gaps that exist in the other major sources of documented technical subprograms: the IBM Scientific Subroutine Package, algorithms published in *Communications of the ACM*, and universities.

X. Current Category Index

The current Category Index has the following entries:
1. curve fitting
2. eigenvalues and eigenvectors
3. equation solving
4. integration by quadrature
5. integration of ordinary differential equations
6. interpolation
7. least-squares estimation
8. linear independence
9. multidimensional equation solving
10. nonlinear equations
11. numerical differentiation
12. one-dimensional equation solving
13. optimization—one-dimensional
14. optimization—multidimensional
15. orthogonalization
16. plotting operations
17. polynomial fitting
18. reformatting
19. regression
20. root-finding
21. scientific subroutine package
22. statistics

XI. Sample Documentation

A sample of the documentation for an interpolation subroutine from the Technical Subprogram Library is given in the following pages.

1. Description

1.1 Scope:
This subprogram performs first, second, or third order Lagrangian interpolation.

1.2 Model:
Given values in an independent variable (X) array and corresponding values in a dependent variable (Y) array this subprogram will interpolate to find a value for the dependent variable corresponding to any value of the independent variable. First, second or third order Lagrangian interpolation is used (see Appendix A). For additional information, consult reference 1, paragraph 4.

1.3 Limitations:
The X array must be in monotonically increasing sequence but need not be equally spaced. The desired value for the independent variable must lie within the range specified in the given X array, or an error will be indicated. The accuracy may suffer close to the ends of the arrays.

2. Subprogram Usage

2.1 General:
The routine is designed to supply an answer whenever the desired value of the independent variable falls within the range supplied to the routine.

2.2 Declaratives:
The following declaratives are needed:

DIMENSION X(n), Y(n)

where n is of sufficient size to contain the necessary tabular information.

2.3 Argument List:
The calling sequence is

```
CALL INTER (NV, IN, X, Y, ARG, ANS, KØDE)
    NV is the number of values in each array.
    IN is the order of interpolation, 1, 2, or 3.
    X is the name of the array of the independent variable
        ordered in increasing sequence.
    Y is the name of the array of the dependent variable.
    ARG is the value of x for which the y value is desired.
    ANS is the desired value of y.
    KØDE returns error information
```

2.4 Error Returns:
Normal output from the program will be accompanied with KØDE = 0.

If the value specified for ARG is below the range given for the X array, the value of ANS will correspond to Y(1) and the value of KØDE will be set to +1. If the value given for ARG is above the range covered by the X array, the value of ANS will correspond to Y(NV) and the value of KØDE will be set to +2. If the interpolation request is not feasible (for example 2nd order interpolation is requested but only 2 points are supplied) then the routine will return the best ANS possible (linear interpolation in this example) and KØDE will be set to 3. In case no points are supplied (NV less than 1) then ANS is set to 0. and KØDE is set to 4.

3. Example of Use

Appendix C illustrates the use of subroutine INTER to interpolate between values in a logarithm table. The sample problem illustrates two possible error codes that can be returned from the subprogram.

4. References

1. Hamming, R. W., Numerical Methods for Scientists and Engineers, McGraw-Hill, New York, 1962, pp. 94-95.

2. Milne, W. E., Numerical Calculus, Princeton University Press, 1949, pp. 83-92.

3. Sterba, V. J., "Private Communication".

149

Appendix A—Algorithm

$X_1, X_2, \ldots, X_n$ — given independent variable points ordered in increasing sequence.

$Y_1, Y_2, \ldots, Y_n$ — corresponding dependent variable points.

X_d — the value of X for which the value of Y is desired.

Y_d — the value of Y at X_d.

i — indicates the location of the smallest value in the X array which is greater than X_d ($X_i > X_d$).

n — number of points in the X and Y arrays.

First Order Lagrangian Interpolation:

One point above and one point below X_d are used:

$$Y_d = (X_d - X_{i-1})(X_d - X_i) \left[\frac{Y_{i-1}}{(X_d - X_{i-1})(X_{i-1} - X_i)} + \frac{Y_i}{(X_d - X_i)(X_i - X_{i-1})} \right]$$

Second Order Lagrangian Interpolation:

Two points above and one point below X_d are used if X_d lies between X_1 and X_2; two points below and one above X_d are used if X_d lies between X_{n-1} and X_n. At other points within the range, the three points will be selected so that the middle point is the closest to X_d. This will result in either two points below and one above or one below and two above being used.

$$S_1 = \frac{Y_{i-1}}{(X_d - X_{i-1})(X_{i-1} - X_i)(X_{i-1} - X_{i+1})}$$

$$S_2 = \frac{Y_i}{(X_d - X_i)(X_i - X_{i-1})(X_i - X_{i+1})}$$

$$S_3 = \frac{Y_{i+1}}{(X_d - X_{i+1})(X_{i+1} - X_{i-1})(X_{i+1} - X_i)}$$

$$Y_d = (S_1 + S_2 + S_3)(X_d - X_{i-1})(X_d - X_i)(X_d - X_{i+1})$$

Third Order Lagrangian Interpolation:

Two points above and two points below of X_d are used unless X_d lies between X_1 and X_2 or X_{n-1} and X_n. In this case, i is defined so that three points above and one below (or 1 above and 3 below) will be used.

$$S_1 = \frac{Y_{i-2}}{(X_d - X_{i-2})(X_{i-2} - X_{i-1})(X_{i-2} - X_i)(X_{i-2} - X_{i+1})}$$

$$S_2 = \frac{Y_{i-1}}{(X_d - X_{i-1})(X_{i-1} - X_{i-2})(X_{i-1} - X_i)(X_{i-1} - X_{i+1})}$$

$$S_3 = \frac{Y_i}{(X_d - X_i)(X_i - X_{i-2})(X_i - X_{i-1})(X_i - X_{i+1})}$$

$$S_4 = \frac{Y_{i+1}}{(X_d - X_{i+1})(X_{i+1} - X_{i-2})(X_{i+1} - X_{i-1})(X_{i+1} - X_i)}$$

$$Y_d = (S_1 + S_2 + S_3 + S_4)(X_d - X_{i-2})(X_d - X_{i-1})(X_d - X_i)(X_d - X_{i+1})$$

Note that if the argument is exactly equal to one of the elements in the X array, the corresponding value in the Y array is returned.

Appendix C

Sample Calling Program

```
C      TEST INTER ROUTINE
       DIMENSION   X(10), Y(10), ANSR(3), PCT(3), KOD(3)
   25 FORMAT(  16H1 ARGU- CORRECT ,3(10H-----ORDER,I2,8H------ ),
      1 /     16H    MENT  ANSWER ,3(20HKOD ANSWER PCT ERR  ))
   30 FORMAT(F7.0,F8.4,3(I4,2F8.4))
       DO 10 I      = 1,10
             X(I)   = 10*I
   10        Y(I)   = ALOG(X(I))
       PRINT 25, (K,K=1,3)
       DO 20 I      = 7,105,7
             ARG    = I
             ANS    = ALOG(ARG)
       DO 15 JP     = 1,3
       CALL INTER (10,JP,X,Y,ARG,ANSR(JP),KOD(JP))
   15        PCT(JP) = 100.*(ANSR(JP) - ANS)/ANS
   20 PRINT 30, ARG,ANS,(KOD(K),ANSR(K),PCT(K),K=1,3)
       STOP
       END
```

ARGU-	CORRECT	-----ORDER 1------			-----ORDER 2------			-----ORDER 3------		
MENT	ANSWER	KOD	ANSWER	PCT ERR	KOD	ANSWER	PCT ERR	KOD	ANSWER	PCT ERR
7.	1.9459	1	2.3026	18.3295	1	2.3026	18.3295	1	2.3026	18.3295
14.	2.6391	0	2.5798	-2.2437	0	2.6144	-0.9357	0	2.6252	-0.5237
21.	3.0445	0	3.0363	-0.2708	0	3.0492	0.1544	0	3.0464	0.0622
28.	3.3322	0	3.3201	-0.3631	0	3.3295	-0.0804	0	3.3350	0.0827
35.	3.5553	0	3.5450	-0.2901	0	3.5531	-0.0633	0	3.5564	0.0304
42.	3.7377	0	3.7335	-0.1114	0	3.7387	0.0267	0	3.7379	0.0064
49.	3.8918	0	3.8897	-0.0543	0	3.8915	-0.0072	0	3.8919	0.0029
56.	4.0254	0	4.0214	-0.0978	0	4.0248	-0.0139	0	4.0255	0.0037
63.	4.1431	0	4.1406	-0.0615	0	4.1435	0.0099	0	4.1432	0.0015
70.	4.2485	0	4.2485	0.0	0	4.2485	0.0	0	4.2485	0.0
77.	4.3438	0	4.3420	-0.0424	0	4.3436	-0.0043	0	4.3438	0.0008
84.	4.4308	0	4.4291	-0.0379	0	4.4310	0.0047	0	4.4308	0.0005
91.	4.5109	0	4.5103	-0.0114	0	4.5109	0.0009	0	4.5108	-0.0003
98.	4.5850	0	4.5841	-0.0190	0	4.5851	0.0027	0	4.5849	-0.0009
105.	4.6540	2	4.6052	-1.0484	2	4.6052	-1.0484	2	4.6052	-1.0484

5.10   The Boeing Library and Handbook of Mathematical Routines

*A. C. R. Newbery**

THE BOEING COMPANY

RENTON, WASHINGTON

The Boeing library and handbook became operational in the summer of 1969, about twenty months after the planning and designing phase was completed. Although the design emerged entirely independently of the similar project at Bell Laboratories (Gentleman and Traub, 1968) the two projects have strong similarities, particularly with regard to motivation, rather less so with regard to implementation.

The Boeing library covers the following areas: (1) programmed arithmetic; (2) elementary functions; (3) polynomials and special functions; (4) ordinary differential equations; (5) interpolation, approximation, and quadrature; (6) linear algebra; (7) probability, statistics, and time series; (8) nonlinear equation solvers. In each area it was our objective to supply Fortran IV decks for a comprehensive set of well-tested programs. We avoided duplication of algorithms except in cases where it could be shown that duplication would significantly strengthen the library. In addition to the decks and writeups, we also put out a Handbook (totaling about 1800 pages) containing information and advice that is likely to be helpful to whomsoever is using the decks. Examples of the kind of information we give under this heading are: (a) how to reformulate problems, e.g., how to trade a lambda-matrix problem for a nonlinear equation; (b) the relative merits of the various matrix decompositions as against matrix inversion; (c) comparative test results on batteries of test problems; (d) advice, empirical and/or theoretical, concerning the interpretation of an output, e.g., the manner in which the norm of a residual vector may be related to the error norm of the "solution" vector to a set of linear equations. Advice on how to select a method for a given problem is sometimes given explicitly in the text. More often (and more effectively) it is given implicitly in so far as no alternative program may be offered for a given problem. We consider it important to give some kind of directive, one way or the other, rather than to present the programmer with a bewildering assortment of alternatives and let him make a random choice.

* Present address: University of Kentucky, Lexington, Kentucky.

Concerning motivation, all the considerations mentioned in Gentleman and Traub (1968) were applicable in our case, especially the need to set uniform standards of quality control and to draw adverse attention to any widely used program which failed to meet these standards. On occasions one might have wished for the authority to search and destroy such programs, but we were under constraint to seek less flamboyant remedies. In addition to the technical motivation, one should not overlook the managerial aspect. A scientific programmer is often an exceedingly conservative individual. It may have been a long series of traumatic experiences that made him that way, but the fact is that he has a healthy distrust of programs which are presented to him from the outside; this distrust is often offset by an exaggerated and misplaced confidence in any deck that has resided in his drawer for a year. If new and improved programs are introduced into an area in a routine manner, it will ordinarily take many months before they begin to find acceptance and to break down the barrier of distrust. On the other hand, if a major package (in our case three-hundred programs) is released at one time, it is then hardly possible for the programing community to give it the silent treatment. In particular, when it has been necessary for the certification sections of the Handbook to "knock" a program which is widely used and almost universally respected, some kind of reaction is necessarily evoked. Most likely the reaction is, "I knew it all along." In our experience we have found it advantageous to put almost all of the library on disk. Making it readily accessible in this way is a big step towards getting it widely used and accepted.

As regards implementation, it appears that Boeing differs from Bell in two main respects. [These remarks are somewhat conjectural, since the speaker's knowledge of the Bell project is restricted to Gentleman and Traub (1968) and to the presentation and discussion in Las Vegas.] Firstly, Boeing's main effort was directed towards the handbook rather than the library. Of approximately 10 man-years spent, only around 20% was devoted to coding algorithms and tests; the handbook, 1800 typed pages, accounted for about 60%, and the remaining 20% went to the "Maintenance Supplement." This supplement, consisting of program documentation is also about 1800 pages, but these pages are relatively quick to write because they largely follow a stereotyped pattern. The other main point of divergence between Boeing and Bell is the fact that Boeing takes a firmly authoritarian view towards languages. There shall be no other language than Fortran IV. The only departures from this policy are with elementary function subroutines and with a few fast-running machine-coded modules like the vector dot-product. In the latter cases we require that there should also be a Fortran-coded standby with identical name and call sequence, so that we can switch machines if necessary with a minimum of dislocation. It is the speaker's view that the authoritarian attitude has paid off for Boeing. We would not have been able to get the system operational—with its 24,000 source-deck cards and extensive documentation—within 20 months of its inception, if we had been obliged to divert some of our resources to provide interfaces with other languages and dialects.

In spite of the foregoing remarks, one should be far more impressed with the similarities than with the dissimilarities of the two systems. The area coverage seems to be almost identical, and there is full agreement on the importance of comparative testing, compatibility of storage conventions, accessibility, and machine independence of the programs. Perfect machine independence may be an unattainable goal, but it is still worth pursuing. Our programs were checked out on the CDC 6600 and have not been extensively tested on other equipment; nevertheless, it has been our policy to disregard any nonstandard permissive features of our compiler, and to concentrate those parameters which are word-length dependent in one place. We even go so far in the latter respect as to provide a single program whose function is to store an assortment of machine constants (e.g., the smallest nonzero floating number) and mathematical constants in single and double precision. If we were to switch machines in such a way that these numbers changed or had different representations, it would be inconvenient to check through for the occurrence of every constant in every program. By storing all the constants in one program we hope to avoid this necessity; all the troubles are concentrated in that one program which would have to be rewritten. The concept has not been subjected to the ultimate test of a machine switch. When the test comes, it may well show some weaknesses, particularly if a different normalization convention is involved, but we are confident that it will help to ease the problem.

In conclusion, it is hoped that this review of the Boeing Library and Handbook will not have created the impression that the whole package is now considered to be in its final "fossilized" form. It is constantly under review and development. When it was described as having been "completed" in 20 months, we meant that we had set ourselves the goal of achieving certain specific capabilities, and when these were achieved the project was considered "complete" in that sense. We did not wish to imply that we had left no more work for other numerical analysts to do. Our program list is given in the Appendix.

REFERENCE

Gentleman, W. M., and Traub, J. H. (1968). The Bell Laboratories Numerical Mathematics Program Library Project, *Proc. ACM Nat. Conf.* pp. 485–490. Brandon Systems Press.

Appendix

The following is a brief description of each program in the library. The "classes" referred to indicate the origin with the following coding:

B	Boeing.
CDC	Control Data 6600 Computer System Programming System/Library Function, 2nd ed. 1964, rev.
H	Adaptation of algorithm published in the Handbook series of "Numerische Mathematik," Springer, New York.
N	Miscellaneous non-Boeing sources.
SSP	Scientific Subroutine Package, IBM Corporation Document 360 A CM 03x.
VIM	Control Data 6600 Series Users Group.
SHARE	IBM Corporation Users Group.

I. Programmed Arithmetic

A. Integer Arithmetic

All the following routines are of class N:

HCF	Finds the highest common factor of two integers
LCM	Finds the least common multiple of two integers
FAFRAC	Adds two numbers of the form a/b and c/d where a, b, c, d are integers, and leaves the answer as a fraction in its lowest terms
FMFRAC	Multiplies two numbers of the form a/b and c/d where a, b, c, d are integers, and leaves the answer as a fraction in its lowest terms
FFRAC	Transforms a vector with fractional components into one with integer components, times a scalar fraction

B. Range Arithmetic
None.

C. Unnormalized Arithmetic
None.

II. Elementary Functions

All the following routines are of class CDC except COSH and CBAREX which are of class B:

SIN, DSIN, CSIN, CØS, DCØS, CCØS, TAN, ASIN, ACØS, ATAN, ATAN2, DATAN, DATAN2	Computes trigonometric functions

SINH, CØSH, TANH	Computes hyperbolic functions

EXP, DEXP, CEXP,
  ALØG, DLØG, CLØG,   Computes exponential and logarithmic functions
  ALØGLO, DLØGLO

SQRT, DSQRT,
  CSQRT, CBRT   Computes roots

(——)   An array of programs to compute $x ** y$ as follows:

x/y	Integer	Real	Double	Complex
Integer	IBAIEX			
Real	RBAIEX	RBAREX	RBADEX	
Double	DBAIEX	DBAREX	DBADEX	
Complex	CBAIEX	CBAREX		

III. POLYNOMIALS AND SPECIAL FUNCTIONS

A. Polynomials

All the following routines are of class B:

ADR CADR	Adds coefficients of like powers of two real or complex polynomials
SBR CSBR	Subtracts coefficients of like powers of two real or complex polynomials
MPYR CMPYR	Finds the product of two real or complex polynomials
LDIV CLDIV	Divides a real or complex polynomial by the linear expression $(X+B)$, where B may be either real or complex
QDIV CQDIV	Divides a real or complex polynomial by the quadratic expression $(X^2 + BX + C)$, where B and C may be real or complex
PDIV CPDIV	Provides the quotient and remainder obtained by dividing one real or complex polynomial by another
EVREAL CØMPEV CCØMPE	Evaluates a real or complex polynomial at a real or complex value of the independent variable
DERIV CDERIV	Computes the coefficients of a polynomial which is the derivative of another real or complex polynomial given the coefficients of the latter
INT CINT	Computes the coefficients of a polynomial which is the integral of another real or complex polynomial given the coefficients of the latter

PTRAN CPTRAN	Effects a coordinate translation such that the real or complex polynomial P(X) becomes the polynomial P(X + T), where T may be either real or complex
REV CREV	Reverses the order of the real or complex polynomial coefficients in an array
SHRINK CSHRINK	Computes the coefficients of the polynomial P(AX) from the coefficients of the real or complex polynomial P(X), where A may be either real or complex
PARFAC	Resolves a rational function into partial fractions; real coefficients, real isolated poles

B. Special Functions

The routines listed below belong to the class specified following their description:

NBESJ	Computes Bessel functions of the first kind for real argument and integer orders (class B)
RBESY	Computes Bessel functions of the second kind for positive real argument and integer orders (class B)
CØMBES	Computes Bessel functions of the first or second kind for complex argument and complex orders (class B)
BESNIS	Computes modified Bessel functions of the first kind for real argument and integer orders (class VIM)
BESNKS	Computes modified Bessel functions of the second kind for real argument and integer orders (class VIM)
BSJ	Computes spherical Bessel functions of the first kind for real argument and integer orders (class VIM)
HANKEL	Evaluates the complex-valued Hankel function of first or second kind for real argument and integer order (class B)
ELF	Computes incomplete elliptic integrals of the first and second kinds (class B)
ELK	Computes complete elliptic integrals of the first and second kinds (class B)
EL3	Computes the incomplete elliptic integral of the third kind (could be used to evaluate the complete elliptic integral of the third kind under certain conditions) (class B)
CEL3	Computes the complete elliptic integral of the third kind (class B)
ERF	Evaluates the error function (class B)
ERFINV	Evaluates the inverse error function (class B)
GAMMA	Evaluates the gamma function (class N)

LØGGAM   Computes the natural logarithm of the gamma function for complex argument (class B)

SICI   Evaluates the sine and cosine integrals (class SSP)
$$\int (\sin t/t)\, dt, \qquad \int (\cos t/t)\, dt$$

AMCØN   Provides a list of machine constants and mathematical constants for the CDC 6600 (class B)

IV. ORDINARY DIFFERENTIAL EQUATIONS

The routines listed below belong to the class specified following their description:

A. Initial Value Problems

BLCKDQ   Multistep method of order 8; starts and restarts at step changes by a modification of Picard's method; a good performer, especially when high accuracy is needed (class B)

DRATEX   Uses the modified Euler method as a basis for a scheme of rational extrapolations (class H)

RKINIT   5th-order Runge–Kutta; error estimate based on comparison of two 4th-order solutions; 5th-order results obtained by an extrapolation process; consistent good performer, especially for low-accuracy work (class B)

B. Boundary Value Problems

BVP   Solves nonlinear problems with general boundary conditions, supplied as functions; iterates by a generalized secant method on initial values that are guessed (class B)

LINBVP   Linear boundary value problems solved by superposition; faster and more reliable than the general programs for the problems it is designed to solve (class B)

These two programs solve n-point boundary problems, and the constraints do not need to be explicit values.

V. INTERPOLATION, APPROXIMATION AND QUADRATURE

A. Interpolation

The routines listed below belong to the class specified following their description:

1. Polynomials

FLGNEW   Constructs nth-degree Lagrangian interpolating polynomial through $(n + 1)$ points (class N)

FHRNEW   Constructs nth-degree Hermitian interpolating polynomial (class N)

HRMT 1	Performs Hermite interpolation at one point (class B)
HRMT 2	Performs Hermite interpolation for several values of independent variable (class B)
TBLU 1 2 3	Lagrangian interpolation in one-, two-, or three-dimensional tables; arbitrary order (class B)
LAGINT	Lagrangian interpolation of arbitrary order using Barycentric formula (class B)
AITKEN	Lagrangian interpolation at one point; uses Aitken's method (class B)

2. *Rational function*

ACFI ATSM	Performs continued fraction interpolation on tabular data using inverted differences; determines number of points required to meet a given accuracy criterion (class SSP)

3. *Trigonometric*

All of the following routines are of class B:

SINSER	Constructs sine series for discrete data; variable spacing
FOURI	Constructs finite Fourier series for N data points
FOURAP	Constructs least-squares Fourier series (equispaced data)
TRGDIF	Constructs differentiated Fourier series from Fourier series
TRGINT	Constructs integrated Fourier series from given series
SIGSMT	Performs σ-smoothing of Fourier series
SINEVL CØSEVL	Evaluates sine and cosine series, respectively

4. *Splines*

All of the following routines are of class B:

CØMCUB	Constructs cubic spline through N points (Monotone abscissas required); second-order continuity
SPLINE	Constructs and evaluates, at user-specified points, the 5th-degree spline through equispaced data points; 4th order continuity
SURFS	Constructs a surface, through a set of points (X_i, Y_i, Z_i), which has a continuous gradient; requires a rectangular grid (X_i, Y_i) as input points
UNCSPL	Constructs nonlinear cubic spline (continuous 2nd derivative) through (X_i, Y_i)

B. *Approximation* (*Exact Data*)

The routines listed below belong to the class specified following their description:

MIGEN   Computes c_i so that

$$\max_{X \varepsilon [X_0, X_n]} \left| f(x) - \sum_{i=1}^{N} c_i g_i(X) \right|$$

is a minimum, where the $g_i(X)$ are continuous functions furnished by the user (class B)

MINRAT   Computes rational minimax approximation of given degree to discrete data set (class B)

CFQME   Constructs minimax polynomial through weighted set of points (class B)

PADE   Constructs Padé approximation (class N)

CHEBAP   Approximates the minimax polynomial for continuous data (class B)

C. Approximation (*Inexact Data*)

The routines listed below belong to the class specified following their description:

1. *Fitting and Smoothing*

(*a*) *Least Squares*

FDLSQ   Constrained derivative (ordinate); least-squares polynomial fit
FCLSQ   (Klopfenstein) (class N)

FLSQFY   Weighted least-squares polynomial curve fit (Forsythe) (class N)

PRONY   Finds exponential approximation (class B)

FITLIN   Straight-line least-squares curve fitting; minimizes distance of data points from straight line (class B)

RATL   Performs least-squares curve fitting with rational functions (class B)

ORTHFT n-dimensional general function weighted least-squares fit (i.e., routine finds "best" linear combination of the given functions) (class B)

(*b*) *Cubic Spline*

SMOCUB   Smooths and fits two-dimensional data using cubic spline (class B)

2. *Smoothing*

SMOOTH   Does M-point least-squares smoothing of degree N (class B)

MILN2   Smooths data by an averaging process (class B)

3. *Enriching*

NRICH   Enriches to a specified chord height tolerance by using interpolating cubics; 1st derivative continuity (class B)

RICH	Enriches to a specified chord height tolerance using an interpolating function which attempts to minimize the ripple in curvature (class B)

D. Quadrature

The routines listed below belong to the class specified following their description:

RØMBG	Romberg integration (an extrapolation technique based on repeated use of the trapezoidal rule) (class N)
GMI	Gauss–Legendre for multiple integrals (class B)
HERMIT	Gauss–Hermite quadrature (class B)
LAGRAN	Quadrature based on Lagrangian interpolation over arbitrarily spaced data (class B)
LAGUER	Gauss–Laguerre quadrature (class B)
LEGEND	Gauss–Legendre quadrature (class B)
PARBL	Simpson's rule on tabular (equispaced) data; provision is made to handle also an even number of data points (class B)
QUAD	Gauss–Legendre; improves capability by judicious division in sub-intervals; error estimate by comparison of different order formulas (class N)
SIMPRC	Simpson's rule (class B)

E. Numerical Differentiation

All the following routines are of class B:

DERIV	Differentiates polynomials
DIFTAB	Computes derivatives at data points or midpoints in equispaced data
LAGDIF	Computes derivatives at or between data points in arbitrarily spaced tabular data

VI. Linear Algebra

A. Inner Product Subroutines

The following subroutines (all of class B) are COMPASS coded Fortran IV callable subroutines which outperform Fortran IV coded counterparts by a time factor of over ten. Since most linear algebra algorithms for evaluating determinants, solving equations, and inverting matrices consist largely of inner products, the time advantage is extremely significant. Fortran IV coded counterparts have nevertheless been written for use on machines other than the CDC 6600.

VIP	Computes the inner product of two vectors

VIPA	Computes the inner product of two vectors and adds the result to a given scalar
VIPS	Computes the inner product of two vectors and subtracts the result from a given scalar
VIPD	Computes the inner product of two vectors using double-precision arithmetic; the result is given as a rounded single-precision number
VIPDA	Computes the inner product of two vectors and adds the result to a given scalar using double-precision arithmetic throughout; the result is given as a rounded single-precision number
VIPDS	Computes the inner product of two vectors and subtracts the result from a given scalar using double-precision arithmetic throughout; the result is given as a rounded single-precision number
CINPRD	Complex inner product using VIPDA or VIPA

B. *Elementary Matrix Operations*

All the following routines are of class B:

MATRIARCH	A package of subroutines for elementary matrix–vector operations; uses rapid inner products where appropriate see Table I

TABLE I

Subroutine Names for Elementary Matrix and Vector Operations
Performed[a]

Operation	Matrix or vector type	
	Real	Complex
Inner product	VIP, VIPA, VIPS, VIPD, VIPDA, VIPDS	CINPRD
Matrix–vector product	FMVX	FMVCX
Transposed matrix–vector product	FMTVX	FMTVCX
Matrix–matrix product	FMMX	
Transposed matrix–matrix product	FMTMX	
Linear combination of two vectors	FCØMB	
Sparse matrix–vector product	SMVX	
Transposed, sparse matrix, vector product	SMTVX	
Vector euclidean norm	FABSV	
Vector normalization	FNØRM1	
Vector orthogonalization	FPUR	
Matrix transpose	FMTR	

[a] All routines listed are of class B.

C. Direct Methods for Determinants, Linear Equations, and Inverses

We have about seventy programs in this area, but the table is not reproduced here. The programs (mostly of class H) rely on decomposition, with optional iterative refinement for equation solving. The list includes programs which are specialized in the direction of complex, banded, symmetric, and overdetermined systems.

D. Algebraic Eigenvalue Problems

The routines listed below belong to the class specified following their description:

1. Real, Symmetric Matrices

EIGCHK   Given an approximate eigenvalue–eigenvector pair of a real, symmetric matrix, and the matrix, and estimates of the closest eigenvalues to the given eigenvalue; to provide error bounds and possible refinement of the eigenvalue (class B)

EIGSYM   Finds all eigenvalues and eigenvectors (optionally a possibly empty subset of the latter); uses TRIDI and VECTØR (see below); also one of SEPAR, SYMQR, or SYMLR, as chosen by the user (class B)

RAYLGH   Given an approximate eigenvector, a relatively more accurate eigenvalue is found by evaluation of the Rayleigh quotient (may be used in an iterative scheme) (class B)

SEPAR   Finds all eigenvalues of a tridiagonal matrix by Sturm sequencing (class B)

TRIDI   Reduces a real, symmetric matrix to tridiagonal form by the Householder reduction; information to reconstruct the matrix and transform eigenvectors are stored into part of the matrix (class B)

VECTØR   Finds $m(\leqq n)$ eigenvectors of a tridiagonal matrix of order n; performs transformation to eigenvectors of the unreduced matrix (class B)

SYMLR   LR method for eigenvalues of tridiagonal symmetric matrix (class B)

SYMQR   QR method for eigenvalues of tridiagonal symmetric matrix (class N)

REDSY1   Programs for finding eigenvalues and vectors for $BA - \lambda I$,
REDSY2 $AB - \lambda I$, $A - \lambda B$; A, B symmetric, B positive definite; each
RECOV1   problem is reformulated into a symmetric eigenvalue problem of
RECOV2   standard form (class H)

2. *General Real Matrices*

DEIG   To solve for the eigenvalues and right eigenvectors of the dynamical system $A\ddot{x} + B\dot{x} + Cx = 0$ where A, B, and C are real, general, matrices (class B)

EIGCØ1   Improves the guess for a real eigenvalue and gives the corresponding eigenvector (class B); uses HSSN, EIGVCH, EIGIMP, and SIMP

EIGIMP   Input a real Hessenberg matrix, and approximations to a real eigenvalue and its eigenvector; refines the eigenvector by inverse iteration (class B)

EIGVCH   Input a real Hessenberg matrix, and an approximation to an eigenvalue; improves the approximate eigenvalue and gives its eigenvector, by Wielandt inverse iteration (class B)

EIG5   Gives some (or all) eigenvalues of a real matrix; the method is by reduction to Hessenberg form and subsequent use of Laguerre's method on its determinant with implicit deflation (class N)

HSSN   Reduces general real matrix to upper Hessenberg by transformations using elementary stabilizing matrices; the transformation is saved in the reduced part of the matrix (class B)

LATNTR   Finds all eigenvalues of a real matrix; reduction to Hessenberg form by Householder's method; double QR-iterations and progressive deflation (class VIM)

SIMP   Transforms eigenvectors of a Hessenberg matrix to those of the original matrix, where the original transformation is stored according to the rules used in HSSN (class B)

3. *Complex Matrices*

QREIGN   Finds all eigenvalues of a complex matrix; reduction to Hessenberg form by Householder's method; single complex QR-iterations with progressive deflation; uses a system of subroutines (class VIM)

TCDIAG   To compute partial or complete eigensystems of Hermitian matrices (class B)

VALVEC   Finds all eigenvalues and eigenvectors (optionally a subset of the latter); reduction to Hessenberg form by elementary stabilized matrices; single QR-iterations and deflations; eigenvectors are found by Wielandt inverse iteration and then transformed back to those of the given matrix (class VIM)

VECORD   To order a set of complex numbers according to magnitude (used by VALVEC) (class B)

VII. PROBABILITY FUNCTIONS (All of Class B)

NAME	DESCRIPTION
PRBUNF	Uniform probability function
PNORM	Normal probability function
PRBEXP	Exponential probability function
PTRNRM	Truncated normal probability function
PLGNRM	Log-normal probability function
PWEBL	Weibull probability function
PGMMA	Gamma probability function
PBETA	Beta probability function
PCHY	Cauchy probability function
PRAYL	Rayleigh probability function
CHIPPB	Chi-square probability function
PFDIST	F probability function
PTDIST	t probability function
PUNFD	Discrete uniform probability function
PBINOM	Binomial probability function
POIS	Poisson probability function
PHYPGE	Hypergeometric probability function
PNBIN	Negative binomial probability function
PGEOM	Geometric probability function
PIUNF	Inverse of uniform probability function
PINORM	Inverse of normal probability function
PIEXP	Inverse of exponential probability function
PITRNM	Inverse of truncated normal probability function
PILGNM	Inverse of log-normal probability function
PIWEBL	Inverse Weibull probability function
PIGAMA	Inverse of gamma probability function
PIBETA	Inverse of beta probability function
PICHY	Inverse of Cauchy probability function
PIRAYL	Inverse of Rayleigh probability function
PICHI	Inverse of chi-square probability function
PIFDIS	Inverse of F probability function
PIT	Inverse of t probability function
PIUNFD	Inverse of discrete uniform probability function
PIBIN	Inverse of binomial probability function
PIPOIS	Inverse of Poisson probability function
PIHYPG	Inverse of hypergeometric probability function

| PINBIN | Inverse of negative binomial probability function |
| PIGEO | Inverse of geometric probability function |

RANDOM NUMBER GENERATORS (All of Class B)

NAME	DESCRIPTION
RAND	Generates uniformly and normally distributed numbers
URAND	Generates uniformly distributed numbers (for STATIC)
NRAND	Generates normally distributed numbers—central limit (for STATIC)
NRML	Generates normally distributed numbers—exact distribution (for STATIC)
EXRAND	Generates exponentially distributed numbers (for STATIC)
IRAND	Generates uniformly distributed integers
XIRAND	Generates uniformly distributed integers (for STATIC)
PORAND	Generates Poisson distributed numbers (for STATIC)

DESCRIPTIVE STATISTICS AND HYPOTHESIS TESTS (All of Class B)

NAME	DESCRIPTION
DSCRPT	Computes mean, variance, standard deviation, skewness, and kurtosis with editing
DSCRP2	Computes median, minimum, maximum, and range
CHIRUD	χ^2 test for runs up and down
CHIRAB	χ^2 test for runs above and below zero
CHIDST	χ^2 distribution test
BRTLTT	Bartlett's test for homogeneity of variances

TIME SERIES (Mostly Class B)

NAME	DESCRIPTION
CORCOV	Autocorrelation or autocovariance coefficients
FILTER	Moving average—autoregressive filter
HARM	Fast Fourier transform

IØ (All of Class B)

NAME	DESCRIPTION
YPLOT	Printer plot of Y_i vs i
XYPLOT	Printer plot of Y_i vs X_i

TRANSFORMATIONS (All of Class B)

NAME	DESCRIPTION
CONRAY	Arithmetic operation—constant and array elements
OPIRAY	Transformation of elements of array
OP2RAY	Arithmetic operation—corresponding elements of two arrays
ZRMN	Subtracts mean from each element of array
DLETE	Deletes specified elements from array

SPECIAL (Mostly Class B)

NAME	DESCRIPTION
BETAR	Incomplete beta ratio
GAMAIN	Incomplete gamma function
HSTGRM	Counts values for histogram
CHSQO	Computes observed value of χ^2
RUNSUD	Counts runs up and down
RUNSAB	Counts runs above and below zero
SUMPS	Double-precision sums of powers of observations
VARORD	Places array elements in increasing order

PROGRAMS (Class B)

STATIC	Main program for statistical processing

VIII. NONLINEAR EQUATION SOLVERS

The routines listed below belong to the class specified following their description:

Polynomial zeros:

CNSLVL	To estimate the error performed in the evaluation of a complex polynomial in the neighborhood of one of its zeros (class B)
CPØLRT	Accelerated gradient technique for finding all zeros of a polynomial with complex coefficients (class B)
HELP	Finds all zeros of a polynomial with complex coefficients; uses the method of Lehmer–Schur to isolate zeros in the complex plane (class SHARE)
MULLP	Muller's algorithm to find all zeros of a polynomial with complex coefficients by parabolic extrapolation (class N)
NSLVL	To estimate the error performed in the evaluation of a real polynomial at a complex point in the neighborhood of one of its zeros (class B)

PROOT Bairstow–Newton algorithm for finding all zeros of a polynomial with real coefficients (class VIM)

Zeros of single arbitrary functions:

ZAFUM Muller's method to find N zeros of an arbitrary complex-valued function of a complex variable (class B)

ZAFUJ Jarratt's method of rational approximation to find N zeros of an arbitrary complex-valued function of a complex variable (class B)

ZAFUR Jarratt's method of rational approximation to find N zeros of an arbitrary real-valued function of a real variable (class B)

ZCØUNT Counts the number of zeros of an arbitrary function inside a given contour in the complex plane (class B)

Solution of systems of m nonlinear equations in n unknowns:

NRSG
PS–1034 Newton–Raphson, steepest gradient (class B)

NONLIQ
PS–945 Secant-type algorithm (class B)

NEWT
PS–871 Newton–Raphson with step-size limitation (class B)

QNWT
PS–1154A Quasi-Newton algorithm (class B)

RQNWT
PS–1195 Quasi-Newton algorithm with random perturbation of initial guess if QNWT fails to obtain solution (class B)

5.11   Software for the Elementary Functions*

W. J. Cody

ARGONNE NATIONAL LABORATORY
ARGONNE, ILLINOIS

I. Introduction

During the past decade our knowledge of how to accurately compute elementary function values has grown considerably. For example, we now understand the principles of argument reduction and recognize the interplay between the design of algorithms and the characteristics of the arithmetic units of computers. Unfortunately, only a few computer manufacturers have exploited this increased knowledge in their software. In all fairness to the manufacturers, while some of this knowledge has been collected in the open literature (e.g., Cody, 1967), much of it has appeared only in bits and pieces, or is regarded as unpublishable folklore by the "experts" in the field. If a criticism is to be levied, it is that few of the manufacturers have called upon the "experts" for advice.

The number of library certification projects listed in the appendix is one indication of the growing awareness by some individuals of the importance of the function library, and also of growing suspicion as to its quality. In many cases their work is known to only a few. The present paper is an attempt to acquaint a wider audience with the activities and results of some of these certification projects, as well as to survey the quality of typical Fortran libraries provided by manufacturers, and of replacement libraries available at some installations.

An in-depth survey of every routine in every library would require more time and space than is available. Instead, we have chosen eight different functions and have examined the corresponding subroutines in the Fortran libraries provided by five computer manufacturers for major machines in their lines. The libraries thus surveyed are those for the CDC 6000 series, the IBM System/360, the PDP-10, the Sigma 7, and the Univac 1107–1108 computers. These libraries are believed to be typical of those used extensively in scientific computation today. The routines surveyed span the spectrum

* Work performed under the auspices of the US Atomic Energy Commission.

from the naive approach of a beginning calculus student to the exquisitely subtle touch of the professional, the latter as evidenced in the new IBM hyperbolic sine routine.

The choice of libraries was dictated partly by the availability of source material and partly by the characteristics and distribution of the computers involved. Formal and informal progress reports of the various certification projects listed in the appendix form an important source of information. For the CDC library they are almost the only source. Listings and library manuals were consulted for the IBM System/360, Sigma 7, and Univac 1107 libraries. Additional sources from the open literature are cited as used.

We believe that the survey accurately reflects the status of these libraries as of at least January 1969, in most cases, and as of January 1970 (OS Release 18) for the IBM machines.

II. Preliminaries

There are three important performance characteristics to be considered for subroutines—reliability, speed of execution, and storage requirements. Reliability includes both numerical accuracy and adequate diagnosis of errors. A reliable routine is one that returns accurate values for all legitimate arguments and flags all other arguments. However, we will ignore error diagnosis in our discussion and think of reliability as meaning numerical accuracy.

In general, it is possible to improve the accuracy of almost any routine at the expense of lower speed and larger storage. The design of a particular routine therefore reflects the relative importance the designer places on these characteristics. The author's personal emphasis has always been on accuracy, provided the cost in speed and storage is not prohibitive. Responsibility for only a fast machine with a large memory, an IBM 360/75 at the moment, has made this approach a reasonable one. On the other hand, a manu-facturer, with his concern about performance on all machines in a line, must judge the relative merits of a routine differently. Routines that are acceptable, even desirable, on our machine might not be appropriate for universal use on the System/360 series. Criticism of existing library routines should not, then, be aimed at marginal deficiencies in some characteristic that can be repaired only at an unacceptable cost in terms of other characteristics, but rather at major deficiencies that must be repaired at any cost, and minor deficiencies that can be remedied cheaply.

Function routines can be divided into three classes: primary, secondary, and management. Primary routines include the basic building blocks of any library. They are completely self-contained routines that rely upon no other routines for computation. They may be frequently called upon by other routines in the library to perform critical computations. Typically, routines for the exponential, logarithm, and sine and cosine are among the primary routines in a library.

Secondary routines do part of the computation themselves, but rely upon primary or other secondary routines for some of the computation. Routines for the inverse trigonometric functions and the hyperbolic functions are frequently in this category.

Management routines are those that rely solely on other routines for computation, merely managing the flow of information from one routine to another. Routines for exponentiation and for certain hyperbolic functions and complex functions (i.e., functions of a complex variable) are usually found in this category.

Our primary yardstick in measuring the performance of subroutines will be their accuracy. There are two major sources of error associated with any function value. The first is *transmitted error*; i.e., error due to a small error in the argument. Consider

$$z = f(x)$$

where $f(x)$ is a differentiable function, and let δz denote the relative error and Δz the absolute error in z. Then

$$\delta z = \Delta z / z \simeq dz/z = \frac{f'(x)}{f(x)} \, dx \simeq x \frac{f'(x)}{f(x)} \delta x \tag{1}$$

The transmitted error δz depends solely on the *inherited error* δx, and not on the subroutine. The second type of error is *generated error*; i.e., the error generated within the subroutine. It includes the error due to the truncation of an essentially infinite process at some finite point, as well as error due to the round-off characteristics of the machine. In particular, it includes the error due to the inexact representation of constants as machine words. The constant π, for example, is irrational, hence cannot be represented exactly by any finite string of bits.

Since subroutines have no control over inherited error, they should be designed to minimize generated error under the assumption that the inherited error is zero; i.e., that the arguments are exact. Similarly, certifications should attempt to measure only generated error. Most of the certifications listed in the appendix use techniques that measure only generated error (Cody, 1969), although it has been necessary in some cases to use techniques similar to those proposed by Hammer (1967), which relax the controls on the inherited error. Our discussion of accuracy will concern only generated error.

III. Primary Routines

The overall quality of a function library can be no better than the quality of its primary routines. We therefore begin our survey by examining routines for the natural logarithm, exponential, sine and cosine, and tangent functions.

The computation of these functions involves a reduction of the argument to some primary interval, followed by the evaluation of an approximation to the function over that interval. With a few exceptions to be noted below, the approximations used in the routines surveyed appear to be accurate enough. In a few cases more economical approximations might have been used; e.g., minimax instead of economized Taylor's series or Padé approximations. The improvements, however, would have been marginal in time and storage, and probably nonexistent in accuracy. Therefore, the quality of almost all of these routines hinges upon the care used in the argument reduction.

For purposes of illustration, consider a typical argument reduction scheme for a sine routine. Let the floating point argument be x and let

$$z = (4/\pi) \cdot |x| = n + f$$

where n is an integer and $0 \leq f < 1$. Then the reduced argument is either

$$g = (\pi/4) \cdot f = |x| - n \cdot (\pi/4) \tag{2}$$

or

$$g' = (\pi/4)(1 - f) = (n + 1)(\pi/4) - |x|$$

depending upon the value of

$$m \equiv n \bmod 8$$

and the main computation will be the sine or cosine of g or g' with appropriate sign. This latter computation can be carried out accurately *provided the reduced argument is accurate*. A loss of accuracy can occur in forming the difference between nearly equal quantities, $|x|$ and an integer multiple of $\pi/4$. If, for example, these two quantities agree for the first k bits, then g may be in error in the last k bits, leading to a large value of δg even though $\delta x \equiv 0$.

If we let .xxxxxxxx represent the computer bits devoted to the normalized fractional part of a floating point number, a typical computation of g is as follows:

$$
\begin{array}{rl}
x = & \text{.xxxxxxxx} \\[2pt]
-n \cdot \pi/4 = & \dfrac{-\,.\text{xxxyyyy}}{.000\text{zzzzz}} \\[6pt]
= & .\text{zzzzz000}
\end{array}
$$

where the renormalization of the intermediate result shifts zero bits into the low-order positions. If we had more precision, the computation would have been

$$x = \quad .xxxxxxxx\ 00000000$$

$$-n \cdot \pi/4 = \frac{-.xxxyyyy\ yyyyyyy}{.000zzzzz\ zzzzzzzz}$$

$$= \quad .zzzzzzzz$$

The extension of x by the appendage of extra zero bits is correct since δx was assumed to be zero; i.e., the original machine representation of x was assumed to be exact. Note that the low-order zero bits in the final result of the original computation should actually be (in general, nonzero) significant bits that depend only upon the product $n \cdot \pi/4$. If these bits can be preserved, then δg will be limited to the error due to machine rounding. Certainly δg will be bounded by 1 ulp (unit in the last place). From Eq. (1) we see that

$$\delta \sin(g) = g \cot(g)\, \delta g \qquad (3)$$

and

$$\delta \cos(g) = -g \tan(g)\, \delta g$$

so the magnitude of δg has a great influence on the accuracy of the computation over the reduced range.

The low-order bits of g can be preserved in several ways. For single-precision floating point routines, the argument reduction could be done in fixed point, with its generally greater significance, or in double-precision floating point, if that is available as a hardware capability. Only a few operations need be performed in the higher precision, so the cost is rather small.

Of course, when double-precision hardware is not available, or the routine is already in double precision, a different approach is necessary. Since $\pi/4$ is a universal constant, it can be written to extra precision as two working precision words:

$$\pi/4 \simeq a \equiv a_1 + a_2$$

where schematically

$$a_1 = .xxxxx000$$

$$a_2 = \quad\quad xxxxxxxx$$

and the floating point exponent of a_2 has been adjusted properly. The essential feature of this decomposition of a is that $n \cdot a_1$ must be *exactly* expressible in the working word length for all values of n such that

$$n \leq \bar{x} \cdot 4/\pi$$

where $\bar{x}$ is the largest legitimate argument for the sine routine. Unless g is extremely small it can then be formed to full precision as

$$x - n \cdot \pi/4 = (x - n \cdot a_1) - n \cdot a_2$$

at a total extra cost of one multiply, one subtraction, and a few storage locations. Although differing in detail, argument reduction schemes for the logarithm, exponential, and tangent subroutines will all benefit from careful handling of constants in a similar manner.

The value of this approach can be illustrated by the following example, which is useful for checking the argument reduction scheme in any sine routine. Let $x = 22.0$. Then

$$g = 0.00885 \ldots$$

indicating a loss of three to four significant decimal places (~ 11 bits) in g unless care is taken. From Eq. (3) we see that

$$\delta \sin(22) \simeq \delta g$$

If the argument reduction scheme is poor, a comparison of the computed value of $\sin(22)$ against published tables should reveal about three fewer significant decimal figures of accuracy than would normally be expected for the floating point hardware involved.

There is one other frequently avoidable source of error associated with base-16 floating point arithmetic—the variable significance of normalized floating point numbers. A normalized floating point number in base 16 has its fractional part in the interval $[1/16, 1)$. Thus, there may be up to three leading zero bits in the fraction. For example,

$$\pi/2 = 1.57 \ldots = 16^1 \cdot (0.098 \ldots)$$

contains three leading zero bits in the fraction. The single-precision fraction on the IBM 360 and the Sigma 7 computers contains 24 bits, of which only 21 can be significant in this case. On the other hand,

$$\pi/4 = 0.707 \ldots = 16^0 \cdot (0.707 \ldots)$$

has no leading zero bits in its representation, hence can be carried to a full 24 bits of precision. However, the computation of $\pi/4$ as $0.5 \cdot (\pi/2)$ results in only 22 bits of accuracy (the 22nd bit of $\pi/4$ is a zero bit, agreeing with the first of the three "garbage" bits shifted in during the computation). With careful planning, computations can often be rearranged to avoid poor normalization of intermediate results, thus preserving significance in the final result. In our trivial example above, we could have retained full significance in our computation of $\pi/4$ by starting from $2/\pi$, which has no leading zero bits in its representation, and using division instead of multiplication.

In general, we may preserve as much as one decimal place of significance by computing $x \cdot \pi/2$ as $x/(2/\pi)$. This saving is nontrivial in any case, and especially so if the arithmetic involved has only seven significant figures to start.

With the above discussion as background, let us now consider the specific routines surveyed.

CDC 6000 Library   Although only meagre source material is available for the CDC library, what is available indicates that CDC argument reduction schemes are quite good. The certification of the single-precision routines reported by Scharenberg (1969) and the single-precision test results of Clark and Cody (1969) indicate no serious problems, and only minor inaccuracies in the sine routine. Rice has reported privately that the original sine routine, which uses an economized Taylor series for the computation for $0 \leq x < \pi/4$, suffers about a 5% loss of four or more bits for x in that range. The replacement of the original approximation with the corresponding minimax approximation from Hart *et al.* (1968) improves the accuracy to a 2% loss of two bits, and no losses of three or more bits. Appropriate statistics may be found in Scharenberg (1969).

Since none of the CDC source material mentions a tangent routine, we assume that the sine and cosine routines are used. This is a poor approach from the user's viewpoint.

Double-precision test results reported by Clinnick (1969) apparently make heavy use of techniques similar to those of Hammer (1967), hence are not as sharp as might be desired. While they do not indicate any serious problems, they are not as reassuring as the single-precision test results. Unpublished probes of the double-precision routines, similar to those of Clark and Cody (1969) for the single-precision routines, suggest that there may be a problem in the logarithm routine for x close to 1. Several cases of four- and five-bit errors were encountered.

IBM 360 Library   The IBM routines surveyed were generally of high quality, but the double-precision sine/cosine and tangent routines still lack precise argument reduction. Hillstrom *et al.* (1970) give specific performance figures. Revised routines have been written by K. Hillstrom at Argonne National Laboratory, with the expected improvement in accuracy and only slight deterioration in speed.

PDP-10 Library   Bennett's survey for the PDP-10 library includes statistics for some replacement routines that he has written. The original library does not contain the careful argument reduction, whereas his replacements do. In the case of the logarithm, for example, he has reduced the maximum relative error for his tests over the interval [0.5, 1.0] from the original

1.521×10^{-4} to 3.641×10^{-8}. The penalty in this case was an execution speed of ~ 300 μsec versus the original ~ 185 μsec. In the case of the exponential, the RMS, and maximum relative errors for the interval [10, 80] were reduced from the original 1.204×10^{-6} and 3.654×10^{-6}, respectively, to 3.211×10^{-7} and 7.633×10^{-7}, respectively, with essentially no loss of speed.

Sigma 7 Library   Routines in this library all lack the careful argument reduction mentioned above. The computation of sin(22) in double precision, for example, is accurate to only 13 significant decimals out of about 16. In addition, few if any attempts are made to avoid the special problems associated with base-16 arithmetic. Thus, the floating point exponential routines use the constant $1/\ln(2)$, with its three leading insignificant bits, in the argument reduction. The only accuracy tests available are the superficial tests of Clark and Cody (1969), which report errors as large as eight bits (out of 56) for the double-precision exponential.

Univac 1107–1108 Library   This library also lacks the careful argument reduction. In addition, the tangent routine is not a primary routine, but rather a management routine calling the sine and cosine routines. Wallace (1969), in his 1108 test results, reports up to 9 out of 27 bits incorrect in the single-precision tangent for x in the interval $(\pi/2 - 0.01, \pi/2)$, and up to five incorrect bits for the interval $(0, 0.2)$. Part of the trouble can be traced to the sine routine where he reports losses of up to six bits in the primary range of computation, $[0, \pi/4)$. Figures given by Devine (1968) for his 1108 tests are in essential agreement.

It appears that the approximation used in the computation of the sine function, an economized Taylor series of degree 7, is not quite accurate enough. Even the minimax approximation of the same form guarantees only 8.49 significant figures of accuracy (Hart *et al.*, 1968), whereas the 27-bit mantissa on the 1108 has an accuracy threshold of 8.43 significant figures. An approximation of one higher degree should improve the sine routine over the primary interval.

IV. Secondary Routines

The secondary routines surveyed were those for the complex absolute value, the arcsine and arccosine, and the hyperbolic sine. Each of the usual algorithms for these particular functions contains one critical computation that is often done poorly.

In the complex absolute value function it is a simple ploy to reduce the possibility of complete destruction of the computation due to underflow or overflow. Let

$$z = x + iy$$

The most obvious computation,

$$|z| = (x^2 + y^2)^{1/2}$$

can lead to underflow or overflow of the intermediate quantities x^2 or y^2 even though $|z|$ is a reasonable number. The usual method for avoiding this is to let

$$v = \max(|x|, |y|)$$

$$w = \min(|x|, |y|)$$

and to compute

$$|z| = v[1 + (w/v)^2]^{1/2} \tag{4}$$

This remedy is not new, but dates to the early days of the IBM 704, at least, and is probably even older than that.

Base-16 machines should use a slightly modified algorithm to avoid normalization problems. The quantity under the radical sign in Eq. (4) will have two to three leading zeros in its floating point representation, since it is of the form $1 + s$, where $0 \leq s \leq 1$, and the square root will definitely have three. A certain amount of scaling is therefore required to insure maximum significance of $|z|$ on a base-16 machine. For example, if Eq. (4) were replaced by

$$|z| = 2v[1/4 + (w/2v)^2]^{1/2} \tag{5}$$

the square root would never have a leading zero bit.

There are several possible algorithms for the computation of the arcsine. The best involves evaluation of a direct approximation to the function for $0 \leq x \leq a < 1$, coupled with appropriate identities such as

$$\arcsin(x) = \pi/4 + \tfrac{1}{2}\arcsin(2x^2 - 1)$$

or

$$\arcsin(x) = \pi/2 - \arcsin(1 - x^2)^{1/2}$$

for arguments not in the primary range. A second algorithm is based upon the relation

$$\arcsin(x) = \arctan[x/(1 - x^2)^{1/2}] \tag{6}$$

and uses a subroutine generally called ATAN2 for evaluating the inverse tangent of a ratio.

Regardless of the algorithm used, some care must be taken when the argument is close to 1. Even if we again assume that x is exact, i.e., is represented exactly as a machine word, it does not necessarily follow that x^2 is exact. Thus any cancellation of leading bits in forming the difference $1 - x^2$ results in incorrect low-order bits. However, the computation can be recast as

$$1 - x^2 = (1 - x)(1 + x)$$

which preserves low-order bits at the cost of one extra operation. On certain machines, because of rounding characteristics, lack of guard characters, etc., it may be necessary to compute

$$1 - x^2 = [(0.5 - x) + 0.5](1 + x)$$

to retain full significance of the results.

Routines for the hyperbolic sine usually make use of the relation

$$\sinh(x) = (e^x - e^{-x})/2 \tag{7}$$

for x not too small in magnitude. Equation (7) is obviously not suitable for computation for small x because of the unavoidable loss of significance in the subtraction. Yet three of the five libraries surveyed used Eq. (7) for all x!

Even for large x the use of Eq. (7) as it stands can lead to subtle problems. Since asymptotically

$$\sinh(x) \sim e^x/2$$

overflow will occur in the exponential routine before it becomes a problem for the hyperbolic sine. If we rewrite Eq. (7) as

$$\sinh(x) = \exp[x + \ln(\tfrac{1}{2})] - 1/[4 \exp[x + \ln(\tfrac{1}{2})]] \tag{8}$$

overflow will not occur prematurely in the exponential routine. However, from Eq. (1) we see that

$$\delta e^z \simeq z \, \delta z \simeq \Delta z \tag{9}$$

Since $\ln(\tfrac{1}{2})$ is not an exact binary number, we must introduce error into

$$z = x + \ln(\tfrac{1}{2})$$

by rounding the result to working precision. When x is large, Δz is also large. The larger x becomes, the larger Δz becomes, and the larger the error in the hyperbolic sine. This trouble can be avoided by revising Eq. (8) so that the argument for the exponential routine is exact whenever x is. Thus, the computation becomes

$$\sinh(x) = (1 + \delta)[\exp(x + \ln v) - v^2/\exp(x + \ln v)] \tag{10}$$

where $1 + \delta = 1/(2v)$, and v is chosen so that $\ln v$ has an exact machine representation with sufficient trailing zero bits so that $x + \ln v$ will be exactly represented whenever x is. Since v will be close to $\frac{1}{2}$, δ will be small. In this manner, the undesirable overflow in the exponential computation can be avoided without introducing new error into the computation.

There is one other advantage to this approach for base-16 arithmetic. For certain intervals, $e^x/2$ is of the form $16^n(1 + \varepsilon)$, while $\sinh(x)$ has the representation $16^n(1 - \sigma)$. Thus, on these intervals the quantity $e^x/2$ has poor normalization (three leading zero bits), while the final result has no leading zero bits, but instead three low-order "gargage" bits as a consequence of the poor normalization of the intermediate computation. With the proper choice of v, $\exp(x + \ln v)$ will have good normalization whenever $\sinh(x)$ does.

CDC 6000 Library   The complex absolute value routines do not avoid premature underflow or overflow. Our source material does not indicate the existence of arcsine or hyperbolic sine routines per se, but does indicate that Eq. (7), or a variant of it, is used for all x in the hyperbolic tangent, complex sine, and complex cosine routines.

IBM 360 Library   All of the routines surveyed were of excellent quality. The complex absolute value and hyperbolic sine routines use Eqs. (5) and (10), respectively, to avoid the problems associated with base-16 arithmetic.

PDP-10 Library   No source material is available for a complex absolute value routine; apparently there is none in the library. Statistics given by Bennett (1969) show serious problems in the arcsine and arccosine routines for arguments close to 1. His replacement routines limit errors to 2 ulp and are about twice as fast as the original routines. Both the hyperbolic sine and hyperbolic tangent routines use Eq. (7) for all arguments. Again, Bennett's replacements are much more accurate than the original routines with only modest penalty (less than 5%) in speed.

Sigma-7 Library   This library contains no routine for the arcsine. The routines for the complex absolute value use Eq. (4) to avoid underflow/overflow problems, and the hyperbolic sine routines use special small argument approximations instead of Eq. (7). However, no special effort is made to avoid the problems associated with base-16 arithmetic, nor is premature overflow prevented in the hyperbolic sine computation.

Univac 1107–1108 Library   The complex absolute value routine in the 1107 library examined did not guard against premature underflow or overflow. The arcsine routine on the 1107 used Eq. (6) together with the ATAN2 routine. Wallace (1969), in his survey of the 1108 library, reports some cases for arguments close to 1 in which only 20 bits (out of 27) were correct for the single-precision arcsine, and some in which only 14 bits were correct for the arccosine.

The computation of the hyperbolic sine is again bad, using Eq. (7) every-where. Test results of Devine (1968), Ng and Lawson (1969), and Wallace (1969) all show the resulting severe loss of accuracy for small x.

V. Management Routines

An almost universal characteristic of management routines is that they can be improved in accuracy by raising their status to that of secondary, or even primary routines. One example of this has already been mentioned in the case of the tangent routines in the Univac library. The penalty associated with this elevation in status is an increase in storage requirements, but a compensating side benefit is frequently an increase in speed. This latter phenomenon is a result of the elimination of calls to other routines with the accompanying linkage overhead, such as saving and restoring registers.

Many of the routines classified as management routines are used in-frequently and are therefore less important than the primary and secondary routines. The exponentiation or power routines, however, are some of the most important in the library, corresponding to the Fortran "**" operator. If x and y are floating point numbers we expect the computation of $x + y$ or $x * y$ to be precise to within 1 ulp regardless of the magnitude of the arguments. It is sometimes shocking to discover that the result of computing $x ** y$ for moderate arguments can have the last 7–10 bits in error! Yet the survey of floating point power routines reported in Clark and Cody (1969) shows that errors of this magnitude are common.

The standard approach to floating point exponentiation involves the computation

$$x ** y = \exp[y \ln(x)] \tag{11}$$

using the primary routines for the exponential and logarithm functions. As Eq. (9) shows, if

$$w = y \ln(x)$$

and

$$z = \exp(w)$$

then

$$\delta z \simeq \Delta w$$

If w is computed in working precision, the word length of the computer assures us that Δw, hence δz, is likely to be large whenever w is large. This phenomenon is independent of the accuracy of the exponential and logarithm routines involved.

As an example, consider the computation of

$$z = 2.0 ** 25.0$$

Then

$$w = 25.0 * \ln(2.0) \simeq 17.3$$

Since the binary representation of w is

$$w = (10001.01 \ldots)_2$$

the first five bits of the floating point fraction representing w are devoted to the integer part of the number. If the floating point fraction contains t bits, then

$$\Delta w = 2^{-t+5}$$

and, from Eq. (9),

$$\delta z \simeq 2^{-t+5}$$

i.e., the last five bits of the floating point representation of z are probably wrong.

To avoid this error, we must obviously compute both $\ln(x)$ and w to higher than working precision, assuming that the arguments x and y are precise. Then, if the argument reduction

$$w = n \cdot \ln(2) + f, \qquad |f| \leq \ln(2)/2$$

is also performed in this higher precision, the final computation

$$z = 2^n \cdot \exp(f)$$

will be essentially as accurate as the exponential computation.

If the working accuracy is single precision, the extra precision required for this approach can be obtained by doing the appropriate steps in either fixed point, with its longer word length, or in double-precision floating point, if that is available. Primary, or self-contained, single-precision routines of this type have been available on the IBM 7094 at the University of Toronto and on the CDC 3600 at Argonne National Laboratory since the early 1960's, and on the IBM 360 at Argonne and the IBM 7094 at the University of Chicago since about 1967.

For double-precision routines, the problem is more difficult since higher than double-precision arithmetic is expensive in terms of time. Clark and Cody (1969) give an economical algorithm that uses only double-precision arithmetic. The corresponding double-precision routine has been available on the IBM 360 at Argonne since early 1969. Similar routines have since been developed by Kuki at The University of Chicago for IBM equipment, and possibly by Bennett at The University of Western Ontario for the PDP-10. These routines are all primary routines, computing the logarithm and exponential functions themselves. They are fast and accurate (errors limited to 1 or 2 ulp until y gets large), but require large storage in comparison to the management routines. Some of the overall storage could be retrieved by reducing the standard exponential and logarithm routines to appropriate entries in the corresponding power routine. Unfortunately, experimental work at The University of Chicago and at Argonne indicates that this would impose an unacceptable penalty of up to a 50% increase in execution time for the logarithm, and about a 10% increase in time for the exponential (Clark *et al.*, Chap. 6).

To the author's knowledge, the only self-contained power routines in existence are those mentioned. With the exception of the CDC 3600 routine, which became standard on that equipment, all of these routines exist at the moment only in private libraries. All of the manufacturer's routines examined as part of this survey were management routines subject to the large errors previously mentioned.

VI. Summary

With one or two exceptions, the surveyed Fortran subroutine libraries for the elementary functions, developed and distributed by the manufacturers, do not reflect the present technology in this field. There is much more to a good function subroutine than the use of a sufficiently accurate and well-conditioned approximation. We have included in our discussion such things as how to reduce arguments properly, how to handle constants, some of the effects of the design of the arithmetic unit, the elimination of undesirable underflow and overflow, and the importance of primary versus secondary or management status of certain routines. In our opinion, good subroutines cannot be designed without considering ideas such as these.

Although there is still room for some improvement, the IBM 360 library stands well above its contemporaries in spite of the added problems associated with accurate computation in base-16 arithmetic. Aside from a few excellent primary routines in the CDC 6000 series library, we found very little in non-IBM libraries that was noteworthy. In a few cases, notably in some of the hyperbolic routines, we found programs that were simply bad. In many cases replacement routines with the desired characteristics already exist in private libraries.

Appendix

Following is an alphabetical list of computing centers that have active or recently completed projects to certify basic Fortran elementary function libraries.

Installation	Individual responsible	Machines involved
Argonne National Laboratory	W. J. Cody	CDC 3600 IBM 360
Jet Propulsion Laboratory	C. L. Lawson	IBM 7094 Univac 1108
Lawrence Radiation Laboratory	M. L. Clinnick	CDC 6600
Lewis Research Center, NASA	R. L. Turner	IBM 7094 IBM 360 PDP-9
Purdue University	M. J. Scharenberg	CDC 6500
Sandia Laboratories	R. D. Halbgewachs	CDC 6600 Univac 1108
The University of Chicago	H. Kuki	IBM 7094 IBM 360
University of Western Ontario	M. Bennett	IBM 7044 PDP-10
University of Wisconsin	W. Wallace	Univac 1108

ACKNOWLEDGMENTS

Much of the material on the idiosyncracies of base-16 arithmetic is an outgrowth of a continuing discourse between R. L. Turner, H. Kuki, and the author. In particular, the discussion of the hyperbolic sine summarizes unpublished joint work with H. Kuki. We also wish to acknowledge the cooperation of the individuals listed in the Appendix, all of whom made results available to the author.

REFERENCES

Bennett, J. M. (1969). Performance of the DEC PDP-10 Fortran Subroutine Library (unpublished report), Computer Science Dept., Univ. of Western Ontario, London, Ontario.

Clinnick, M. L. (1969). Preliminary Report on Study of Fortran Library Functions, *Comput. Center Newsletter* **6**, No. 1, Lawrence Radiation Laboratory, Berkeley, California.

Clark, N. W., and Cody, W. J. (1969). Self-Contained Exponentiation, *AFIPS Conf. Proc.* **35**, 701–706.

Cody, W. J. (1967). The Influence of Machine Design on Numerical Algorithms, *AFIPS Conf. Proc.* **30**, 305–309.

Cody, W. J. (1969). Performance Testing of Function Subroutines, *AFIPS Conf. Proc.* **34**, 759–763.

Devine, C. J., Jr. (1968). Accuracy of Single Precision Fortran Functions, JPL Sect. 314 Tech. Memo. No. 209. JPL, Pasadena, California.

Hammer, C. (1967). Statistical Validation of Mathematical Computer Routines, *AFIPS Conf. Proc.* **30**, 331–333.

Hart, J. F., Cheney, E. W., Lawson, C. L., Maehly, H. J., Mesztenyi, C. K., Rice, J. R., Thacher, H. C., Jr., and Witzgall, C. (1968). "Computer Approximations." Wiley, New York.

Hillstrom, K. E., Clark, N. W., and Cody, W. J. (1970). IBM OS/360 (Release 18) Fortran and PL/I Library Performance Statistics, Argonne National Laboratory Rep. ANL-7666, Argonne, Illinois.

Ng, E. W., and Lawson, C. L. (1969). Accuracy of Double Precision Fortran Functions, Jet Propulsion Laboratory Sect. 314 Tech. Memo. No. 214. JPL. Pasadena, California.

Scharenberg, M. (1969). Report on Analysis of Fortran Elementary Functions, Purdue Computer Science Center Newsletter, June–July, 1969. Purdue Univ., Lafayette, Indiana.

Wallace, W. (1969). Fortran V Library Functions Reference Manual for the 1108, Univ. of Wisconsin, Madison, Wisconsin.

5.12 Mathematical Function Subprograms for Basic System Libraries—Objectives, Constraints, and Trade-Off

H. Kuki

THE UNIVERSITY OF CHICAGO

CHICAGO, ILLINOIS

I. Objectives

Libraries of programs we discuss here are those collections of elementary functions which are considered so basic that it has become customary to be supplied by manufacturers of computers along with a selection of language processors. Since these programs are specifically intended for the general use, one wishes to set down a list of objectives and priorities among them which would satisfy most users all the time and all the users most of the time. This is not very easy to do. What is given in the following are the general guidelines this author has been following over the years.

These programs are intended for heavy use, and therefore careful preparation and rigorous product tests are required not only from their values by themselves but also from an economic point of view. Development cost per program may be relatively high, but cost per usage is very low. Another consideration is their general use. When a set of programs is to be used by a great variety of users, some of them requiring strict accuracy, others not, then such programs are required to aim at satisfying the most demanding users.

Important conditions that must be met include the following:

(a) reliability—the program must produce acceptable answers for all legal argument values;

(b) domain—legal domain should include virtually all those arguments whose function values are representable in the number system of the given computer;

(c) contingencies—proper diagnostics should be given for those arguments which are outside the legal domain;

(d) features—those features that either are required by the host system or enhance its capabilities should be provided;

(e) accuracy—a rigid accuracy standard which is commensurate with the precision of the given computer must be maintained;

(f) speed—optimal speed under the above constraints should be aimed at;

(g) size—the smaller the better after the above requirements are met.

To produce optimal programs in this regard, it is necessary to take advantage of the peculiarities of the computer for which the library is designed. First, part of the above requirements are machine dependent. Accuracy and domain requirements are stated in terms of the given machine. Second, the repertoire of the given instruction set must be utilized to the best advantage. This leads to the use of the assembler language.

On the other hand, when it is required that a single program library is required to service a family of compatible computers of varying performance, we are faced with some difficulty since a code which is optimal for one model of the line is often far from optimal for another model. For example, a single library serves all the models of IBM 360 line computers for the support of Fortran. The timing figures shown in Table I are excerpted from the IBM reference manual (1969).

TABLE I

EXECUTION SPEED

Function	Model 65 (μsec)	Model 85[a] (μsec)	Model 91 (μsec)
EXP	85	14	16
SQRT	58	10	9
DEXP	145	18	13
DSQRT	92	16	8
CEXP	282	47	55
CSQRT	207	43	33

[a] With the fast-multiply feature.

Performance ratio here between Model 65 and Model 85 reflects the ratio of the basic instruction speeds. It is suspected that the failure of these programs to perform well on the 91 is largely due to branching instructions and data access delay. Programs tailor-made for the 91, relieving traffic jams of these sorts, would outperform the standard programs very easily.

A normal solution of this dilemma is to aim at a code which is on the average efficient and also which does not do too much damage to any model of the line.

II. Choice of Programming Language

We have referred earlier to the choice of the assembler language. No serious objection has been raised so far against use of the assembler code for basic systems libraries. Yet, in view of the increasing trend toward use of higher-level languages for general programming, it is worthwhile to examine this question more closely. In addition to those cited earlier, reasons favoring the assembler code for basic libraries include the following:

(a) At the current state of compiler art, codes generated by compilers cannot compete with the assembler codes in terms of economy of space and speed.

(b) Reduction algorithms of several elementary functions are intimately related to the internal number representation.

(c) It is easier to attain razor-sharp accuracy with the assembler code.

(d) Since a program optimized for a given system is most likely far from optimal for other systems, an unacceptable degree of compromise in quality would be imposed if we were to make the code transportable and permanent. Cost of repeated development is well justified by the heavy use of these programs during the life span of the computer system for which they are written.

To illustrate these points, three Fortran codes were prepared. These codes implement standard algorithms without the kind of refinements discussed in later sections. These codes were compiled by the S/360 H compiler using optimization level 2, and execution time figures cited below are for the 360/65 G computer.

Figure 1 gives a code for the exponential subroutine aiming at the basic accuracy of 10^{-7}. This code requires a multiplication by 2^n, where n is an

```
C
C     EXPONENTIAL FUNCTION      RELATIVE ERROR < 10**-7 IN BASE RANGE
C
C     EXP(X) = 2**(X*LOG2(E)) = 2**(N-R) = 2**N * 2**(-R)
C
      FUNCTION EXP(X)
      Y = X*1.4426950E0
      N = Y
      IF (Y) 20,20,10
10    N = N+1
20    Y = Y-N
      Y = ((((Y*0.93891E-3 + 0.9188698E-2)*Y + 0.55279830E-1)*Y
     :       + 0.24017224E0)*Y + 0.69314226E0)*Y + 0.99999993E0
      EXP = Y*(2.0**N)
      RETURN
      END
```

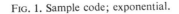
FIG. 1. Sample code; exponential.

integer. An assembler code for a binary machine would accomplish this very economically by manipulation of the internal number representation. To do this without use of internal number representation would require repeated multiplications. The given Fortran code obtains the value 2^n by a call to A $**$ I subroutine. The compiled code, exclusive of the A $**$ I subroutine, requires 452 bytes of storage, and 128 μsec. Computation of A $**$ I for this application can take up to 135 μsec in addition. In comparison, an assembler code for an essentially identical algorithm requires only 184 bytes of storage and 86 μsec.

```
C
C     LOGARITHMIC FUNCTION        RELATIVE ERROR < 10**-7
C
C     LOG(X)  = N*LOG(2) + LOG(R)      WHERE X = R*2**N, 0.5 =< R < 1.0
C     WE ASSUME EXISTENCE OF A BUILT-IN FUNCTION 'NSCALE' WHICH
C        OBTAINS THE BINARY EXPONENT N OF X
C
      FUNCTION  ALOG(X)
      N = NSCALE(X)
      Y = X*2.0**(-N)
      A = 1.0
      IF (Y-0.707107) 10,20,20
   10 A = 0.5
      N = N-1
   20 Y = (Y-A)/(Y+A)
      Z = Y*Y
      Y = Y*((7*0.41517739E0 + 0.66644073E0)*Z + 2.0000009E0)
      ALOG = N*0.6931471R5E0 + Y
      RETURN
      END
```

FIG. 2. Sample code; logarithm.

Figure 2 gives a sample code for the logarithmic subroutine of similar accuracy. Here the situation is even worse. A typical algorithm such as the one used here requires decomposition of the argument x as a product $x = 2^n \cdot r$, where $0.5 \leq r < 1$. It is very expensive to carry this out without recourse to the internal number representation. Therefore we assume existence of a built-in function which bridges this gap in the host language. The function identified as NSCALE produces an integer such that $2^{n-1} \leq x < 2^n$. After this expediency, we still need recourse to the A $**$ I subroutine to obtain the value $r = x/2^n$. The compiled code requires 458 bytes of storage, not counting these two subroutines. Execution time ranged up to total 291 μsec. Assembler code implementation of the same algorithm requires 184 bytes of storage all contained, and 83 μsec of execution time.

Figure 3 gives a sample code for the sine/cosine subroutine which does not depend critically on the number representation. Therefore this example is a more typical one in reflecting the current state of compiling art. The compiled code requires 712 bytes of storage and 167 μsec. Assembler code implementation of the same algorithm requires 200 bytes of storage and 76 μsec.

```
C
C      SINE/COSINE FUNCTION      RELATIVE ERROR < 10**-7 IN BASE RANGE
C
C      LET |X| = N*PI/2 + R  WHERE -PI/4 < R <= PI/4
C      IF COSINE, RAISE N BY 1.    IF SINE AND X<0, RAISE N BY 2
C      ANSWER IS  SIN(N*PI/2)*COS(R) + COS(N*PI/2)*SIN(R)
C
       FUNCTION  SIN(X)
       L = 0
       IF (X) 10,20,20
10 L = 2
       GO TO 20
C
       ENTRY     COS(X)
       L = 1
20 Y = ABS(X)*0.63661977E0
       N = Y+0.5
       Y = 2*(Y-N)
       N = N+L
       Z = Y*Y
       IF (N-2*(N/2)) 30,30,40
30 Y = (((-0.35950E-4*Z + 0.2490001E-2)*Z - 0.80745433E-1)*Z
     :      + 0.78539816E0)*Y
       GO TO 50
40 Y = ((-0.31972RE-3*Z + 0.15349684E-1)*Z - 0.30842417E0)*Z
     :      + 0.99999997E0
50 IF (N-4*(N/4)-2) 70,60,60
60 Y = -Y
70 IF (L-1) 80,90,80
80 SIN = Y
       RETURN
90 COS = Y
       RETURN
       END
```

FIG. 3. Sample code; sine/cosine.

III. Systems Specifications

These include what were listed earlier as considerations for contingences and features. First, programs in a standard library should be equipped to handle any input argument. Excluded from this requirement are those quantities that are incapable of being produced by the host system. In the case of most Fortran systems, unnormalized quantities are not supposed to be produced. Therefore, Fortran library programs are usually not protected against unnormalized arguments. If one is encountered, the result is quite possibly unpredictable. Since the host system pledges not to produce such quantities, the cost of protection would seem unjustifiable. Except for this, programs in the basic library should be able to handle any argument, either to reject it with a proper message or to produce an acceptable answer.

Next, programs should behave consistently with the conventions of the host system. One typical example is the handling of exponent overflows and underflows. Many systems treat overflows and underflows with different weights, treating underflows as the lesser evil. In case of the IBM S/360, underflow interrupts can be suppressed by choice of a program mask, but overflow interrupts cannot be suppressed. This lack of symmetry is a sensible one, since in knowing this one can formulate his computation in such a way as to take advantage of it—namely, to arrange the algorithm in such a way that underflowed quantities are conveniently ignorable. On the other hand, in general computations, we cannot ignore underflows always. Therefore users of the system should be provided with an appropriate warning facility for such contingencies. And such a facility should be respected by programs in the basic library.

The rules which the author found feasible to enforce in his work are as follows:

(a) Intermediate underflows—those underflows which do not affect the final result—are either avoided completely, or allowed to occur in such a way as not to cause unwanted warnings to users. Possibility of such underflows is quite common. The example of the sine/cosine code in Fig. 3 would cause intermediate underflows when executed in a 360 computer, if the argument is less than 10^{-37} in magnitude.

(b) Intermediate or premature overflows are avoided completely. If we let one occur, it will, in general, affect the result drastically.

(c) Result underflow cases are accompanied by an underflow warning if users had elected to have such messages. Otherwise the result is quietly set to 0.

(d) Result overflow cases are usually specifically screened out with proper diagnostic messages. In some complex-valued function subroutines, this was found to be too cumbersome to accomplish. In such cases, processing was left to the standard exponent overflow facility.

Implementing such rules is not always a trivial task. A typical example of cases when special care is needed is complex arithmetic; in particular, the complex divide subroutine. A straightforward algorithm for this operation is

$$\frac{a + bi}{c + di} = \frac{ac + bd}{c^2 + d^2} + \frac{bc - ad}{c^2 + d^2} i \tag{1}$$

Denoting the largest number allowable in the system by Ω, this algorithm will cause an exponent overflow when $|c + di| > \sqrt{\Omega}$, or, alternatively, when $|a + bi| \cdot |c + di| > \sqrt{2} \cdot \Omega$, even if the absolute value of the result is very much less than Ω. Similarly the probability of premature underflow/divide check is quite high. In other words, this algorithm is applicable only for a drastically small subset of the full legal range. Another difficulty to watch for is the possible loss of accuracy due to partial underflows. Referring to Eq. (1), if the product bd underflows it may seriously affect the accuracy of $ac + bd$. Thus it is highly desirable to apply an exponent scaling to the operands.

The cost of the scaling operation is not necessarily trivial. Scaling requires restoration of scale at the end and also proper processing of the result overflow/underflow contingencies. In some cases it is possible to avoid premature underflows and overflows without resorting to a direct scaling. An example of such an approach is the Release 18 version of the complex square root, S/360 Fortran Library [IBM Systems Reference Library (1969)]. A flow diagram of this program is given in Fig. 4, to be examined by interested

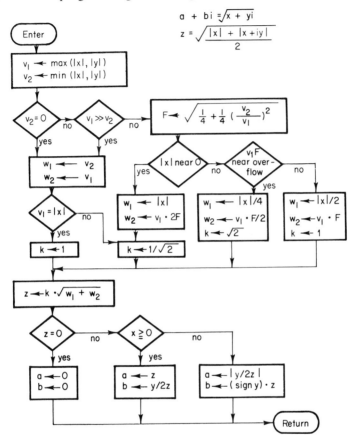

$$a + bi = \sqrt{x + yi}$$

$$z = \sqrt{\frac{|x| + |x + iy|}{2}}$$

FIG. 4. A flow chart for complex square root.

readers. The quantity z in the diagram is equal to $[\frac{1}{2}(|x| + |x + iy|)]^{1/2}$. The main task is to compute this quantity without causing irregularities. Also this flow was designed to enhance the last digit accuracy.

IV. Standard Reference for Accuracy

Before we discuss accuracy we must decide what we regard as the exact answer to a given problem, since accuracy is a measure of deviation of the obtained answer from the exact one. The author, among others, has been using the following standard:

(a) The value of an input argument is that value for which the argument in its machine representation stands as it is passed on to the subroutine. No allowance is made for conversion errors, accumulation of errors from prior processing, or even the minimal rounding. The machine representation is the only information the subroutine receives. No information relating to the past history of the argument value is passed on to the subroutine. Here we are not concerned with the significance arithmetic and other arithmetics of similar purposes.

(b) The infinite-precision function value corresponding to this interpretation of the argument value is taken to be the exact answer. Therefore, even the minimal round-off error to fit this exact answer to the precision of the given machine is accounted as an error.

Some authors do not follow the second rule, and take the best answer within the given machine precision system to be the exact function value, and evaluate errors as deviations from this value. The difficulty of this scheme is that there is no place to account for the inevitable minimal rounding. Consider the following example. Let two subroutines SQRT and SQUARE be of such quality as to produce always the rounded best answers. Then, according to the alternative standard, these routines do not generate any errors. Take three machine numbers x_1, x_2, and x_3 which are consecutive in their machine representations. In other words the distances between x_1 and x_2 and between x_2 and x_3 are minimal in the given precision. Taking the square root reduces the scaled distances to a half. Therefore the distance between $\sqrt{x_1}$ and $\sqrt{x_3}$ is very close to the minimal increment for their machine representations. Thus SQRT(x_1) and SQRT(x_3) will be consecutive to each other, and SQRT(x_2) will be obliged to equal one of them. Follow SQRT with the squaring operation SQUARE and we will get only two distinct values among three answers: SQUARE(SQRT(x_i)), $i = 1, 2, 3$. Then, for at least one of x_i's, the result is not equal to its original value. Here the user is helpless. He did not contribute any error. Neither SQRT nor SQUARE admits any error.

Adoption of the suggested standard eliminates this and other similar difficulties. Users are interested in the amount of the total error, including the minimal round-off. Although this minimal round-off may seem small by itself, it may be considerably magnified by ensuing computations. Thus, this error should be accounted for, and should be included as part of the inherited error in the next stage of computation.

V. Effect of an Argument Error

Let x be the argument and $y = f(x)$ be the exact function value in the sense of the above standard. Let Δx be the absolute error and δx the relative error of x, both inherited from prior computations. Let Δy be the absolute error and δy the relative error in y, both due to Δx. Then

$$\Delta y = (df/dx)\, \Delta x \qquad (2)$$

$$\delta y = (x/y)(df/dx)\, \delta x \qquad (3)$$

If there was no other error in the computation, except for the minimal conversion error, then $\delta x \leqq 2^{-30}$ for a 30-digit binary representation. This may be regarded as small, but, depending on the function f and the argument value x, the factor $(x/y)(df/dx)$ may be quite large. In other words, errors due to the minimal indeterminancy of the argument can be quite large.

Because of its versatility, scientific computations are usually carried out in the floating point arithmetic, and we are here concerned with this arithmetic. Since floating point arithmetic is also a fixed precision arithmetic, relative errors are usually of more concern to us than the absolute error. Fixed precision means that there is an upper limit to the relative accuracy, and the system is capable of carrying up to this upper limit. Thus attaining this upper-limit accuracy becomes the natural goal of those who are engaged in preparing function subroutines. However, it must be remembered that this preference is related to the capability of the machine in solving isolated and localized segments of user problems. Depending on the context in which these problem segments are found, other errors such as absolute errors are often more relevant than relative errors. This choice also depends on the quality of the argument and other considerations. For instance, the author finds the absolute error to be, in general, more useful and relevant than the relative error in evaluating the performance of sine function subroutines for arguments outside the principal range.

VI. Errors Due to Straight Coding

According to the standard of Sect. IV, errors implicit in the argument are the responsibility of the prior computations. Therefore their effect on the accuracy has nothing to do with the performance of a subroutine. However, Eqs. (2) and (3) are related to the way some errors are generated within a subroutine.

If we denote by $\bar{\Delta}y$ and $\bar{\delta}y$ the absolute error and the relative error, respectively, generated within the subroutine, then the total errors in the answer are $\bar{\Delta}y + \Delta y$ and $\delta y + \bar{\delta}y$, respectively. The ideal is to aim at $\bar{\Delta}y = \bar{\delta}y = 0$, or rather $\bar{\delta}y$ smaller than the minimal rounding error of the precision, according to our standard. Unfortunately, in many cases this is not easy to accomplish. The reason is as follows. A subroutine requires a series of steps to be executed in sequence. Typically, earlier steps involve reduction of the argument to basic cases, and later steps involve computation of the function value for the reduced argument. A slight error, which is in general unavoidable, will be magnified, if it was committed at an earlier stage, just as an argument error is magnified. For instance, a slight round-off error at the reduction stage will be magnified by the factor $(x/y)(df/dx)$ as its contribution to the generated error $\bar{\delta}y$. Such is usually the main cause of large generated errors in otherwise sound subroutines.

Take, for instance, the algorithm of Fig. 3 for computation of $\sin(x)$. In the figure, the three statements starting at the label 20 comprise the reduction stage. The task of reduction here is to decompose x as $x = (\pi/2)(n + r)$, where n is an integer and $-0.5 \leq r < 0.5$. Then $\sin(x)$ is, except for the sign, equal either to $\sin[(\pi/2)r]$ or to $\cos[(\pi/2)r]$, depending on whether n is even or odd. Thus accuracy of the answer is decided by the accuracy of the reduced argument r. How accurate is the latter? This quantity is obtained as $(2/\pi)x - n$. Using a straightforward coding such as shown in Fig. 3, the round-off in this computation introduces an absolute error approximately equal to $n \cdot 2^{-p}$, where p is the binary precision of the machine. Since $\sin[(\pi/2)r] \approx 2r$, one would expect an absolute error of $n \cdot 2^{-p+1}$ in the answer from this source. This is the same as the effect of the minimal round-off error in the argument. Since $\sin(x)$ becomes 0 periodically, this means that the generated relative error will become infinite periodically.

From this discussion, it becomes clear that for most functions a straightforward reduction generates an error approximately equal in magnitude to the effect of the minimal round-off error in the argument.

VII. Techniques of Reducing Generated Errors

It is possible to reduce generated errors of the type mentioned above with some additional codes in most instances. The basic technique is the use of guard digits at crucial points of computation. Fixed point computation was used profitably when the precision of the fixed point representation is several digits longer than that of the floating point. For single-precision subroutines, a limited use of double-precision computation allows a very accurate reduction with only modest increase in cost. However, when we run out of the backup precision such as the double precision, use of guard digits becomes rather costly.

Some progress has been made during the last three years in this regard. One of the earliest applications of this new technique can be found in the DEXP subroutine of the Release 18 S/360 Fortran (1969). The basic idea of the reduction scheme used in this program is due to Turner (1967). The problem was not whether we can generate extra accuracy; this can always be done at some cost. The aim is to accomplish this at a minimal expense.

Let us examine the code of Fig. 3 again. It would be too expensive to carry out the entire reduction scheme in a simulated extra precision. Reduction called for decomposition of x as $(\pi/2)(n + r)$, where n is an integer and $-0.5 \leq r < 0.5$. In order to get a very accurate r, x must be multiplied with $2/\pi$ in a simulated arithmetic, and the long product must be carefully decomposed into its integer part and its fraction part. However, reexamination of the whole algorithm reveals that the requirement $-0.5 \leq r < 0.5$ is not an absolute one. This requirement can be violated somewhat and r can be somewhat outside this range. The important requirement is that $(\pi/2)(n + r)$ must agree with x for several more digits than the working precision of the machine arithmetic. From this the following strategy emerges. Multiply x by $2/\pi$ in the machine arithmetic and extract the integer part n. By means of a phased reduction, subtract an extra-precision product $(\pi/2) \cdot n$ from x to obtain $(\pi/2)r$, which is used as the reduced argument. This technique turns out to be far less expensive than a full extra-precision reduction.

An example of a rather involved application of similar techniques can be found in Clark *et al.* (Chap. 6).

VIII. Two Levels of Accuracy Objectives

For programs in the standard library, it is highly desirable virtually to eliminate generated errors. It is true that actual benefits users get from such refinements are largely psychological. For, what percentage of users use uncontaminated quantities as arguments? Even an input which has just been read into the computer had already suffered from conversion round-off. In view of this, is it not sufficient if generated errors are kept at the level of the effect of the minimal round-off in the argument? Is it wise to make users pay for refinements which may not materially benefit them? Arguments along this line were advanced previously by the author [Kuki (1967)]. Obviously, this is a question of trade-off. It is the question of how much extra cost is involved in obtaining these refinements.

There are some tangible benefits from these refinements. The following example is due to Kahan (1967). The function $f(x) = (\log(1 + x))/x$ is a well-defined continuous function of x in $(-1, \infty)$, if we define $f(0) = 1$. Moreover the factor $(x/y)(dy/dx)$ appearing in Eq. (3) is $1/(y + xy) - 1$, which is quite tame except when x is very near -1. In other words, this function deserves very accurate computation for x in $(-0.8, \infty)$, say. However, it is out of the question to expect an accurate result if we compute

$f(x)$ as defined above. For x near 0, a substantial part of x will be chopped away when the sum $1 + x$ is formed, and this loss will be reflected directly in the accuracy. This is true regardless of the quality of the logarithmic subroutine, since the loss occurred outside the subroutine. On the other hand, suppose we apply the following maneuver. Let $y = 1 + x$, and compute $f(x) = \log(y)/(y - 1)$ if $y - 1 \neq 0$, and set $f(x) = 1$ if $y - 1 = 0$. With a computer equipped with a guard digit for subtraction, it turns out that this computation produces accurate results if and only if the logarithmic routine is meticulously accurate to the last digit.

Also, occasionally users are dealing with floating point quantities that happen to be exact; for instance, integers. For such cases, benefits of the above type of refinements can be real. For example, the assembler code for the power routine in Clark *et al.* (Chap. 6) produces the exact answer a^b always, if all a, b, and a^b are integral quantities that can be exactly expressible in the given precision. This is not possible without heavy use of guard digits and the rounding operation in the subroutine.

The most important benefit is psychological, however. Users would like to treat programs in the basic library as a set of black boxes which can be entirely relied upon. They wish to exclude these programs from consideration as a possible source of inaccuracy in their computation. If they can trust these subroutines to be accurate virtually to the last digit, then they feel secure. It is not easy to explain the reasonableness of the relaxed standard. If a user problem is such that errors caused by sound straightforward subroutines are not acceptable, then it is a very good indication that they should switch to a higher precision. Meticulously prepared subroutines of the same precision probably would not help him. However, users do not wish to listen to a qualified characterization of program performance that requires careful interpretation. They are already harassed by enough troubles of other kinds. They want a simple and straightforward answer. At least this is the attitude of the vocal minority with whom the author has come in contact. It is suspected that the silent majority is similarly disposed.

The trade-off point between cost and refinements is a matter of individual judgment. This author feels quite comfortable in increasing execution time by 10% and storage cost by 15% to attain a virtual last-digit accuracy. Actually, it is fun to devise a scheme whose cost is within this limit. On the other hand, if the cost exceeds this limit substantially, he is in favor of settling with the more relaxed standard.

ACKNOWLEDGMENT

The author wishes to express his deep gratitude to W. J. Cody, of Argonne National Laboratory for his constant encouragement, and to Miss Louise Smith for her editorial assistance and typing. Also, coefficients in Figs. 2 and 3 are taken from Hart *et al.* (1968).

REFERENCES

Hart, J. F., Cheney, E. W., Lawson, C. L., Maehly, H. J., Meztenyi, C. K., Rice, J. R., Thatcher, H. C. Jr., and Witzgall, C. (1968). "Computer Approximations." Wiley, New York.

IBM Systems Reference Library (1969). FORTRAN IV Library: Mathematical and Service Subprograms, Form C28–6818. IBM Corp., New York.

Kahan, W. (1967). Private communication. Univ. of California, Berkeley.

Kuki, H. (1967). Comments on the ANL evaluation of OS/360 Fortran Math Function Library, SHARE Secretary Distribution 169, 12 July 1967.

Turner, L. R. (1967). Private communication. Lewis Research Center, NASA, Cleveland.

5.13 On Writing an Automatic Integration Algorithm*

Carl de Boor

PURDUE UNIVERSITY

LAFAYETTE, INDIANA

The task at hand is the construction of an algorithm which has:

(1) *Input*: (a) two real numbers a, b; (b) access to a procedure for evaluating a given real-valued function $f(x)$ for every x between a and b; (c) a desired relative error tolerance RERR; and (d) a desired absolute error tolerance AERR;

and which produces:

(2) *Output*: an estimate EST for the number $\int_a^b f(x)\,dx$, which satisfies

$$\left| \int_a^b f(x)\,dx - \text{EST} \right| \leqq \max\left(\text{AERR}, \text{RERR} \left| \int_a^b f(x)\,dx \right|\right)$$

(3) Further, this output should be produced *efficiently*, say, in as few function evaluations as possible.

(4) Finally, this output should be produced *reliably* and without fail.

Of course, such an algorithm cannot be written. For, the estimate EST will have to be based on the value of the integrand $f(x)$ at *finitely* many points, i.e., on information about $f(x)$ which can be altered arbitrarily without a change in the integral. To say it differently, the estimate EST for $\int_a^b f(x)\,dx$ must be based in information about $f(x)$ which in no way distinguishes $f(x)$ from a host of other functions on which the functional of integration from a to b can take on any value we like. In short, as Golomb and Weinberger (1959) pointed out some time ago, an estimate for $\int_a^b f(x)\,dx$ based solely on the value of $f(x)$ at finitely many points is meaningless.

Once the prospective algorithm writer has recovered from this mathematical blow, he will, of course, realize that this last argument can be turned around. No matter how we calculate EST from the value of $f(x)$ at certain finitely many points, there will be quite a few functions agreeing with $f(x)$ at these points for which EST is the exact value of the integral. Hence, no

* This work was supported in part by NSF Grant GP-05850.

matter what kind of quadrature algorithm one writes, there will be quite a few functions which this algorithm integrates satisfactorily. To give an extreme example, the exceedingly simple algorithm which merely returns zero as the estimate integrates all odd functions on $[-1, 1]$ correctly.

We will call the collection of all quadrature problems—specified by a, b, f, AERR, RERR—which are solved satisfactorily by a given quadrature algorithm, the domain of validity (or effectiveness), or simply *the domain of the algorithm*. A more realistic description of the task at hand would read: Construct an algorithm with input (1) and output (2) as specified above such that the efficiency requirement (3) is satisfied (or at least not ignored), while the reliability requirement (4) is replaced by:

(4′) The domain of the algorithm should include most of the quadrature problems commonly solved at a typical computer installation; also, the algorithm should recognize failure, in some sense.

Useful effectiveness theorems for quadrature algorithms—of the kind Hull (1968) proved for certain methods in differential equations—will have to wait, I think, until new ways of classifying functions according to properties of structural complexity have been developed. For, any proof that a given quadrature problem involving a certain function $f(x)$ is in the domain of a given quadrature algorithm will have to use, in some sense, properties of $f(x)$ which allow the conclusion that the behavior of $f(x)$ on the interval of integration can be fully understood by looking at the value of $f(x)$ at finitely many points. Clearly, the customary mathematical classification according to the number of continuous derivatives is of no help in such arguments.

Since it is at present impossible to describe in useful and precise terms the domain of any quadrature algorithm, it has become standard practice to write such an algorithm somehow, guided by certain vague ideas or by the desire to have certain very specific sample problems included in the domain. During and after the writing, much time is then spent to ascertain which problems with known solutions are, in fact, contained in the domain of the algorithm. And a list of such problems, at times together with a list of problems not in the domain, is then taken as an indication of what the domain of the algorithm might be.

In the remainder of this paper, I want to discuss several of these vague ideas which are at present used in constructing quadrature algorithms, at times arguing that one should be preferred to others and adding one or two of my own. For a thorough discussion of a wide variety of existing automatic integration algorithms, the reader is referred to the excellent chapter on that topic in Davis and Rabinowitz (1967).

Existing quadrature algorithms can be roughly classified as being either *adaptive* or *fixed*, depending on whether or not the choice of points at which the integrand is evaluated is based on the nature of the integrand. In a fixed scheme, the integral is estimated from the value of the integrand at a predetermined set of points. In an adaptive scheme, the interval of integration is subdivided somehow into a set of subintervals

$$[a, x_1], [x_1, x_2], \ldots, [x_n, b]$$

using small intervals where the integrand is difficult and large intervals where the integrand is easy to handle. In each of these subintervals, some fixed integration scheme is used to estimate the integral, and the results are summed to give the final estimate. Although I will give below other reasons for using an adaptive scheme rather than a fixed one, the primary attraction of adaptive schemes is clearly economy. Since I consider the sparing use of function evaluations an important characteristic of a "good" automatic integration algorithm, I will restrict further discussion to adaptive integration schemes.

The potential for economy inherent in an adaptive scheme can, of course, be realized only if the division points $x_1, \ldots, x_n$ are chosen "right." These points are usually picked during integration, as more and more information about the integrand becomes available. There are, at present, two schools of thought on how this should be done.

One school holds that the calculation of $\int_a^b f(x)\, dx$ is a rather simple initial-value problem. Consequently, one traverses the interval $[a, b]$ from left to right, choosing the next point x_{i+1} based on the experience gained during the successful integration over the subinterval $[x_{i-1}, x_i]$, $i = 1, 2, \ldots$, until finally the point b is reached.

I think that numerical integration is tough enough as it stands and does not need the additional difficulties of initial-value problems, such as the problem of how to choose the first point x_1, or the problem with what might be called "Achilles' disease": As one approaches a difficult spot in the integrand, a singularity or near-singularity, from the left, the steps taken become evermore cautious and small, and, in some cases, the algorithm simply gives up right at the trouble spot's doorstep. In watching such a performance at a teletype, one begins to sympathize with those Greeks who were persuaded that Achilles could not overtake the tortoise.

A second school first attempts integration with a fixed scheme over the entire interval $[a, b]$. If things work out, fine; if not, the interval is subdivided into two or three subintervals of equal size, and each of these is now treated separately in the same way. In this fashion, one may end up subdividing many times in regions where the integrand is difficult, while using relatively large subintervals, and consequently few function evaluations, in regions where the integrand is nice.

I belong to this school, but wish to propose a modification in the order in which subintervals are usually looked at. Assume for definiteness that a rejected interval is subdivided into *two* subintervals of equal length. Then one concentrates now on the left subinterval, subdividing it further, if necessary. In any event, the other, right subinterval, is examined only after the entire left subinterval has been handled successfully. In this way, each subinterval is treated separately, usually with no knowledge of the behavior of the integrand in the remaining subintervals. The modification is simple: before concentrating on the left subinterval, check the right subinterval. If things work out, add the estimate found to the total estimate; if not, merely mark the right subinterval for further subdivision. Then proceed to work on the left subinterval.

The effect of this modification is best illustrated by an example. Suppose that the integrand $f(x)$ is a very nice function on the interval of integration $[0, 1]$ except for a jump at $\frac{1}{3}$; see Fig. 1. The first attempt to integrate over

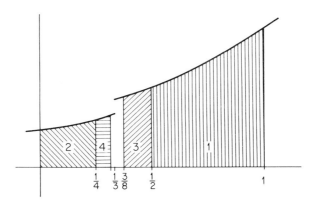

FIG. 1.   Integrating a singularity into a corner.

the entire interval $[0, 1]$ fails, but integration over the right subinterval $[\frac{1}{2}, 1]$ succeeds, leaving the interval $[0, \frac{1}{2}]$ to work on. Again, integration over $[0, \frac{1}{2}]$ fails, as does integration over the right subinterval $[\frac{1}{4}, \frac{1}{2}]$, which is now marked for further subdivision. But the left subinterval $[0, \frac{1}{4}]$ is handled successfully, leaving the subinterval $[\frac{1}{4}, \frac{1}{2}]$. This subinterval was marked for further subdivision; trying its right subinterval $[\frac{3}{8}, \frac{1}{2}]$ succeeds, leaving the subinterval $[\frac{1}{4}, \frac{3}{8}]$ as the only one to worry about. As this process continues, an estimate for the integral over most of the interval $[0, 1]$ is obtained together with an estimate of the error so far committed. Hence, error requirements for the estimate on the remaining small subinterval can be set in a realistic way, which usually means that the estimate can be quite rough without affecting the accuracy of the estimate for the whole interval of integration. More importantly, as the algorithm labors mightily on the one small subinterval left, it is, by this very fact, aware that it must be working on some trouble spot, and hence can test various hypotheses about the kind of trouble present. Should the integrand react positively to one of these tests, a fixed integration scheme designed to integrate functions with precisely this difficulty can then be applied to finish off this small subinterval quickly.

We come now to the central problem of adaptive, and indeed of all automatic, integration: given an estimate EST for the integral $\int_0^h f(x)\,dx$, how do we decide that this estimate satisfies

$$\left| \int_0^h f(x)\,dx - \text{EST} \right| \leqq \varepsilon$$

for some prescribed ε? As was pointed out at the beginning, this question can never be decided rigorously, if the only evidence about the integrand available consists of the value of the integrand at certain finitely many points. Hence, however the decision is made, it is guaranteed to be wrong sometimes. It is also clear that the batting average of any given decision procedure can always be improved simply by examining the integrand at additional points, thus increasing the reliability or the domain of the algorithm while decreasing its efficiency or economy. When faced with this choice between reliability and efficiency, the algorithm writer might find the following remarks helpful.

In the absence of a generally accepted list of problems which "ought" to be contained in the domain of a general purpose integration algorithm, it is easy to become preoccupied with pathological sample problems while working on such an algorithm. Having managed to include yet another weird integrand among those correctly integrated by the algorithm is sure to generate pride and joy in the algorithm designer. But if this feat was achieved at the cost of lowering the overall efficiency of the algorithm even by as little as 5%, it was probably not worth it, since the average user of such an algorithm most likely never has seen nor ever will see such an integrand.

It is often possible to increase reliability at little or no increase in cost merely by examining the available evidence about the integrand more carefully. Bounds on the error of a particular estimate EST for $\int_0^h f(x)\,dx$ are (usually) arrived at in some systematic way, typically by making some assumptions about the (local) behavior of the integrand. Although it is usually not possible to *prove* that these assumptions are valid for the integrand under consideration, it is often possible to *test* the validity of these assumptions by checking whether or not the available data are consistent with these assumptions. The calculated bound on the error is then accepted only if the available data passes such test or tests. Looked at this way, numerical integration consists in the execution of a sequence of numerical experiments on the integrand designed to ascertain one particular property, its integral over some given interval.

Whether or not a particular decision procedure can be made more reliable this way depends in an essential way on the assumptions on which it is based. Consider, for example, the following rather common technique for obtaining an estimate EST for $\int_0^h f(x)\,dx$ and a bound for its error: one calculates two estimates EST_r and EST_s which satisfy

$$\int_0^h f(x)\,dx = \text{EST}_r + C_r h^{r+1} f^{(r)}(\xi_r)$$

$$\int_0^h f(x)\,dx = \text{EST}_s + C_s h^{s+1} f^{(s)}(\xi_s)$$

given that $f(x)$ is sufficiently smooth, with C_r and C_s some constants, ξ_r and ξ_s some points in $[0, h]$, and r and s two integers. Several of the existing quadrature algorithms take

$$r < s$$

and then employ a decision procedure which seems to be based on the following argument: we don't know what $f^{(r)}$ or $f^{(s)}$ look like nor do we know ξ_r or ξ_s; hence, let's forget about them. We are left with the information that EST_s is a higher-order estimate than is EST_r, therefore, a much better estimate than EST_r, therefore

$$\left| \int_0^h f(x)\,dx - \text{EST}_r \right| \approx \left| \text{EST}_s - \text{EST}_r \right|$$

hence, EST_s being better than EST_r, we certainly have

$$\left| \int_0^h f(x)\,dx - \text{EST}_s \right| < \left| \text{EST}_s - \text{EST}_r \right|$$

Whatever the precise assumptions on $f(x)$ might be which would make this argument rigorous, it would seem extremely difficult to devise procedures for testing the validity of these assumptions.

By contrast, consider the specific example of the compound or composite trapezoid rule. If T_n denotes the compound trapezoid rule estimate for $\int_0^h f(x)\,dx$ based on 2^n subintervals of equal length, then, for a sufficiently smooth $f(x)$, one has

$$\int_0^h f(x)\,dx = T_n + 2^{-2n}C + o(2^{-2n}h^2)$$

where C depends on $f(x)$ and h, but not on n. Hence, if h is small enough or n large enough, then the error in the estimate T_n is about $2^{-2n}C$; i.e.,

$$\int_0^h f(x)\,dx = T_m + 2^{-2m}C, \qquad m = n - 1, n$$

which we can solve for $2^{-2n}C$ to get

$$2^{-2n}C = (T_n - T_{n-1})/3$$

This is, of course, the well-known argument which starts off Romberg integration and leads to the statement: With

$$T_{n,1} = T_n + (T_n - T_{n-1})/3$$

we have

$$\left| \int_0^h f(x)\,dx - T_{n,1} \right| \leq |T_n - T_{n-1}|/3$$

provided h is small enough or n is large enough.

Ignoring this last proviso is, I think, as bad as ignoring those derivatives in the earlier "argument." But, in this case, we can test the validity of such an assumption as follows [see Lynch (1967)]. If it is true that

$$\int_0^h f(x)\,dx = T_m + 2^{-2m}C, \qquad m = n-2, n-1, n$$

then it would follow that

$$R_n = \frac{T_{n-1} - T_{n-2}}{T_n - T_{n-1}} = \frac{3.2^{-2n+2}C}{3.2^{-2n}C} = 2^2 = 4$$

and this condition is very easy to test for. In the same vein, one can test the other columns of the T-table generated in Romberg integration, looking for a ratio of about 2^{2k} in the kth column, $k = 1, 2, \ldots$. The resulting fixed iterative scheme for estimating $\int f(x)\,dx$ might be called *cautious Romberg integration*. It is used as the fixed scheme in the adaptive quadrature algorithm CADRE discussed in Chapter 7 of these proceedings.

Similar tests can be applied in conjunction with a variety of decision procedures which are based on the assumption that the "dominant" error term is, in fact, dominant. Since this assumption can be expected to be satisfied only for sufficiently small h, the resulting fixed schemes will have to be used adaptively. To put it differently, the capability of employing such cautious fixed schemes successfully is as good a reason as any to prefer adaptive schemes to fixed schemes.

In the case of cautious Romberg integration, a somewhat unexpected benefit accrues. A significant number of commonly used integrands are not smooth enough to permit the error expansion for the compound trapezoid rule used above. Still, the error can be expanded somehow in terms of n and h. Thus, to take a very simple example, one gets [see Lyness and Ninham (1967)], with $\alpha \neq 0$,

$$\int_0^h x^\alpha\,dx = T_n + 2^{-(1+\alpha)n}C + o((2^{-n}h)^{1+\alpha})$$

Hence, in this case, the ratio R_n approaches $2^{1+\alpha}$ as n becomes large (rather than 4). Such a behavior of the ratio R_n can, of course, be detected and leads to the estimate

$$T_{n,1} = T_n + (T_n - T_{n-1})/(2^{1+\alpha} - 1)$$

with the error bound

$$\left| \int_0^h x^\alpha \, dx - T_{n,1} \right| \leq |T_n - T_{n-1}|/(2^{1+\alpha} - 1)$$

Again, if $f(x)$ has a jump discontinuity in $[0, h]$, but is otherwise "nice", then

$$|R_n| \cong 2$$

a fact easily detected and acted upon.

The two remaining major difficulties in the construction of adaptive integration algorithms are (1) assignment of error requirements for subintervals and (2) the integration of functions which are "noisy" relative to the given (hence, unrealistic) error requirements. Specifically, one might be forced to subdivide the original interval excessively in the sense that the noise in the calculated function values becomes the dominant feature of the integrand on the resulting small subinterval. This problem seems to have been tackled so far only by Lyness (1969). Extending the technique used by Lyness a little bit, one might try to combat this difficulty as follows. Given that a cautious integration scheme is used on subintervals, one declares the dominant behavior of the integrand on a given subinterval to be "noise" if (1) all tests for regular behavior fail and (2) on some previously examined (larger) subinterval containing the given subinterval the integrand was found to satisfy one of the tests for regular behavior, further subdivision having been forced only because the resulting bound on the error was too large. Once "noise" is recognized, one merely accepts the estimate for the given subinterval, whether or not the local error requirement is satisfied, and issues a warning to the user of the algorithm concerning this fact.

The assignment of error requirements for subintervals is an even knottier problem. Recall that the final estimate EST for $\int_a^b f(x) \, dx$ should satisfy

$$\left| \int_a^b f(x) \, dx - \text{EST} \right| \leq \max\left(\text{AERR}, \text{RERR} \left| \int_a^b f(x) \, dx \right| \right)$$

where AERR and RERR are given positive numbers. This is somewhat at variance with current practice, which would replace

$$\text{RERR} \left| \int_a^b f(x) \, dx \right| \quad \text{by} \quad \text{RERR} \int_a^b |f(x)| \, dx$$

Although the latter is easier to check for and satisfy than the former, it strikes me as an undesirable requirement from the user's viewpoint. A user specifying a certain relative error requirement usually has in mind a certain number of correct significant digits. Since he usually does not know $\int_a^b |f(x)| \, dx$, he will find it impossible to express his wish for k correct digits in terms of the input RERR, whenever the integrand changes sign in $[a, b]$.

Given the error requirement as stated above, use of an adaptive scheme makes it necessary to translate this requirement into an error requirement for the estimate to be obtained in the various subintervals. This is usually done by prorating the requirement. Thus with

$$E = \max\left(\text{AERR}, \text{RERR} \; \middle| \; \text{current best estimate for } \int_a^b f(x) \, dx \; \middle| \right)$$

the absolute error tolerance ε for the current subinterval of length h is simply set at

$$\varepsilon = Eh/|\, b - a\,|$$

This procedure is found to lead to overly strict requirements for "very small" subintervals. It might be better to make the requirement slightly tighter for "large" subintervals, thereby gaining the freedom to deal with less care with possibly difficult "small" but less important subintervals. In any event, a rigorous procedure such as prorating ignores completely the possibility that errors in various subintervals may well cancel out, hence leads often to very conservative appraisals of the actual error committed. This seems to be the price one has to pay for using an adaptive scheme.

In conclusion, I wish to acknowledge the many discussions I had with Robert E. Lynch on cautious integration schemes. I acquired from him the conviction that an error estimate for the approximate answer to a given numerical problem should be accepted only if tests show the assumptions on which the estimate is based to be consistent with all the available numerical evidence about the problem.

REFERENCES

Davis, P. J., and Rabinowitz, P. (1967). "Numerical Integration," Blaisdell, Waltham, Massachusetts.

Golomb, M., and Weinberger, H. F. (1959). Optimal approximation and error bounds, *In* "On Numerical Approximation" (R. Langer, ed.), pp. 117–190. University of Wisconsin Press, Madison, Wisconsin.

Hull, T. E. (1968). The effectiveness of numerical methods for ordinary differential equations, *Stud. Numerical Analysis* **2**, 114–121.

Lynch, R. E. (1967). Generalized trapezoid formulas and errors in Romberg quadrature, Blanch anniversary volume, Aerospace Research Laboratories, Office of Aerospace Research, United States Air Force, pp. 215–229.

Lyness, J. N., and Ninham, B. W. (1967). Numerical quadrature and asymptotic expansions, *Math. Comp.* **21**, 162–178.

Lyness, J. N. (1969). Notes on the adaptive Simpson quadrature routine, *J. Assoc. Comp. Mach.* **16**, 483–495.

5.14 Experience and Problems with the Software for the Automatic Solution of Ordinary Differential Equations*

C. W. Gear†

STANFORD UNIVERSITY

STANFORD, CALIFORNIA

I. Introduction

In this paper the experience of a small group in the implementation of some mathematical software will be discussed. When we talk about Mathematical Software we are necessarily talking about the total problem-solving process. It is no longer sufficient to consider only the neat theoretical solution of a nicely stated problem; we must examine all the dirty details. It is this attitude which, I think, separates the computer scientist from the mathematician when both call themselves numerical analysts. It is necessary to investigate all phases of the problem-solving process in order to develop adequate numerical methods, not only because of software implications but also because the methods needed are not apparent until the whole process has been understood. We hope to convince you of this by means of an example from our own experience.

For some time we have been concerned with the implementation of program packages for the integration of ordinary differential equations that are designed to be used by the nonprogrammer and the nonnumerical analyst for the reasonably automatic solution of a problem specified in a fairly natural way; that is, in a language that is close to the language usually used to describe the source problem. In this paper we will discuss the objectives of such programs, objectives that have changed during the course of the investigations, and the problems that have been encountered with the implementation and use of the packages.

The three packages that will be discussed are: "An Interactive System for the Numerical Integration of Ordinary Differential Equations" [Gear

---

* Research supported in part by the U.S. Atomic Energy Commission under grant U.S. AEC AT (11–1) 1469.

† Present address: University of Illinois, Urbana, Illinois.

(1966)]; "A Batch System for the Automatic Integration of Differential Equations" [Dill *et al.* (1968)]; and, yet to be completed, "Network Analysis and Simulation System" [Gear (1969)], which have an interactive graphical interface with the user. The first package also represented the first experiment by this group. The second was a derivative (or should we say integral) of it, as it allowed for a broader class of equations (including stiff problems), was far more automatic, and included some symbolic analysis. The third package allows for algebraic equations as well as differential equations, and is an attempt to use an external language that is closer to the natural language of an engineer than the language of differential equations used in the first two packages.

Throughout this paper we use the word "package" to refer to a collection of programs. The word "system" will be used for a set of equations—or the physical system which they model.

A. The Goals of Automatic Software for Numerical Problems

The adjective automatic refers to two distinct phases of the problem-solving process. The first is the automatic analysis of the source language of the problem to reduce it to the object language required as input by the numerical package. This might be the compilation of subroutines for a numerical integrator from the differential equations written in an algebraic language, or the analysis of a network to derive such equations and then to compile them. The second is the automatic selection of the method to be used. This might be as simple as the automatic selection of step size if the class of methods is restricted to a single integration technique with variable step size, or it could involve the selection from among a large class of methods depending upon various characteristics of the problem that may be determined in earlier stages of analysis.

The goals of the packages we are concerned with are the same as we believe the goals of any large numerical package should be.

Goal I They should provide one of three answers to the user: (a) an answer within the requested accuracy; (b) the statement that the desired accuracy cannot be achieved, together with whatever numerical information was obtained an an estimate of its accuracy; or (c) an incorrect answer without an indication that the desired accuracy was not achieved (this cannot be avoided in some problems such as integration).

The last case should not occur more frequently than is likely to happen if other available and practical approaches to the problem solution had been used. (One can make a statement like that, happy that it can never be verified in practice. The important point is that either fully automatic software or analysis by users will fail sometimes. Not all problems could, or should, find their way to the desk of a numerical analyst.)

Goal 2   They should provide a user interface in a problem-oriented language—that is one in which the details of the problem can be clearly stated, and which is suitable for automatic analysis of the problem to determine the method to be used. The user should not have to provide any information such as the method of solution which is not essential to the statement of the problem.

Goal 3   They should solve the problem in a time that is not large compared to the best times that might be achieved by a direct approach involving programing by someone familiar with current, but not innovative, techniques. (Obviously, a package is not going to invent new approaches to problems, although it may exercise a better selection from among existing techniques.) The solution time can afford to be a large multiple of the best method for short problems, but as the problem becomes more and more complex, the package should not be too much slower than of the best method.

Several things are immediately evident when we consider a relatively sophisticated automatic program:

(1) It will certainly perform badly on simple tasks compared to a simple method because it will spend too much computer time selecting a method compared to the trivial amount required for the solution.

(2) Except for trivial problems it will not be possible to guarantee results because the state of the art of numerical analysis does not yield a priori error bounds that are practical; that is, they are so large that an unreasonable amount of computation would be necessary to achieve sufficient guaranteed accuracy.

(3) It will not be able to compete in computation time with a program that has been hand tailored for a much smaller class of problems.

However, these drawbacks are not necessarily serious when we consider the alternatives open to users. Simple methods must typically be coded directly by the user. Although not a difficult task, it requires additional skills of the user (how to program, knowledge of when the method is adequate, etc.), so that the increased total solution time offsets the decreased computer time (which is not a large item anyway). Most users will not be able to derive more satisfactory error bounds for a sophisticated problem than the package, so the user may go away with the wrong answers in either case. Finally, packages that are hand tailored for small classes of problems have a limited application (as we have found out) and tend to limit the user in the scope of his investigation. Woe to the user who has a problem beyond the scope of the available ones. The time loss in this case may far outweigh the gains on other occasions.

Problems involving ordinary differential equations were chosen solely because of this writer's interest in them; but it is hoped that the experiments that are being performed will provide some guidelines for the implementation of packages in other areas. In many ways, the situation in numerical packages is ideal with respect to the attainment of goal 3. Typically, the source need only be analyzed once prior to the numerical reevaluation of the resultant object language many times. This is particularly true in differential equations, but is also valid for algebraic and matrix problems. Because of this, a fairly large amount of programming and computer time can be invested in the initial analysis in order to determine the best way of performing the numerical tasks. In ordinary differential equations, the derivatives may be evaluated on the order of 10,000 times. The compiler can take 10,000 times as long to compile its object code as to execute it (which time would allow a large amount of optimization of the object code) and the resulting package would still only run twice as slow as the best code written by a programmer.

B. Techniques That Must Be Considered in Automatic Numerical Software

One of the most serious problems facing the implementer of automatic software is the fact that successful packages will have to employ techniques from many fields. Many earlier packages have suffered from the problem that they were designed by persons expert in some of the areas but relatively nonprofessional in others. For example, one only has to look at the number of network analysis and simulation programs that used Runge-Kutta as the standard integration technique. (Runge-Kutta was described to me as an "industry standard" by one simulation professional. Indeed, many new integration techniques are compared with Runge-Kutta—and they naturally look very good on all but the most trivial of problems.)

Efficient and flexible packages are going to have to call on methods from the fields of

(1) symbolic analysis (e.g., differentiation, simplification, etc.),
(2) artificial language translation (particularly compiling optimized object codes),
(3) "system programming" (providing good interactive interfaces to the user and making full use of the file storage media provided by the system),
(4) data structure theory (to store and manipulate the source language and its intermediate forms in an efficient way; e.g., sparse techniques should be used to handle large arrays with that property),

in addition to the use of sensible numerical methods and the application of techniques suitable for particular problems; e.g., graph theory for network analysis.

Practitioners of numerical analysis typically have not been involved in this sort of large-scale endeavor, although there is every reason to believe that they would be as blind to the importance of the other areas to the complete solution as we have complained that others are to the importance of numerical analysis. In the past we have frequently reduced the problem to some canonical form and then produced good methods and theorems for that form. For example, all papers or books that discuss the integration of ordinary differential equations look at the problem $y' = f(y, t)$ or minor variants. While it is true that almost all initial-value problems of interest can be formally put in such a form, it ignores the fact that in some practical applications as much human or computer resources may be spent on reducing the problem to such a form as are spent in the solution. This is the problem we will examine in detail.

II. The Implementation and Use of Automatic Packages

The three packages we are going to discuss form a progression starting from a relatively simple system for "textbook"-style problems to what is planned as a very comprehensive system for a class of application problems. In this section we will give a very brief description of the packages.

A. An Interactive Terminal Package

This package described in Gear (1966) allowed the user to input a series of differential equations, each typed in a Fortran-like form as

$$Y_i'' = \text{(expression involving other Y's, their derivatives, the independent variable T, and other parameters)}$$

i could range from 0 to 9 and differentiation was indicated by the prime operator, up to fourth derivatives being allowed. The equations had to be in a canonical form in which the highest derivative of each variable used appeared only on the left-hand side. After specifying the equations, the user was asked a series of questions in which he had to specify the initial values, where to print, what to print or plot, and when to stop. In addition, he had to either specify the step size (fixed) or give an error parameter, and he had the option to specify the order of the method to be used. (If this was not specified, eight, the maximum available, was used.) The motivation behind the package was that the user would be able to observe the results of two or three choices of orders and errors and probably extrapolate from that to select the method parameters that would provide reasonable results.

The package consisted of a compiler and a numerical integrator plus a small section concerned with the interaction. It should first be admitted that it never progressed beyond the status of a demonstration program, partly because it was written in the assembly language of a one-of-a-kind computer (the ILLIAC II) with limited use outside of our own group and partly because there were few problems that would fit within its constraints. Consequently, we should be careful about extrapolating too much from the experience, which was mainly a result of demonstrating it to visitors and contracting agencies! However, even in its relatively short operating life, one noticed a tendency of the user to avoid the choice of parameters and use the default conditions if they were available. Since it was possible to repeat an integration with a different error control parameter by simply typing "ERROR = nnnnn,REPEAT", the easiest way (from the user's point of view) was to execute two runs with different error requests and compare the results. Unless the integration time was particularly lengthy there was no incentive to experiment with different orders. For a general problem it is most unreasonable to expect the user to have any idea what is a reasonable order or step size.

B. A FULLY AUTOMATIC PACKAGE

The second package, called ODESSY [described in Dill et al. (1968)], was implemented to run as a batch program on any (large) IBM 360. The choice of batch was not fundamental to the approach, and the package could readily be adapted to be run in an interactive environment. However, it is almost impossible to find two installations with the same requirements for an interactive package, and the desire was to make the package available to as many installations as possible to get experience with both the approach and the qualities of the numerical techniques involved. Secondly, from what little experience had been gained from the first package, it was felt that the user did not, in fact, want to interact in order to control the method of integration, but only in order to vary the problem being solved. By restricting ourselves to a batch environment, we had to face up to the problems of making the choice of all the parameters completely automatic, and to find ways to handle all eventualities that could reasonably be due to a bad choice of method on our part.

The input to this package was not basically different from the first. A system of up to 40 differential equations (now restricted to be of first order because of some numerical considerations) could be input symbolically in a

Fortran-like form. However, they still had to be in the canonical form with the derivative on the left-hand side, and the right-hand sides derivative-free. The remaining input consisted of only that essential to the statement of the problem; that is, the initial values, where and what was to be printed, when the integration was to stop, and an error control parameter. The numerical method used was still a multistep type similar to that in the first package, but it now used a set suitable for stiff equations [Gear (1968b)] rather than the Adams methods used before. The order was now variable and selected by the method dynamically along with the step size to try and reduce the computation time [Gear (1968a)]. Because stiff equations were to be considered, it was necessary to compute approximations to the Jacobian matrices of the equations. Since the input was symbolic, this was done by symbolic differentiation during the compilation process. A two-pass compiler produced two machine language subroutines, one of which evaluated the derivatives from the input equations, the other evaluated the Jacobians. These were both called by the integration program repeatedly during execution, so some attention was paid to producing code that made efficient use of the four floating point registers available in the IBM 360.

The program made an initial attempt to use the interval specified for printing as the basic step, since this was the only scale available in the time direction. It would then reduce the step as necessary to keep its estimate of the error within the desired bound. In order to decide on the step size and order of method to use it would calculate estimates of various derivatives from the solution. One problem that was observed was that on occassions these estimates would not behave as might be expected. For example, if a third-order method was currently being used, estimates of $h^p y^{(p)}$ were computed, where h was the current step size and p was 3, 4, or 5. If a step failed because the estimated error based on these values was too large, it was repeated with a smaller h or different-order method. The new estimates of $h^p y^{(p)}$ should be scaled by the pth power of the reduction in h (approximately). On occassions they did not behave at all like that, so that the error test again failed. In that case, one more try was made, and then it was assumed that the information being saved about the earlier values of the solution had accumulated so many errors that it was unreliable. This information was discarded, and the integration was started afresh from the most recently computed step. This would frequently get through the region that gave rise to the problem with no apparent ill effects. However, occasionally the problem would recur several times, at which point we stated to the user that the problem was too difficult for the package to solve (which was certainly true, although it left open the question of why).

The only explanation we can suggest for this type of behavior is that it is a form of instability that can arise in all of the extra data that are accumulated about the solution and the behavior of the method. A numerical process for step-by-step computation, such as an integration program, which in one step processes q pieces of data to produce q new pieces of data, will have to concern itself with the eigenvalues of a q by q matrix. (The error amplification matrix for small perturbations.) A fixed-step, fixed-order multistep method can be completely analyzed, at least for linear problems [Henrici (1962), Chap. 5], but an automatic method is saving and modifying many additional pieces of data; for example, the order, the step, and the error estimates. Each of these can give rise to another eigenvalue that, if it gets out of the unit circle, can cause instability problems. One possible solution if instability is sensed is to perform an operation for which most of the eigenvalues are zero: namely, setting most of the information to zero!

C. An Interactive Simulation and Modeling Package

A brief preliminary description of the purpose of this package is available in Gear (1969). It is currently under development. Whereas the previous packages were intended for any user who had a set of differential equations, or could reduce his problem to such a set, this package attempts to accept the problem in a language closer to the original problem. The penalty of this approach is to restrict its potential class of users, its advantages are that it is no longer necessary that the user perform the initial translation step, a step which both deterred potential users and lost information that might have been useful in the solution process.

The source language consists of a set of "element" definitions and a network that uses these elements, together with the specification of input functions to the network. The output is the time behavior of the network. Elements consist of a number of terminals with associated variables and sets of differential and algebraic equations relating these variables. Elements may be connected to the nodes of a network by their terminals. There are some implied algebraic equations that terminal variables satisfy when they are connected together. These are equivalent to the Kirchoff laws. The user specifies the number of terminals, the number of variables, and the relations that they satisfy on each element so that there are no restrictions such as limit typical network analysis or simulation programs. The network that he subsequently draws using instances of these elements is equivalent to a system of differential and algebraic equations which must be integrated.

The stages in this process are as follows:

(1) Process the element definitions to check for well formedness.

(2) Process the network definitions to generate the equations relating the terminal variables at the nodes, and to generate sets of equations for each

instance of each element used. (This is a recursive operation because elements may be defined by networks.)

(3) Perform symbolic manipulation to: (a) eliminate uninteresting variables and equations (e.g., those that are constant, or trivial functions of other variables); (b) categorize the remaining variables according to the manner in which they are used (the numerical program needs to know which are only used linearly, which are functions of time only, and which appear with derivatives or nonlinearly).

(4) Compile a set of subroutines that evaluate the residuals of the equations that are supposed to be zero.

(5) Compile code that affects the premultiplication of a vector by the inverse of the Jacobian matrix. There are two sections to this code: the first is used whenever the values of the Jacobian change; the second performs the actual premultiplication using values generated by the first. A sparse factored LU method is used [see Tinney (1967)].

(6) Execute the numerical integrator that calls on these sets of compiled code.

III. Problems and Some Solutions

The problems can be put into three classes: usage, numerical, and software. The attempt to solve a problem in one area has usually caused problems in the others, so in this section we will attempt to trace the development of the latest package as a series of reactions to problems that arose.

A. Usage

One factor that limited the utility of the first two packages was that they required the input to be in the "text book" form for an ordinary differential equation; namely, $y' = f(y, t)$, where y and f could be vectors. (A certain number of additional variables were allowed and could be computed by arithmetic assignment statements during the integration, but that feature did not materially improve the range of application.) Because of this, two classes of potential users were excluded: (1) those which needed to execute complex procedures in order to find the derivatives, and (2) those which could only convert their problem into the standard form with considerable effort. Even among those problems that were already in the right form, we found that the limitations on the number of variables and the fact that the package would not interface to other packages prevented some interesting problems from being run. In fact, the major successes of the numerical part of the package, which was the first version of our automatic order and step control package for-stiff equations, were in situations in which the outer layers were stripped off and replaced with code hand tailored to the situation. For example, the front end has been replaced in one installation with a program that accepts chemical reaction equations and produces the corresponding differential equations. It was these factors that caused us to consider moving closer to the source language in the latest package. The attempt now is to develop what might be considered a compiler–compiler system for problem-oriented languages; that is, a system which will allow one user to specify the basic units (elements) so that another user can model his problem.

B. Numerical Problems

The system specified by the user will contain differential and algebraic equations. In future expansions it is also desirable that arbitrary programs (in any procedure-oriented language locally available) be allowed both to specify functions used in the equations and to provide overall control of the sequencing of a series of simulations. For some time we attempted to set restrictions on the definition of the equations associated with elements such that the equations of the system were already in an explicit form with the derivatives appearing only on the left-hand side. It soon became apparent that this was not reasonable. If, for example, one of the elements was defined with the properties of a capacitor, it would have generated a differential equation equivalent to $v' = i/c$, where v is the voltage across it. (It might not appear in this form, but it would be equivalent.) If two such elements were connected in parallel, the result would be the system of three equations (or equivalent)

$$v_1' = i_1/c_1, \qquad v_2' = i_2/c_2, \qquad v_1 = v_2$$

It is evident that we cannot view the first two equations in the usual way of differential equation solvers since they do not determine v_1 and v_2; rather the three of them determine the two equations

$$v_1' = i_1/c_1, \qquad i_2 = c_2{}^*i_1/c_1$$

Packages designed to analyze electrical networks have no difficulty with this problem since they are programmed to perform such eliminations. However, in order to do this it must be known what type of elements are allowed at the time they are programmed. The purpose of the package we are building is to allow arbitrary elements.

We next attempted to find ways of eliminating the redundant variables from the system of equations to put them into an explicit form. That failed for several reasons. One is that the equations could be nonlinear in the derivatives (although this could be disallowed if necessary). This would mean that the derivatives could only be calculated in terms of the function's values at integration time by a time-consuming iterative process. A second is that even for linear equations the sparseness would be destroyed. Since the hope is to handle large networks of elements which are only loosely interconnected (as usually happens), it is important to be able to exploit the sparseness at integration time. A third reason is that in nonlinear networks the coefficients of the derivatives, even if they appear linearly, may be changing, so there is no guarantee that a numerically stable inversion to put the equations in explicit form can be performed before the size of the coefficients is known. We were fortunate that we stumbled onto the solution to these problems while considering another aspect of the numerical problem discussed below.

We had also to be concerned with the solution of simultaneous algebraic equations during the course of the numerical integration. At the time that it was planned to convert the equations into the form

$$\mathbf{y}' = \mathbf{f}(\mathbf{y}, \mathbf{z}, t), \qquad \mathbf{0} = \mathbf{g}(\mathbf{y}, \mathbf{z}, t)$$

we were considering ways in which the algebraic variables $\mathbf{z}$ could be determined efficiently from the functions $\mathbf{g}$ when the values of $\mathbf{y}$ were known. A Newton-like method was being considered in which the first approximation was to be predicted from the values at previous time steps. A reason for this (in addition to the fact that it had some hope of being reasonably fast) was that it resembled the technique to be used for the differential equations so that the same code could be used for both. Because most networks give rise to stiff equations, methods suitable for stiff equations were to be used. It had been shown [Gear (1968b)] that a multistep predictor–corrector, in which the corrector equations are solved by a Newton-like method, could be put in the form

$$\mathbf{y}^i_{n,(0)} = B\mathbf{y}^i_{n-1} \tag{1}$$

$$\mathbf{y}^i_{n,(m+1)} = \mathbf{y}^i_{n,(m)} - \mathbf{c} \sum_j W_j{}^i F^j(\mathbf{y}_{n,(m)}) \tag{2}$$

where $\mathbf{y}_n{}^i$ is the set of information saved for the ith component of the system of equations at time t_n, and F^i represent the differential equation written in the form

$$F^i(\mathbf{y}) = \mathbf{y}^i|_1 - hf^i(\{\mathbf{y}^j|_0\}, t)$$

It is assumed that the zeroth component of $\mathbf{y}_n{}^i$, written $\mathbf{y}_n{}^i|_0$ (numbering of vectors starting from 0), is an approximation to the value of y^i at t_n, while $\mathbf{y}_n{}^i|_1$ is an approximation to hy'^i at t_n. The matrix B represents the extrapolation process which predicts the values at t_n from the information at t_{n-1}. It involves no derivative evaluations. The correction process is iterative. The vector $\mathbf{c}$ depends on the coefficients of the method while the matrix W is given by

$$W = [c_1 - hc_0 \, \partial f/\partial y]^{-1}$$

where c_0 and c_1 are the first two components of the vector $\mathbf{c}$. If the matrix W is calculated exactly at each corrector iteration, it corresponds to using the Newton method to solve the corrector. If W is taken as I/c_1 we get the conventional corrector iteration (which will not converge for stiff equations), while in-between cases correspond to various Newton-like iterations. We have been keeping W fixed until the order of the method changes (which changes $\mathbf{c}$) or until corrector convergence could no longer be achieved in three steps. It was then reevaluated in terms of the most recent function values available.

It is also apparent that the solution of the algebraic equations can be handled by similar means. For variables that only appear algebraically, the vector $\mathbf{y}_n$ still represents that information being saved from step to step. The matrix B is the extrapolation process based on polynomial extrapolation in our case, while the corrector iteration is the Newton or Newton-like iteration to solve the algebraic equation. The function F^i is

$$F^i(\mathbf{y}) = g^i(\{\mathbf{y}^j|_0\}) = 0$$

while W is given by

$$W = [c_0 \, \partial g/\partial y]^{-1}$$

for the Newton iteration. The vector $\mathbf{c}$ can be chosen according to the components of $\mathbf{y}_n$. Thus, if $\mathbf{y}_n = (y_n, y_{n-1}, y_{n-2}, \ldots)^T$ (the previous history), then $\mathbf{c}$ is $(1, 0, 0, 0, \ldots)^T$ for the Newton iteration.

The information that is saved from step to step can be subject to a linear transformation without changing the characteristics of the method (although it will change both the round-off properties and the computational time). Since the purpose of the saved information is to perform a polynomial prediction of the values at the next step in the case of both differential variables and algebraic variables, we could save any set of information that serves to determine that polynomial. This could be the previous function values as suggested for the algebraic variables, or a combination of these and their derivatives at previous points as is conventionally done in predictor–corrector methods, or the values of the derivatives of the polynomial at the current time as is done in our package at the moment. (These may not be the most efficient computationally for large systems, although they do have about the best round-off properties and make the estimation of derivatives for error control most easy.) If both the algebraic and differential processes are converted into this form by a linear transformation, the effect is to change $\mathbf{c}$. Thus if the linear transformation to take the saved information $\mathbf{y}_n$ into the set of derivatives

$$\mathbf{a}_n = (y_n, hy_n', h^2 y_n^{(2)}, \ldots, h^k y_n^{(k)}/k!)^T$$

is given by

$$\mathbf{a}_n = T\mathbf{y}_n$$

the method can be obtained by premultiplying Eqs. (1) and (2) by T to get

$$\mathbf{a}^i_{n,(0)} = TBT^{-1}\mathbf{a}^i_{n-1} \tag{3}$$

$$\mathbf{a}^i_{n,(m+1)} = \mathbf{a}^i_{n,(m)} - T\mathbf{c}\sum_j W^i_j\, F^j(\mathbf{a}_{n,(m)}) \tag{4}$$

If the method used for the differential equations is one of the stiffly stable methods given in Gear (1968b), then the $T\mathbf{c}$ obtained after this transformation is identical to the $T\mathbf{c}$ obtained after the transformation of the algebraic equations to the similar form. This is because the limit of the stiff equation

$$y' = \lambda f(y, t)$$

as λ goes to $-\infty$ is the algebraic equation $0 = f(y, t)$, and the stiff methods were chosen to have all zero parasitic eigenvalues in this limit. The Newton method for algebraic equations gives new values that do not depend on the old values once the iteration has converged; so viewed as a step-by-step process it will also have zero eigenvalues. Since it can be shown that the eigenvalues uniquely determine the $\mathbf{c}$ if the matrix B is such that the extrapolation is of maximum order [Gear (1967)], the vectors from the two processes will be identical.

Since the methods for the two processes are identical, we do not care whether or not we are solving differential or algebraic equations. It now only remains to remark that the iterative process (4) converges to a solution of

$$F(\mathbf{y}_n) = 0 \tag{5}$$

if it converges at all, to realize that we do not need to put the equations into an explicit form. Since the solution of iteration (4) is such that

$$\mathbf{a}_n = \mathbf{a}_{n,(0)} + \mathbf{l}b \tag{6}$$

where $\mathbf{l} = T\mathbf{c}$ and b is a scalar, we ask what is the solution of Eq. (5) by a Newton method, which is of the form (6), and find that it is given by iteration (4) with W redefined as

$$W = [\partial F^i/\partial y^j l_0 + \partial F^i/\partial y'^j l_1]^{-1}$$

We are then solving algebraic equations for those variables y^j for which $\partial F/\partial y'^j = 0$. Thus, all that the integration package needs is a subroutine which will evaluate the residuals of the equations F^i, be they differential or algebraic, and a subroutine which will premultiply these residuals by the matrix W.

Since W is the inverse of a sparse matrix, this inverse must not be calculated explicitly; rather a factored form is used. The original plan was to symbolically differentiate the original equations to develop this matrix, as was done in ODESSY. There were two reasons for this. One was a reluctance to rely on numerical differentiation; the second was a plan to attempt to do part of the matrix inversion symbolically, since many problems involving large networks give rise to many elements that are small integers like 1 and 2 or simple symbols for which exact (symbolic) arithmetic can be performed. Currently we do not plan any symbolic differentiation for several reasons. It was always realized that it was not going to be possible to symbolically differentiate a function that was user defined by a section of code or graphically defined by a curve, so it was always going to be necessary to resort to numerical differentiation. Secondly, it was found in some experiments on the 360 that single precision in the matrix W was adequate for double-precision elsewhere, so that relatively inaccurate numerical differences were sufficient. (In fact, this was found when we prepared a temporary version using numerical differentiation to avoid preparing the symbolic differentiator in the initial package. Saved by serendipity!)

We are now starting tests of code which will determine the sparseness properties of the Jacobians numerically. The aim is to determine by numerical differentiation which elements of the Jacobian are zero, which are small integers, and which are arbitrary. The functions will be evaluated for somewhat random data (none of the variables will be integral) and then differenced. Information from an earlier symbolic stage indicates which variables are known to appear only linearly, which will tell us which of the arbitrary derivatives are constant.

This information is fed to a sparse matrix inverter which compiles code to generate the factored form of the inverse from the elements of the matrix and code to multiply F by that inverse. It uses the information about the nature of the elements to determine a pivot strategy. The strategy selects pivots to try and minimize the number of additional nonzeros introduced. When several choices are equally good, it selects the pivot that leads to the least arithmetic at execution time. Thus, constants are favored over variables, while small integers are even better.

We have yet to face up to the problem of selecting pivots also for numerical stability. In nonlinear problems, the Jacobians will change and it is likely that a single pivot strategy will not be stable over the whole range. A proposed solution is to compile tests into the object code that detect unacceptable instability and invoke a new compilation of the inversion code using the latest numerical values to help in pivot determination. Fortunately, great numerical accuracy in W is not important. In many situations 10% error may be acceptable.

C. SOFTWARE PROBLEMS

Software maintenance has caused the greatest problems in the second package ODESSY. Because it was an experimental package, little effort was put into the finishing touches that should probably take more time than planning, coding, and debugging. Consequently, when the original programmers left, modification or correction became an increasingly difficult task. This is no news to anybody associated with large codes. The problem is to find a way to develop packages such as this experimentally in an environment in which each new package does not require starting from scratch. One approach is to use an automatic software writing package such as a compiler–compiler. However, the range of types of software to be produced currently preclude such an approach. To our knowledge there are no automatic packages that can produce code for applications as diverse as numerical analysis, symbolic manipulation, and data structure manipulation (as is involved in the input–output sections of the simulation package which interfaces to a display).

The maintenance problem also affects the ability of a research group (or production group) to react quickly to new approaches which require adding new programs to an existing package. The solution will almost certainly have to be to demand greater modularity of software components and to be less flexible in the interfaces between them. Since it is not possible to predict the range of new ideas that may lead to new packages to be added to the system, we feel that an important requirement will be to make the data that is exchanged at interfaces contain its own description. If each subpackage checks the description of the data it is handed and generates a description of its output, it will at least be possible to check for compatibility between new and old subpackages dynamically.

IV. Future Plans for the Package

The development of the current package has been (and still is) a dynamic process, with the objectives and the methods changing as implementation proceeds. It is sometimes difficult to decide whether or not to stop all changes in order to get a working package as quickly as possible, or whether to pursue all new approaches that suggest themselves. If the software were to be better modularized (which would require making some decisions on standard data description formats) we might be in a better position to react to changes. However, there is some reluctance to start making such decisions until one version of the package has been completely implemented and many of the potential problems are understood.

Many of the parameters used in the integration package were either heuristically obtained, or based on some tests of various cases. An example is the choice of the number of corrector iterations that are used before it is decided that convergence has not occurred. Three are currently used, based on some numerical experiments (Ratliff, 1968). However, this choice depends on the ratio of the time taken to evaluate the residuals of the equations to the time taken to evaluate and invert the Jacobian. In a large package it would be plausible to perform an execution of each of these steps in order to time them for the system being studied, and then to choose the appropriate value of the parameter.

As the package evolves, it will allow for a number of different types of analysis (for example, eigenvalue determination of the Jacobian, which is a small signal response analysis, or steady state determination, which is the solution of the algebraic equations obtained when all of the derivatives are set to zero). Consequently, there will be much more software involved than may be needed for any particular application. Furthermore, in a typical operating environment, the user will be calling on a standard set of elements that have previously been defined. It would be more efficient to perform the analysis of these elements once at the time they are defined in order to produce a package for the particular user that is stripped of the unused features (the compiler–compiler approach).

In the present version of the numerical section, the variables that only appear linearly are handled separately because it can be shown [Gear (1970)] that no additional corrector iterations are required when the prediction step is omitted. Consequently the vector y_n of saved information for these variables contains only one component. An alternative approach is to eliminate these variables earlier by solving an appropriate system of linear equations. This will usually reduce the sparseness so its net effect will be to increase solution time. However, for problems with sufficiently few nonlinear or differential variables, the resulting system would be so small that sparseness would be unimportant. In future versions we would like to make the choice of performing this elimination automatically based on timing measurements.

REFERENCES

Dill, C., Ellis, C. A., Gear, C., and Ratliff, K. (1968). The Automatic Integration Package for Ordinary Differential Equations. Univ. of Illinois at Urbana, Dept. of Comp. Sci. File No. 779.

Gear, C. (1966). Numerical Solution of Ordinary Differential Equations at a Remote Terminal, *Proc.* 1966 *ACM Nat. Conf.*, pp. 43–49.

Gear, C. (1967). The Numerical Integration of Ordinary Differential Equations, *Math. Comp.* **21**, 145–156.

Gear, C. (1968a). The Control of Parameters in the Automatic Integration of Ordinary Differential Equations, Univ. of Illinois at Urbana, Dept. of Comp. Sci. File No. 757.

Gear, C. (1968b). The Automatic Integration of Stiff Ordinary Differential Equations, Information Processing 68 (Morrell, ed.), Vol. 1, pp. 187–193. North Holland Publ., Amsterdam, 1969.

Gear, C. (1969). An Interactive Graphic Modeling System, Univ. of Illinois at Urbana, Dept. of Comp. Sci. Rep. No. 318.

Gear, C. (1971). The Simultaneous Solution of Differential and Algebraic Equations, *IEEE Trans. Circuit Theory* **CT18** (1).

Henrici, P. (1962). "Discrete Variable Methods in Ordinary Differential Equations." Wiley, New York.

Ratliff, K. (1968). A Comparison of Techniques for the Numerical Integration of Ordinary Differential Equations, Univ. of Illinois at Urbana, Dept. of Comp. Sci. Rep. No. 274.

Tinney, W. F. (1967). Direct Solution of Sparse Network Equations by Optimally Ordered Triangular Factorization, *Proc. IEEE* **55**, 1801–1809.

5.15 Comparison of Numerical Quadrature Formulas*

D. K. Kahaner

LOS ALAMOS SCIENTIFIC LABORATORY

UNIVERSITY OF CALIFORNIA

LOS ALAMOS, NEW MEXICO

I. Introduction

Algorithm testing and certifying has recently been the subject of several attempts at definition and unification [Fosdick and Usow (1969); Lyness (1970)]. Until guidelines are established, potential algorithm users must be content with individual experiences reflecting other people's preferences and requirements. This is particularly true with subroutines which require user-provided function subprograms. In this paper we compare eleven different one-dimensional numerical quadrature codes for speed and accuracy on a particular collection of integrands. The goal is selection of a routine for inclusion in a library package for use by many other scientists. Experience at Los Alamos has shown that most users would prefer a single quadrature code that is both simple to use and reliable. Programs for single integrals usually run quickly (a few seconds or less on a CDC 6600)[1] and users are willing to sacrifice maximum speed in favor of the considerations above.

If an integral is troublesome, either because it is being evaluated unreliably or because its evaluation adds too much time to the total program, an "expert" will probably be consulted for something "better" than the standard library subroutine. It is at this point that more complicated codes should enter. Faced with a large selection of quadrature codes in a program library, most users are reluctant to choose those with long calling sequences, especially if that involves setting unfamiliar parameters on which convergence depends. Consequently, in designing a test of quadrature codes, this author placed

* This research was supported by the U.S. Atomic Energy Commission under Contract W-7405-ENG-36.

[1] Machine accuracy parameter is 0.71×10^{-14}.

230   D. K. KAHANER

strong emphasis on reliability and simplicity. Readers whose goals differ from these should be particularly wary of the author's conclusions.

II. Quadrature Codes

Eleven quadrature codes were tested. All the codes were written in Fortran IV. Some are listed in Appendix II, and references to the others are provided. All the codes were written as Fortran functions and all contain as arguments the interval end points, the name of the function to be integrated (user provided as a Fortran function), a parameter indicating the accuracy[2] required in the approximate integration, and an output parameter giving the number of function evaluations required by the program to estimate the integral. Several of the codes provided by others to the author have additional parameters. The utility of these was not tested. From the point of view of the author all these codes have the same simple calling sequence mentioned above. Consequently some of these routines may not be providing maximum possible information, either because some output parameters were suppressed or because some input parameters have been fixed in the code to "nominal" values. All of the codes have been modified by this author to exit when the number of function evaluations becomes excessive; in this case, 5000 or more. What follows is a short description of each code and reference to its appearance in the literature.

G 96: This is a standard 96-point Gaussian quadrature code. Consequently the input parameter indicating required accuracy is not used.

ROMB, HAVIE, SHNK: These codes are all based on extrapolation from the trapezoidal rule. ROMB is a standard Romberg quadrature code [Bauer *et al.* (1963)]. HAVIE is an improved version with better error estimation characteristics [Havie (1966); Kubik (1965)]. SHNK is a nonlinear extrapolation based on the ε-algorithm of Wynn (1956). In all these codes the number of function evaluations for the trapezoidal rule is 2, 4, 8, 16, ...

SIMPSN, QNC7, QUAD, SQUANK, GAUSS, QABS, RBUN: These are all examples of iterative-adaptive codes [Davis and Rabinowitz (1967)]. SIMPSN is the standard Fortran version of adaptive Simpson's rule [Davis and Rabinowitz (1967)]. SQUANK is an improved version [Lyness (1969, 1970)] with several interesting innovations, including some protection from round-off error.[3] QNC7 and QUAD are patterned after SIMPSN with two major differences: (1) interval bisecting rather than trisecting is used, and (2) Simpson's rule is replaced by higher-order Newton–Cotes formulas,

---

[2] Absolute accuracy.
[3] The author wishes to thank Dr. J. Lyness for the opportunity to use his code.

seven points in QNC7, and ten points in QUAD [Kahaner (1969)]. QABS[4] [O'Hara and Smith (1968)] uses a combination of Romberg and Curtis Clenshaw quadrature. GAUSS is an interval bisecting adaptive code that uses five- and seven-point Gauss quadrature to obtain independent estimates of the integral on each subinterval. RBUN is an adaptive code using Romberg integration[5] [Bunton et al. (1969); Bunton (1970)].

III. Functions

Twenty-one functions were used as integrands. They are provided as tested in Appendix II. None of the functions incorporates a variable para-meter, and they are essentially all different, although many have features in common. They were selected as being *representative of those functions used repeatedly at the computer installation most familiar to the author* (*Los Alamos Scientific Laboratory*).

IV. Tests

Each quadrature code was asked to integrate each function to accuracies[6] of 10^{-3}, 10^{-6}, and 10^{-9}. The computations were performed on a CDC 6600.[7] Appendix IV contains the detailed results for each function and each quadrature code.[8] Appendix III contains a summary. In summarizing these results a basic decision had to be made on what to do with codes that returned rapidly, but with inaccurate answers; e.g., not within the required tolerance. To simply omit them would, in the author's opinion, underemphasize their volatility, whereas including them in unmodified form was probably in-

[4] The author wishes to thank Dr. H. O'Hara for the opportunity to use his code (called SPLITABS by O'Hara).

[5] The author wishes to thank Mr. W. Bunton for the opportunity to use his code. In the form given by Bunton (1970) this code is known as ROMBS. We have changed its name to emphasize some differences in our version. The most important change was the requirement that three input parameters in ROMBS be fixed in our test code. These are

$$HSTAR \equiv (B - A)/4$$
$$HMIN \equiv HSTAR * 1.E\text{-}10$$
$$HMAX \equiv HSTAR$$

These parameters control the starting, minimum, and maximum step size.

[6] (See Footnote 2.)

[7] (See Footnote 1.) Storage requirements. The extrapolation codes (ROMB, HAVIE, SHNK) could be coded to run in under 500 words. The shortest adaptive code was GAUSS (254 words) which saves no function evaluations. Of the other adaptive codes, RBUN used 521 words, the others about twice that. G96 needed but 246 words.

[8] The times are in seconds and represent the average of five repetitions. Accuracy is to thousandths of a second.

correct too. In such cases it was decided to reset the time and function evaluation count to be greater than comparable values for any code that integrated correctly. This rather arbitrary process tends to penalize routines for being unreliable. In any case this did not amount to a large fraction of the cases tested, except for the 96-point Gauss code, which has no provision for reducing the error, and for SQUANK, which liked to attribute discontinuities and high oscillation to round-off. In such cases it returned quickly with a value that is "best possible" before round-off sets in. The originator of the code, J. Lyness, states that it should only be used for $C^{(4)}$ functions. Consequently, it does not satisfy the conditions set forth in the introduction. It was included anyway because of its widespread interest.

V. Summary

A. RELIABILITY

The average reliability for all tolerances was

QUAD	90%	SQUANK	79%	SHNK	60%
QABS	89	ROMB	79		
RBUN	87	HAVIE	78		
SIMPSN	87	G 96	76		
QNC7	84	GAUSS	73		

Two routines, SQUANK and RBUN, in addition to being reliable, also were too accurate the fewest number of times. All the codes except ROMB and G 96 integrated about the same number of functions correctly at the highest and lowest tolerance.

B. SPEED

For the tolerance of 10^{-3}, the median time per integral was about the same for all codes except GAUSS and HAVIE, whose times were substantially higher. One routine, SQUANK, integrated the most functions with fewest function evaluations and the most functions in minimum time. As expected, it was inaccurate on most of the functions without four continuous derivatives.

SHNK, while the least reliable, had the smallest median time per integral. This reflects the fact that successful integrations were very fast while unsuccessful ones were slow.

At smaller tolerances the best times were recorded by G 96 and SHNK and the higher-order adaptive routines QABS, QNC7, and QUAD.

The worst times were recorded by ROMB and HAVIE.

VI. Conclusion

Based on the constraints noted in the introduction and the results of this particular test, the author believes that a high-order adaptive quadrature code, such as QUAD, QABS, and QNC7, would be most appropriate for use as a general purpose subroutine in a system library. These routines combine desired reliability with good computational speed.

Two other adaptive routines, RBUN and GAUSS, are somewhat slower, and the latter also seems less reliable.

The two codes ROMB and HAVIE based on Romberg integration do not seem competitive, but perhaps they could be improved by a different number of subintervals for the trapezoidal rule.

Several other routines, such as SQUANK, SHNK, and G 96, seem appropriate for special situations. For example, the last two might be effectively employed in a parameter study where one could test their reliability on an integrand of the same general type that one wishes to examine.

VII. Multiple Integrals

It should be noted that these tests are for single integrals only. The problem of approximation of multiple integrals is one that must be studied separately. At this installation most multiple integrals involve decaying exponentials over semi-infinite regions, and Gauss–Laguerre quadrature has been satisfactory for this type of problem. Other users may also find that general purpose quadrature schemes are inappropriate for the multiple integrals that appear at their computing center.

VIII. Appendixes

Appendix I lists the test functions, the interval of integration, and the exact integral. The latter was computed by using several independent quadrature codes and computing the integrals double precision until all agreed to at least nine figures.

Appendix II lists the quadrature codes not available in the literature or whose coding is not entirely routine[9]. The routines omitted are

SIMPSN	QABS	G 96
SQUANK		ROMB
HAVIE		
RBUN		

---

[9] SHNK has some recent modifications. These do not change the test results significantly, but make the code more reliable.

The first four have been published. QABS is available from the author.[10] The last two are trivial to code.

Appendix III summarizes the detailed output which follows it. For each tolerance 10^{-3}, 10^{-6}, and 10^{-9}, five statistics are computed for each quadrature rule.

(1) Minimum function evaluations: For how many of the 21 functions did each code compute the integral within the specified tolerance with fewer function evaluations than any other routine. Thus, for TOLERANCE = 10^{-3}, SQUANK computed 8 of 21 integrals correctly with fewer function evaluations than the other codes. Ties are allowed.

(2) Minimum time: For how many of the 21 functions did each code compute correctly in the least time?

(3) Median time per integral (in seconds).

(4) Functions integrated to tolerance: How many of the 21 functions were integrated to at least the required tolerance?

(5) Functions integrated to 1/100 tolerance: How many of the 21 functions were integrated to at least 10^{-2} tolerance?

Note comments in Sect. IV.

---

[10](See Footnote 4.)

APPENDIX I

```
          INTEGRANDS USED IN QUADRATURE CODE TESTS
      FUNCTION F1 F(X)
C         INTERVAL IS (0,1)
C         INTEGRAL IS 1.7182818284
      F1F =EXP(X)
      RETURN
      END
      FUNCTION F2F(X)
C         INTERVAL IS (0,1)
C         INTEGRAL IS .7
      F2F =0
      IF ( X .GE. .3 ) F2F = 1.
      RETURN
      END
      FUNCTION F3 F(X)
C         INTERVAL IS (0,1)
C         INTEGRAL IS .66666666667
      F3F =SQRT(X)
      RETURN
      END
      FUNCTION F4 F(X)
C         INTERVAL IS (-1,1)
C         INTEGRAL IS .47942822668S
      F4F =23./50.*(EXP(X)+EXP(-X))-CCS(X)
      RETURN
      END
      FUNCTION F5 F(X)
C         INTERVAL IS (-1,1)
C         INTEGRAL IS 1.58223296373
      F5F =1./(X*X*X*X+X*X+.9)
      RETURN
      END
      FUNCTION F6 F(X)
C         INTERVAL IS (0,1)
C         INTEGRAL IS .4
      F6F =X**(3./2.)
      RETURN
      END
      FUNCTION F7F(X)
C         INTERVAL IS (0,1)
C         INTEGRAL IS 2.
      F7F=0.
      IF ( X .GT. 0. ) F7F = 1. / SQRT(X)
      RETURN
      END
      FUNCTION F8 F(X)
C         INTERVAL IS (0,1)
C         INTEGRAL IS .866972987339S
      F8F =1./(1.+X*X*X*X)
      RETURN
      END
      FUNCTION F9 F(X)
C         INTERVAL IS (0,1)
C         INTEGRAL IS 1.15470066904
      F9F =2./(2.+SIN(10.*3.14159*X))
      RETURN
      END
      FUNCTION F10F(X)
```

APPENDIX I (cont.)

```
C              INTERVAL IS (0,1)
C              INTEGRAL IS .69314718056
       F10F =1./(1.+X)
       RETURN
       END
       FUNCTION F11F(X)
C              INTERVAL IS (0,1)
C              INTEGRAL IS .37988549930417
       F11F =1./(1.+EXP(X))
       RETURN
       END
       FUNCTION F12F(X)
C              INTERVAL IS (0,1)
C              INTEGRAL IS .77750463411
       F12F =1.
       IF ( X .GT. 0. ) F12F = X/(EXP(X) -1.)
       RETURN
       END
       FUNCTION F13F(X)
C              INTERVAL IS( .1,1)
C              INTEGRAL IS .9098645256*E-02
       F13F =100.
       IF ( X .GT. 0. ) GO TO 10
       RETURN
   10  F13F =SIN(100.*X*3.14159)/(3.14159*X)
       RETURN
       END
       FUNCTION F14F(X)
C              INTERVAL IS (0,10)
C              INTEGRAL IS .50000021116E
       F14F =SQRT(50.)*EXP(-50.*3.14159*X*X)
       RETURN
       END
       FUNCTION F15F(X)
C              INTERVAL IS (0,10)
C              INTEGRAL IS 1.00
       F15F =25.*EXP(-25.*X)
       RETURN
       END
       FUNCTION F16F(X)
C              INTERVAL IS (0,10)
C              INTEGRAL IS .4993638C287
       F16F =50./((3.14159)*(2500.*X*X+1.))
       RETURN
       END
       FUNCTION F17F(X)
C              INTERVAL IS (.01, 1)
C              INTEGRAL IS .11213956963
       F17F = 50.
       IF ( X .GT. 0. ) GO TO 10
       RETURN
   10  F17F =(SIN(50.*3.14159*X))**2/((3.14159*X)**2
      1*50)
       RETURN
       END
       FUNCTION F18F(X)
C              INTERVAL IS (0,3.1415927)
C              INTEGRAL IS .83867632338
       F18F = COS( COS(X)+3.*SIN(X)+2.*COS(2.*X)+SIN(2.*X)+3.*COS(3.*X)
      1 +2.*SIN(2.*X))
       RETURN
```

APPENDIX I (cont.)

```
      ENC
      FUNCTION F19F(X)
C           INTERVAL IS (0,1)
C           INTEGRAL IS  -1.
      F19F=0
      IF(X .GT. 1.0E-15) F19F = ALOG(X)
      RETURN
      END
      FUNCTION F20F(X)
C           INTERVAL IS (-1,1)
C           INTEGRAL IS 1.564396443
      A=1.005
      F20F=1./(X*X+A)
      RETURN
      END
      FUNCTION F21F(X)
C           INTERVAL IS (0,1)
C           INTEGRAL IS .2108027354
      HSECF(Z)=2./(EXP(Z)+EXP(-Z))
      F21F=(HSECF(10.*(X-.2)))**2+(HSECF(100.*(X-.4)))**4+(HSECF(1000.
     1 *(X-.6)))**6
      RETURN
      END
```

Appendix II

```
      FUNCTIONQNC7(FOF,Y1,Y2,FERR, KOUNT)
C         NUMERICAL QUADRATURE BASED ON SEVEN POINT NEWTON COTES RULE
C         USED ADAPTIVELY
C         WRITTEN BY DAVID KAHANER, LOS ALAMOS SCIENTIFIC LABORATORY
      DIMENSION FOT(20),F1T(20),F2T(20),F3T(20),
     1F4T(20),F5T(20),F6T(20),F(13),W(4),LEG(20),
     2CXT(20),XMT(20),ART(20),EPST(20),
     3ESTT(20),SUM1(20)
      W(1)=41./840.
      W(2)=216./840.
      W(3)=27./840.
      W(4)=272./840.
      A=Y1
      B=Y2
      EPS=FERR
      DA=B-A
      AREA=1.
      EST=1.
      L=1
      DO10I=1,13,2
   10 F(I)=FOF(A+DA*(I-1)/12.)
      KOUNT=7
    1 DX=DA/2.
      XM=A+DX
      DO11I=2,12,2
   11 F(I)=FOF(A+DA*(I-1)/12.)
      KOUNT=KOUNT+6
      QNC7=0
      IF (KOUNT .GT. 5000) RETURN
      ESTL=(W(1)*F(1)+W(2)*F(2)+W(3)*F(3)
     1+W(4)*F(4)+W(3)*F(5)+W(2)*F(6)
     2+W(1)*F(7))*DX
      ESTR=(W(1)*F(7)+W(2)*F(8)+W(3)*F(9)
     1+W(4)*F(10)+W(3)*F(11)+W(2)*F(12)+W(1)*F(13)
     2)*DX
      AREA=AREA-ABS(EST)+ABS(ESTL)+ABS(
     1ESTR)
      SUM=ESTL+ESTR
      IF(ABS(EST -SUM)-EPS*AREA)2,2,3
    2 IF(EST-1.)6,3,6
    3 IF(L-20)5,6,6
    5 L=L+1
      LEG(L)=2
      FOT(L)=F(7)
      F1T(L)=F(8)
      F2T(L)=F(9)
      F3T(L)=F(10)
      F4T(L)=F(11)
      F5T(L)=F(12)
      F6T(L)=F(13)
      CXT(L)=DX
      XMT(L)=XM
      ART(L)=AREA
      EPST(L)=EPS/1.4
      ESTT(L)=ESTR
      DA=DX
      F(13)=F(7)
      F(11)=F(6)
      F(9)=F(5)
      F(7)=F(4)
      F(5)=F(3)
      F(3)=F(2)
      EST=ESTL
      EPS=EPST(L)
      GOTO1
    6 IF(LEG(L)-2)9,7,7
    7 SUM1(L)=SUM
```

APPENDIX II (cont.)

```
      LEG(L)=1
      A=XMT(L)
      CA=CXT(L)
      F(1)=F0T(L)
      F(3)=F1T(L)
      F(5)=F2T(L)
      F(7)=F3T(L)
      F(9)=F4T(L)
      F(11)=F5T(L)
      F(13)=F6T(L)
      AREA=ART(L)
      EST=ESTT(L)
      EPS=EPST(L)
      GCTO1
    9 SUM=SUM1(L)+SUM
      L=L-1
      IF(L-1)111,111,6
  111 QAC7=SUM
      RETURN
      END
      FUNCTION GAUSS( A, B, FUNC, RE, KOUNT)
C         NUMERICAL QUADRATURE  BASED ON FIVE AND SEVEN PCINT GAUSS
C         QUACRATURE USED ADAPTIVELY
C         WRITTEN BY J. MELENDEZ AND D. KAHANER   LOS ALAMCS SCI LAB
      DIMENSION X5(2),W5(3),X7(3),W7(4)
      DATA(X5(I),I=1,2)/4.69100770306668E-2,.230765349947158/
      DATA(W5(I),I=1,3)/.118463442528094,.239314335249683,.284444444444
     1/
      DATA(X7(I),I=1,3)/2.544604382E621E-2,.1292344072003C2,.29707742431
     11301/
      DATA(W7(I),I=1,4)/6.4742483084434E-2,.139852695744638,.19091502525
     12558,.208979591836734/
      I=0
      GAUSS=0
      XM=B
      A1 = A
      KCUNT =0
    1 I = I+1
      IF ( I .GE. 48   ) GO TO 1C
      B1 = XM
   20 C=B1-A1
      XM=.5*(A1+B1)
      FM=FUNC(XM)
      SUM1 = W5(3)*FM
      CC 2 J=1,2
      AC = C * X5(J)
      X1 = A1 + AC
      X2 = B1 - AC
    2 SUM1=SUM1+W5(J)*(FUNC(X1)+FUNC(X2))
      SUM2 = W7(4)*FM
      CC 3 J=1,3
      AC = C * X7(J)
      X1 = A1 + AC
      X2 = B1 - AC
    3 SUM2=SUM2+W7(J)*(FUNC(X1)+FUNC(X2))
      KCUNT = KOUNT + 11
      IF(ABS(SUM2-SUM1) .GT. RE*SUM2*2**(-I/2)) GO TO 1
   10 GAUSS= GAUSS+ C*SUM2
      IF (KOUNT .GT. 5000) RETURN
      IF(B1.EC.B) RETURN
      A1=B1
      I = I - 1
      B1=A1+2.*C
      IF(B1.LE.B) GO TO 20
      B1 = B
      GC TO 20
   50 RETURN
      END
```

APPENDIX II (cont.)

```
      FUNCTION QUAD (FOF,Y1,Y2,FERR, KOUNT)
C        NUMERICAL QUADRATURE BASED ON TEN POINT NEWTON COTES RULE
C        USED ADAPTIVELY
C        WRITTEN BY DAVID KAHANER, LOS ALAMOS SCIENTIFIC LABORATORY
      DIMENSION FOT(20),F1T(20),F2T(20),F3T(20),F4T(20),F5T(20),F6T(20)
     1,F7T(20),F8T(20),F(19),W(5),LEG(20),DXT(20),XMT(20),ART(20),EPST(2
     20),ESTT(20), SUM1(20),FST(20)
      W(1)=2857./8960C.
      W(2)=15741./8960C.
      W(3)= 1080./8960C.
      W(4)=19344./8960C.
      W(5)= 5778./8960C.
      A=Y1
      B=Y2
      EPS=FERR
      DA=B-A
      AREA=1.
      EST=1.
      L=1
      DC 10 I=1,19,2
   10 F(I)=FOF(A+DA*(I-1)/18.)
      KOUNT = 10
    1 DX=DA/2.
      XM=A+DX
      DO 11 I=2,18,2
   11 F(I)=FOF(A+DA*(I-1)/18.)
      KOUNT = KOUNT +9
      QUAD = 0
      IF (KOUNT .GT. 5000) RETURN
      ESTL= W(1)*F(1)+W(2)*F(2)+W(3)*F(3)+W(4)*F(4)+W(5)*F(5)+W(5)*F(6)
     1 +W(4)*F(7)+W(3)*F(8)+W(2)*F(9) +W(1)*F(10)
      ESTL=ESTL*DX
      ESTR= W(1)*F(10)+W(2)*F(11)+W(3)*F(12)+W(4)*F(13)+W(5)*F(14)
     1+W(5)*F(15)+W(4)*F(16)+W(3)*F(17)+W(2)*F(18)+W(1)*F(19)
      ESTR=ESTR*DX
      AREA= AREA- ABS(EST)+ABS(ESTL)+ABS(ESTR)
      SUM= ESTL+ESTR
      IF(ABS(EST-SUM)-EPS*AREA) 2,2,3
    2 IF(EST-1.)6,3,6
    3 IF(L-20)5,6,6
    5 L=L+1
      LEG(L)=2
      FOT(L)=F(10)
      F1T(L)=F(11)
      F2T(L)=F(12)
      F3T(L)=F(13)
      F4T(L)=F(14)
      F5T(L)=F(15)
      F6T(L)=F(16)
      F7T(L)=F(17)
      F8T(L)=F(18)
      F9T(L)=F(19)
      DXT(L)=DX
      XMT(L)=XM
      ART(L)=AREA
      EPST(L) = EPS/1.4
      ESTT(L)=ESTR
      DA=DX
      F(19) =F(10)
      F(17) = F(9)
      F(15) = F(8)
      F(13) = F(7)
      F(11) = F(6)
      F( 9) = F(5)
      F( 7) = F(4)
      F( 5) = F(3)
      F( 3) = F(2)
      EST = ESTL
      EPS = EPST(L)
      GC TO 1
```

APPENDIX II (cont.)

```
    6 IF(LEG(L)-2)9,7,7
    7 SUM1(L)=SUM
      LEG(L)=1
      A=XMT(L)
      CA=CXT(L)
      F(1)=F0T(L)
      F(3)=F1T(L)
      F(5)=F2T(L)
      F(7)=F3T(L)
      F(9)=F4T(L)
      F(11)=F5T(L)
      F(13)=F6T(L)
      F(15)=F7T(L)
      F(17)=F8T(L)
      F(19)=F9T(L)
      AREA = ART(L)
      EST = ESTT(L)
      EPS = EPST(L)
      GC TO 1
    9 SUM =SUM1(L) +SUM
      L=L-1
      IF(L-1) 111,111,6
  111 QUAD = SUM
      RETURN
      END
      FUNCTION SHNK(FOF,Z,B,E,KOUNT)
C-----------------EPSILON ALGORITHM----------------------------------
C----------------------DAVID K. KAHANER-----------------------------
C-------------------VERSION NUMBER 2--------------------------------
C-----------------WRITTEN DECEMBER 1970-----------------------------
C-----------------LOS ALAMOS SCIENTIFIC LABORATORY-----------------
      DIMENSION EP(20,20)
      H=B-Z
      K=C
      KOUNT=2
      EP(1,1)=H/2*(FOF(Z)+FOF(B))
      NMIN=3
    4 K=K+1
      H=H/2
      S=0.
      M=2**(K-1)
      DC 1 J=1,M
      AJ=2*J-1
    1 S=S+FOF(Z+AJ*H)
      KOUNT=KOUNT+M
      EP(K+1,1)=EP(K,1)/2+H*S
      SHNK=(4.*EP(K+1,1)-EP(K,1))/3.
      IF(KOUNT .GT. 5000) RETURN
      TTEST=ABS(EP(K+1,1)-EP(K,1))+ABS(EP(K+1,1)-EP(K-1,1))+ABS(EP(K,1)
     .  -EP(K-1,1))
      IF(K.GE.NMIN .AND. TTEST .LT. 1.0E-11) RETURN
      DC 2 L=1,K
      LL=L+1
      KK=K+1-L
      IF(L .EQ. 1) GO TO 10
      EP(KK,LL)=EP(KK+1,L)-EP(KK,L)
      IF(EP(KK,LL).NE. 0)EP(KK,LL)=1./EP(KK,LL)+EP(KK+1,L-1)
      GC TO 2
   10 EP(KK,LL)=EP(KK+1,L)-EP(KK,L)
      IF(EP(KK,LL) .NE. 0)EP(KK,LL)=1/EP(KK,LL)
    2 CONTINUE
      IF((K/2)*2 .EQ. K) GO TO 12
      SHNK=EP(2,K)
      IF(ABS(EP(1,K)-SHNK) .LE. E*ABS(SHNK) .AND. K .GE. NMIN )RETURN
      GC TO 4
   12 SHNK=EP(1,K+1)
      IF(ABS(EP(3,K-1)-SHNK) .LE. E*ABS(SHNK) .AND. K .GE. NMIN)RETURN
      GC TO 4
      END
```

SUMMARY OF DETAILED TEST OUTPUT FOR TOLERANCE OF 10**(-3)

	CUAD	SIMPSN	GAUSS	QNC7	QABS	RC*B	G´96	S*NK	RBUN	SQUANK	HAVIE
MINIMUM FUNCTION EVALUATIONS	0	1	1	0	0	1	1	2	7	8	
MINIMUM TIME	0	3	7	0	4	1	2	2	7	5	
MEDIAN TIME PER INTEGRAL	.0076	.0052	.0356	.0060	.0048	.0044	.0064	.0028	.0064	.0048	.018
FUNCTIONS INTEGRATED TO TOL	20	19	16	18	19	19	18	13	15	17	1
FUNCTIONS INTEGRATED TO 1/100 TOL	15	16	13	15	16	16	17	8	8	8	1

SUMMARY OF DETAILED TEST OUTPUT FOR TOLERANCE OF 10**(-6)

	CUAD	SIMPSN	GAUSS	QNC7	QABS	RQMB	G 96	SHNK	RBUN	SQUANK	HAVIE
MINIMUM FUNCTION EVALUATIONS	1	1	3	2	1	0	4	2	4	2	
MINIMUM TIME	2	1	4	2	3	0	6	2	2	2	
MEDIAN TIME PER INTEGRAL	.0104	.0120	.0508	.0056	.0084	.0696	.0072	.0092	.0332	.0156	.070
FUNCTIONS INTEGRATED TO TOL	19	19	16	18	18	16	16	12	18	17	1

FUNCTIONS INTEGRATED TO 1/100 TOL

	QUAD	SIMPSN	GAUSS	QNC7	QABS	RCMB	G 96	SHNK	RBUN	SQANK	HAVIE
	17	16	15	15	17	14	14	8	12	7	1

SUMMARY OF DETAILED TEST OUTPUT FOR TOLERANCE OF 10**(-9)

	QUAD	SIMPSN	GAUSS	QNC7	QABS	RCMB	G 96	SHNK	RBUN	SQANK	HAVIE
MINIMUM FUNCTION EVALUATIONS	3	1	4	2	1	1	6	4	0	0	1
MINIMUM TIME	3	1	3	2	2	0	8	3	0	0	1
MEDIAN TIME PER INTEGRAL	.0208	.0556	.0676	.0180	.0176	.2268	.0076	.0172	.0932	.0916	.126
FUNCTIONS INTEGRATED TO TOL	18	17	14	17	19	15	14	13	18	16	1
FUNCTIONS INTEGRATED TO 1/100 TOL	17	7	10	16	17	14	13	9	12	13	1

Appendix IV

DETAILED TEST OUTPUT FOR TOLERANCE OF 1C**(-3)

FUNCTION NUMBER 1

	QUAD	SIMPSN	GAUSS	QNC7	QABS	RCMB	G 96	SHNK	RBUN	SQUANK	HAVI
ERROR	.14E-13	.91E-07	.28E-13	.36E-13	.22E-12	.36E-13	.19E-12	.14E-07	.86E-06	.14E-C7	.23E-0
TIME IN SECONDS	.002	.001	.001	.CC2	.CC1	.002	.005	.002	.001	.CC1	.00
FUNCTION EVALUATIONS	37	19	11	25	13	17	96	17	5	5	

FUNCTION NUMBER 2

	QUAD	SIMPSN	GAUSS	QNC7	QABS	RCMB	G 96	SHNK	RBUN	SQUANK	HAVI
ERROR	.69E-04	.23E-05	.71E-14	.1CE-C3	.57E-04	.19E-02	.54E-02	.70E+00	.13E-1C	.18E-02	.29E-0
TIME IN SECONDS	.008	.005	.051	.CC6	.CC6	.011	.004	.274	.018	.CC2	.03
FUNCTION EVALUATIONS	163	115	1518	121	141	257	96	8193	271	25	102

FUNCTION NUMBER 3

	QUAD	SIMPSN	GAUSS	QNC7	QABS	RCMB	G 96	SHNK	RBUN	SQUANK	HAVI
ERROR	.10E-03	.79E-05	.25E-03	.68E-C4	.3CE-05	.13E-03	.11E-06	.54E-C5	.41E-06	.32E-C2	.16E-0
TIME IN SECONDS	.003	.004	.001	.CC3	.CC4	.004	.004	.002	.017	.CC1	.00
FUNCTION EVALUATIONS	55	55	11	45	77	65	96	17	211	5	6

FUNCTION NUMBER 4

	QUAD	SIMPSN	GAUSS	QNC7	QABS	RCMB	G 96	SHNK	RBUN	SQUANK	HAVI
ERROR	.21E-13	.40E-06	.18E-13	.11E-11	.14E-1C	.78E-10	.60E-13	.64E-C5	.13E-02	.2CE-C5	.13E-0
TIME IN SECONDS	.005	.003	.002	.CC4	.CC2	.003	.012	.003	.001	.CC2	.00

FUNCTION EVALUATIONS

FUNCTION NUMBER 5

METHOD	ERRCR	TIME IN SECCNDS	FUNCTION EVALUATIONS
QUAD	.78E-09	.002	37
SIMPSN	.11E-05	.001	19
GAUSS	.15E-04	.001	11
QNC7	.41E-C7	.CC2	25
QABS	.79E-06	.CC1	13
RCMB	.66E-05	.002	17
G 96	.14E-11	.004	96
SHNK	.46E-C4	.001	17
RBUN	.25E-C3	.001	11
SQUANK	.25E-03	.CC1	5
HAVI	.18E-0	.00	1

FUNCTION NUMBER 6

METHOD	ERRCR	TIME IN SECCNDS	FUNCTION EVALUATIONS
QUAD	.91E-06	.005	37
SIMPSN	.10E-04	.002	19
GAUSS	.36E-05	.002	11
QNC7	.27E-C5	.CC3	25
QABS	.51E-C6	.CC2	13
RCMB	.86E-05	.003	17
G 96	.99E-11	.011	96
SHNK	.59E-06	.003	17
RBUN	.30E-03	.001	5
SQUANK	.54E-04	.CC2	5
HAVI	.77E-0	.00	5

FUNCTION NUMBER 7

METHOD	ERRCR	TIME IN SECCNDS	FUNCTION EVALUATIONS
QUAD	.38E-03	.024	361
SIMPSN	.48E-05	.016	235
GAUSS	.14E-07	.055	1023
QNC7	.48E-C3	.C17	241
QABS	.1CE+04	.CC8	133
RCMB	.20E+01	.431	8193
G 96	.90E-02	.005	96
SHNK	.44E-04	.003	33
RBUN	.10E-04	.015	211
SQUANK	.17E-C2	.C1C	1C5
HAVI	.1CE+0	.40	819

FUNCTION NUMBER 8

METHOD	ERRCR	TIME IN SECCNDS	FUNCTION EVALUATIONS
QUAD	.60E-12	.002	
SIMPSN	.32E-06	.001	
GAUSS	.69E-07	.001	
QNC7	.19E-C5	.CC2	
QABS	.23E-08	.CC1	
RCMB	.14E-07	.002	
G 96	.10E-12	.003	
SHNK	.16E-05	.002	
RBUN	.55E-C2	.001	
SQUANK	.78E-C6	.CC1	
HAVI	.81E-0	.00	

APPENDIX IV (cont.)

FUNCTION EVALUATIONS

	QUAD	SIMPSN	GAUSS	QNC7	QABS	RCMB	G 96	SHNK	RBUN	SQUANK	HAVI
FUNCTION EVALUATIONS	37	19	11	25	13	17	96	17	5	5	

FUNCTION NUMBER 9

	QUAD	SIMPSN	GAUSS	QNC7	QABS	RCMB	G 96	SHNK	RBUN	SQUANK	HAVI
ERROR	.58E-09	.72E-04	.51E-01	.85E-05	.48E-06	.42E-05	.32E-07	.65E-04	.12E-03	.13E-03	.15E+0
TIME IN SECONDS	.012	.013	.066	.008	.012	.004	.007	.002	.008	.008	.00
FUNCTION EVALUATIONS	145	163	1023	57	149	33	96	17	79	81	

FUNCTION NUMBER 10

	QUAD	SIMPSN	GAUSS	QNC7	QABS	RCMB	G 96	SHNK	RBUN	SQUANK	HAVI
ERROR	.39E-13	.30E-06	.20E-10	.42E-10	.50E-09	.14E-08	.17E-12	.57E-06	.27E-04	.72E-06	.74E-0
TIME IN SECONDS	.002	.001	.001	.001	.001	.001	.003	.002	.001	.001	.00
FUNCTION EVALUATIONS	37	19	11	25	13	17	96	17	5	5	

FUNCTION NUMBER 11

	QUAD	SIMPSN	GAUSS	QNC7	QABS	RCMB	G 96	SHNK	RBUN	SQUANK	HAVI
ERROR	.18E-14	.48E-08	0.	.18E-13	.26E-12	.38E-12	.57E-13	.27E-08	.19E-06	.25E-08	.20E-0
TIME IN SECONDS	.003	.001	.001	.002	.001	.002	.005	.002	.001	.001	.00
FUNCTION EVALUATIONS	37	19	11	25	13	17	96	17	5	5	

FUNCTION NUMBER 12

	QUAD	SIMPSN	GAUSS	QNC7	QABS	RCMB	G 96	SHNK	RBUN	SQUANK	HAVI
ERROR	.21E-13	.16E-08	.71E-14	.21E-13	.11E-13	.18E-13	.12E-12	.16E-09	.10E-07	.15E-09	.65E-0
TIME IN SECONDS	.003	.002	.001	.002	.001	.002	.006	.002	.001	.001	.00
FUNCTION EVALUATIONS	37	19	11	25	13	17	96	17	5	5	

FUNCTION NUMBER 13

QUAD	SIMPSN	GAUSS	QNC7	QABS	RCMB	G 96	S+NK	RBUN	SQLANK	HAVI
ERRCR .11E-07	.91E-C2	.15E+0C	.12E+CC	.28E-07	.27E-08	.59E-12	.91E-C2	.39E-05	.23E-04	.91E-0
TIME IN SECCNDS .080	.002	.084	.CC5	.C49	.082	.008	.624	.055	.C51	.00
FUNCTICN EVALUATIONS 865	19	1067	45	573	1025	96	8193	522	425	

FUNCTION NUMBER 14

QUAD	SIMPSN	GAUSS	QNC7	QABS	RCMB	G 96	S+NK	RBUN	SQLANK	HAVI
ERRCR .11E-08	.52E-07	.18E-C9	.24E-C7	.18E-08	.13E-06	.31E-11	.50E+C0	.28E-05	.12E-C6	.96E-1
TIME IN SECCNDS .011	.009	.08C	.CC5	.CC8	.073	.007	.0C2	.006	.CC6	.02
FUNCTICN EVALUATIONS 127	103	1023	57	85	1025	96	17	51	45	25

FUNCTION NUMBER 15

QUAD	SIMPSN	GAUSS	QNC7	QABS	RCMB	G 96	S+NK	RBUN	SQLANK	HAVI
ERRCR .19E-06	.37E-05	.25E-1C	.13E-C6	.79E-08	.35E-05	.32E-12	.10E+C1	.25E-04	.13E-04	.20E-0
TIME IN SECCNDS .008	.007	.057	.CC6	.CC6	.030	.006	.444	.046	.CC5	.05
FUNCTICN EVALUATIONS 109	103	1022	85	85	513	96	8193	527	52	102

FUNCTION NUMBER 16

QUAD	SIMPSN	GAUSS	QNC7	QABS	RCMB	G 96	S+NK	RBUN	SQLANK	HAVI
ERRCR .91E-08	.35E-05	.13E-10	.25E-C8	.93E-07	.30E-04	.37E-05	.50E+00	.39E-04	.88E-05	.18E-1
TIME IN SECCNDS .008	.006	.036	.CC6	.C05	.072	.004	.273	.006	.CC4	.06
FUNCTICN EVALUATIONS 163	115	1022	121	1C9	2049	96	8193	87	52	204

FUNCTION NUMBER 17

	QUAD SIMPSN	GAUSS	QNC7	QABS	RCMB	G 96	SPNK	RBUN	SQLANK	HAVI
ERROR	.41E-04 .83E-03	.21E-02	.11E-C2	.5CE-03	.16E-08	.16E-10	.11E+00	.49E-04	.75E-C3	9CE-0
TIME IN SECONDS	.029 .014	.075	.C16	.014	.083	.008	.634	.012	.CC7	.04
FUNCTION EVALUATIONS	307 151	1023	165	149	1025	96	8193	105	57	51

FUNCTION NUMBER 18

	QUAD SIMPSN	GAUSS	QNC7	QABS	RCMB	G 96	SPNK	RBUN	SQLANK	HAVI
ERROR	.76E-04 .23E-05	.16E-02	.38E-C6	.29E-06	.16E-06	.90E-12	.19E-03	.65E-02	.15E-04	.49E-0
TIME IN SECONDS	.020 .038	.263	.C24	.021	.036	.025	.018	.012	.016	.01
FUNCTION EVALUATIONS	73 139	1023	65	77	129	96	65	35	53	6

FUNCTION NUMBER 19

	QUAD SIMPSN	GAUSS	QNC7	QABS	RCMB	G 96	SPNK	RBUN	SQLANK	HAVI
ERROR	.41E-05 .37E-05	.21E-11	.32E-C5	.26E-04	.82E-03	.68E-04	.10E+01	.56E-05	.88E-C3	.10E+0
TIME IN SECONDS	.024 .014	.068	.C18	.014	.268	.006	.523	.022	.CC5	.49
FUNCTION EVALUATIONS	307 175	1023	217	181	4097	96	8193	211	45	819

FUNCTION NUMBER 20

	QUAD SIMPSN	GAUSS	QNC7	QABS	RCMB	G 96	SPNK	RBUN	SQLANK	HAVI
ERROR	.61E-09 .79E-05	.11E-04	.25E-C7	.66E-06	.51E-05	.22E-12	.50E-04	.26E-03	.26E-C3	.55E-0
TIME IN SECONDS	.002 .002	.001	.CC2	.CC1	.001	.004	.CC2	.001	.CC1	.00
FUNCTION EVALUATIONS	37 31	11	25	13	17	96	17	11	5	1

248

FUNCTION NUMBER 21

	QUAD	SIMPSN	GAUSS	QNC7	QABS	RCHB	G 96	SHNK	RBUN	SQLANK	HAVI
ERRCR	.11E-02	.11E-02	.14E-01	.11E-C2	.11E-02	.20E-05	.24E-02	.15E-01	.12E-02	.15E-C1	.11E-0
TIME IN SECCNDS	.032	.032	.24C	.C24	.019	.962	.023	.004	.016	.CC5	.06
FUNCTION EVALUATIONS	127	127	1023	57	77	4097	96	17	65	17	25

DETAILED TEST OUTPUT FOR TOLERANCE CF 1C**(-6)

FUNCTION NUMBER 1

	QUAD	SIMPSN	GAUSS	QNC7	QABS	RCHB	G 96	SHNK	RBUN	SQLANK	HAVI
ERRCR	.14E-13	.11E-08	.28E-13	.36E-13	.22E-12	.36E-13	.19E-12	.14E-C7	.22E-05	.22E-C5	.22E-0
TIME IN SECCNDS	.002	.0C4	.0C1	.CC2	.C01	.C02	.005	.CC2	.002	.CC2	.00
FUNCTION EVALUATIONS	37	55	11	25	13	17	96	17	21	17	1

FUNCTION NUMBER 2

	QUAD	SIMPSN	GAUSS	QNC7	QABS	RCHB	G 96	SHNK	RBUN	SQLANK	HAVI
ERRCR	.72E-08	.38E-10	.71E-14	.1CE-C6	.56E-C7	.70E+00	.54E-02	.70E+00	.13E-1C	.18E-C2	.1CE+0
TIME IN SECCNDS	.017	.011	.051	.C12	.011	.280	.003	.274	.018	.CC2	.23
FUNCTION EVALUATIONS	361	235	1518	241	261	8193	96	8193	271	25	819

FUNCTION NUMBER 3

	QUAD	SIMPSN	GAUSS	QNC7	QABS	RCHB	G 96	SHNK	RBUN	SQLANK	HAVI
ERRCR	.88E-08	.68E-08	.25E-12	.6CE-C8	.59E-08	.26E-06	.11E-06	.49E-08	.34E-08	.2CE-C7	.27E-0
TIME IN SECCNDS	.013	.011	.045	.01C	.008	.178	.004	.0C4	.032	.CC5	.16
FUNCTION EVALUATIONS	217	199	1023	157	145	4097	96	65	422	1C5	409

APPENDIX IV (cont.)

FUNCTION NUMBER 4

QUAD	SIMPSN	GAUSS	QNC 7	QABS	RCMB	G 96	SMK	RBUN	SQLANK	HAVI
ERROR										
.21E-13	.50E-08	.18E-12	.11E-11	.8CE-13	.78E-10	.60E-13	.25E-11	.13E-0?	.1CE-C7	.13E-0
TIME IN SECONDS										
.005	.008	.0C2	.CC4	.C03	.003	.012	.0C9	.001	.CC4	.00
FUNCTION EVALUATIONS										
37	55	11	25	25	17	96	65	5	25	

FUNCTION NUMBER 5

QUAD	SIMPSN	GAUSS	QNC 7	QABS	RCMB	G 96	SMK	RBUN	SQLANK	HAVI
ERROR										
.78E-09	.13E-09	.14E-08	.36E-1C	.4CE-11	.83E-10	.14E-11	.31E-C7	.10E-07	.3CE-C5	.55E-0
TIME IN SECONDS										
.002	.008	.037	.CC3	.C02	.004	.003	.004	.004	.CC5	.00
FUNCTION EVALUATIONS										
37	163	1022	45	49	65	96	65	55	65	6

FUNCTION NUMBER 6

QUAD	SIMPSN	GAUSS	QNC 7	QABS	RCMB	G 96	SMK	RBUN	SQLANK	HAVI
ERROR										
.29E-07	.52E-08	.16E-12	.15E-C7	.5CE-C9	.47E-07	.99E-11	.32E-C9	.65E-05	.71E-C7	.76E-0
TIME IN SECONDS										
.010	.012	.121	.CC8	.CC8	.017	.012	.009	.058	.CC5	.01
FUNCTION EVALUATIONS										
73	91	1022	61	65	129	96	65	382	25	12

FUNCTION NUMBER 7

QUAD	SIMPSN	GAUSS	QNC 7	QABS	RCMB	G 96	SMK	RBUN	SQLANK	HAVI
ERROR										
.38E-03	.15E-04	.14E-07	.48E-C3	.1CE+C4	.2CE+01	.90E-02	.18E-C7	.61E-C5	.15E-C4	.1CE+0
TIME IN SECONDS										
.024	.068	.055	.017	.CC6	.431	.006	.008	.051	.1C5	.40
FUNCTION EVALUATIONS										
361	1027	1022	241	85	8193	96	129	587	1152	819

FUNCTION NUMBER 8

QUAD	SIMPSN	GAUSS	QNC 7	QABS	RCMB	G 96	SMK	RBUN	SQLANK	HAVI

	QUAD	SIMPSN	GAUSS	QNC7	QABS	RCMB	G 96	SHNK	RBUN	SQLANK	HAVI
ERROR	.60E-12	.56E-09	.70E-12	.19E-09	.11E-10	.11E-09	.10E-12	.16E-05	.16E-10	.37E-07	.32E-0
TIME IN SECONDS	.002	.004	.034	.001	.001	.002	.004	.001	.003	.002	.00
FUNCTION EVALUATIONS	37	67	1023	25	25	33	96	17	41	29	3

FUNCTION NUMBER 9

	QUAD	SIMPSN	GAUSS	QNC7	QABS	RCMB	G 96	SHNK	RBUN	SQLANK	HAVI
ERROR	.79E-09	.87E-10	.51E-01	.93E-06	.11E-09	.53E-10	.32E-07	.82E-08	.24E-05	.13E-06	.42E-0
TIME IN SECONDS	.033	.071	.066	.025	.025	.011	.007	.006	.028	.040	.00
FUNCTION EVALUATIONS	397	871	1023	265	313	129	96	65	267	377	3

FUNCTION NUMBER 10

	QUAD	SIMPSN	GAUSS	QNC7	QABS	RCMB	G 96	SHNK	RBUN	SQLANK	HAVI
ERROR	.39E-13	.37E-08	.20E-10	.42E-10	.50E-09	.14E-08	.17E-12	.12E-07	.20E-08	.37E-08	.25E-0
TIME IN SECONDS	.002	.002	.001	.001	.001	.002	.003	.002	.002	.001	.00
FUNCTION EVALUATIONS	37	55	11	25	13	17	96	33	31	21	1

FUNCTION NUMBER 11

	QUAD	SIMPSN	GAUSS	QNC7	QABS	RCMB	G 96	SHNK	RBUN	SQLANK	HAVI
ERROR	.18E-14	.48E-08	0.	.18E-13	.26E-12	.38E-12	.57E-13	.27E-08	.25E-08	.25E-08	.25E-0
TIME IN SECONDS	.003	.001	.001	.002	.001	.002	.006	.002	.002	.001	.00
FUNCTION EVALUATIONS	37	19	11	25	13	17	96	17	11	9	

FUNCTION NUMBER 12

	QUAD	SIMPSN	GAUSS	QNC7	QABS	RCMB	G 96	SHNK	RBUN	SQLANK	HAVI
ERROR	.21E-13	.16E-08	.71E-14	.21E-13	.11E-13	.18E-13	.12E-12	.16E-09	.10E-07	.15E-09	.15E-0

FUNCTION NUMBER 13

	QUAD	SIMPSN	GAUSS	QNC7	QABS	RCPB	G 96	SHNK	RBUN	SQLANK	HAVI
TIME IN SECCNES	.003	.002	.001	.CC2	.CC1	.002	.006	.002	.001	.CC1	.00
FUNCTICN EVALUATIONS	37	19	11	25	13	17	96	17	5	5	
ERRCR	.10E-10	.91E-02	.15E+0C	.55E-11	.38E-1C	.18E-11	.59E-12	.91E-C2	.59E-C9	.65E-C6	.18E-0

FUNCTION NUMBER 14

	QUAD	SIMPSN	GAUSS	QNC7	QABS	RCPB	G 96	SHNK	RBUN	SQLANK	HAVI
TIME IN SECCNES	.152	.002	.084	.144	.126	.160	.008	.623	.270	.294	.15
FUNCTICN EVALUATIONS	1639	19	1067	1525	1449	2049	96	8193	2382	2545	204
ERRCR	.80E-10	.62E-09	.18E-09	.81E-C9	.86E-1C	.82E-10	.31E-11	.50E+00	.72E-08	.22E-C7	.52E-1

FUNCTION NUMBER 15

	QUAD	SIMPSN	GAUSS	QNC7	QABS	RCPB	G 96	SHNK	RBUN	SQLANK	HAVI
TIME IN SECCNES	.014	.030	.080	.C12	.CC9	.143	.007	.002	.01C	.C2C	.03
FUNCTICN EVALUATIONS	163	331	1023	133	109	2049	96	17	91	165	51
ERRCR	.95E-10	.52E-08	.25E-1C	.46E-C9	.33E-1C	.17E-11	.32E-12	.10E+C1	.11E-05	.54E-C7	.65E-0

FUNCTION NUMBER 16

	QUAD	SIMPSN	GAUSS	QNC7	QABS	RCPB	G 96	SHNK	RBUN	SQLANK	HAVI
TIME IN SECCNES	.010	.023	.058	.C1C	.CC9	.115	.005	.444	.355	.C2C	.11
FUNCTICN EVALUATIONS	145	343	1023	133	133	2049	96	8193	4117	213	204
ERRCR	.12E-09	.18E-07	.13E-1C	.68E-C9	.23E-C5	.5CE+00	.37E-05	.50E+00	.10E-06	.15E-C6	.68E-1

	QUAD	SIMPSN	GAUSS	QNC7	QABS	RCPB	G 96	SHNK	RBUN	SQLANK	HAVI
TIME IN SECCNES	.009	.024	.036	.CC5	.CC6	.276	.003	.273	.01C	.C2C	.12

FUNCTION NUMBER 17

	QUAD	SIMPSN	GAUSS	QNC7	QABS	RCMB	G 96	S+NK	RBUN	SQLANK	HAVI
FUNCTION EVALUATIONS	181	511	1023	181	145	8193	96	8193	141	273	409
ERROR	.32E-10	.82E-07	.21E-02	.11E-02	.51E-10	.40E-11	.16E-10	.11E+00	.60E-07	.33E-03	.66E-0
TIME IN SECONDS	.094	.206	.079	.036	.073	.163	.008	.634	.081	.082	.08

FUNCTION NUMBER 18

	QUAD	SIMPSN	GAUSS	QNC7	QABS	RCMB	G 96	S+NK	RBUN	SQLANK	HAVI
FUNCTION EVALUATIONS	1009	2275	1023	385	829	2049	96	8193	697	697	102
ERROR	.57E-11	.56E-08	.16E-02	.79E-05	.13E-11	.76E-10	.90E-12	.70E-08	.80E-08	.55E-05	.21E-0
TIME IN SECONDS	.055	.150	.262	.050	.056	.070	.025	.270	.058	.050	.07

FUNCTION NUMBER 19

	QUAD	SIMPSN	GAUSS	QNC7	QABS	RCMB	G 96	S+NK	RBUN	SQLANK	HAVI
FUNCTION EVALUATIONS	199	547	1023	181	205	257	96	1025	195	301	25
ERROR	.57E-06	.11E-07	.21E-11	.83E-06	.10E+04	.10E+01	.68E-04	.10E+01	.51E-08	.71E-06	.10E+0
TIME IN SECONDS	.029	.038	.068	.020	.008	.532	.006	.524	.041	.027	.50

FUNCTION NUMBER 20

	QUAD	SIMPSN	GAUSS	QNC7	QABS	RCMB	G 96	S+NK	RBUN	SQLANK	HAVI
FUNCTION EVALUATIONS	361	499	1023	241	105	8193	96	8193	403	257	819
ERROR	.61E-09	.18E-11	.65E-05	.30E-12	.50E-13	.20E-10	.22E-12	.24E-10	.67E-08	.20E-07	.33E-1
TIME IN SECONDS	.002	.008	.035	.002	.002	.004	.003	.004	.002	.004	.00

FUNCTION NUMBER 21

	QUAD	SIMPSN	GAUSS	QNC7	QABS	RCMB	G 96	S+NK	RBUN	SQLANK	HAVI
FUNCTION EVALUATIONS	37	163	1023	45	49	65	96	65	45	45	3

APPENDIX IV (cont.)

	QUAC	SIMPSN	GAUSS	QNC7	QABS	RCMB	G 96	SMNK	RBUN	SQLANK	HAVI
ERRCR	.11E-02	.19E-08	.14E-C1	.11E-C2	.11E-C2	.21E+C0	.24E-02	.21E+C0	.26E-06	.11E-C2	.67E-0
TIME IN SECCNDS	.063	.170	.24C	.C52	.C48	1.918	.022	1.914	.088	.C5C	.94
FUNCTION EVALUATIONS	253	691	1023	2C5	197	8193	96	8193	327	185	409

DETAILEC TEST OUTPUT FOR TOLERANCE CF 1(**(-5))

FUNCTICN NUMBER 1

	QUAC	SIMPSN	GAUSS	QNC7	QABS	RCMB	G 96	SMNK	RBUN	SQLANK	HAVI
ERRCR	.14E-13	.14E-10	.28E-13	.36E-13	.28E-13	.36E-13	.19E-12	.99E-13	.57E-13	.92E-13	.13E-1
TIME IN SECCNDS	.003	.010	.001	.CC2	.CC2	.CC2	.005	.0C4	.00E	.CC6	.CO
FUNCTICN EVALUATIONS	37	163	11	25	25	17	96	65	73	65	1

FUNCTICN NUMBER 2

	QUAC	SIMPSN	GAUSS	QNC7	QABS	RCMB	G 96	SMNK	RBUN	SQLANK	HAVI
ERRCR	.72E-08	.38E-10	.71E-14	.1CE-C6	.55E-1C	.7CE+00	.54E-02	.70E+C0	.13E-1C	.18E-C2	.1CE+0
TIME IN SECCNDS	.017	.010	.051	.C12	.C16	.280	.004	.274	.015	.CC2	.23
FUNCTICN EVALUATIONS	361	235	1518	241	381	8193	96	8193	271	25	819

FUNCTICN NUMBER 3

	QUAC	SIMPSN	GAUSS	QNC7	QABS	RCMB	G 96	SMNK	RBUN	SQLANK	HAVI
ERRCR	.29E-11	.16E-10	.25E-12	.6CE-11	.44E-11	.67E+C0	.11E-06	.12E-11	.64E-12	.14E-11	.1CE+0
TIME IN SECCNDS	.021	.050	.045	.C17	.C16	.352	.004	.012	.121	.C42	.32
FUNCTICN EVALUATIONS	361	883	1023	285	289	8193	96	257	1595	513	819

FUNCTICN NUMBER 4

	QUAC	SIMPSN	GAUSS	QNC7	QABS	RCMB	G 96	SMNK	RBUN	SQLANK	HAVI

Comparison of numerical integration methods (Functions 5–8)

	SIMPSN	QUAD	GAUSS	QNC7	QABS	RCMB	G 96	SHNK	RBUN	SQLANK	HAVI
FUNCTION NUMBER 5											
ERROR	.21E-13	.11E-13	.20E-12	.11E-11	.2CE-13	.25E-13	.60E-13	.59E-13	.87E-12	.25E-11	.33E-1
TIME IN SECONDS	.005	.045	.126	.CC4	.C11	.CC5	.012	.017	.022	.C17	.CO
FUNCTION EVALUATIONS	37	331	1023	25	85	33	96	129	135	1C5	3
FUNCTION NUMBER 6											
ERROR	.15E-11	.71E-14	.14E-08	.15E-11	.16E-11	.14E-11	.14E-11	.94E-11	.87E-11	.17E-11	.62E-1
TIME IN SECONDS	.004	.023	.038	.CC5	.CC8	.CC7	.004	.010	.017	.C21	.CO
FUNCTION EVALUATIONS	73	487	1023	57	181	129	96	257	235	285	12
FUNCTION NUMBER 7											
ERROR	.50E-11	.91E-11	.16E-13	.42E-11	.51E-12	.46E-10	.99F-11	.62E-11	.12E-12	.18E-11	.74E-1
TIME IN SECONDS	.021	.056	.122	.C18	.C18	.243	.011	.016	.214	.C25	.23
FUNCTION EVALUATIONS	163	427	1023	133	137	2049	96	129	1423	161	204
FUNCTION NUMBER 8											
ERROR	.38E-03	.15E-C4	.14F-C7	.46E-C3	.1CE+C4	.2CF+C1	.90E-C2	.17E-C9	.61E-C5	.1CE+C4	.1CE+0
TIME IN SECONDS	.046	.280	.055	.C42	.CC6	.431	.005	.C15	.21C	.45C	.40
FUNCTION EVALUATIONS	685	4279	1023	565	89	8193	96	257	2467	5CC1	819
ERROR	0.	.14E-11	.70E-12	.12E-11	.71E-14	.15E-12	.10E-12	.13E-C9	.1CE-1C	.20E-11	.28E-1

Appendix IV (cont.)

	QUAD	SIMPSN	GAUSS	QNC7	QABS	RCMB	G 96	SINK	RBUN	SQLANK	HAVI
TIME IN SECNDS	.004	.021	.034	.CC4	.CC4	.CC4	.003	.003	.005	.C11	.00
FUNCTION EVALUATIONS	73	463	1023	73	97	65	96	65	125	145	6
FUNCTION NUMBER 9											
ERRCR	.33E-12	.15E-10	.51E-C1	.41E-11	.17E-12	.33E-11	.32E-07	.46E-11	.17E-05	.18E-1C	.16E-0
TIME IN SECNDS	.063	.322	.066	.CEC	.C7C	.020	.007	.019	.093	.26E	.01
FUNCTION EVALUATIONS	757	3967	1023	657	893	257	96	257	882	2525	12
FUNCTION NUMBER 10											
ERRCR	.39E-13	.13E-1C	.71E-13	.17E-11	.16E-12	.96E-13	.17E-12	.13E-C9	.53E-13	.2EE-12	.35E-1
TIME IN SECNDS	.002	.012	.032	.CC2	.CC2	.CC4	.003	.003	.008	.CC7	.00
FUNCTION EVALUATIONS	37	271	1023	37	49	65	96	65	122	97	6
FUNCTION NUMBER 11											
ERRCR	.18E-14	.17E-11	.18E-13	.18E-13	0.	.12E-13	.57E-13	.76E-13	.59E-12	.6CE-12	.14E-1
TIME IN SECNDS	.003	.010	.0C1	.CC2	.CC2	.CC3	.006	.005	.004	.CC3	.00
FUNCTION EVALUATIONS	37	151	11	25	25	33	96	65	39	32	1
FUNCTION NUMBER 12											
ERRCR	.21E-13	.19E-10	.71E-14	.21E-13	.11E-13	.18E-13	.12E-12	.24E-11	.21E-13	.28E-13	.24E-1
TIME IN SECNDS	.003	.004	.001	.CC2	.CC1	.CC2	.006	.0C3	.004	.CC3	.CO
FUNCTION EVALUATIONS											

256

FUNCTION NUMBER 13 (continued)

	QUAD	SIMPSN	GAUSS	QNC7	QABS	RCMB	G 96	S+N	RBUN	SQLANK	HAVI
ERRCR	.56E-12	.91E-C2	.15E+CC	.56E-12	.57E-12	.53E-12	.59E-12	.91E-02	.10E+04	.1CE+C4	.15E-1
TIME IN SECCNS	.257	.447	.084	.251	.277	.316	.008	.624	.564	.571	.30
FUNCTION EVALUATICNS	2773	5003	1067	2C73	3197	4097	96	8193	5001	5CC1	409

FUNCTION NUMBER 14

	QUAD	SIMPSN	GAUSS	QNC7	QABS	RCMB	G 96	S+N	RBUN	SQLANK	HAVI
ERRCR	.75E-13	.35E-11	.18E-C9	.18E-12	.71E-13	.6CE-13	.31E-11	.50E+C0	.22E-1C	.57E-12	.52E-1
TIME IN SECCNS	.023	.102	.08C	.C22	.C21	.283	.007	.0C2	.025	.C52	.03
FUNCTION EVALUATICNS	253	1123	1023	241	245	4097	96	17	255	757	51

FUNCTION NUMBER 15

	QUAD	SIMPSN	GAUSS	QNC7	QABS	RCMB	G 96	S+N	RBUN	SQLANK	HAVI
ERRCR	.15E-12	.12E-10	.25E-1C	.5CE-12	.5CE-13	.56E-12	.32E-12	.10E+C1	.10E+04	.65E-11	.14E-1
TIME IN SECCNS	.015	.100	.057	.C18	.C18	.227	.006	.444	.434	.1C6	.21
FUNCTION EVALUATICNS	217	1483	1023	241	281	4097	96	8193	5001	1145	409

FUNCTION NUMBER 16

	QUAD	SIMPSN	GAUSS	QNC7	QABS	RCMB	G 96	S+N	RBUN	SQLANK	HAVI
ERRCR	.33E-11	.78E-10	.13E-1C	.29E-11	.94E-12	.50E+C0	.37E-05	.50E+C0	.14E-11	.37E-11	.68E-1
TIME IN SECCNS	.017	.115	.035	.C2C	.C18	.276	.003	.272	.09€	.117	.12
FUNCTION EVALUATICNS	343	2467	1023	357	397	8193	96	8193	1435	1645	409

FUNCTION NUMBER 17

	QUAD	SIMPSN	GAUSS	QNC7	QABS	RCMB	G 96	S+N	RBUN	SQLANK	HAVI
ERRCR	.33E-11	.11E+00	.21E-02	.36E-C3	.33E-11	.37E-11	.16E-10	.11E+C0	.42E-11	.10E+C4	.32E-1

FUNCTION NUMBER 17 (continued)

	QUAD	SIMPSN	GAUSS	QNC7	QABS	RCMB	G 96	SINK	RBUN	SQLANK	HAVI
TIME IN SECONDS	.187	.452	.075	.126	.178	.322	.008	.634	.316	.580	.16
FUNCTION EVALUATIONS	1999	5003	1022	1345	2025	4097	96	8193	2741	5001	204

FUNCTION NUMBER 18

	QUAD	SIMPSN	GAUSS	QNC7	QABS	RCMB	G 96	SINK	RBUN	SQLANK	HAVI
ERROR	.10E-11	.16E-10	.16E-02	.86E-12	.11E-11	.72E-12	.90E-12	.19E-11	.21E-11	.18E-10	.79E-1
TIME IN SECONDS	.095	.798	.262	.114	.160	.137	.025	1.071	.224	.475	.13
FUNCTION EVALUATIONS	343	2923	1022	405	585	513	96	4097	753	1585	51

FUNCTION NUMBER 19

	QUAD	SIMPSN	GAUSS	QNC7	QABS	RCMB	G 96	SINK	RBUN	SQLANK	HAVI
ERROR	.57E-06	.12E-08	.21E-11	.85E-06	.10E+04	.10E+01	.68E-04	.10E+01	.21E-05	.19E-08	.10E+0
TIME IN SECONDS	.033	.170	.068	.034	.007	.531	.006	.523	.155	.202	.49
FUNCTION EVALUATIONS	415	2203	1022	421	85	8193	96	8193	1571	1585	819

FUNCTION NUMBER 20

	QUAD	SIMPSN	GAUSS	QNC7	QABS	RCMB	G 96	SINK	RBUN	SQLANK	HAVI
ERROR	.36E-13	.14E-13	.65E-09	.28E-13	.57E-13	.20E-12	.22E-12	.24E-10	.19E-11	.12E-11	.60E-1
TIME IN SECONDS	.003	.022	.034	.005	.006	.006	.003	.004	.012	.018	.00
FUNCTION EVALUATIONS	73	487	1022	57	145	129	96	65	185	245	6

FUNCTION NUMBER 21

	QUAD	SIMPSN	GAUSS	QNC7	QABS	RCMB	G 96	SINK	RBUN	SQLANK	HAVI
ERROR	.10E-09	.10E-09	.14E-01	.98E-10	.10E-09	.21E+00	.24E-02	.21E+00	.20E-05	.16E-05	.10E+0
TIME IN SECONDS	.172	.926	.240	.178	.155	1.941	.023	1.916	.291	.445	1.89
FUNCTION EVALUATIONS	685	3751	1022	705	633	8193	96	8193	1075	1657	819

REFERENCES

Bauer, F. L., Rubishauer, H., and Striefel, E. (1963). New Aspects in Numerical Quadrature. *In* "Experimental Arithmetic, High Speed Computing Mathematics," pp. 199–218. American Mathematical Society, Providence, Rhode Island.

Bunton, W. (1970). ROMBS is available in various forms from Computer Software Management and Information Center (COSMIC), Barrow Hall, Univ. of Georgia, Athens, Georgia. Program Nos. NPO-11295, NPO-11296, and NPO-11297.

Bunton, W., Diethelm, M., and Haigler, K. (1969). Romberg Quadrature Subroutines for Single and Multiple Integrals, Jet Propulsion Laboratory Rep. TM-324-221.

Davis, P., and Rabinowitz, P. (1967). "Numerical Integration," pp. 162, 198. Ginn (Blaisdell), Boston.

Fosdick, L., and Usow, K. (1969). Proposed Certification Projects and activities, and Guidelines for Evaluating an Algorithm for Publication, presented at SIGNUM Subroutine Certification Meeting, August, 26, 1969.

Havie, T. (1966). On a Modification of Romberg's Algorithm, *BIT* 6, 24–30.

Kahaner, D. (1969). Comparison of Numerical Quadrature Formulas, Los Alamos Scientific Lab. Rep. No. LA-4137.

Kubik, R. (1965). Algorithm 257, Havie Integrator, *Comm. ACM* 8, 381.

Lyness, J. (1969). Notes on the Adaptive Simpson Quadrature Routine, *JACM* 16, 483–495.

Lyness, J. (1970). Testing and Comparing Subroutines Which Require User-Provided Function Subprograms. Private communication.

Lyness, J. (1970). Algorithm 379 SQUANK (Simpson Quadrature Used Adaptively, Noise Killed), *Comm. ACM* 13, No. 4.

O'Hara, H., and Smith, F. (1968). The Evaluation of Definite Integrals by Interval Subdivision, Nat. Bur. of Standards, Rep. N69-11541.

Wynn, P. (1956). On a Device for Computing the $e_m(S_n)$ Transformation, *Math. Tables Aids Comp.* 10, 91–96.

5.16 Evaluation of NAPSS Expressions Involving Polyalgorithms, Functions, Recursion, and Untyped Variables*

L. R. Symes†

PURDUE UNIVERSITY

LAFAYETTE, INDIANA

I. Introduction

The Numerical Analysis Problem Solving System (NAPSS) project has been undertaken at Purdue University to design an on-line interactive system for solving numerical problems [Rice and Rosen (1966); Roman and Symes (1968)]. The system is designed to accept input in a language [Symes and Roman (1968)] which is closely akin to normal mathematical notation. This implies that a wide variety of operands and operators may appear in an expression, and thus the evaluation of expressions becomes a major component of the software for NAPSS. The user may manipulate quantities other than scalars [Symes (1970)]: e.g., arrays, functions, and arrays of functions. Some of the more complex operators which may appear in an arithmetic expression invoke polyalgorithms [Rice (1968)] to perform the operation.

In addition, clerical statements such as those used to declare variables are removed from the NAPSS language. Instead, variables are contextually declared when they are assigned values. Therefore the type of a variable can change dynamically throughout a user's program, further complicating arithmetic expression evaluation.

This paper is primarily concerned with the problem of how algebraic expressions involving the various operands and operators permitted in the language are evaluated.

II. Types of Expressions

Rather than present a detailed description of NAPSS language [Symes and Roman (1969)], we describe a sampling of the allowable expressions. They

* Research supported in part by NSF Grant GP-05850.
† Present address: University of Saskatchewan, Regina, Saskatchewan, Canada.

permit the direct manipulation of numeric scalars, vector arrays, symbolic and tabular functions, and variables which denote symbolic expressions. The user need not worry about the type or mode of the operands, as long as the expression is mathematically correct.

The number of allowable types of expressions is numerous because of the wide variety of forms appearing in mathematics. An expression type is included in NAPSS for many of the common mathematical forms. This removes from the problem solver the burden of learning a strange notation. He already knows mathematical notation, so the closer NAPSS is to this the easier the system will be for him to use.

The NAPSS language is linear, as opposed to mathematics' two-dimensional notation. The transformation between the two, however, is straightforward. The transformation is aided by the inclusion of several special characters; for example: $\int$ for integration, $|\ |$ for absolute value, and $'$ for differentiation and transposition. But to permit the use of standard terminals and to aid the goal of machine independence for the system, the number of characters in the NAPSS language is limited to 63. This results in some nonstandard appearances. For example, $\partial f(x, y)/\partial g \mid_{y=3}^{x=2}$ is written DER(F(X, Y)/ (Y) | X ← 2, Y ← 3) in NAPSS. Blanks are significant in NAPSS to allow for implied multiplication (e.g., 2A1, A2 + C, and A 1 + AB C mean $2 * A1$, $A2 + C$, and $A * 1 + AB * C$, respectively).

There are several methods for constructing vectors and arrays in arithmetic expressions:

(i) $(1, -3, 2, 6, -10)$

(ii) $(1, 2, \ldots, 20)$

(iii) (1 FOR 20 TIMES)

(iv) $(2 + I \uparrow 3$ FOR I ← 1 TO N BY 3)

(v)   ([0:5], 1 TO 6)

(vi)   ([1, 1:11], 3.5 TO 4.5 BY 0.1)

(vii) (3.5 TO 4.5 BY 0.1)$'$

(viii) $([-1:3, 4], (1$ FOR 4 TIMES), $(-2, -1.75, \ldots, -1.25)$, (3 TO 6), $(-10, -20, -30, -40))$

The first five examples are vectors, which are considered to be column vectors in NAPSS. The lower index bounds of the first four vectors is 1 by default. Nonstandard ranges are indicated by square brackets, and thus the index range of the fifth vector is from 0 to 5. Vectors six and seven are identical row vectors. The eighth example is a square array with the row

index ranging between -1 and 3 and the column index between 1 and 4
The resulting array is:

$$\begin{bmatrix} 1 & 1 & 1 & 1 \\ -2 & -1.75 & -1.50 & -1.25 \\ 3 & 4 & 5 & 6 \\ -10 & -20 & -30 & -40 \end{bmatrix}$$

From a numerical array a single element, a row, a column, or any arbitrary
contiguous subarray may be extracted. If A is a two-dimensional array with
the first subscript ranging from -3 to 3 and the second from 0 to 3, then
A[0, *] denotes the zero row of A, and A[-1:2, 1] denotes the column
vector consisting of the third through sixth elements of the first column of
A.

Arithmetic expressions which yield array results may be subscripted in the
same fashion as variables. For example, $(A * B + E)[I, J]$ and $(A \uparrow 2)$
[I1 : 12, *] are both valid expressions.

NAPSS also permits arrays of functions to be manipulated an element
at a time: $2f'(3.5)[1, 3]$ is the NAPSS equivalence of $2f'_{1,3}(3.5)$.

Several examples of arithmetic expressions and assignment statements
appear below:

   (i)   DOTPRODUCT $\leftarrow$ | D[2, *]' | | D[2, *] |

  (ii) EQUALSVAR $= K * (1, 1 + H/K, \ldots, 1 + N * H/K)$

 (iii) $L(X) \leftarrow A\ X \uparrow 2 + B\ X + C$

 (iv) $E(X) = A\ X \uparrow 2 + B\ X + C$

  (v) $W \leftarrow A * (\int F(X, Y), (Y \leftarrow 0\ TO\ 1)) - G'(X)$

 (vi) $V(X) \leftarrow (1/(1 + X), 1/(2 + X), 1/(3 + X), 1/(4 + X))$

(vii) HILBERT $\leftarrow ([4, 4]\ V(1), V(2), V(3), V(4))$

(viii) DINVERSE $\leftarrow (DIAG(A) * IDENT(N)) \uparrow -1$

 (ix) $JACOBI(X) \leftarrow X + DINVERSE * (B - A\ X)$

  (x) $MDOMAIN(X) \leftarrow X - 1, (X > 1) \leftarrow X + 1, (X < -1) \leftarrow 0$

The left arrow ($\leftarrow$) indicates that the arithmetic expression on the right is
to be evaluated and its value is to be assigned to the variable on the left,
similar to what Fortran's equals sign ($=$) signifies. The equals sign has the
more usual mathematical meaning. Statement (ii) means EQUALSVAR is
equivalent to the expression $K * (1, 1 + H/K, \ldots, 1 + N * H/K)$. Values are
only substituted for the variables in the expression to the right of the equals
sign when a value is needed for the variable on the left. Variables defined
to the left of an equals sign are referred to as equals variables and variables

defined to the left of left arrow are called left-arrow variables, or simply variables.

Statements (iii) and (iv) illustrate a similar structure for functions. Functions are called equals functions or left-arrow functions according to how they are defined.

Statement (v) shows how a polyalgorithm may appear in an expression. Since X is not a variable of integration its current value is used.

Statements (vi) and (vii) combine to construct a 4 by 4 Hibert matrix, and show that functions with scalar arguments can return array values.

Statements (viii) and (ix) construct a vector-valued function with a vector argument that can be used to solve the system of linear equations $AX = B$ by the Jacobi method.

Statement (x) illustrates that functions may be defined to have different values on different domains. If there is no domain specified with the last definition of the function, this definition is used, whenever the point of evaluation does not lie in any of the explicitly stated domains.

III. Basic Construction of the Interpreter

NAPSS source text is transformed by a compiler into an internal text. The internal text generated for arithmetic expressions is a form of three-address code.

During compilation a name control block is created for each user variable. At this time the name control block is used as a name table entry. It contains the name of the variable and some basic attributes as to how it appears in the program.

The name control block is used during execution to hold values, pointers to values, and definitive attribute information for the variable. The name control block of a numeric scalar contains the actual value of the variable. If the name control block denotes something else, it contains a pointer to where the values are stored and information such as bounds, type of variable, number of dimensions, and number of arguments.

There is a fixed set of name control blocks used for storing temporary results during the evaluation of arithmetic expressions. They are similar to a user variable name control block except for the name field and are called temporary name control blocks.

The memory that a NAPSS program has is made up of a few pages of real memory which reside in core and a larger number of pages of virtual memory which reside in secondary storage and are brought in and out of real memory. As the number of user variables increases, the name table size is dynamically increased by obtaining a page at a time from real memory.

To aid the goal of machine independence for the system, almost all of the system is written in Fortran. The machine-dependent operations are restricted to "black-box" type modules coded in assembly language.

IV. Normal Arithmetic Expressions with Nonrecursive Operands

The flow of control in the arithmetic expression evaluator for expressions which do not involve recursive variables, function evaluations, or calls on polyalgorithms is given in Fig. 1.

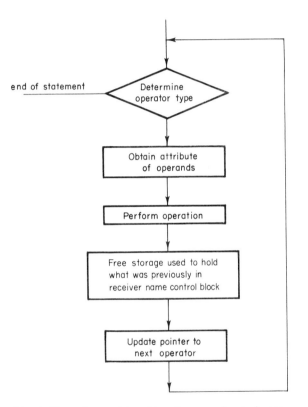

FIG. 1. Flow of control in arithmetic expression evaluator.

The operators are tested for in a fixed order, with the most frequently occurring ones first.

The attribute or type of an operand must be determined at execution time because attributes are not associated with variables during compilation. They are associated during execution time and may dynamically change during the execution of the program.

The fact that NAPSS gives the user the option of declaring some attributes and having the rest associated contextually complicates the attribute field of a name control block. Thus it contains a set of flags from which an attribute number is decoded. At the same time that the attribute of a variable is determined, necessary pointers are obtained from the name control block so that the variable may be used as an operand.

To reduce the work necessary to obtain the attribute of an operand, a look-ahead scheme is used where possible. If the result of an operation is an operand of the next operator then the attribute of that operand is flagged as being known. This scheme, even though only local, is quite useful, for frequently the result of the previous operation is an operand of the next operator.

There are three types of numeric scalars in NAPSS: real single precision, real double precision, and complex single precision. Integers are stored internally as real numbers and converted to the nearest integer.

With only three types of numeric scalars the number of addition routines needed to permit all possible combinations of operands is 3^2. If a fourth data type, double precision complex, were added the number of routines needed would be 4^2 or an increase of 77%. For this reason double-precision complex numbers are not now provided in NAPSS.

NAPSS does not use 3^2 routines for each of the basic binary arithmetic operators but rather only 3. This is achieved by converting one of the operands to match the attribute of the other. The scalar operands are placed in a work area before the operations are performed. The conversion is performed during transfer to the work area by zeroing a word when necessary.

The number of routines needed to perform the various operations cannot be reduced for array arithmetic to the same extent as for scalar arithmetic. This is because of the time needed to convert one operand to match the other and the increase in memory required to hold the operands. The number of routines needed to perform the binary array operations is 3^2 for multiplication and 2×3 for addition and subtraction. The number of routines needed to perform addition and subtraction is reduced more than for multiplication by taking into account the similarity between data types.

Arrays are stored in a random file and are brought into memory only when needed. The empty records in this file are chained together so that when a record is requested and the file is full the user can be asked to free a variable holding an array to allow his program to continue.

The computations of array arithmetic are performed in an area called the work pool, which is a dynamic area of real memory. When an array operation is to be performed, enough space is assigned to the work pool to hold the operands and resulting arrays. This may require removing pages from real memory. When a page of real memory is assigned to the work pool it is removed from real memory until the work pool is explicitly reduced in size.

The result of the array operation is not immediately put out in the random file with the other arrays. Rather the work pool remains intact with the operands and the result left in it. When the next array operation occurs the work pool is checked to see if it is empty; if not, the operands are compared with the contents of the work pool. If the result of the previous array operation is an operand of the present array operation then the result array need only be written out into the array file if it is an operand of a future operation.

The work pool is completely emptied at the end of each statement. Therefore, the process of optimizing the manipulation of arrays is only performed locally. The reason for this is that the work pool is used to manipulate other data types in addition to arrays.

When performing array arithmetic the system checks to see if the operands are conformal. The values of the index bounds of the operands do not affect the operations if the number of elements in the corresponding dimensions agrees. For example, it is illegal to multiply two row vectors or to add a row vector and a column vector. The system does not attempt to determine what the user intended in these situations. Rather it gives an error message, and asks the user to clarify the meaning of the statement.

The index bounds of a result array take their values from the bounds of the operand arrays. There is one exception to this. When two arrays are added or subtracted and their index bounds are not identical, the lower bounds of the result array are set to one.

If the result of an array operation is a one-element array, it is not treated as an array by the system but is stored as a scalar.

A temporary variable may be assigned several values during the evaluation of an arithmetic expression. This would pose no problem if all the results were scalars, because scalar values are stored in the name control blocks. However, the name control blocks for other data types only contain pointers to where the values are stored. This causes the problem of when to free the storage used to hold temporary results. Storage can be returned to the system periodically using a garbage collection scheme, or storage can be returned immediately, at the point it is no longer referenced.

The second alternative is used by the NAPSS interpreter and storage is freed immediately after a new value is assigned to the temporary variable, thereby permitting an operation to have the same temporary variable as an operand and as a result. This scheme has two main advantages: first, the type of storage to be freed is known at this point; second, the time required to free storage is uniformly consumed. This is of importance, since the system is intended for use in an on-line incrementally executing mode.

The arithmetic expression evaluator is called from various places in the interpreter and not just to evaluate arithmetic expressions appearing to the right of assignment statements. For this reason, and to facilitate recursion, the result of an evaluation is associated with a fixed temporary name control block. The results of every arithmetic expression evaluation may be obtained from this temporary name control block by whatever portion of the interpreter requested the evaluation.

The name control block which receives the result of an arithmetic expression evaluation is only used to pass the value along to whatever portion of the interpreter invoked the arithmetic expression evaluator. Thus, the storage associated with its previous value is not returned to the system. If the storage associated with the result temporary name control block were freed each time a new value is associated with it, storage would be returned which may now be associated with a user variable or which has already been freed by some other portion of the interpreter.

V. Evaluation Arithmetic Expression with Recursive Operands

The occurrence of an equals variable in an arithmetic expression causes the arithmetic expression evaluator to recurse. The recursion needed to evaluate equals variables is limited to one routine, the master controller. This routine is responsible for determining what the next operator is, what the attributes of the operands are, and what routine is to be invoked to perform the operation. The routines which perform the various operations expect to receive pointers to where the actual values of the operands may be obtained. This implies that the equals variable's expression must be evaluated before calling the operator routine.

When recursion occurs the text for the current arithmetic expression is written out onto a sequential file along with a group of variables that must be saved for the interpreter, and all the temporary name control blocks except for the temporary name control block used to hold the final result of an arithmetic expression evaluation. All of these variables are equivalenced to one contiguous area so that they may be manipulated as a unit. A flag is set in the interpreter's recursive variable area just before the push-down of storage is performed. This flag is used to return to the point in the master controller where recursion occurred after the symbolic variable's expression has been evaluated.

Because of the manner in which storage associated with temporary variables is freed, all temporary variables are set to undefined after the push-down area has been written out. This allows them to be reused during the evaluation of the new expression without the danger of freeing storage which was associated with the temporary variables on the previous level. After the internal text for the new expression is read in and the necessary pointers are adjusted, control is transferred to the main entry point of the master controller to begin execution. This new expression may also contain symbolic variables; if so, the process is repeated.

The compiler does not check for symbolic definitions which yield non-terminating definitions. This is the responsibility of the arithmetic expression evaluator during execution. The statement $A = A + B$ and the statements $A = B + C$ and $B = A + D$ both cause this situation to occur. The interpreter could check for the occurrence of this when the assignment of an equals variable is made or could keep a list of what variables have caused recursion and check this before each recursion to eliminate the possibility of infinite recursion. Neither of these methods are used, however, in NAPSS because each require extensive checking be performed, and for the vast majority of cases this is unnecessary. Instead a limit has been placed on the depth of recursion. If the arithmetic expression evaluator attempts to recurse past this limit, the user is given an error message indicating that the recursion depth has been exceeded and suggesting that the definitions of the symbolic variables which caused the initial recursion is inconsistent.

The result from the evaluation of an equals variable is put in the temporary name control block used to receive the result of all arithmetic expression evaluations. This name control block is fixed in the compiler and the interpreter and is the only temporary name control block which is not in the push-down area.

Before the push-down area can be restored and execution of the original expression resumed, a pass must be made through the other temporary name control blocks to free any storage that is associated with them. If this were not done this storage space would be lost since garbage collection is not used.

All temporary name control blocks need not be checked during the freeing process because the compiler assigns the temporary variables in a linear fashion and reuses them as soon as their results are no longer needed. Thus the interpreter need only scan them until the first name control block is encountered which is still marked as undefined.

After all the temporary variables are freed, the push-down area is restored and the name control block containing the result of the equals variable expression is copied into a special temporary name control block which is used only for the values of symbolic variables. This is done to permit both operands of an operator to be symbolic. The special name control block is used in place of the symbolic variable in the evaluation of the original arithmetic expression.

To avoid needless recursions to evaluate the same symbolic variable, a local check is made to determine if any of the other operands of the current operator are the same variable. If any of them are, the special name control block is substituted in the arithmetic expression for them also.

There is a problem associated with the use of the work pool and recursion. If there are any arrays in the work pool when a symbolic variable is encountered, the work pool must be emptied. This saves the temporary result array which had resided only in the work pool in the array file. Were this not done and the symbolic expression to be evaluated involved any array arithmetic, this result array would become associated with a temporary name control block on the wrong level.

If an error occurs while evaluating the expression for a symbolic variable the storage associated with the temporary variable name control blocks on the difference levels must be freed. This is not necessary if the arithmetic expression evaluator is at level zero when the error occurs, because in this case the normal freeing mechanism frees the storage associated with temporary variable the next time the arithmetic expression evaluation is called. However, when an error is detected on a nonzero level, the storage associated with all temporary variables is freed a level at a time until level zero is reached. Information about what caused the error and at what level it occurred is saved before the recursion levels are rolled back, so that an error message can be given the user by the section of the interpreter which initially called the arithmetic expression evaluator.

VI. Evaluation of Arithmetic Expressions Involving Symbolic Functions

The independent variables in the definition of a symbolic function should be place-markers, to insure that the values of the variables used as independent variables in a function definition will not be changed when the function is evaluated.

When a symbolic function is defined in NAPSS the compiler replaces all references to the N independent variables in the expressions defining the function with references to the first N temporary name control blocks. Therefore when the function is evaluated the name control blocks which represent the value at which the function is to be evaluated are copied onto the first N temporary name control blocks.

This can not be done directly for two reasons. First, the function evaluation may appear at any point in an arithmetic expression and therefore some values may already reside in the first N temporary name control blocks. Second, one or more of the actual parameters may be arithmetic expressions which have been evaluated and had their results put in some of the temporary name control blocks.

These problems cause the arithmetic expression evaluator to recurse before the actual argument name control blocks are copied into the temporary name control block and force the use of a temporary area to collect the parameter name control blocks.

The appearance of an equals variable as an actual parameter is not handled in the same fashion as other types of parameters. Its name control block is not directly copied onto the corresponding temporary name control block. If it were this would cause the arithmetic expression evaluator to recurse each time this parameter appears in the text for the function. Since the value of the equals variable cannot change during the evaluation of the function, this is avoided by having the arithmetic expression evaluator recurse and evaluate the equal variable before its name control block is copied into the corresponding temporary name control block. Thus, the name control block for the result of evaluating the equals variable is used in place of the name control block of the equal variable itself.

After all the name control blocks of the arguments are in the temporary area used to collect them, the arithmetic expression evaluator recurses and the actual parameter name control blocks are copied onto the first N temporary name control blocks.

Before evaluation commences the function is checked to see if it is a left-arrow or equals function. If it is a left-arrow function then all nonparameter variables appearing in the function text had their values fixed when the function assignment was made. To fix the value of these variables when the function assignment was performed a copy of each of their name control blocks and associated storage was created. Thus to evaluate a left-arrow function these local name control blocks are brought into the name table area and pointers are adjusted so that these variables are referenced while the function is being evaluated.

If the function to be evaluated is an equals function all nonparameter variables appearing in the function text are not fixed when the function assignment is made, but assume their current value when the function is evaluated. Thus no local name control blocks are associated with equals functions.

The point at which the function is to be evaluated is checked to see if the function is defined at this point. The check is performed by evaluating the boolean expressions associated with the various definitions of the function. The boolean expressions are evaluated in the order the user has stated them. When no boolean expression appears with a definition the function is assumed to have this definition everywhere or everywhere else, depending on whether or not other definitions with associated boolean expressions precede it.

After the result of a function, evaluation is put in the temporary name control block which receives the result of all arithmetic expression evaluations, the arithmetic expression evaluator returns to the level at which the function invocation occurred.

The process of returning to the level in the arithmetic expression evaluator at which the function invocation occurred is similar to what occurs when returning from the evaluation of an equals variable. The only difference is the freeing of the temporary name control blocks before the recursion area is restored. All of the temporary name control blocks may not be freed as they were after the evaluation of an equals variable, because to evaluate a function the first N temporary name control blocks were used to hold copies of the parameter name control blocks.

The copy of the actual name control block for the parameter is flagged when it is put into the corresponding temporary name control block, so that when the temporary name control blocks are freed the ones used to hold parameters will not be freed. However, one type of temporary name control block used to hold a parameter is not flagged and must have its associated storage freed. This is the temporary name control block used to hold the value of a parameter which corresponds to an equals variable. Since the equals variable is evaluated before the evaluation of the function, the only name control block pointing to the value of the equals variable is the temporary name control block used as parameter.

If an error occurs during the evaluation of a function, the arithmetic expression evaluator saves information as to what caused the error and at which level it occurred and returns to level zero as it does when an error occurs during the evaluation of an equals variable.

VII. Evaluation of Arithmetic Expressions with Polyalgorithm Calls

A polyalgorithm is formed by grouping several numerical procedures and a supervisor into a single procedure for solving a specific problem. The polyalgorithm combines the various methods along with the strategy for their selection and use into a single method which is relatively efficient and very reliable.

The appearance of either an integral or a derivative in an arithmetic expression causes the arithmetic expression evaluator to invoke a polyalgorithm to perform the operation. Although the polyalgorithm contains its own supervisor, it requires the arithmetic expression evaluator to evaluate the function involved. Therefore the process of evaluating an integral or derivative of a function is recursive. It is also considerably more complicated than evaluation of an equals variable or a function. In the later two cases only the master controller of the arithmetic expression evaluator itself is involved; here the arithmetic expression evaluator and a polyalgorithm are involved. In addition, since the polyalgorithm may require that the value of the function involved be computed repeatedly, the normal process of function evaluation, which is itself recursive, cannot be used in this case for practical reasons.

When a derivative or integral appears in an arithmetic expression being evaluated all the arguments required by the polyalgorithm, such as number of derivatives, integral bounds, or point of differentiation, are evaluated in the arithmetic expression evaluator before the polyalgorithm is invoked. The values of these parameters are passed to the polyalgorithm initially so that the arithmetic expression evaluator need only be reentered from the polyalgorithm to evaluate the function involved.

Before the polyalgorithm is called the arithmetic expression evaluator recurses as it does when evaluating a function. The text of the function involved in the operation is placed in the appropriate place in the interpreter for evaluation. All parameters necessary for evaluation are also set up except for filling in the temporary name control block which corresponds to the variable of differentiation or integration. Thus when the polyalgorithm needs to evaluate the function, all that remains to be done is supply the value of this point.

When the polyalgorithm is called from the arithmetic expression evaluator and a value of the function involved is needed, the arithmetic expression evaluator must be returned to, or must be called from the polyalgorithm. If the polyalgorithm calls the arithmetic expression evaluator, the address where the arithmetic expression evaluator was initially called from would be destroyed. If the polyalgorithm returns to the arithmetic expression evaluator, this would create problems in the organization of the polyalgorithm. For if the point at which the function must be evaluated is several routines removed from the original call on the polyalgorithm, all of these calls would have to be retraced for each evaluation of the function, or the polyalgorithm would have to be reorganized.

To avoid both of these problems direct transfers are used to transfer control between the arithmetic expression evaluator and the polyalgorithm after the polyalgorithm is initially entered. This method of transferring between routines is accomplished by the use of assigned go to statements in each of the routines.

When the polyalgorithm completes its work it returns to the arithmetic expression evaluator normally. The arithmetic expression evaluator then restores itself to the level at which the integral or derivative occurred. The process of freeing storage associated with temporary name control blocks and the popping up of the recursive area is similar to what is done after the evaluation of a function.

If an error occurs which causes the polyalgorithm to terminate evaluation it returns to the arithmetic expression evaluator as if the evaluation was successful but with an error flag set. The arithmetic expression evaluator returns to level zero as is done when an error occurs during an equals variable or a function. The actual message is issued by the routine which initially called the arithmetic expression evaluator.

REFERENCES

Rice, J. R., and Rosen, S. (1966). NAPSS—A Numerical Analysis Problem Solving System, Proc. ACM Natl. Conf. 21st, Los Angeles, 1966, ACM Publ. P-66, p. 51.

Rice, J. R. (1968). On the Construction of Polyalgorithms for Automatic Numerical Analysis, "Interactive Systems for Experimental Applied Mathematics" (M. Klerer and J. Reinfelds, eds.), p. 301. Academic Press, New York.

Roman, R. V., and Symes, L. R. (1968). Implementation Considerations in a Numerical Analysis Problem Solving System, "Interactive Systems for Experimental Applied Mathematics" (M. Klerer and J. Reinfelds, eds.), p. 400. Academic Press, New York.

Symes, L. R., and Roman, R. V. (1968). Structure of a Language for a Numerical Analysis Problem Solving System, "Interactive Systems for Experimental Applied Mathematics" (M. Klerer and J. Reinfelds, eds.), p. 67. Academic Press, New York.

Symes, L. R., and Roman, R. V. (1969). Syntactic and Semantic Description of the Numerical Analysis Programming Language (NAPSS). Purdue Univ. Tech. Rep., CSD TR 11, revised.

Symes, L. R. (1970). Manipulation of Data Structures in a Numerical Analysis Problem Solving System, *Proc. Spring Joint Comp. Conf., AFIPS* Vol. 36, p. 157.

5.17 Toward Computer-Aided Production of Software for Mathematical Programming

*R. Bayer**

BOEING SCIENTIFIC RESEARCH LABORATORIES

SEATTLE, WASHINGTON

I. Introduction and General Problem

This paper describes an approach toward solving a specific case of a rather general problem in the production of software. As a rule, solution methods for computational problems become more data dependent as they are being moved closer to a machine executable form. The published mathematical analysis of a problem often disregards questions of data structuring and data representation, and describes an abstract solution method. When presenting an explicit algorithm, there might be some consideration of how the algorithm depends on the data involved. In a high-quality computer program, however, written with an effort to minimize storage requirements and run-time, one generally finds that the algorithmic part is completely tied in with and highly dependent on data declarations, data access methods, and data representations. In the course of this development the program realizing an algorithm becomes quite unreadable and the mathematical core of the algorithm is being increasingly obscured. Any changes in the structure or representation of the data make it necessary to modify the program extensively or to rewrite it completely, even if the underlying algorithm remains basically the same. This situation is typical for many application areas, where ideally the same algorithm should be useful for solving a large class of problems, as, for example, in control theory, file processing, network analysis, data retrieval, and matrix calculations.

It is clearly desirable to overcome this wide discrepancy between a mathematical algorithm and a high-quality computer program. Ideally one should be able to start with a less involved description of a computational process in which the algorithmic logic and the special properties of data are fairly well separated and clearly visible. Then the power of the computer should be used

* Present address: Purdue University, Lafayette, Indiana.

to automatically generate an efficient program tailored to the special structure of a problem [Bayer *et al.* (1968)].

Little is known how to reach this goal, in general, but progress seems possible for special application areas. We have developed one approach which looks promising. Whether it is susceptible to generalization in other application areas is an open question.

II. Matrix Calculi for Mathematical Programming

The special application area considered is mathematical programming. The typical data structures involved are large rather sparse matrices with an elaborate substructure of zero and nonzero blocks. Three *concatenation operators* serve to describe such matrices conveniently: namely, horizontal "$\rightarrow$", vertical "$\downarrow$", and diagonal "$\searrow$" concatenation. They are explained by the obvious illustrations in Fig. 1 with A, B, C, D, E, and F being matrices or arbitrary matrix valued expressions.

FIG. 1. Concatenations of matrices.

Submatrices are extracted by selecting arbitrary sequences of rows and columns from a matrix. This is accomplished by introducing sets of indices, called *ranges*, which are finite sequences of positive integers, and the *extraction operators* "$\lhd$" for extracting rows and "$\rhd$" for extracting columns. Rows and columns are indexed by consecutive integers starting with 1. Denoting ranges by $[r_1, r_2, \ldots, r_l]$, submatrices can be formed from rows of a matrix M, e.g., $[2, 3, 2] \lhd M$, and from columns of M, e.g., $M \rhd [4, 2]$, as illustrated in Fig. 2.

$$M = \begin{pmatrix} 1 & 3 & 2 & 4 \\ 2 & 7 & 6 & 5 \\ 2 & 1 & 8 & 9 \end{pmatrix}; \quad [2,3,2] \lhd M = \begin{pmatrix} 2 & 7 & 6 & 5 \\ 2 & 1 & 8 & 9 \\ 2 & 7 & 6 & 5 \end{pmatrix}; \quad M \rhd [4,2] = \begin{pmatrix} 4 & 3 \\ 5 & 7 \\ 9 & 1 \end{pmatrix}$$

FIG. 2. Submatrices of rows and columns.

In addition there are the usual matrix operators: $+$, $-$, $/$, and $\times$ are for elementwise addition, subtraction, division, and multiplication, respectively;

$\otimes$ is for matrix multiplication; := is the assignment sign which we treat like an operator; and ^{T} is the matrix transposition.

To illustrate how efficient programs depend on the structure of the data involved we consider the assignment

$$A := B \otimes (C \twoheadrightarrow D) \tag{1}$$

This is a concise and clear description of the task to be performed. In writing a program to carry out this computation one would observe that the right-hand expression can be written as a concatenation resulting in the assignment

$$A := (B \otimes C) \twoheadrightarrow (B \otimes D)$$

which can now be split up into the two assignments

$$A_1 := B \otimes C; \qquad A_2 := B \otimes D$$

where A_1 and A_2 are certain submatrices of A which can be described by the column extractions $A \rhd R_1$ and $A \rhd R_2$, respectively, using certain ranges R_1 and R_2. Now the two assignments

$$A \rhd R_1 := B \otimes C; \qquad A \rhd R_2 := B \otimes D \tag{2}$$

are computationally equivalent to the original assignment in the sense that performing them has the same effect as performing (1). Translating (2) into a piece of Algol or Fortran program results in two separate sets of loops reflecting directly the structure of the data, namely of the concatenated matrix $C \twoheadrightarrow D$.

The transformations performed in this example are aimed at optimizing the evaluation of expressions. We found two characteristics indicating that an expression is in a form suitable for evaluation on a computer:

(i) All concatenations $\downarrow$, $\twoheadrightarrow$, and $\searrow$ are outermost operators.

(ii) All submatrix extraction operators $\rhd$ and $\lhd$ are innermost; where "outermost" and "innermost" refer to the nested structure of the completely parenthesized expression. Expressions having both characteristics are said to be in *normal form*.

Intuitively, the reasons that normal form expressions are suitable for evaluation are as follows:

(i) Moving concatenations in expressions toward the outside corresponds to moving logical tests—in our example, whether to select a column from C or from D to form an inner product with a certain row from B—outside of loops, and to perform the tests at run-time only once outside of instead of many times inside of the loops.

(ii) Moving submatrix extractions inside corresponds to performing the operations only on the necessary submatrices, instead of performing them on the whole matrices and then selecting a submatrix from the obtained result. Thus machine operations and possibly auxiliary storage are generally saved by moving "$\triangleright$" and "$\triangleleft$" inside as indicated by the following example:

$$(B \otimes C) \triangleright S \equiv B \otimes (C \triangleright S)$$

Rules can be found for transforming expressions so that they come closer to being in normal form. In the following examples of transformation rules the expressions enclosed in brackets are range expressions whose operators are explained in Sect. III. These rules mean that an expression to the left of "$\equiv$" is computationally equivalent to the expression to the right of "$\equiv$" which is closer to being in normal form.

$R1$:

$$A + (B \rightarrow C) \equiv A \triangleright [B\#] + B \rightarrow A \triangleright [C\# + \rho(B\#)] + C$$
$$A + (B \downarrow C) \equiv [\#B] \triangleleft A + B \downarrow [C\# + \rho(B\#)] \triangleleft A + C$$
$$A + (B \searrow C) \equiv ([\#B] \triangleleft A \triangleright [B\#] + B \downarrow [\#C + \rho(\#B)]$$
$$\triangleleft A \triangleright [B\#]) \rightarrow ([\#B] \triangleleft A \triangleright [C\# + \rho(B\#)]$$
$$\downarrow [\#C + \rho(\#B)] \triangleleft A \triangleright [\#C + \rho(\#B)] + C)$$

$R2$:

$$A \otimes (B \rightarrow C) \equiv A \otimes B \rightarrow A \otimes C$$
$$A \otimes (B \downarrow C) \equiv A \triangleright [\#B] \otimes B + A \triangleright [\#C + \rho(\#B)] \otimes C$$
$$A \otimes (B \searrow C) \equiv A \triangleright [\#B] \otimes B \rightarrow A \triangleright [\#C + \rho(\#B) \otimes C$$
$$(A \rightarrow B) \otimes C \equiv A \otimes [A\#] \triangleleft C + B \otimes [B\# + \rho(A\#)] \triangleleft C$$
$$(A \downarrow B) \otimes C \equiv A \otimes C \downarrow B \otimes C$$
$$(A \searrow B) \otimes C \equiv A \otimes [A\#] \triangleleft C \downarrow B \otimes [B\# + \rho(A\#)] \triangleleft C$$

What one hopes for is to find a set of transformation rules and a suitable class of normal form expressions, such that every properly formed expression can automatically be transformed into a computationally equivalent expression in normal form. In the matrix calculus as described so far this is the case. The details of this development are described in Bayer and Witzgall (1968, 1970).

In order to obtain the desired descriptive power several extensions of the matrix calculus considered so far are necessary. They are primarily:

(i) introduction of *lists* of matrices and ranges up to arbitrary levels; i.e., lists of lists, etc.;

(ii) extension of operators to list operators like $*+*$, $*+$, and $+*$;

(iii) extension of matrix operators ω to become cumulative operators $\int \omega$ over all elements of a list; e.g., in order to add up or to concatenate all matrices of a list.

Some of these generalizations are indicated by the illustrations in Fig. 3 where K, L, and A are lists of matrices, and M is a simple matrix.

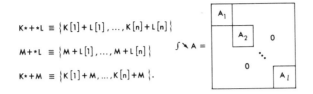

$$K*+*L \equiv \{K[1]+L[1], \ldots, K[n]+L[n]\}$$

$$M+*L \equiv \{M+L[1], \ldots, M+L[n]\}$$

$$K*+M \equiv \{K[1]+M, \ldots, K[n]+M\}.$$

Fig. 3. Lists, list operators, and cumulative operators.

It is also necessary, of course, to generalize normal form expressions and transformation rules to cover the extensions to lists and list expressions. These extensions are omitted in this paper, but they are described in considerable detail in Bayer and Witzgall (1968, 1970).

III. Ranges and Range Manipulation

As a consequence of matrix manipulations, rather complicated operations are necessary on ranges. As indicated before a *range* is a (possibly empty) sequence of positive integers:

$$R = [r_1, r_2, \ldots, r_l]; \quad r_i > 0, \quad i = 1, 2, \ldots, l$$

A range S consisting of the consecutive integers from i to j is also abbreviated as

$$S = [i:j]; \quad i, j > 0$$

and is called an *interval.*

We found it useful to consider the following types of special ranges:

monotonic: the elements of the range are nondecreasing;
interval: the range is of the form [i:j];
initial: the range is of the form [1:j];
permutation: a permutation of an initial range; e.g., [3, 1, 2];
nonrepetitive: no two elements of the range are the same.

Some types allow more economical representations and faster evaluation than in the general case.

All operations necessary on ranges can be described using the following operators:

∘   Range composition, similar to function composition; e.g.,
    $[7, 8, 9, 5] \circ [4, 2] = [5, 8]$.

γ   A monadic sorting operator; e.g., $\gamma[7, 3, 1, 4, 3] = [1, 3, 3, 4, 7]$.

α   A monadic permutation operator yielding for an arbitrary range R a
    permutation such that $(\gamma R) \circ (\alpha R) = R$; e.g., $\alpha[7, 3, 1, 4, 3] = [5, 2, 1, 4, 3]$.

Applied to an arbitrary matrix expression E to yield $\#E$, the row
    range of the value of E, and $E\#$, the column range of the value of E.

ρ   An integer function yielding the length of a range; i.e., $\rho R =$ the
    number of elements of R.

ι   Yielding an initial range; namely, $\iota R = [1 : \rho R]$.

∪   Range concatenation (not set union); e.g.,
    $[1, 2, 5] \cup [2, 3] = [1, 2, 5, 2, 3]$.

+   A shift operator adding a positive integer to all elements of a range;
    e.g., $[3, 7] + 2 = [5, 9]$.

μ   A shift operator subtracting a positive integer from all elements of a
    range and deleting nonpositive elements; e.g., $[3, 7, 9, 5] \mu 5 = [2, 4]$.

η   Restriction operator deleting all elements of a range which are
    greater than a given integer, e.g. $[3, 4, 7, 5] \eta 5 = [3, 4, 5]$.

If for the moment we consider ranges as integer vectors, then we can give a
concise definition of our range operators in terms of Iverson's APL operations
as described in Pakins (1968).

Range operations		Equivalent APL expressions
$\#E$	=	$\iota(\rho E)[1]$
$E\#$	=	$\iota(\rho E)[2]$
ρR	=	ρR
ιR	=	$\iota \rho R$
$R + n$	=	$R + n$
$R \cup S$	=	R, S
$R \circ S$	=	$R[S]$
$R \mu n$	=	$((R > n)/R) - n$
$R \eta n$	=	$(R \leq n)/R$
γR	=	$R[\blacktriangle R]$
αR	=	$\blacktriangle \blacktriangle R$

These operators need not necessarily be accessible to the programmer as part of a language but are only needed in the machine for manipulation purposes. They were selected with only two goals in mind: namely,

(i) to provide just as much descriptive power as needed;

(ii) to allow fast evaluation, the slowest being the sorting and permutation operators requiring both on the order of (n log n) computer operations, where n is the length of the range involved.

As a consequence of the operator $\#$ the range expressions obtained through the matrix manipulation may contain arbitrarily complicated matrix expressions as subexpressions. It is not acceptable, of course, to first evaluate matrix expressions in order to obtain ranges needed to evaluate the same matrix expressions more efficiently. It is therefore necessary to eliminate matrix operators from range expressions. This can easily be done by repeated application of a set of transformation rules like the following:

$$\#(A + B) \equiv \#A \quad or \quad \#B$$
$$\#(A \otimes B) \equiv \#A$$
$$(A \searcher B)\# \equiv \iota((A\#) \cup (B\#))$$
$$\#(R \lhd A) \equiv \iota R$$

Notice that this is again an example, quite simple in this case, of the situation described in Sect. II: There is a set of properly formed expressions, namely all range expressions containing possibly matrix-valued subexpressions. A subclass of these expressions, which we can consider as normal form expressions, are the range expressions free of matrix operators. A set of rules allows the transformation of arbitrary range expressions into computationally equivalent normal forms.

After the elimination of matrix operators, however, range expressions can be simplified even further using some of the special properties of range operators. We found between forty and fifty simplification rules like the following for range expressions:

$$\alpha\gamma R \equiv \iota R$$
$$\iota(R \circ S) \equiv \iota S$$
$$(\#A) \circ R \equiv R$$
$$R \circ (S \cup T) \equiv (R \circ S) \cup (R \circ T)$$

Unfortunately, there is no obvious property of range expressions to single out a subclass of simplified normal forms. Such a subclass can be implicitly specified, however, as those expressions to which no further simplification rules are applicable. The completeness problem, namely whether all expressions can be transformed into normal form, then becomes a termination problem, namely whether for every range expression (without matrix operators) there is a sequence of simplifications which terminates in the sense that no further simplifications are applicable to the resulting expression.

Our simplification rules have no obvious property which would settle the termination problem: Not all simplification rules physically shorten expressions, and in some pathological cases a simplified expression might require more machine operations than the unsimplified expression. We were able to prove, however, that simplification always terminates if the rules are applied in a certain way. These results and various generalizations involving lists of matrices and ranges will be described in detail in a forthcoming technical report.

Let us now look at a particular example. Transforming the matrix expression

$$(M \lhd (E \downarrow F)) \otimes (A + (B \searrow C))$$

into normal form one obtains as a subexpression the following range expression:

$$\#([\#C + \rho(\#B)] \lhd A \rhd [C\# + \rho(B\#)] + C) + \rho(\#([\#B]$$
$$\lhd A \rhd [C\# + \rho(B\#)]))$$

After removing all matrix operators from it one obtains

$$\iota[\#C + \rho(\#B)] + \rho(\iota[\#B])$$

Further simplification finally leads to

$$\#C + \rho(\#B)$$

We are now in a position to give an example of a complete manipulation. Consider the assignment

$$A := ((B + C) \searrow D) \otimes (E \rightarrow F)$$

After transforming it into normal form one obtains four assignments containing very complicated range expressions. Removing matrix operations from the range expressions and then simplifying range expressions one obtains the following four assignments, range expressions being enclosed in brackets:

$$[\#B] \lhd A \rhd [E\#] := (B + C) \otimes [B\#] \lhd E$$
$$[D\# + \rho B\#] \lhd A \rhd [E\#] := D \otimes [D\# + \rho B\#] \lhd E$$
$$[\#B] \lhd A \rhd [F\# + \rho E\#] := (B + C) \otimes [B\#] \lhd F$$
$$[D\# + \rho B\#] \lhd A \rhd [F\# + \rho E\#] := D \otimes [D\# + \rho B\#] \lhd F$$

These then yield the following piece of Algol program, in which "rows" and "columns" are functions resulting in the number of rows or columns, respectively, of a matrix:

for i := 1 *step* 1 *until* rows(B) *do*
for j := 1 *step* 1 *until* columns(B) *do*

AUX[i, j] := B[i, j] + C[i, j];

for i := 1 *step* 1 *until* rows(B) *do*
for j := 1 *step* 1 *until* columns(E) *do*

  begin S := 0;
    for k := 1 *step* 1 *until* columns(B) *do*
    S := S + AUX[i, k] × E[k, j];
    A[i, j] := S *end*;

for i := 1 *step* 1 *until* rows(D) *do*
for j := 1 *step* 1 *until* columns(E) *do*

  begin S := 0;
    for k := 1 *step* 1 *until* columns(D) *do*
    S := S + D[i, k] × E[columns(B) + k, j];
    A[i + rows(B), j] := S *end*;

for i := 1 *step* 1 *until* rows(B) *do*
for j := 1 *step* 1 *until* columns(F) *do*

  begin S := 0;
    for k := 1 *step* 1 *until* columns(B) *do*
    S := S + AUX[i, k] × F[k, j];
    A[i, columns(E) + j] := S *end*;

for i := 1 *step* 1 *until* rows(D) *do*
for j := 1 *step* 1 *until* columns(F) *do*

  begin S := 0;
    for k := 1 *step* 1 *until* columns(D) *do*
    S := S + D[i, k] × F[k + columns(B), j];
    A[i + rows(B), j + columns(E)] := S *end*

IV. Some Language Design and Implementation Problems

A. PROCEDURES AND PROGRAM MANIPULATION

We have seen that a piece of program as it will eventually be executed on a computer may heavily depend on the data involved. This causes difficulties with the concept of a closed subroutine or procedure, since some of the data, namely the actual parameters, may be different for each call. Several solutions for the problem are conceivable:

(i) Restricting the class of expressions acceptable as actual parameters so that the manipulation of a procedure body will not depend on the actual parameters.

(ii) Treating procedures as open procedures or macros. In this case a copy of the procedure is substituted into the program for each procedure call, the actual parameters replacing the formal parameters. Thus actual parameter dependent manipulation of procedures is possible without limitations.

(iii) Performing the program manipulation dynamically at execution time. Thus for each procedure call the procedure body would first be manipulated depending on the actual parameters, and then it would be executed.

Unfortunately, any single one of these three approaches has serious drawbacks: Manipulation of closed procedures at compile time is too restrictive, whereas open procedures require often unnecessary repetition of code and in general do not allow recursion. Manipulation at execution time makes the distinction between open and closed procedures disappear and allows recursion and parameter-dependent manipulation. It is, however, quite inefficient, causes redundancy of manipulations and code substitutions, and requires interpretation of the program instead of translation.

A quite satisfactory solution is achieved by providing both open and closed procedures in the language. Open procedures can then be used to the extent made necessary by the manipulation depending on the form of actual parameters, and closed procedures provide the economy in code and the power of recursion. Even more powerful would be a language facility for specifying for each individual procedure call, whether it should be treated as open or closed.

The following examples illustrate the advantages of using a proper combination of open and closed procedures:

open procedure $Q(x)$; $\langle$procedure body of $Q\rangle$;

$$Q(A); \quad Q(B \to C); \quad Q(D \to E)$$

Obviously Q should be open since for two calls manipulation depending on the actual parameters $B \to C$ and $D \to E$ is necessary. Both calls, however, will duplicate essentially the same insertion of code and the same manipulation. One code insertion and one manipulation of the procedure Q could be avoided by introducing a closed procedure P, whose body simply consists of a call of Q, as follows:

open procedure   Q(X); ⟨procedure body of Q⟩;
closed procedure P(Y, Z); Q(Y → Z);
             Q(A); P(B, C); P(D, E)

B. A COMPILE TIME FACILITY FOR PROGRAM ADAPTABILITY

As indicated before computations are performed on matrices with an elaborate substructure which changes frequently during the development and the use of a program. Therefore a high degree of program adaptability is important.

A highly structured matrix can be written as a complicated expression of course. Using this expression explicitly throughout the program, however, requires many tedious and error-prone manual changes whenever the structure of the matrix changes, which is expected to happen frequently.

A considerable improvement is achieved if such expressions can be referred to by simple names, and if the compiler substitutes the corresponding expressions for all occurrences of the name. Such names for expressions can be introduced by declarative "let-statements"; for example,

$$\textit{let } A \textit{ be } (B \to C) + D \searrow E.$$

Any change in the structure of A can then be accommodated by simply modifying the definition of A, but the rest of the source program stays intact. The probably quite extensive "clerical" changes resulting in the object program are achieved by the compiler via substitution of the new expression for A, symbolic manipulation of the new source program, and finally translation. The power of the let-statement could be increased, of course, by using parameters in it.

Although the let-statement may resemble open procedures or functions, it is quite different from both. In contrast to the body of an open procedure, the substituted expression is manipulated and optimized together with the context into which it was substituted. In that sense it is more general and powerful than open procedures. It is more restricted than procedures or functions, however, in the sense that only a very small part of the language, namely expressions, can be used in it.

Together with the concept of symbolic program manipulation the let-statment provides the high degree of program flexibility and adaptability needed for applications in mathematical programming.

C. TYPES, SUBSCRIPTS, ARRAY BOUNDS, LOOPS

The use of structured entities like arrays and lists requires the specification of sets of indices or names which are used to denote the individual elements of such entities. The index or name sets associated with a structured entity remain constant throughout its lifetime.

In order to allow subscript checking and to detect logical inconsistencies in the program at compile-time, it is useful to explicitly introduce certain index sets, e.g., the row and column indices of matrices, as separate entities called *domains* or *set theoretic types* [Reynolds (1969)], into the program. Such domains are not variables and remain constant throughout their scope. It is then possible to restrict the acceptable values of index variables or ranges to such domains. Many otherwise necessary checks at run-time can then be replaced entirely by syntactic checks at compile-time or by less costly run-time checks for the proper binding of variables.

Considerable work on this subject has recently been done by Reynolds (1969), Hoare (1969), and Wirth (1969) under a very general approach using set theoretic types. Here this work is primarily of interest with regard to indexing, naming, and control variables of loops.

To illustrate the problem let us consider two simple examples:

Example I   Let R and C be previously declared domains and assume that the matrix A and the range M are declared as follows:

$$matrix \ A \ rows \ R \ columns \ C$$
$$range \ M \ from \ R$$

indicating that the elements of M must be from the domain R, the same as the rows of A. Then it is not necessary to check the range elements of M in the expression $M \lhd A$ at run-time since, according to information available at compile-time, the elements of M are restricted to the domain R. Checks at run-time may only be necessary in some cases, but not in all, when the value of M is changed.

Example 2   Consider the following matrix declarations and matrix multiplication:

$$matrix \ A \ rows \ R \ colums \ S;$$
$$matrix \ B \ rows \ S \ columns \ C;$$
$$matrix \ D \ rows \ R \ columns \ C;$$
$$for \ i \in R \ do$$
$$for \ j \in C \ do$$
$$begin \ D[i, j] := 0;$$
$$for \ k \in S \ do$$
$$D[i, j] := D[i, j] + A[i, k] \times B[k, j]$$
$$end$$

Assuming that the loop control variables i, j, and k may not be changed inside the loops, it can be recognized at compile-time that all subscripts will be acceptable, and no checks at run-time are necessary at all.

Initial experience with sample programs indicates that with these techniques at least in the area of linear programming the bulk of all subscript checking can be done at compile-time.

D.   CONTROL OF DATA TRAFFIC IN SEVERAL LEVELS OF STORAGE

In the applications we are concerned with computations require large volumes of data. Consequently, a considerable data traffic must be maintained between the main store and various backup stores. The time spent on a computation may largely depend on how well and efficiently this data traffic can be controlled. Various techniques are in use differing mainly in the amount of explicit control exercised by a user program over the data traffic. Two extremes are:

(i) the program handles its own data movements, trying to utilize possible overlaps between moving data and processing;

(ii) all data traffic is handled by the operating system using automatic segmentation and paging techniques.

Either scheme has serious drawbacks and seems unsuitable for applications in mathematical programming. The reason is that generally programs are logically structured in such a way that I/O occurs primarily at the beginning or at the end of blocks. Furthermore it is often predictable which data will be needed in main store next and which are no longer needed, a situation quite different from the usual paging environment.

We are therefore considering language facilities, described in more detail in Bayer (1968), to describe data movements easily and to control them through the dynamic structure of the program. Basically the scheme is a generalization of the block and declaration concepts of Algol 60. The block concept with the implied stack mechanism for storage allocation is extended to encompass not only main store, but several levels of storage.

The scheme is related to that described by Hoare (1968) for file processing applications, but there are several modifications. The basic principles and mechanisms are:

*P*1: The lifetime of a variable, of its value, and of the storage allocation for it is strictly determined by the block in which it is declared, and is exactly the same as the dynamic lifetime of that block. This is in contrast to *owns* in Algol 60 and similar constructions.

*P*2: A declaration not only defines a variable, but it also specifies a storage medium on which space should be allocated for it.

*P*3: There are two kinds of declarations: (i) *Direct declarations*: They are very similar to those familiar from Algol 60 with the addition that they also specify a storage medium. (ii) *Copy declarations*: They are the main tool for specifying data movements. They define a new variable whose structure and possibly value (initialization) are copied from some other, already declared, variable for which storage usually is allocated on a different medium. Several versions of copy declarations are available in order to initialize variables at the entry to a block or to save their value on leaving a block. Of course, copy declarations also specify a storage device for the new variable.

The following consequences of these principles are immediate:

*C*1: The movement of data between storage media and the allocation and release of space on all storage media is easily controlled via the block structure of the program.

*C*2: The strict enforcement of block structure (no *owns*) allows allocation and release of space on all storage media by the very efficient stack mechanism. In particular, storage fragmentation, which is undesirable even in a virtual store [Fenichel and Yochelson (1969)], and garbage collection are avoided on all storage media.

Example   The following example is typical for a temporary "replacement" of the variable A. The effect is similar to a page replacement.

> *begin declare drum* A;
>   *declare* B;
>
> *begin copy* A *from* A;
>   operate on A, B *end*;
>
> *begin declare* C;
>   operate on B, C *end*;
>
> *begin copy* A *from* A;
>   operate on A, B *end*
>
> *end*

Omission of a storage specification in a declaration causes space allocation in main store.

E. Implementation of Symbolic Manipulation

Let us consider the implementation of symbolic manipulation for expressions of matrices and of lists of level one (excluding cumulative list operators).

In this case there is only a finite number of transformation rules [Bayer and Witzgall (1968, 1970)], about 150 to 200, of the kind presented as examples in Sect. II. They can easily be stored in the main memory of a computer. Some of these rules are quite complicated, and the manipulation task may look rather formidable at first glance.

Assume for the purpose of this discussion that before we start manipulating a program has been parsed and is represented as a tree describing its syntactic structure.

Example Consider the tree for the expression $A \otimes (B \rightarrow C)$ in Fig. 4. It is important to observe here that the tree can be traversed in a certain way; namely, from a given node we can look directly up and down all arcs incident with that node. Now consider a very simple manipulation example; namely, application of the rule

$$A \otimes (B \rightarrow C) \equiv (A \otimes B) \rightarrow (A \otimes C)$$

This rule corresponds to the tree transformation shown in Fig. 4.

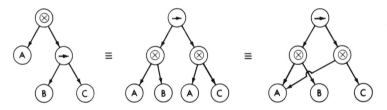

FIG. 4. Tree and acyclic graph representation of the transformation $A \otimes (B \rightarrow C)$ $\equiv (A \otimes B) \rightarrow (A \otimes C)$.

Instead of representing the transformed tree explicitly as a tree and repeating the potentially very complicated expression A twice, it can also be represented as a directed graph without cycles as illustrated in Fig. 4. This has several advantages:

(i) The symbolic manipulation can be done largely by fast pointer manipulation instead of rewriting entire strings, namely parts of the program. It also leads to a more compact representation of the transformed program.

(ii) Many common subexpressions are identified automatically; thus, work can be saved during translation and execution time by normalizing and evaluating common subexpressions only once.

Basically the same manipulation techniques can be used to transform matrix expressions, to remove matrix operators from range expressions, and to simplify range expressions.

F. PARALLELISM IN BLOCK PREPARATION

The *preparation* of a block B for execution consists of allocating storage for all variables declared in B and initializing them, typically by bringing data from a backup store into main store. If possible, blocks should be prepared in parallel with the processing of other blocks. To some degree this is feasible. The statement

prepare B

indicates that the block labeled B should be prepared for future processing.

Let us first indicate how parallel block preparation could be implemented. Any implementation scheme should be simple, but the main criteria for an acceptable scheme shall be that it avoid storage fragmentation and the need for garbage collection. Consider an example with the block structure, also represented in the obvious way as a tree, as in Fig. 5. Also shown are three

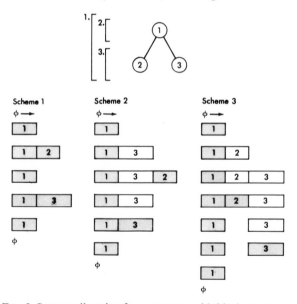

FIG. 5. Storage allocation for a program with block structure.

different storage allocations, the arrow indicating that storage is being filled toward the right, and ϕ indicating an empty storage. The storage is displayed at successive times during execution, shaded areas meaning that a block has been entered, nonshaded areas indicating that a block has been prepared, but not entered yet. Scheme 1 is the stack scheme commonly used in Algol 60 implementations which does not allow any overlap between processing and block preparation. In scheme 2, block 3 is prepared while the computation for block 1 is performed, then block 2 is entered, and finally block 3. Observe that no storage fragmentation is caused, although the blocks are prepared and executed in different orders. Scheme 3 causes storage fragmentation and would not be acceptable by our rules.

Now consider a store in which several stacks are available; e.g., the two stacks S4 and S5. Initializing and executing blocks in the same order as in scheme 3, but using both stacks S4 and S5 simultaneously as shown in Fig. 6 avoids storage fragmentation.

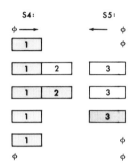

Fig. 6. Allocation avoiding storage fragmentation.

The difference between this and scheme 3 may seem very superficial, but it is not. Main stores of computers can be viewed as two stacks: one running from the lower part of storage upward, and one running from the higher part downward, with all of free storage being one contiguous area in between. Now it is clear that scheme 3 fragments free storage whereas the two stacks 4 and 5 together do not.

We are interested in general conditions which allow parallel block preparation to some extent and still ensure that an acceptable implementation is possible. The following conditions are sufficient (they are not necessary):

R1: A block can be prepared only if the immediately surrounding block has been entered.

R2: A block B can be prepared on a stack S only if the last block allocated on S surrounds (not necessarily immediately) B.

R3: If a block is entered, but has not been prepared yet, it can be allocated on any stack.

In the example in Fig. 7, blocks are entered in the order in which they are numbered, and a storage allocation is shown satisfying our conditions.

The conditions for preparing blocks enforce the proper use of the statement by the familiar scope rules for the label B, just as in the statement *go to* B.

prepare B

When a block is prepared via a proper *prepare* statement, all the global quantities which may be necessary for the declarations in that block are available.

Observe that in preparing procedures it is necessary to use the label of the particular procedure call, and not of the procedure declaration.

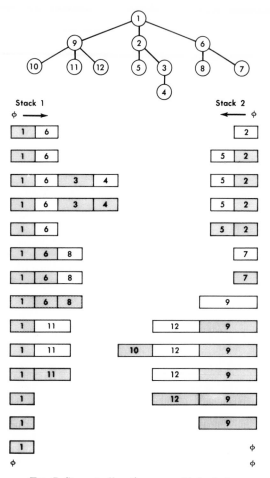

FIG. 7. Storage allocation on multiple stacks.

ACKNOWLEDGMENTS

Many of the ideas presented in this paper were arrived at in close cooperation with C. Witzgall. I also want to thank D. MacLaren, E. McCreight, and J. Reynolds for stimulating discussions on the subject.

REFERENCES

Bayer, R. (1968). Multilevel Storage Control in Programming Languages. Document D1–82–0795, Boeing Scientific Research Laboratories, Seattle, Washington.

Bayer, R., and Witzgall, C. (1968). A Data Structure Calculus for Matrices. Document D1–82–0726, Boeing Scientific Research Laboratories, Seattle, Washington.

Bayer, R., and Witzgall, C. (1970). Some Complete Calculi for Matrices. *Comm. ACM* **13**, 223.

Bayer, R., Bigelow, J. H., Dantzig, G. B., Gries, D. J., McGrath, M. B., Pinsky, P. D., Schuck, S. K., and Witzgall, C. (1968). MPL Mathematical Programming Language. Rep. CS 119, Computer Science Dept., Stanford Univ., Stanford, California.

Fenichel, R. R., and Yochelson, J. C. (1969). A LISP Garbage-Collector for Virtual Memory Computer Systems. *Comm. ACM* **12**, 611.

Hoare, C. A. R. (1968). Data Structures in Two-Level Store. *Invited Papers IFIP Congr. 68, Edinburgh*, 1968 p. 111.

Hoare, C. A. R. (1969). Private communication.

Pakins, S. (1968). APL/360. Science Research Associates, Chicago.

Reynolds, J. (1969). A Set-Theoretic Approach to the Concept of Type. Applied Mathematics Div., Argonne Nat. Lab., Argonne.

Wirth, N. (1969). Private communication.

5.18 Software for Nonnumerical Mathematics

J. E. Sammet

IBM CORPORATION

CAMBRIDGE, MASSACHUSETTS

I. Introduction

For a number of years there has been an increasing utilization of digital computers in doing nonnumerical mathematics.[1] In this context the term "nonnumerical mathematics" covers three broad categories: (1) theorem proving, (2) "pure" mathematics, and (3) formula manipulation. A discussion of tools for categories (1) to (3) is also included.

The term "software" in some contexts refers to generalized programs which can be used for a variety of problems; in a broader sense it means any program. For this paper, the broad meaning is being taken, so that programs which have been used to solve a single unique problem in nonnumerical mathematics can be discussed or mentioned.

To clarify the subject matter of the paper a little bit further, it is desirable to amplify the meaning of the three main areas that are included. *Theorem proving* means the use of the computer to construct or check a formal proof of a theorem; the latter can exist in any domain, including formal logic. In fact, most of the items in the theorem proving category relate to proving theorems in logic, since this is the area in which most of the work has been done.

The *"pure" mathematics* category involves both intent and methodology. There is some work, e.g., in group theory, geometry, and topology, that has been done on a computer in a symbolic fashion. There is still more work that has been done in these fields, or in analytic number theory, where the computer was really doing numeric computation, but for the purpose of obtaining results or information in technical fields that at least currently can be classified as pure mathematics. These have been included, based on "intent" rather than methodology.

The category of *formal algebraic manipulation*, which is often abbreviated as *formula manipulation*, pertain to the use of a computer to manipulate mathematical expressions in a formal manner. In some cases, the manipulation may involve only numeric computations, as is often true

---

[1] *Note added in proof:* The Second Symposium on Symbolic and Algebraic Manipulation sponsored by the ACM Special Interest Group on Symbolic and Algebraic Manipulation (SIGSAM) held March 23–25, 1971, provided a major forum for the presentation of work in this area. None of the developments reported there could be included in this paper which was written in 1970. A copy of the Proceedings may be purchased from ACM Headquarters, 1133 Avenue of the Americas, New York, N. Y. 10036.

with polynomial handling, whereas in other cases abstract symbol manipulation may be used. Here again, the intent of the routine is important in determining its applicability for inclusion.

In some sense one can say that the digital computer itself contributed to an initially slow development in its use for nonnumerical mathematics. Prior to the existence of computers, many mathematical problems (e.g., involving integration, differential equations) were either solved analytically or not solved at all. With the advent of accurate high-speed computation it rapidly became clear that, for problems not susceptible to analytic solution, useful answers could be obtained numerically. Thus, numerical analysis which had been relatively unimportant prior to the existence of computers became a major activity, and the analytic techniques were relegated to courses in mathematics unrelated to computers. However, the increasing sophistication in knowledge of how to use computers for symbol manipulation is beginning to redress this balance.

Although a computer was used for formal differentiation as early as 1953 by Kahrimanian (1954) and Nolan (1953), and theorems in the propositional calculus were proved on a computer in 1956 by Newell, Shaw, and Simon [see Newell *et al.* (1956, 1957)], it was not until the early and mid 1960's that there began to be significant software development for nonnumerical mathematics. As one minor measure of this (using documented—although not necessarily published—programs as a criteria), an annotated bibliography by this author published in the summer of 1966 [Sammet (1966a)] contained approximately 300 items. A revised version of this [Sammet (1968b)], covering the period *through* 1966, contained about 80 additional items. Three lists of additional items [Wyman (1968, 1969, 1970], covering through early 1970, bring the total to around 550 items. While this number may be small in comparison with other fields and/or uses of computers, the growth rate is significant.

There are very different reasons for the increased computer utilization of the three areas being covered. In the case of formula manipulation, it has become obvious that there are a large number of problems requiring algebraic manipulation that are very tedious, time consuming, error prone, and straightforward. These characteristics make computer solution both necessary and desirable. On the other hand, the increased use of computers to prove theorems and do mathematics is partially because these areas have an exactly opposite characteristic—namely, they are creative and nonstraightforward (although they may be tedious and prone to error). The people who are interested in gaining insight into how the human being solves problems have been very interested in these domains, and have used these subjects as vehicles for work in artificial intelligence.

Surveys of the fields of formula manipulation and parts of the fields of theorem proving and pure mathematics have been thoroughly documented by this author and others. For this reason, it is not practical for this paper to provide a complete survey of the software for nonnumerical mathematics. However, a sketch of the earlier activities will be given, and appropriate cross-references will be shown in each section, so that the emphasis will be on more recent activities.

Section II covers formula manipulation. Section II,A provides some introductory comments and a brief history for that area. Section II,B discusses technical issues, while II,C deals with current systems. Some of the theoretical mathematical work is mentioned in II,D and an indication of some of the applications is given in II,E.

Section III discusses theorem proving, with introductory remarks in III,A with III,B devoted to logic, while III,C covers theorem proving in "pure" mathematics, and III,D describes some miscellaneous work.

Section IV describes work done in "pure" mathematics, with IV,A providing an introduction, IV,B discussing number theory and combinatorial analysis, and IV,C describing abstract algebra.

Section V discusses very briefly some of the tools used in developing software for nonnumerical mathematics, while Section VI provides some thoughts on future developments. A lengthy reference list is provided.

II. Formula Manipulation

A. INTRODUCTORY COMMENTS

Although this author was one of the first users of the term "formula manipulation," she would concede that it is actually inaccurate because we tend to deal with mathematical *expressions* rather than *formulas* in a formal sense. That is, we normally think of a digital computer as producing a number when we write something of the form $Y = (A - B) * (A + B)$, where it is assumed that A and B are numbers. When we deal with formal algebraic manipulation (which is a more accurate term) we really mean that writing $Y = (A - B) * (A + B)$ will yield the result $A^2 - B^2$. The fundamental difference is that in the numeric case, as exemplified by Fortran and Algol, all of the variables take on only numeric values and all partial and final results are numbers. In the formula manipulation case, we assume that variables stand for themselves or for other mathematical expressions, and we are able to perform on them the same kinds of operations that are done in high school algebra or in calculus (e.g., formal differentiation).

This author has already written two fairly lengthy papers surveying the field of formula manipulation [Sammet (1966b, 1967)]. It is unnecessary to repeat all of that material here merely to provide a complete self-contained survey. Instead, a sketchy history of the major developments will be given, and concentration will be aimed at some of the broader issues and an indication of the current state of the art, with particular emphasis on achievements over the past four to five years.

The first known use of a computer to do what can reasonably be called formal mathematics or formula manipulation was the differentiation programs written independently in 1953 by Kahrimanian (1954) and Nolan (1953). In the late 1950's and the early 1960's the emphasis was on the development of individual packages and subroutines either to solve a particular application [e.g., Herget and Musen (1959)] or to carry out a particular facet of formula manipulation, e.g., differentiation [see Hanson et al. (1962)].

In the summer of 1962, this author (assisted by R. G. Tobey) decided that what was really needed was a system wherein the formula manipulation capability was added to an existing numerical language so that the user had the facility of doing both nonnumeric and numeric computations within the same program. The base language chosen was Fortran, and the system —known as FORMAC—was implemented as an experiment on the 7090/94. Completely independently, A. J. Perlis reached a similar conclusion on the need for combining formula manipulation capabilities (as well as string and list processing facilities) with a numeric language, and he and graduate students at Carnegie-Mellon implemented a system called Formula Algol on the G20. Although that system had some very interesting ideas and concepts, it has been dropped and therefore will not be discussed here any further. A detailed description is given in Perlis et al. (1966); a brief description of the language is in Sammet (1969, Chap. VIII). Another major development was the ALPAK system of subroutines for handling polynomials and rational functions [see Brown et al. (1963)]. Both ALPAK and FORMAC were used for numerous applications; e.g., for ALPAK see Takacs (1963) and for FORMAC see Tobey (1966).

All those systems were designed primarily for use in a batch environment, although a subset of FORMAC was put under MIT's Compatible Time Sharing System (CTSS) in 1965 [see Bleiweiss et al. (1966), and Crisman (1965), respectively]. The first attempt known to this author at an on-line system was the Magic Paper [see Clapp and Kain (1963)], which was never really completed as originally described. The first completed system designed primarily for on-line use was MATHLAB [see Engleman (1965)] running on the IBM 7030 and under CTSS. MATHLAB was also the first system to include in the language such higher-level commands as "integrate" and "solve," but handled only polynomials and rational functions.

The other major landmark in the early 1960's was the integration program called SAINT developed by Slagle (1963). This program did formal integration about as well as many freshmen in calculus classes.

A number of other individual programs and systems were under development in the mid-60's. For a broad overview of the work under development in the mid-60's, see the *Proceedings of the Symposium on Symbolic and Algebraic Manipulation*, which appeared as the August 1966 issue of the *Communications of the ACM* (Vol. 9, No. 8) and the proceedings of the conference held under IFIP auspices at Pisa in September 1966 [Bobrow (1968)]. See also this author's bibliographies [Sammet (1966a, 1968b)].

The activities which were mentioned in this section represent the outline of the major, or "first-generation," software for formula manipulation.

B. TECHNICAL ISSUES

This section will discuss the technical issues in formula manipulation. The main categories are

1. general purpose system versus individual capabilities,
2. level of language and system,
3. level of capability and type of user,
4. general capabilities,
5. specific capabilities.

(This list and outline follow the framework used for Sammet (1967). The reader interested in more details is referred to that paper, since only highlights will be sketched here.)

1. *General Purpose System versus Individual Capabilities*

As indicated in the introduction, the term software tends to be used most often for generalized systems rather than for individual programs with a single specific objective. However, to the extent that the term applies to both, this is a major issue that must be dealt with and there are as many different views on this as there are people. The arguments for general purpose versus special purpose in the formula manipulation context are about the same as those in any other framework.

2. *Level of Language and System*

In this context one must consider whether or not there is going to be a separate package of routines for use in several problems, whether a package of subroutines is to be embedded in some existing system, whether some particular higher-level language is to be augmented by the formula manipulation capabilities, or whether there is to be a specially created higher-level language. There are illustrations of all of these, with a specific differentiation program illustrating the individualized system, ALPAK representing the package of subroutines augmenting an existing (assembly) system, FORMAC representing language additions to an existing higher-level language, and lastly MATHLAB or REDUCE as stand-alone higher-level languages.

3. *Level of Capability and Type of User*

It is extremely important to determine whether the potential user is to be a mathematician, a systems programmer, a moderately capable scientific programmer, etc. This is a major factor in determining the level of language and system. For example, although the scientific programmer accustomed to Fortran is used to using ** for exponentiation, the mathematician might find that intolerable.

4. *General Capabilities*

In considering *any* kind of software for nonnumerical mathematics, the first question to be asked is what types of expressions are to be handled. There are three broad categories: The first and simplest is that of polynomials, which may be in one or many variables. The second, which often is a relatively simple extension, is the ability to handle rational functions, i.e., the quotient of two polynomials. The third is the broad facility to handle any mathematical expression containing the elementary functions. As a minor deviation from this last, we can consider whether or not the system has the capability of handling formulas instead of, or in addition to, handling expressions. It is relatively simple for a system handling only expressions to manipulate formulas, and conversely; however, some differing language elements are required.

A second major factor pertains to the type of arithmetic. While numerical analysts are very conscious of round-off, they may be less conscious of the difficulties accruing in a formula manipulation system if there is no rational arithmetic available. For example, if one looks at the expression

$$0.5X + 0.16667X^2 + 0.08333X^3 + 0.05X^4$$

it would be impossible to tell where it came from without trying to get the rational equivalent of the decimal fractions. Even by representing it in the form

$$\frac{X}{2} + \frac{X^2}{6} + \frac{X^3}{12} + \frac{X^4}{20}$$

it is still hard to tell that it represents the following expression:

$$\frac{X}{1 \cdot 2} + \frac{X^2}{2 \cdot 3} + \frac{X^3}{3 \cdot 4} + \frac{X^4}{4 \cdot 5}$$

The next broad capability is one which generates more heat than almost any other, and all too frequently more heat than light. This pertains to the subject of simplification. In the most elementary sense, this means the concept of transformations that the system can or does make—either independently or under user control—to convert from one form to another. The most obvious and elementary of these transformations is doing the obvious and appropriate manipulations with zeros and ones; e.g.,

$$A + 0 \to A$$
$$A^1 \to A$$

etc. A second level of complexity involves combining like terms; e.g.,

$$2X + 3X \to 5X$$
$$\frac{3Y^2}{6XY} \to \frac{Y}{2X}$$

Some systems (e.g., FORMAC) have the philosophy that this will be done automatically for the users. It is fairly obvious that the user must control some of the expansion and factoring; i.e., the user and not the system must decide whether an expression should be in the form $A(B + C)$ or $AB + AC$. On the other hand, Fenichel (1966) takes the view in his system FAMOUS that nothing should be done unless directly requested by the user. While I tend to favor the automatic transformations on zeros and ones and combining of like terms to be done automatically by the system, it is nevertheless necessary to admit that there are a few cases in which this hinders the intent of the individual. For example, if the system changes

$$\frac{(X - 1)^2}{(X + 1)(X - 1)}$$

to

$$\frac{X - 1}{X + 1}$$

some meaning may be lost. Readers interested in pursuing the philosophical or detailed aspects of simplification further are referred to the following sources: Tobey et al. (1965), Moses (1967, p. 57ff), Hearn (1969), and Krugman and Myszewski (1969).

The concept of pattern matching becomes important in a formula manipulation system because it affects not only the kind of simplification that can be done but also plays a major role in certain specific capabilities, e.g., integration. As the simplest but most obvious illustration, the ability to find an expression of the form $SIN^2 X + COSINE^2 X$ is fairly simple. However, to be able to also recognize something of the form A COSINE^2 Y + A SIN^2 Y, where A and Y can be any legal expressions whatsoever, is far from trivial.

Of interest primarily to the specialist or those dealing primarily with complex numbers is the issue of whether they can be handled directly or must be programmed separately.

Probably the most important characteristic of the general capabilities other than the types of expression and simplification is the issue of the environment; more specifically, whether the system allows for man–machine interaction. While a number of problems have been successfully solved using batch systems such as FORMAC and ALPAK, it is nevertheless true that in the long run an on-line system is far more important for nonnumerical mathematics than for numeric. The primary reason is that in the numeric case the developer of the problem has some general understanding of what his results may be, or at least the ranges within which they fall. Thus, although he may not know whether his answer will be between 1 and 100 or between 100 and 1000, etc., if these distinctions are meaningful to him, he is able to program ahead of time any necessary branches depending on the size of the numbers. That this is manifestly impossible in the formula manipulation case should be obvious; if not, try to predict the general format of the derivative of the following expression:

$$\frac{X(X - 1)^2}{(X + 1)(X^2 + 1)}$$

I do not accept the contention of some people that any formula manipulation system *must* be interactive to be effective; on the other hand, I certainly do feel that an "ideal" system—which does not exist and is not on the drawing boards—must have an on-line capability with the facility of immediately shifting into a batch environment as soon as the user desires. The actual physical format of the input and output is crucial in this situation as well as in the batch, and this will be discussed later under input/output.

5. Specific Capabilities

The first significant specific capability in a formula manipulation system is differentiation. Fortunately, this is sufficiently simple that it is often assigned as an exercise in symbol manipulation classes. The trick is not merely to do the differentiation, however, but to determine the types of simplification that will be involved and the interactions with the remainder of the system.

Certainly one of the most exciting specific capabilities is that of integration. In the case of polynomials this is sufficiently trivial that it is not of any significance. However, once we get beyond that the matter becomes one of strong interest. Earlier mention was made of the pioneering work of Slagle

(1963) in developing SAINT. It was several years later that Manove *et al.* (1968) developed an integration package for rational functions within the MATHLAB system. This was then followed by SIN (Symbolic Integrator) of Moses (1967) which used the MATHLAB work. The work of Moses is significant in the development of software for nonnumerical mathematics because, instead of using primarily the general heuristic techniques that Slagle used, he used algorithms to dispose of very many of the easier cases and resorted to heuristics only in the more difficult problems. SIN is used in another routine called SOLDIER, which solves differential equations. Integration is one of the areas in which the theoretical mathematical results play a primary role, and more will be said about this in Sect. II,D.

A very important specific capability is that of factoring. For a long time the primary technique was some modification of the Kronecker algorithm. However, more recent work by Berlekamp (1967) [which is also described in Knuth (1969)] indicates a better algorithm.

Of primary concern to any user is the form of input/output and editing. The two most significant pieces of work that have been done along these lines are the CHARYBDIS routines within MATHLAB [see Millen (1968)] and the scope routines developed by Martin (1967) in the Symbolic Mathematical Laboratory. In the former, the routine will display on a typewriter the output in a "reasonable" two-dimensional form where this actually has a very specific technical meaning. The work done by Martin actually caused specific symbols (e.g., summation sign) to be shown in a natural format on the scope.

On the other hand, there remains the problem of input. Here one must separate this into two subproblems—one deals with two-dimensional input and the other with handwriting, although I would contend that neither is unique to software for nonnumerical mathematics. That is, the ability to read handwritten and/or two-dimensional input is important regardless of whether one is doing numeric or nonnumeric computations. Specific equipment and typewriters have been used in the past for the numeric case [see Wells (1961); or Klerer and Grossman (1967)]. Unfortunately, none of the people interested in software for nonnumerical mathematics have taken these facilities of the numeric systems and adopted them for the nonnumeric case. If the user writes

$$ Y = \frac{A_{i+1} - B_{i,j}}{A_{x+3,y}} $$

he really doesn't care on a local basis whether that is actually numeric or nonnumeric. He is primarily concerned with his ability to write it in that fashion. The people concerned with input in the nonnumeric software seem to have jumped over the specialized typewriter and gone immediately to the handwriting situation [see Anderson (1968), Bernstein and Williams (1969), and Blackwell and Anderson (1969)].

One important specific capability is the ability to compare expressions. The difficulty is that there are many forms of the expression involved; e.g., the expressions $(A + B)^2$ and $A^2 + 2AB + B^2$ are certainly equivalent mathematically but they are not identical. The user needs the capability of specifying whether he wants mathematical equivalence or whether he requires identity. Unfortunately, it has been shown mathematically that the problem of equivalence is, in general, unsolvable; i.e., it is impossible to determine, in general, whether a given expression is identically equal to zero, and this of course means that one cannot always tell whether or not two expressions are equivalent. Of major concern to numerical analysts in this context is the consideration of the coefficients and tolerances in determining equivalences. If rational numbers are not used, and we have the expressions $x + y$ and $0.9999x + 0.9995y$, we might want to specify the tolerance for each coefficient before we would say that they were equivalent.

The concept of automatic substitution (also called unraveling) is a philosophical question pertaining to how much control the user should have over his expressions as the computation proceeds. If he writes $A = B - C$, $D = C + A$, then the user may or may not want the equivalent of A substituted in the expression to yield $D = B$.

One of the major specific capabilities is that dealing with polynomial manipulation. This is of great importance because a very large number of practical applications involve polynomials, or at most rational functions. The implementation is very much easier and very much more efficient if one knows that only these kinds of expressions are allowed. Among the earliest practical software for polynomials and rational functions was ALPAK [see Brown et al. (1963)]. More recently the SAC system of Collins (1968a, b) has provided other facilities along with the polynomial manipulation, e.g., Collins and Horowitz (1970). He has also developed a series of increasingly better algorithms for finding the greatest common divisor of polynomials in one variable. The most recent appears to be in Collins (1969).

There are numerous other specific capabilities and factors in implementation whose discussion is beyond the scope of this paper [see Sammet (1967)].

C. Current Systems

This section discusses briefly some of the current and/or major systems; namely, PL/I-FORMAC, REDUCE, SAC-1, IAM, SYMBAL, and work at IBM Research Center. This is by no means a complete survey nor a complete description. In particular, no comments will be made about the following systems: ALPAK [see Brown et al. (1963)], ALTRAN [see McIlroy and Brown (1966), Brown (1966)], PM [see Collins (1966)], 7090 FORMAC [see Sammet and Bond (1964), FORMAC (1965)], Formula Algol [see Perlis *et al.* (1964, 1966)], Symbolic Mathematical Laboratory [see Martin (1967)], MATHLAB [see Engelman (1965, 1969)], and the work of van de Riet (1968), since they have been amply documented or are not really, in significant current use. The language facilities of ALTRAN, FORMAC Formula Algol, Symbolic Mathematical Laboratory, and MATHLAB are described in Sammet (1969, Chap. VIII).

The PL/I-FORMAC is "second-generation" formula manipulation software in the sense that it is based on concepts developed in the 7090 FORMAC, but has major conceptual improvements. The change to PL/I as a base language allows the user full access to all of the facilities of PL/I, together with the capabilities needed for formula manipulation; numerical and symbolic computations can still be intermingled in the same program. It is implemented on the IBM System/360. It should be noted that PL/I-FORMAC is batch oriented and general purpose in the sense that it allows the user to include the elementary functions in his expressions. This has the advantage of not requiring the user to do anything special to include them, but has the disadvantage of providing less efficient handling of polynomials and rational functions. For full details on the system see the manual by Tobey *et al.* (1969). One of the more interesting and unusual applications of PL/I-FORMAC has been the work involving the theorem of Pappus in projective geometry [see Cerutti and Davis (1969)].

A system called REDUCE developed by Hearn has been implemented on several machines, including IBM System/360, PDP-10, and the CDC 6400. It was written in LISP and is usable in both batch and on-line modes, with more current use of the latter form. It has undergone several versions; its primary motivation initially was as a tool for the solution for problems in physics, but it is actually far more general purpose than that. The basic elements that the user can operate on are rational functionals of polynomials. While the user can introduce elementary functions, he must write some extra source language statements to do so. The language style is similar to Algol 60, but is not actually an extension of it. There are some convenient facilities for introducing new operators and for controlling the substitution process. Further information can be found in Hearn (1968, 1969).

The SAC-1 polynomial system developed by G. E. Collins is a series of subroutines available under Fortran for manipulating polynomials and rational functions. It is written in Fortran, except for a few primitive routines which must be coded in machine language, and has been implemented on at least the CDC 1604, 3600, 6400, IBM System/360, GE 645, and UNIVAC 1108. The user invokes the facilities of this system by means of CALL statements in Fortran programs. More than any other system, this is based on some highly theoretical work, including the development of an excellent algorithm for finding the greatest common divisor of two polynomials, which is a major problem in all such systems. In addition, considerable work has been done in terms of giving theoretical maximum computing times for differing subprograms. A series of unpublished reports [e.g., Collins (1968a, 1969b), Collins and Horowitz (1970)] describe the system.

The IAM (Interactive Algebraic Manipulation) system has recently been developed by ADR/Computer Associates, Inc., in Wakefield, Mass. This was designed from the beginning as an on-line system. It includes provisions for execution of single statements, and also collection of statements to permit execution of a program with loops. In addition to allowing the elementary functions, it contains an integrate command, as well as a command to solve for a single variable in a relatively simple equation. Further information about IAM appears in Wheelock (1970) and Krugman and Myszewski (1969).

Two systems strongly related to Algol 60 are SYMBAL, developed by Engeli (1969), and the work of van de Riet (1968). The SYMBAL system (implemented on the CDC 6600) is not a direct extension of Algol 60 but the van de Riet work is. Among the interesting facilities of SYMBAL are specific operators for sums and products of series.

A very recent system is an experimental interactive system developed at the IBM Research Center in Yorktown Heights [see Blair *et al.* (1970)]. This system is truly a conglomerate since it incorporates in one package a number of individual facilities that have been programmed in LISP by a variety of different people. In particular, it includes: (1) rational functional manipulation and simplification, as well as the differentiation from RE-DUCE; (2) symbolic integration (SIN); (3) polynomial factorization, solution of linear differential equations, direct and inverse symbolic Laplace transforms (from MATHLAB); and (4) two-dimensional output on both the IBM 2741 terminal and the IBM 2250 display (using the picture display from the Symbolic Mathematical Laboratory on the scope, and the CHARYBDIS program from MATHLAB on either the terminal or the scope). There are other features in the system. The language is not related to any existing one, but merely contains facilities felt by the developers to be natural to the mathematicians. Probably the most interesting or novel facility in this system is the ability for the user to define new syntax in terms of existing facilities.

D. THEORETICAL MATHEMATICAL WORK

It is undoubtedly a sign of the maturity of the development of software for nonnumerical mathematics that theoretical mathematics is beginning to play a major role. In the early 1960's much of the emphasis and concentration was on the implementation and details of programming. Some small attention was paid to the mathematical foundations, but the emphasis was on the programming techniques used. It is now becoming increasingly clear that the theoretical developments in mathematics are playing a major role from several points of view. In the first place, they are determining some theoretical constraints on what can and can not be done, so that the implementers do not spend large amounts of time attempting to produce or improve something which is theoretically impossible. The best example of this involves the result of Richardson (1966), that it is recursively unsolvable to determine in general whether an expression involving elementary functions is zero. This means that, in general, it is impossible to tell when two expressions are equivalent. Further work in this area was done by Caviness (1970).

Another area on which mathematical results are playing a major role is in the development of new machine-oriented algorithms for integration, finding greatest common divisors, and for the factoring of polynomials. Work in these areas has been done, respectively, by Risch (1968, 1969), Collins (1967, 1969), Berlekamp (1967), and Knuth (1969).

In other words, the development of efficient software for formula manipulation systems will in the long range depend at least as heavily upon the use of appropriate mathematical algorithms and techniques as it will upon appropriate programming techniques. Such questions as the best way of storing expressions internally, canonical forms, methods of simplification, etc., are no longer solely dependent upon effective and efficient programming techniques. Many of the theoretical issues involved with polynomials are covered (although not necessarily with this motivation) in Knuth (1969).

E. Applications

Applications can be classified in many ways, with the two major breakdowns being the mathematical techniques used (e.g., differential equations, matrices, Laplace transforms, numerical analysis), and the specific scientific source of the problem (e.g., astronomy, physics, meteorology). There are a number of people who have written specific individualized programs to solve a particular problem. For historical reasons, it is important to cite one of the very first of these, namely Boys et al. (1956). Very brief descriptions of a number of specific applications are given in Sect. 4.2 of Sammet (1967).

It is worth noting that, as of this time, apparently only three *systems* have been used to any significant degree to obtain meaningful results: FORMAC (in both the Fortran 7090 and the PL/I 360 versions), ALPAK-ALTRAN, and REDUCE. A number of applications for the 7090 FORMAC are described in Tobey (1966). More recent applications are described in

Neidleman (1967), Howard and Tashjian (1968), and Cerutti and Davis (1969). Some of the more recent references for ALPAK-ALTRAN are Russell (1969) and Gey and Lesser (1969); for examples of the use of REDUCE see Hearn *et al.* (1969) and Aldins *et al.* (1969).

Although systems such as those described in Sect. II,C are actually useful in doing numerical analysis and can be used for investigation of theoretical work in numerical analysis itself, there appears to have been very little of this done.

It must be noted that the biggest problem in actual usage of these systems is the need for more storage. Expressions grow extremely large during the calculations, and sometimes overflow before they can be simplified to an equivalent small form. This author has coined a new version of Parkinson's law to describe this chronic problem: "Expressions expand to exceed the available storage space." Other problems are discussed in Sammet (1968a).

To provide the reader with a concrete illustration of formula manipulation systems, Fig. 1 (p. 309) shows a simple but realistic use of PL/I-FORMAC.

III. Theorem Proving

A. INTRODUCTORY COMMENTS

The subject of proving theorems on a computer can be divided or classified in a number of different ways. The first and most obvious is the domain in which the theorems are being proved. For example, programs have been written to prove theorems in logic (both the propositional and predicate calculus), in a large portion of plane geometry, in number theory, in group theory, and even to prove trigonometric identities (which admittedly is a simple and special case). Another way of classifying the work in theorem proving is by whether it is heuristic or algorithmic. Again, work has been done in both of these areas. A third means of classification is the contrast between general and specific. In the former case, a system of axioms and rules of inference are established, and the machine (i.e., computer program) may be instructed to derive the proofs of a large number of stated theorems. On the other hand, there are cases in number theory where a very particular theorem can only be proved by the computer, simply because the human being is incapable of generating the number of paths and steps required. Still another classification is the difference between demonstrating the validity of a theorem (where computation can be used to test all cases) and actually providing the steps in the proof. Finally, in the broad concept of theorem proving, we can distinguish between checking proofs and actually generating them. The distinction is sometimes a subtle one, because checking a proof implies that somebody feels that a sequence of steps is in fact a proof; however, by rigorous standards it may not really be, although the missing

```
INPUT TO FORMAC PREPROCESSOR
  TAYLOR:   PROC OPTIONS(MAIN);
            /* THIS PROGRAM CAN COMPUTE THE TAYLOR SERIES FOR ANY NUMBER OF
            FUNCTIONS AROUND ANY POINT.  FOR THIS PROGRAM THE FUNCTIONS
            EXP(X)*SIN(X)  AND  SIN(X)*LOG(X+1) WERE EXPANDED TO A
            20'TH DEGREE TAYLOR SERIES POLYNOMIAL AROUND THE POINT X=0.
            IF IT HAD BEEN DESIRED, THE FUNCTIONS, THE POINTS OF
            EXPANSION, AND THE NUMBER OF TERMS COULD ALL HAVE BEEN
            READ IN FROM DATA CARDS     */
            DCL ENTRY(BIN FIXED(31),BIN FIXED(31));
            FORMAC_OPTIONS;
            LET ( CH=CHAIN(  EXP(X)*SIN(X)  ,   SIN(X)*LOG(X+1)   ));
            OPTSET(NOINT;LINELENGTH=72);
            LET(A=0);
            N=20;
            DO NUMBER=1 TO NARGS(CH);
                 LET(FX = ARG( "NUMBER" , CH ));
                 LET(TAYLOR = EVAL(FX,X,A));
  LOOP:          DO I= 1 TO N;
                    LET(I="I");
                    LET( DFX = DERIV(FX,X,I));
            LET ( TAYLOR = TAYLOR + (X-A)**I * EVAL(DFX,X,A) / FAC(I) );
                    END LOOP;
            PRINT_OUT(FX;TAYLOR);
            PUT SKIP(5);
            END TAYLOR;
```

OUTPUT

```
            X
FX = #E   SIN ( X )
------------------
                     2              3              5              6
TAYLOR =  X / 1! + 2 X   / 2! + 2 X   / 3! - 4 X   / 5! - 8 X   / 6!
-------------------------------------------------------------------------
     7              9              10              11              13
 - 8 X   / 7! + 16 X   / 9! + 32 X    / 10! + 32 X    / 11! - 64 X
-------------------------------------------------------------------------
              14               15               17               18
 / 13! - 128 X    / 14! - 128 X    / 15! + 256 X    / 17! + 512 X
-------------------------------------------------------------------------
              19
 / 18! + 512 X    / 19!
------------------------
```

```
FX = SIN ( X ) LN ( X + 1 )
---------------------------
                 2              3              4              5              6
TAYLOR = 2 X   / 2! - 3 X   / 3! + 4 X   / 4! - 20 X   / 5! + 110 X
-------------------------------------------------------------------------
            7              8              9                 10
 / 6! - 651 X   / 7! + 4520 X   / 8! - 36000 X   / 9! + 322618 X
-------------------------------------------------------------------------
              11                 12                 13
 / 10! - 3213595 X    / 11! + 35226860 X    / 12! - 421419492 X    /
-------------------------------------------------------------------------
              14                 15
 13! + 5463436134 X    / 14! - 76301056755 X    / 15! + 1142009233872
-------------------------------------------------------------------------
  16                 17                         18
 X   / 16! - 18236159031584 X    / 17! + 309463272791538 X    / 18!
-------------------------------------------------------------------------
                 19                      20
 - 5561354285804115 X    / 19! + 105510576441518164 X    / 20!
-------------------------------------------------------------------------
```

FIG. 1.

steps may be quite easy to fill in. This is the situation that arises in most mathematical textbooks or papers when they use the phrase "it is obvious that" or "it follows immediately from." There is a significant difference between trying to write a program to fill in a small but logically necessary number of steps between each major statement and asking a computer to generate an entire proof from scratch.

In the current environment where man–machine interaction is common, it is quite reasonable to consider cases in which the human and computer work together to generate proofs (or theorems) which are beyond the individual capability of either alone. Even in a batch environment there are cases in which the algorithm for finding a proof is well defined but the speed and accuracy of a computer are needed to carry out all the steps. In at least one case (the resolution principle) a technique was developed by Robinson (1965) in response to a need for a method which would reduce the number of combinations involved, but which also involved more inferences than the human could comprehend in a single step. On the other hand, suggestions or guidance from the human may reduce computation time or help produce more efficient proofs.

There is at least one case—discussed in Sect. III,C—in which the computer (although aided by a human) generated a proof of a new theorem, i.e., one which was not known to be true ahead of time.

In Slagle and Bursky (1968), they provide a framework for a multipurpose heuristic program, in which the user must describe the particular theorem-proving domain, including methods of transforming propositions to be proved into subpropositions which might be easier to prove.

Some—although not all—of the motivation for doing work in theorem proving has been because of its implications for artificial intelligence. Thus, many of the workers in that field are using theorem proving as a vehicle for obtaining a better understanding of how the human mind works and solves its problems.

We shall first discuss theorem proving done purely in the domain of logic, then in geometry, then in pure mathematics, and finally in some other areas (including number theory).

B. Logic

This section is quite brief and does not supply much technical detail. It is meant primarily to serve as a guide to the kinds of programs which have been developed. For a more thorough coverage of the technical issues and the whole subject area, see Cooper (1966) and the papers of Robinson (1967, 1969); the last has an excellent bibliography. For an annotated bibliography through 1966, see Sammet (1968b).

The earliest work on theorem proving appears to be that of Newell et al.

(1957) in their development and implementation of the Logic Theory Machine. The objective in this work was to investigate ways in which the computer program could learn to prove theorems in the propositional calculus. The "Principia Mathematica" of Russell and Whitehead was the source of the axioms and postulates. One of the important points made in that paper was the existence of the so-called British Museum algorithm, which said essentially that, although one could create any proof desired by generating "all possible sequences of statements," the time for this was incomprehensibly large and therefore heuristics were needed. However, Wang (1960a, b) actually exhibited a specific algorithm that made it possible to generate all of the proofs for the propositional calculus theorems of Russell and Whitehead in "Principia Mathematica" in a matter of minutes. This has since become known as Wang's algorithm and has been programmed by others in a variety of languages.

In the more difficult area of the predicate calculus, early work was done by Gilmore (1959, 1960). In the latter, he showed that a sentence of quantification theory was logically true without actually proving it. The algorithm in Wang (1960a) also took a large step forward in proving many of the theorems of the predicate calculus. Various difficulties and many improvements of this type of theorem proving are described in Davis and Putnam (1960), Davis et al. (1962), and Prawitz et al. (1960).

In Davis (1963) there is a good explanation of basic concepts in proof procedures and he evaluated the techniques which were then in use. He then described an improved procedure and applied it to the field of group theory. More recent techniques are discussed in Loveland (1968).

In Robinson (1963), techniques are described in which some of the combinatorial explosion difficulties encountered in many proof construction procedures are overcome. One of the techniques used is to allow the user to supply useful information. By permitting the user to supply a set of premises which can be redundant, the computer was able to prove the theorem that "if A is a prime number then the square root of A is irrational." A formulation of the first-order predicate calculus specifically designed to be used in computer theorem-proving programs was given in Robinson (1965). This paper introduced a new inference principle called the resolution principle; it is machine oriented, in the sense that it may permit a single inference that is far beyond any person's ability to grasp in a single instance. This contrasts with previous work in which it was tacitly assumed that each step in a proof had to be understandable by a human being. The statement is made by Wos et al. (1964) that substitution of the rules of inference of the type proposed by Robinson can produce a reduction of the combinatorial explosion by a factor in excess of 10^{50}. Papers by Wos et al. (1965) and Luckham (1968) indicate its effectiveness in proving simple theorems in group theory.

C. THEOREM PROVING IN "PURE" MATHEMATICS

One of the very interesting and early programs developed was for proving theorems in plane geometry; see Gelernter (1959) and Gelernter et al. (1960). The program proves a significant number of interesting theorems in plane geometry involving parallel lines, congruence, and equality and inequality of segments and angles. The two main devices used are heuristics and the drawing of diagrams; the latter is done only to guide the search for a proof and in no way constitutes a part of it.

The area that lends itself most easily to theorem proving seems to be group theory. A heuristic program developed specifically for this purpose is ADEPT [see Norton (1966)]. Almost 100 theorems were proved, including several nontrivial ones; e.g., if K and H are normal subgroups of G then G/H is isomorphic to (G/K)/(H/K). In the previous section, mention was made of several programs which had proved theorems in group theory, using methods developed for proving theorems in logic. The primary difference is that in ADEPT a specific representation was chosen for group elements and theorems, whereas in the logic case the formulation was basically independent of the domain in which the theorems were being proved.

A completely different approach was taken in the series of computer programs called SAM I through SAM V (Semi-Automated Mathematics), described in Guard et al. (1969), and SAM VI [see Bennett et al. (1968)]. To quote from the introduction to Guard et al. (1969): "Semiautomated mathematics is an approach to theorem proving which seeks to combine automatic logic routines with ordinary proof procedures in such a manner that the resulting procedure is both efficient and subject to human intervention in the form of control and guidance." While this obviously places considerably less burden upon the computer program, it also represents a fruitful direction for future practical work. The system involves the use of a CRT display. Some of the early work in the SAM series also involved proving theorems in logic, but the emphasis has been more mathematical than the work discussed in Sect. III,B. Perhaps the most interesting result of this work is the fact that the system and human together solved an open problem in the field of lattice theory. A lemma was proved which led immediately to the proof of a theorem involving modular lattices whose truth or falsity had previously been unresolved.

Another interactive system has been developed [see Allen and Luckham (1969)] and used to prove some mathematical results concerning questions of dependence in the axiomatization of ternary boolean algebra. The authors indicate that the degree of difficulty is similar to trigonometric identities.

Some theorem proving in the area of number theory is described by

Lehmer *et al.* (1962) and Lehmer (1963). He indicates that there is a primary difference between the type of theorem proving done in number theory and most of the other theorem-proving work in mathematics. In the number theory situation, an attempt is often made only to prove a single specific theorem which may deal with a particular case; there is no attempt at generality. User hints and information are welcomed or mandatory. Furthermore, these theorems tend to be of a very restricted nature, namely involving number theoretic concepts in which the computer must do a vast amount of arithmetic, as well as a very large amount of bookkeeping for the tree of possible alternatives in the steps of a proof. Stated another way, theorem proving in the area of number theory tends to use the computer as a number cruncher, whereas theorem proving in other aspects of mathematics, and in logic, tends to use the computer as a symbol manipulator.

D. MISCELLANEOUS

In McCarthy (1962), he introduced the notion of using computer programs to check theorems and even engineering systems. He introduces: (1) the concepts of the use of decision methods being implied in statements such as, "it can be verified easily that . . .," (2) the idea that a proof may really be a set of directions for generating a proof rather than a complete proof itself, and (3) the possibilities that heuristics can be allowed in generating the proofs. These concepts, plus additional ones along similar lines, were actually implemented by Abrahams (1963) using LISP on the IBM 7090. Although most of the testing and actual proof checking centered on the propositions in "Principia Mathematica," two textbook proofs (one involving operations on finite sets and one concerned with recursion induction) were considered.

Work by Travis (1964) described a program on the Philco 2000, which proves theorems by efficiently utilizing previously proved theorems. A special problem domain was devised for exploration with the program. Heuristics were definitely used.

Finally, work in proving identities in trigonometry with a learning mechanism is described by Johnson and Holden (1964) and Holden (1967).

A small but interesting FORMAC program was written to do mathematical induction. Although the program, published in Sammet and Bond (1964), proves specifically (and only) that the sum of the first N integers is $N(N + 1)/2$, the technique is applicable to all formulas of this kind; in fact, the approach is applicable to any use of mathematical induction for which the manipulation could be defined in terms of FORMAC commands.

While not involving mathematics, it is worth noting that considerable work has been done in recent years in the area of proving programs correct. A bibliography is given by London (1970) and a brief survey is provided by London (1968).

IV. Pure Mathematics

A. INTRODUCTORY COMMENTS

In attempting to discuss the software and use of a computer for "pure mathematics," it is obviously necessary to understand what is meant by the latter term. Unfortunately, it is ill-defined, and attempts to produce a definition usually result in something as unproductive as the statement that pure mathematics is what is done by pure mathematicians. An excellent discussion of the problem is given by Lehmer (1963 and 1966). He distinguishes between results, theorems, and theories by defining a result as a statement about a finite set, a theorem as a statement involving an infinite class, and a theory as a collection of theorems and results that are related in some aesthetically satisfactory way. One of the simplest uses of a computer for pure math is to discover a counterexample for a conjectured theorem, particularly in number theory. Lehmer cites as an example that the conjecture that

$$2 + 1, \quad 2^2 + 1, \quad 2^{2^2} + 1, \quad 2^{2^{2^2}} + 1, \quad \ldots$$

are all primes was proved false by the SWAC in 1954, when it found a factor of the fifth term of this sequence. In a number of cases, particularly in problems of this kind, the computer is really acting as an assistant to the human being. However, we are also concerned with whether or not the computer can do new and original work in the same sense that people can. Just as it is hard in the field of artificial intelligence to determine whether a computer has behaved intelligently, so it is hard to determine what is new and original and mathematically "pure" on a computer. However, there is an easier criterion in the latter case than in the former; if a theorem that was unknown to people is both discovered and proved on a computer, then it seems fair to say that the computer has done original work in mathematics. (See, for example, the work done by the SAM system mentioned in Sect. III,C.) The level of originality is a matter of value judgment, just as it is for work done by humans.

It can be said that there are really four things that can be done for pure mathematics or pure mathematicians on a computer. First, the computer can provide empirical material from which the mathematician can either build theories or make conjectures. Second, the computer can be used to disprove conjectures. Third, the computer can prove conjectures and/or theorems. (The distinction being made here is that the theorem is actually known to be true and the machine proof might be used only for the purpose of obtaining a better or different proof; in the case of a conjecture, the machine proof would establish it as a theorem.) Section III,C mentioned a conjecture in lattice theory which was essentially proved by the computer–human system called SAM. Fourth, the computer can create new theorems; right now this appears beyond the scope of current or even far-reaching state of the art, except in the most trivial sense.

It is not practical to cover in any technical depth the work that has been done in the area of pure mathematics. Only a very brief survey stressing the highlights and an indication of what is happening will be given here. Appropriate existing survey papers will be delineated. The bibliography given for this area is a sampling rather than a complete listing.

The areas of mathematics that are involved seem to fall into the following categories: analytic number theory, group theory, other algebraic structures, topology, and geometry and trigonometry. In the latter two cases, the work has been done primarily for the purpose of proving theorems, and so it is discussed in Sects. III,C and III,D.

As an illustration of major interest in this broad area, a conference on computational problems in abstract algebra was held in England in September 1967; see Leech (1968). Meetings of the American Mathematical Society in March 1966 and March 1970 indicate the interest of even the "pure" mathematicians.

The best single continuing source of material for work in this general area is the periodical, *Mathematics of Computation*, formerly called *Mathematical Tables and other Aids to Computation*. Another useful source is *Mathematical Algorithms*, which is a journal published by MIT Press, 1966–1968, but apparently discontinued. It contains three broad categories of which one is mathematical results obtained wholly or partially by computer. Purportedly, the standards of acceptance are those of any other mathematics journal except that a computer must be used. While this is not a large journal, its existence indicates positive use of computers to do nonnumerical mathematics.

A number of specific articles are listed in the author's annotated bibliography [first version—Sammet (1966a); second version—Sammet (1968b)].

B. NUMBER THEORY AND COMBINATORIAL ANALYSIS

To illustrate some of the types of problems involved, it is worth mentioning briefly some work done in 1953 on the SWAC [see Lehmer (1956)]. The first problem handled involved a conjecture of Polya that states, in effect, that there are more integers having an odd number than an even number of prime factors. This was investigated for numbers less than 800,000, but apparently the program did not improve an already known result. A research problem in which the computer did prove very helpful involved quadratic residues and some formulas relating to them; in this case the computer actually proved some rather complicated results.

Paige and Tompkins (1960) describe a technique for generating a pair of latin squares of order 10. The authors show some useful methods that permit the rejection of isomorphs of situations previously analyzed; this rejection concept is very important for saving both time and space in problems of this kind. In Parker (1963), more progress is made towards solving the problem. In this particular case, the authors state an important problem in finite affine planes, which is equivalent to a problem in latin squares.

Various combinatorial problems and how to program them effectively are described in Lehmer (1960). Ten different methods of automatically generating permutations are given. Another place in which work is done in preventing the analysis or generation of isomorphic cases more than once is in Swift (1960). Efficient handling of this problem is important in areas such as the listing of latin squares, semigroups of fixed order, finite projective planes, etc. Different types of problems in combinatorial analysis and analytic number theory are described in the following items in Sammet (1968b): (BRILJ640), (COHNH620), (COMES550), (HAYAH650), (HALLM550), (KELLH630), (TAUSO600), (WALKR600). A useful survey of the work in combinatorial analysis is given by Hall and Knuth (1965).

Some theorem proving has been done in the area of number theory. This is mentioned briefly in Sect. III,C.

C. ABSTRACT ALGEBRA

Although considerable work has been done to construct on a computer various algebraic structures and substructures, such as finite fields, division algebras, and groups, most of the work has been done in group theory. According to Cannon (1969, p. 3): "Probably the first reference to computing with groups is . . . by M. H. A. Newman, . . . 1951 . . . ." Much of the work involves enumeration of cosets with emphasis on developing efficient programs and machine-oriented algorithms. Another major area is the work done in Germany by Neubüser and colleagues in developing subgroup lattices; unfortunately the original source papers are in German. A very complete survey (including an extensive bibliography) is given by Cannon (1969); however, the reader is cautioned that a fairly thorough knowledge of group theory is needed to fully appreciate that paper. The programs in the broad area of group theory have been developed on a wide variety of machines and using various tools ranging from coding in assembly language to the use of Algol. In many cases problems of efficient storage utilization preclude the use of list processing techniques and require assembly language coding.

A practical method of constructing Galois fields of characteristic 2 of large orders is given by Swift (1960). Determination of division algebras with 32 elements is described by Walker (1963), and the generation of Veblen-Wedderburn systems is described by Kleinfeld (1960).

Some interesting comments are made by Dade and Zassenhaus (1963) in considering the structure of the class semigroup formed by the fractional ideals of an order in an algebraic number field E. They found that some of the mathematical hypotheses that were being used were so difficult to program that they had to be weakened. Actually, by doing this they were helped to obtain a better mathematical result.

Work involving Witt vectors is described in Duby (1968).

In a recent paper [Belinfante and Kolman (1969)] mention was made of a Fortran program to carry out representations of Lie algebras. The paper discusses some of the basic tools needed to perform calculations on the finite-dimensional representations of a complex simple Lie algebra on a computer.

Maurer (1968) describes a program which finds the Galois group of a polynomial of degree less than five over the rationals or over an arbitrary finite field.

Of primarily pedagogical or tutorial interest is the system by Maurer (1966) designed specifically for handling finite groups, rings, fields, semigroups, and vector spaces. This is done primarily through the use of Cayley tables, Upon request, the program will do such things as generate the Cayley table of the alternating group on a given number of letters, or of the symmetric group on the given number of letters, or of a cyclic group of a given order, or various kinds of semigroups, right ideals, or all normal subgroups, etc. It can also test whether a semigroup is a group, or whether a group is Abelian, or whether a particular mapping is an isomorphism, etc. There are separate commands for each of these capabilities, and these commands can be written in a fairly natural language; e.g., GENERATE RIGHT IDEAL $3 FROM ELEMENT $2 OF $1 or GENERATE SET $2 OF ALL NORMAL SUB-GROUPS OF $1, etc., where the $N represents specific names. An additional feature of the system is its ability to accept theorems or conjectures, and prove (or disprove) them for the file of information currently available to the system. Thus, the input statement IF THE GROUPS G1 AND G2 ARE ABELIAN, THEN THE DIRECT PRODUCT OF G1 AND G2 IS ALSO ABELIAN is tested against the entire file and either is shown to be valid or else a counterexample is given.

D. MISCELLANEOUS

Several other types of pure mathematical work have been done on a computer, and are mentioned here briefly.

Work on specifying projective planes by Hall is described in Hall et al. (1956, 1959) and Knuth (1965). MacLaren (1965) described the problem of finding the homotopy groups of manifolds. He programmed a small part of this using COGENT. See also work in topology by Cohn (1969).

Morris (1966) has implemented a model for a finite collection of algebras on the computer.

Work has been done on proving trigonometric identities; this work could probably be mentioned either in this section or in Sect. III. The latter has been chosen since the techniques used seem more in line with that type of work.

Using Formula Algol, Iturriaga (1967) developed heuristic programs to compute nontrivial limits of functions.

V. Tools for Developing Nonnumerical Mathematics Software

A. INTRODUCTORY COMMENTS

This section is devoted to a brief discussion of the tools that have been found to be either necessary or useful in the development of software—both individual programs and systems—for nonnumerical mathematics. The main emphasis is on the languages; the two main streams of activity here are list processing and string processing. The latter is considerably less important in this context, although the existing string-processing systems can, and have been, used for certain types of nonnumerical work.

Although it is not the intent of this paper to try to provide rigorous definitions of the terms *list* and *string processing*, it is worth giving at least one description of the fundamental difference between them. A list essentially refers to the way in which data is stored in a computer, and list processing is concerned with how to get at items of data stored (or to be put) on the list. On the other hand, a string is simply a sequence of data that can have any type of storage representation. In particular, a string can be stored in a list form, and often is. String and list processing are used both for manipulating specific pieces of information based on their content, or handling information based on its position and relatively independent of its content.

This section is in no way a complete representation of the past and present state of the art in this area, since list-processing routines and applications lurk behind every door. A good comparison of COMIT, IPL-V, LISP 1.5, and SLIP with respect to several factors is given by Bobrow and Raphael (1964). Descriptions of all the languages cited in Sect. V can be found in Sammet (1969).

B. LIST PROCESSING

Of all the work described in this paper, the most widely known and documented is that of list processing. Some readers may even ask why a section on list processing is being included in a paper devoted to software for nonnumerical mathematics. There are two reasons for this inclusion. First, list processing is a very fundamental technique in doing nonnumerical mathematics. This is evidenced by the fact that the earliest and most basic work on list processing was that of Newell *et al.* (1956, 1957) and was developed in response to a need shown in their attempt to do theorem proving (discussed in Sect. III,B). Similarly, any manipulation of algebraic expressions other than polynomials clearly requires list processing. An illustration on this point is given in Wilkes (1965). A second reason for including this discussion is that some of the nonnumerical mathematics software has specifically used some of these list-processing systems, and this has significantly affected the results—both good and bad. Thus, while list processing is a tool, rather than the objective itself, it is of sufficient importance to warrant its discussion here.

The fundamental importance of list processing is well known in those applications where the amount of storage to be allocated for a particular table or type of data is unknown at the beginning of the program, and changes radically during the course of the program. Furthermore, not only can the amount of storage vary, but insertion and deletion of items from the list become very nontrivial. In a system where the data is stored sequentially, the insertion or deletion of items from the list requires movement of the data that either precedes or follows the item in question, whereas in a list-structured system one need only change a few addresses. The fundamental principle of implementing a list or list structure is that the items need not be stored consecutively in memory. In the broad sense, the implementation of a list involves two elements for each item of data: one element is, of course, the data itself, but the second is a pointer to the next element in the list. If only a list (sometimes called a simple list, or a chain) is involved, then only one pointer is needed. However, if one talks about a list structure, in which an element of data can itself be a list (and is then called a sublist), then clearly two pointers are needed: one is to the next item in the basic list, and the other is to the heading of the sublist.

Two main characteristics of list-processing systems are needed for discussion or comparison of them: one is the type of language involved, and the second is the method of implementation and the type of list structures actually permitted. With regard to the type of language involved, the earliest form is roughly equivalent in style and format to an assembly language for list-processing operations. The whole IPL family is the best illustration of this. A second category involves the addition of subroutines to an existing language. SLIP is the most widely used example of this type, but is not a language itself. The third language approach is to include list processing as a basic part of the language, as in PL/I. Finally, it is possible to develop a completely independent language, although these tend to be woefully weak in doing any numeric computations. This approach is exemplified by LISP and L^6, and also by IPL.

With regard to the implementation characteristics, one needs to consider whether or not they permit common sublists (or, for that matter, sublists at all), how the free list (i.e., available storage) is handled both in terms of obtaining storage space from there and returning unneeded memory positions to there (i.e., garbage collection), the amount of symmetry and/or threading and/or extra pointing that is used, and finally whether tape or some external storage media can be used.

Because these systems are so well documented, only a few words will be given here with emphasis on their use in developing software for non-numerical mathematics. IPL-V is an outgrowth of the earlier IPL languages that were developed by Newell, Shaw, and Simon, culminating in IPL-V, which has been implemented on many machines and is defined in Newell *et al.* (1964). Among the capabilities available in IPL-V are instructions to copy lists, create lists, insert symbols, search lists, erase lists, etc. The primary use in the domain of this paper has been in the theorem-proving work described in Sect. III,B. There is very little current use of IPL-V.

LISP 1.5 [see McCarthy *et al.* (1962)] is the most elegant in concept and structure of the list-processing systems, although in this author's opinion it is harder to learn. A program in LISP is basically a function or a sequence of functions to be evaluated, followed by an executable statement called "evaluate." All the data is in the form of symbolic expressions built up from atomic symbols. A program itself is considered data. LISP has been the most widely used system for developing formula manipulation software. As discussed in Sect. II,C, a "conglomerate" has been developed primarily by using appropriate portions of many existing LISP-based systems. Among the systems or individual programs pertaining to nonnumerical mathematics written using LISP are SAINT, SIN, REDUCE, MATHLAB, Symbolic Mathematical Laboratory, ADEPT, and the "conglomerate" at the IBM Research Center. These were all mentioned or described earlier in this paper.

SLIP [see Weizenbaum (1963)] has been used to develop several small formula manipulation systems. It is not known to this author whether L^6 [see Knowlton (1966)] has been used for this purpose.

C. String Processing and Pattern Matching

The fundamental characteristic of a string-processing system is its ability to look for a pattern in a string of data and take action with respect to this pattern. The user generally has control over the pattern matching, but not over the way in which the information is stored internally. The main developments in this area have been COMIT [MIT (1961a, b)] and SNOBOL [Griswold *et al.* (1968)]. Facilities providing good string handling were embedded in LISP in a system called CONVERT [see Guzman and McIntosh (1966)]. The reason for the importance of the string processing is that some of the problems in formula manipulation involve pattern matching, which usually requires some type of string processing.

SNOBOL is the major string-processing language in use as of this writing. SNOBOL made a number of obvious and needed improvements to COMIT, and is still growing in power. SNOBOL has been used in several (undocumented) programs for nonnumerical mathematics in addition to the trigonometry identity-proving work mentioned in Sect. III,D.

A version of AMBIT [see Christensen (1965)] called AMBIT/L has been used to implement the formula manipulating system IAM discussed in Sect. II,C. AMBIT/L is a pattern replacement language with data structures similar to those of LISP.

VI. Future Developments—A Scientific Assistant

Many people are trying to develop systems which they consider "better" in some sense than those existing currently. This is both necessary and desirable because the current state of the art does not permit us to develop one system which is equally useful to all people. Moreover, I have not seen any purported list of the components of such an "ideal" system. I will now attempt to provide a list, with the full realization that others may find elements that I have missed. I would like to call this collection of features a "Scientific Assistant" which would be useful to mathematicians, physicists, engineers, and any others in the "scientific" category. The concept of a broad-based powerful system for some of those user classes is not new. Other people have used different terms for similar ideas; e.g., Rochester refers to an "Automatic Handbook," Minsky uses "Mathematical Assistant," Sibley talks about "Engineering Assistant," and Martin and Moses use the phrase "Mathematical Laboratory." The name is not important; what is significant is that each person is trying to indicate the need for a computer system which assists the engineer, scientist, or mathematician in doing the creative work which we expect of humans, while the computer does the drudgery. Unfortunately, the easy part of this problem is to describe what is needed; the hard part is to do it efficiently.

A *fundamental premise* in such a Scientific Assistant is that everything must be done efficiently. To the extent that this can not be done, the trade-offs and sacrifices made in developing such a system will obviously slant its effectiveness towards different classes of users.

The primary purpose of this section is to list components of a Scientific Assistant, rather than to describe in any detail the meaning of the individual elements. The five main components of a Scientific Assistant are: language, environment, facilities, algorithm and table retrieval, and theorem proving.

A. LANGUAGE

1. *Natural to User*

The language should be natural to the user, including the ability to input material for stating a problem in the form found in normal textbooks or mathematical proofs. This specifically includes normal two-dimensional notation (see also Sect. VI,B). Output should be equally natural.

2. *Extensible*

The user must have the ability to define new syntax, new semantics, and reformulate existing syntax. Specifically, he must be able to add new data types and provide for such things as noncommutative operators. Furthermore, the user must be able to specify low-level operations (e.g., precision) when he desires.

3. *High Level*

The primitives should be high level and thus be as nonprocedural as possible. For example, the user should be able to say "INTEGRATE" or "INVERT MATRIX" or "CALCULATE THE DERIVATIVES OF THE FOLLOWING FUNCTIONS AND EVALUATE THEM FOR $X = 1, 2, ..., 9$."

4. *Existing Facilities*

All facilities of existing languages, including loop control, interaction between numeric and nonnumeric data, access to complex data structures (including lists), etc., should be included.

B. ENVIRONMENT

1. *Interactive (with Compatible Batch)*

Obviously, the system must be interactive, but it must also provide a good batch-handling facility. Furthermore, the language and other system features must be compatible between the interactive and batch versions so that experimentation and debugging carried out using the on-line facilities can be relegated to the batch-handling for lengthy processing.

2. *Display Scope and Typewriter*

Both typewriters and display scopes should be provided so that the user may choose the one must suited to his particular purpose. Both a keyboard and light pen should be available with the scope.

3. *Two-Dimensional Input and Output*

Both the input and output of mathematical expressions must be two-dimensional, with normal mathematical notation used on both the typewriter and scope.

4. *Handwritten Input*

Input should be allowed in normal handwritten form.

C. Facilities

1. *Numeric*

Clearly fixed, floating point, and complex arithmetic are needed. However, the most essential feature is rational arithmetic to arbitrary precision. Automatic conversion from one unit to another (e.g., feet to inches) should be included. Control of round-off, precision, etc., must be available to the user who wants them.

2. *Nonnumeric*

This represents the formula manipulating capabilities. Subpackages for handling polynomials and rational functions (including factoring and finding greatest common divisors) with maximum efficiency is required, but the ability to allow expressions with all the elementary functions, and to switch back and forth between the two, is needed. Both complex and real formal variables should be handled. Specific subroutines or subpackages are needed for at least differentiation, integration, and algebraic and differential equation solving. The system must provide automatic simplification to the extent of handling ones and zeros and combining like terms, but the user must have the ability to override all or part of these manipulations or to allow noncommutative or nonassociative operations. The user also must have control over expansion, factoring, and substitution, although default conditions can be supplied for normal use. Other formula manipulating facilities from existing or future systems should be included, e.g., matching expressions.

D. Algorithm and Table Retrieval

The information which is now stored in handbooks for physics, chemistry, etc., should be stored and easily accessible to the user. This includes tables, definitions, formulas, etc. In addition, algorithms and procedures for carrying out technical tasks, e.g., designing circuits, bridges, etc., must be stored and retrievable.

E. THEOREM DERIVING AND PROVING

While perhaps of specialized use only to people working in "pure" mathematics, the system should have capabilities for allowing people to both derive and prove theorems. This requires appropriate internal formats for the types of data involved, and inclusion of techniques for theorem proving. This can also be used for checking derivations of formulas.

REFERENCES

Abrahams, P. (1963). Machine Verification of Mathematical Proof. Rep. No. P-AA-TR-(0045), (Ph.D. Thesis—MIT). ITT, International Electric Corp., Paramus, New Jersey.

Aldins, J., Brodsky, S., Dufner, A., and Kinoshita, T. (1969). Photon–Photon Scattering Contribution to the Sixth Order Magnetic Moments of the Muon and Electron. *Phys. Rev. Lett.* **23.**

Allen, J., and Luckham, D. (1969). An Interactive Theorem-Proving Program. Artificial Intelligence Project Memo AIM-103. Stanford Univ., Computer Science Dept., Stanford, California. (To appear in "Machine Intelligence 5," Edinburgh Press.)

Anderson, R. H. (1968). Syntax-Directed Recognition of Hand-Printed Two-Dimensional Mathematics. *In* "Interactive Systems for Experimental Applied Mathematics" (M. Klerer and J. Reinfelds, eds.), pp. 436–459. Academic Press, New York.

Belinfante, J. G., and Kolman, B. (1969). An Introduction to Lie Groups and Lie Algebras, With Applications. III. Computational Methods and Applications of Representation Theory. *SIAM Rev.* **11,** 510–543.

Bennett, J., Guard, J., Haydock, R., Oglesby, F., Paschke, W., and Settle, L. (1968). CRT-Aided Semi-Automated Mathematics. Contract AF F-19628-67-C-0100, Air Force Cambridge Research Labs., Bedford, Massachusetts. Applied Logic Corp., Princeton, New Jersey.

Berlekamp, E. R. (1967). Factoring Polynomials over Finite Fields. *Bell System Tech. J.* **46,** 1853–1859.

Bernstein, M. I., and Williams, T. G. (1969). A Two-Dimensional Programming System. *Proc. IFIP Congress 1968* (A. J. H. Morrell, ed.), Vol I, pp. 586–592. North-Holland Publ., Amsterdam.

Blackwell, F. W., and Anderson, R. H. (1969). An On Line Symbolic Mathematics System Using Hand Printed Two Dimensional Notation. *Proc. ACM Nat. Conf., 24th,* pp. 551–557.

Blair, F. W., Griesmer, J. H., and Jenks, R. D. (1970). An Interactive Facility for Symbolic Mathematics. Rep. RC-2766. IBM, T. J. Watson Research Center, Yorktown Heights, New York.

Bleiweiss, L., Bond, E., Cundall, P. A., and Hirschkop, R. (1966). A Time-Shared Algebraic Desk Calculator Version of FORMAC. Technical Rep. No. TR00.1415. IBM, Systems Development Div., Poughkeepsie, New York.

Bobrow, D. G. (ed.) (1968). "Symbol Manipulation Languages and Techniques." North-Holland Publ. Amsterdam.

Bobrow, D. G., and Raphael, B. (1964). A Comparison of List-Processing Languages. *Comm. ACM* **7,** 231–240.

Boys, S. F., Cook, G. B., Reeves, C. M., and Shavitt, I. (1956). Automatic Fundamental Calculations of Molecular Structure. *Nature* **178,** No. 4544, 1207–1209.

Brown, W. S. (1966). A Language and System for Symbolic Algebra on a Digital Computer. *Proc. IBM Scientific Computing Symp. Computer-Aided Experimentation,* 320–0936–0, pp. 77–114. IBM Corp., Data Processing Div., White Plains, New York.

Brown, W. S., Tague, B. A., and Hyde, J.`P. (1963). The ALPAK System for Non-Numerical Algebra on the Digital Computer. *Bell System Tech. J.* Part I, **42,** 2081–2120; Part II, **43,** 785–804; Part III, **43,** 1547–1562.

Cannon, J. J. (1969). Computers in Group Theory: A Survey. *Comm. ACM* **12,** 3–12.

Caviness, B. F. (1970). On Canonical Forms and Simplification. *J. ACM* **17,** 385–396.

Cerutti, E., and Davis, P. J. (1969). FORMAC Meets Pappus. *Math. Mon.* **76,** 895–905.

Christensen, C. (1965). Examples of Symbol Manipulation in the AMBIT Programming Language. *Proc. ACM Nat. Conf.* pp. 247–261.

Clapp, L. C., and Kain, R. Y. (1963). A Computer Aid for Symbolic Mathematics. *Proc. Fall Joint Comp. Conf.* **24,** 509–517.

Cohn, H. (1969). Some Computer-Assisted Topological Models of Hilbert Fundamental Domains. *Math. Comput.* **23,** 475–487.

Collins, G. E. (1966). PM, A System for Polynomial Manipulation. *Comm. ACM* **9,** 578–589.

Collins, G. E. (1967). Subresultants and Reduced Polynomial Remainder Sequences. *J. ACM* **14,** 128–142.

Collins, G. E. (1968a). The SAC-1 Polynomial System. Tech. Ref. No. 2. Univ. of Wisconsin Computing Center, Madison, Wisconsin.

Collins, G. E. (1968b). The SAC-1 Rational Function System. Technical Rep. No. 8. Univ. of Wisconsin Computing Center, Madison, Wisconsin.

Collins, G. E. (1969). Computing Time Analysis for Some Arithmetic and Algebraic Algorithms. *In Proc.* 1968 *Summer Inst. Symbolic Mathematical Computing* (R. G. Tobey, ed.), Rep. No. FSC 69–0312, pp. 195–231. IBM, Cambridge, Massachusetts.

Collins, G. E., and Horowitz, E. (1970). The SAC-1 Partial Fraction Decomposition and Rational Function Integration System. Tech. Rep. No. 80. Univ. of Wisconsin, Computer Sciences Dept., Madison, Wisconsin.

Cooper, D. C. (1966). Theorem-Proving in Computers, *In* "Advances in Programming and Non-Numerical Computation" (L. Fox, ed.), pp. 155–182. Pergamon Press, Oxford.

Crisman, P. A. (ed.) (1965). "The Compatible Time-Sharing System, A Programmer's Guide," 2nd ed. MIT Press, Cambridge, Massachusetts.

Dade, E. C., and Zassenhaus, H. (1963). How Programming Difficulties Can Lead to Theoretical Advances. *In* "Experimental Arithmetic, High Speed Computing and Mathematics," *Proc.* 15*th Symp. Appl. Math.* **15,** 87–94. American Mathematical Soc., Providence, Rhode Island.

Davis, M. (1963). Eliminating the Irrelevant from Mechanical Proofs. *In* "Experimental Arithmetic, High Speed Computing and Mathematics," *Proc.* 15*th Symp. Appl. Math.* **15,** 15–30. American Mathematical Soc., Providence, Rhode Island.

Davis, M., Logemann, G., and Loveland, D. (1962). A Machine Program for Theorem-Proving. *Comm. ACM* **5,** 394–397.

Davis, M., and Putnam, H. (1960). A Computing Procedure for Quantification Theory. *J. ACM* **7,** 201–215.

Duby, J. J. (1968). Sophisticated Algebra on a Computer—Derivatives of Witt Vectors. *In* "Symbol Manipulation Languages and Techniques" (D. G. Bobrow, ed.), pp. 71–85. North-Holland Publ., Amsterdam.

326 J. E. SAMMET

Engeli, M. (1969). Formula Manipulation—The User's Point of View. *In* "Advances in Information System Science" (J. Tou, ed.). Plenum Press, New York.

Engelman, C. (1965). MATHLAB—A Program for On-Line Machine Assistance in Symbolic Computations. *Proc. Fall Joint Comp. Conf.* **27**, 413–422.

Engelman, C. (1969). MATHLAB 68. *Proc. IFIP Congress* 1968 (A. J. H. Morrell, ed.), Vol. I. pp. 462–467. North-Holland Publ., Amsterdam.

Fenichel, R. R. (1966). An On-Line System for Algebraic Manipulation. MAC-TR-35 (Ph.D. thesis, Harvard University). MIT, Project MAC, Cambridge, Massachusetts.

FORMAC (1965). FORMAC (Operating and User's Preliminary Reference Manual), No. 7090R2IBM0016. IBM Program Information Dept., Hawthorne, New York.

Gelernter, H. (1959). Realization of a Geometry Theorem Proving Machine. *Proc. Int. Conf. Inform. Proc.* pp. 273–282. Unesco House, Paris.

Gelernter, H., Hansen, J. R., and Loveland, D. (1960). Empirical Explorations of the Geometry Theorem Machine, *Proc. Western Joint Comp. Conf.* **17**, 143–149.

Gey, F. C., and Lesser, M. B. (1969). Computer Generation of Series and Rational Function Solutions to Partial Differential Initial Value Problems. *Proc. ACM Nat. Conf.* 24th, pp. 559–572.

Gilmore, P. C. (1959). A Program for the Production from Axioms, of Proofs for Theorems Derivable Within the First Order Predicate Calculus. *Proc. Int. Conf. Inform. Proc.*, pp. 265–273. Unesco House, Paris.

Gilmore, P. C. (1960). A Proof Method for Quantification Theory—Its Justification and Realization. *IBM J. Res. Develop.* **4**, 28–35.

Griswold, R., Poage, J., and Polonsky, I. (1968). "The SNOBOL4 Programming Language." Prentice-Hall, Englewood Cliffs, New Jersey.

Guard, J., Oglesby, F., Bennett, J., and Settle, L. (1969). Semi-Automated Mathematics. *J. ACM* **16**, 49–62.

Guzman, A., and McIntosh, H. (1966). CONVERT. *Comm. ACM* **9**, 604–615.

Hall, M. Jr., and Knuth, D. E. (1965). Combinatorial Analysis and Computers. *Amer. Math. Monthly* **72**, No. 2, Part II, 21–28.

Hall, M. Jr., Swift, J. D., Walker, R. J. (1956). Uniqueness of the Projective Plane of Order Eight. *Math. Tables and Other Aids to Comput.* **10**, 186–194.

Hall, M. Jr., Swift, J. D., Killgrove, R. (1959). On Projective Planes of Order Nine. *Math. Tables and Other Aids to Comput.* **13**, 233–246.

Hanson, J. W., Caviness, J. S., and Joseph, C. (1962). Analytic Differentiation by Computer. *Comm. ACM* **5**, 349–355.

Hearn, A. (1968). REDUCE: A User-Oriented Interactive System for Algebraic Simplification. *In* "Interactive Systems for Experimental Applied Mathematics" (M. Klerer and J. Reinfelds, eds.), pp. 79–90. Academic Press, New York.

Hearn, A. (1969). The Problem of Substitution. *In Proc.* 1968 *Summer Inst. Symbolic Mathematical Computing* (R. G. Tobey, ed.), Report No. FSC 69–0312, pp. 3–19. IBM, Cambridge, Massachusetts.

Hearn, A. C., Kuo, P. K., and Yennie, D. R. (1969). Radiative Corrections to an Electron–Positron Scattering Experiment, *Phys. Rev.* **187**.

Herget, P., and Musen, P. (1959). The Calculation of Literal Expansions. *Astronom. J.* **64**, No. 1266, 11–20.

Holden, A., and Johnson, D. (1967). The Use of Imbedded Patterns and Canonical Forms in a Self-Improving Problem Solver. *Proc. ACM Nat. Conf.*, 22nd, pp. 211–219.

Howard, J. C., and Tashjian, H. (1968). An Algorithm for Deriving the Equations of Mathematical Physics by Symbolic Manipulation. *Comm. ACM* **11**, 814–818.

Iturriaga, R. (1967). Contributions to Mechanical Mathematics. (Ph.D. thesis.) Carnegie–Mellon Univ., Pittsburgh, Pennsylvania.

Johnson, D., and Holden, A. (1964). A Problem-Solving Machine with the Capacity to Learn from its Experience. *Simulation* 3, 71–76.

Kahrimanian, H. G. (1954). Analytical Differentiation by a Digital Computer. *In* Symposium on Automatic Programming for Digital Computers, May, pp. 6–14. Office of Naval Research, Dept. of the Navy.

Kleinfeld, E. (1960). Techniques for Enumerating Veblen–Wedderburn Systems. *J. ACM* 7, 330–337.

Klerer, M., and Grossman, F. (1967). Further Advances in Two-Dimensional Input-Output by Typewriter Terminals. *Proc. Fall Joint Comp. Conf.* 31, 675–687.

Knowlton, K. (1966). A Programmer's Description of L^6, Bell Telephone Laboratories' Low-Level Linked List Language. *Comm. ACM* 9, 616–625.

Knuth, D. E. (1965). A Class of Projective Planes. *Amer. Math. Soc.* 115, 541–549.

Knuth, D. E. (1969). "The Art of Computer Programming," Volume 2. Addison–Wesley, Reading, Massachusetts.

Krugman, E., and Myszewski, M. (1969). Canonical Form in a System for Automatic Algebraic Manipulation, Rept. No. CA 6912–0931. Applied Data Research, Wakefield, Massachusetts.

Leech, J. (ed.) (1968). "Proceedings of Conference on Computational Algebra." Pergamon Press, Oxford.

Lehmer, D. H. (1960). Teaching Combinatorial Tricks to a Computer. *In* "Combinatorial Analysis," *Proc. of Symposia in Applied Mathematics* 10, 179–193. American Mathematical Soc., Providence, Rhode Island.

Lehmer, D. H. (1963). Automation and Pure Mathematics. *In* "Applications of Digital Computers" (W. Freiberger and W. Prager, eds.), pp. 219–231. Ginn, Boston, Massachusetts.

Lehmer, D. H. (1966). Mechanized Mathematics. *Bull. Amer. Math. Soc.* 72, 739–750.

Lehmer, D. H., Lehmer, E., Mills, W. H., and Selfridge, J. L. (1962). Machine Proof of a Theorem on Cubic Residues. *Math. Comput.* 16, 407–415.

Lehmer, E. (1956). Number Theory on the SWAC. *In* "Numerical Analysis," *Proc. Symp. Appl. Math. Amer. Math. Soc.* 6, 103–108. McGraw-Hill, New York.

London, R. L. (1968). Computer Programs Can Be Proved Correct. *To appear in Proc. Symp. Formal Systems and Non-Numerical Problem Solving by Computers* held at Case Western Reserve Univ., November, 1968.

London, R. L. (1970). Bibliography on Proving the Correctness of Computer Programs. *To appear in* "Machine Intelligence 5." Edinburgh Univ. Press, Edinburgh, Scotland.

Loveland, D. W. (1968). Mechanical Theorem-Proving by Model Elimination. *J. ACM* 15, 236–251.

Luckham, D. (1968). Some Tree-Paring Strategies for Theorem Proving. *In* "Machine Intelligence 3" (D. Michie, ed.), pp. 95–112. American Elsevier, New York.

MacLaren, D. (1965). Notes on the Machine Computation of a Spectral Sequence. Argonne National Laboratory (unpublished).

Manove, M., Bloon, S., and Engelman, C. (1968). Rational Functions in MATHLAB. *In* "Symbol Manipulation Languages and Techniques" (D. G. Bobrow, ed.), pp. 86–102. North-Holland Publ., Amsterdam.

Martin W. A. (1967). Symbolic Mathematical Laboratory. MAC-TR-36. (Ph.D. Thesis.) MIT, Project MAC, Cambridge, Massachusetts.

Maurer, D. (1966). Computer Experiments in Finite Algebra. *Comm. ACM* 9, 598–603.

Maurer, D. (1968). The Uses of Computers in Galois Theory. *In Proc. Conf. Computational Algebra* (J. Leech, ed.). Pergamon Press, Oxford.

McCarthy, J. (1962). Computer Programs for Checking Mathematical Proofs. *In* "Recursive Function Theory," *Proc. of Symposia Pure Math.* **5,** 219–229. American Mathematical Society, Providence, Rhode Island.

McCarthy, J., Abrahams, P., Edwards, D., Hart, T., and Levin, M. (1962). LISP 1.5 Programmer's Manual. Computation Center and Research Laboratory of Electronics, MIT, Cambridge, Massachusetts.

McIlroy, M. D., and Brown, W. S. (1966). The ALTRAN Language for Symbolic Algebra on a Digital Computer. Bell Telephone Laboratories, Murray Hill, New Jersey (unpublished).

Millen, J. K. (1968). CHARYBDIS: A LISP Program to Display Mathematical Expressions on Typewriter-Like Devices. *In* "Interative Systems for Experimental Applied Mathematics" (M. Klerer and J. Reinfelds, eds.), pp. 155–163. Academic Press, New York.

MIT (1961a). An Introduction to COMIT Programming. The Research Laboratory of Electronics and the Computation Center, MIT, Cambridge, Massachusetts.

MIT (1961b). COMIT Programmers' Reference Manual. The Research Laboratory of Electronics and the Computation Center, MIT, Cambridge, Massachusetts.

Morris, A. H., Jr. (1966). Models for Mathematical Systems. U.S. Naval Weapons Laboratory.

Moses, J. (1967). Symbolic Integration. MAC-TR-36. (Ph.D. Thesis.) MIT, Project MAC, Cambridge, Massachusetts.

Neidleman, L. D. (1967). An Application of FORMAC. *Comm. ACM* **10,** 167–168.

Newell, A., Shaw, J. C., and Simon, H. A. (1957). Empirical Explorations of the Logic Theory Machine. *Proc. Western Joint Comp. Conf.* pp. 218–230.

Newell, A., and Simon, H. A. (1956). The Logic Theory Machine—A Complex Information Processing System. *IRE Trans. Inform. Theory* **IT-2,** No. 3, 61–79.

Newell, A., Tonge, F., Feigenbaum, E., Green, B. F., Mealy, G. H., eds. (1964). "Information Processing Language V Manual," 2nd ed. Prentice-Hall, Englewood Cliffs, New Jersey.

Nolan, J. (1953). Analytical Differentiation on a Digital Computer. (M.A. Thesis.) MIT, Cambridge, Massachusetts

Norton, L. M. (1966). A Heuristic Program for Proving Theorems of Group Theory. MAC-TR-33. (Ph.D. Thesis.) MIT, Project MAC, Cambridge, Massachusetts.

Paige, L. J., and Tompkins, C. B. (1960). The Size of the 10×10 Orthogonal Latin Square Problem. *In* "Combinatorial Analysis," *Proc. Symposia Appl. Math.* **10,** 71–83. American Mathematical Society, Providence, Rhode Island.

Parker, E. T. (1963). Computer Investigation of Orthogonal Latin Squares of Order Ten: *In* "Experimental Arithmetic, High Speed Computing and Mathematics," *Proc. 15th Symp. Appl. Math.* **15,** 73–81. American Mathematical Society, Providence, Rhode Island.

Perlis, A. J., and Iturriaga, R. (1964). An Extension to ALGOL for Manipulating Formulae. *Comm. ACM* **7,** 127–130.

Perlis, A. J., Iturriaga, R., and Standish, T. A. (1966). A Definition of Formula ALGOL. Carnegie Institute of Technology, Pittsburgh, Pennsylvania.

Prawitz, D., Prawitz, H., and Voghera, N. (1960). A Mechanical Proof Procedure and its Realization in an Electronic Computer. *J. ACM* **7,** 102–128.

Richardson, D. (1966). Some Unsolvable Problems Involving Functions of a Real Variable. (Ph.D. Thesis.) Univ. of Bristol, Bristol, England.

Risch, R. (1968). On the Integration of Elementary Functions Which Are Built Up Using Algebraic Operations. Rep. SP-2801SDC. , Santa Monica, California.

Risch, R. (1969). The Problem of Integration in Finite Terms. *Trans. AMS* **139,** 167–189.

Robinson, J. A. (1963). Theorem-Proving on the Computer. *J. ACM* **10**, 163–174.

Robinson, J. A. (1965). A Machine-Oriented Logic Based on the Resolution Principle. *J. ACM* **12**, 23–41.

Robinson, J. A. (1967). A Review of Automatic Theorem-Proving. *In* "Mathematical Aspects of Computer Science," *Proc. Symposia Appl. Math.* **19**, 1–18. American Mathematical Society, Providence, Rhode Island.

Robinson, J. A. (1969). An Overview of Mechanical Theorem Proving. *To appear in Proc. Symp. Formal Systems and Non-Numerical Problem Solving by Computers* held at Case Western Reserve Univ., November 1968.

Russell, L. (1969). Linear Circuit Analysis by Symbolic Algebra. *Proc. ACM Nat. Conf. 24th*, pp. 573–586.

Sammet, J. E. (1966a). An Annotated Descriptor Based Bibliography on the Use of Computers for Non-Numerical Mathematics. *Comput. Rev.* **7**, B1–B32.

Sammet, J. E. (1966b). Survey of Formula Manipulation, *Comm. ACM* **9**, 555–569.

Sammet, J. E. (1967). Formula Manipulation by Computer. *In* "Advances in Computers" (F. Alt and M. Rubinoff, eds.), Vol. 8, pp. 47–102. Academic Press, New York.

Sammet, J. E. (1968a). Problems and Future Trends in Formal Algebraic Manipulation. *In* "Symbol Manipulation Languages and Techniques" (D. G. Bobrow, ed.), pp. 55–63. North-Holland Publ., Amsterdam.

Sammet, J. E. (1968b). Revised Annotated Descriptor Based Bibliography on the Use of Computers for Non-Numerical Mathematics. *In* "Symbol Manipulation Languages and Techniques" (D. G. Bobrow, ed.), pp. 358–484. North-Holland Publ., Amsterdam.

Sammet, J. E. (1969). "Programming Languages: History and Fundamentals." Prentice-Hall, Englewood Cliffs, New Jersey.

Sammet, J. E., and Bond, E. (1964). Introduction to FORMAC. *IEEE Trans. Electron. Comput.* **EC-13**, 386–394.

Slagle, J. R. (1963). A Heuristic Program that Solves Symbolic Integration Problems in Freshman Calculus. *J. ACM* **10**, 507–520.

Slagle, J. R., and Bursky, P. (1968). Experiments with a Multipurpose, Theorem-Proving Heuristic Program. *J. ACM* **15**, 85–99.

Swift, J. D. (1960). Construction of Galois Fields of Characteristic Two and Irreducible Polynomials. *Math. Comput.* **14**, No. 70, 99–103.

Takacs, L. (1963). A Single-Server Queue with Feedback. *Bell System Tech. J.* **42**, No. 2, 505–519.

Tobey, R. G. (1966). Eliminating Monotonous Mathematics with FORMAC. *Comm. ACM* **9**, 742–751.

Tobey, R. G. (ed.) (1969). *Proc. 1968 Summer Inst. Symbolic Mathematical Computing.* Rep. No. FSC 69-0312. IBM, Cambridge, Massachusetts.

Tobey, R. G., Bobrow, R. J., and Zilles, S. (1965). Automatic Simplification in FORMAC. *Proc. Fall Joint Comput. Conf.* **27**, 37–52.

Tobey, R. G., Baker, J., Crews, R., Marks, P., Victor, K., Lipson, J., and Xenakis, J. (1969). PL/I-FORMAC Symbolic Mathematics Interpreter. 360D-03.3.004. IBM Corp., Contributed Program Library, Program Information Dept.

Travis, L. (1964). Experiments with a Theorem-Utilizing Program. *Proc. Spring Joint Comput. Conf.* **25**, 339–358.

van de Riet, R. P. (1968). Formula Manipulation in ALGOL 60, Parts 1 and 2. Tracts 17 and 18. Mathematisch Centrum, Amsterdam.

Walker, R. J. (1963). Determination of Division Algebras with 32 Elements. *In* "Experimental Arithmetic, High Speed Computing and Mathematics," *Proc. 15th Symp. Appl. Math.* **15**, 83–85. American Mathematical Society, Providence, Rhode Island.

Wang, H. (1960a). Toward Mechanical Mathematics. *IBM J. Res. Develop.* **4,** 2–22.

Wang, H. (1960b). Proving Theorems by Pattern Recognition I. *Comm. ACM* **3,** 220–234.

Weizenbaum, J. (1963). Symmetric List Processor. *Comm. ACM* **6,** 524–544.

Wells, M. B. (1961). MADCAP—A Scientific Compiler for a Displayed Formula Textbook Language. *Comm. ACM* **4,** 31–36.

Wheelock, B. (1970). An Initial Session with IAM. Rep. No. CA–7003–3011. Applied Data Research, Wakefield, Massachusetts.

Wilkes, M. V. (1965). Lists and Why they are Useful. *Comput. J.* **7,** No. 4, 278–281.

Wos, L., Carson, D., and Robinson, G. (1964). The Unit Preference Strategy in Theorem Proving. *Proc. Fall Joint Comput. Conf.* **26,** Part I, 615–621.

Wos, L., Robinson, G., and Carson, D. (1965). Efficiency and Completeness of the Set of Support Strategy in Theorem Proving. *J. ACM* **12,** 536–541.

Wos, L., Robinson, G., and Carson, D. (1966). Automatic Generation of Proofs in the Language of Mathematics. *Proc. IFIP Congr.* 65 (W. Kalenich, ed.), pp. 325–326.

Wyman, J. (1968). Addition No. 3 to Sammet (1968b). *SIGSAM Bull.* No. 10, Oct. (unpublished).

Wyman, J. (1969). Addition No. 4 to Sammet (1968b). *SIGSAM Bull.* No. 12, July (unpublished).

Wyman, J. (1970). Addition No. 5 to Sammet (1968b). *SIGSAM Bull.* No. 15, July, 1970.

Good General Sources

(1966) *Comm. ACM* **9.**

(1967) Mathematical Aspects of Computer Science, **19,** *Proc. Symposia Appl. Math.* American Mathematical Society, Providence, Rhode Island.

Braffort, P., and Hirschberg, D. (eds.) (1963). "Computer Programming and Formal Systems." North-Holland Publ., Amsterdam.

Davis, M. (1958). "Computability and Unsolvability." McGraw Hill, New York.

5.19 Continuous Distribution Sampling: Accuracy and Speed

W. H. Payne and T. G. Lewis

WASHINGTON STATE UNIVERSITY
PULLMAN, WASHINGTON

I. Introduction

The probability integral transformation assures that if X is sampled from cumulative probability distribution (cdf) G, then $U = G(X)$ is uniformly distributed. Thus if cdf G is to be sampled on a computer a pseudorandom number is computed and $G^{-1}(U) = X$ evaluated. The value X is sampled from G or, when a pseudorandom number U is used, X is said to have a pseudo-G distribution.

Computations of $G^{-1}(X)$ present a problem. Often direct analytic inversion of G is impossible or impractical. Several alternatives to avoid direct computation have been developed. Numerical inverse interpolation of exact G, numerical inverse interpolation in an approximation to G, or numerical inverse approximation to G^{-1} have been used. Generalized rejection techniques can be used to sample from G without computation of G^{-1} or even performing the integration on the density of G. The method of mixtures (or composition) splits the probability density of G into a sum of simple probability densities. A method of subterfuge can be used to create a process distributed similar to G.

It is shown that convergence to uniformity of probability of occurrence of the low-order digits is the appropriate accuracy measure for distribution sampling not pointwise convergence. Further, it is proved that the jth digit of a random number sampled from a differentiable cumulative distribution F is uniformly distributed in the limit as $j \to \infty$. Based on this fact a sampling algorithm generates bits from high to low order. Convergence to the limit is rapid so only a few high-order bits need be generated using tables of conditional probabilities. A uniform random number is attached to form the low-order digits of the binary number sampled from F. This "conditional bit" sampling algorithm is nearly equivalent to the Los Alamos binary directed table search sampling technique. Both the conditional bit and Los

Alamos algorithms are combined to yield the most efficient, to date, general sampling algorithm. ASA standard Fortran code for these algorithms is given.

Consider sampling from the normal distribution with expected value μ and standard deviation σ. Define

$$\Phi(X) = \frac{1}{(2\pi)^{1/2}\sigma} \int_{-\infty}^{X} \exp[-(t-\mu)^2/2\sigma^2]\, dt \tag{1}$$

and $\phi(X)$ to be the same integral except the lower limit equal to 0 rather than $-\infty$. For all of the sampling methods discussed in this paper $\mu = 0$ and $\sigma = 1$, since if X is sampled from the normal $\mu = 0$ and $\sigma = 1$, then $BX + A$ is from a normal with $\sigma = B$ and $\mu = A$.

Nearly all of the above methods have been used to produce normal numbers [Muller (1959)]. For (almost) direct analytic inversion of the normal distribution, Box and Muller developed a method based on the classical technique for integration of $\Phi(X)$ between $-\infty$ and ∞ to obtain unity [Box and Muller (1958), Kronmal (1964), Muller (1957)]. Although very accurate, the disadvantage of this method is that two pseudorandom numbers, a sine, a cosine, a natural logarithm, and a square root must be computed.

A Padé-type or rational approximation to transform a uniform deviate to a normal deviate is not popular because faster procedures requiring less memory locations are available [Hastings (1955)].

The central limit approach approximates a normal number by summing N uniform random numbers (subterfuge method). It requires little memory space, is moderately fast, and portable, but produces numbers restricted to $\pm(3N)^{1/2}$ standard deviations of the mean. Teichroew approximated this distribution ($N = 12$) by curve fitting (Chebychev polynomials), but the normal numbers are restricted to $\pm 4\sigma$ [Donnelly (1969), Muller (1959), Teichroew (1953)].

Several composition procedures exist which are generally very fast and accurate [Marsaglia and Bray (1964), Marsaglia et al. (1964), Muller (1957), Naylor et al. (1966)]. Approximating functions over selected intervals are successively sampled and then summed. These procedures are best coded in assembler language and are sufficiently detailed to hinder portability. Because of large memory requirements they are restricted to applications where speed is essential.

A strict rejection technique attributed to Muller (1959) and von Neumann (1959) is slow because of poor sampling efficiency of uniform numbers. Batchelor's method utilizes advantages of both composition and rejection techniques by splitting the normal probability density function into two parts which are sampled by rejection [Tocher (1963)]. The algorithm compares favorably in speed and memory requirements with other methods. However, calculation of two logarithms and three random numbers is repeated until the rejection process obtains an acceptable number (acceptance rate of 78%).

Improvements to the above methods have been made. For instance, rejection sampling has been used to replace sine/cosine calculations in the Box–Muller method [Bell (1968)]. Also, the central limit theorem approach appears to be particularly well suited to highly parallel computers like ILLIAC IV, where many random numbers can be simultaneously generated and then summed [Kuck (1968), Winje (1969)]. We present a new analysis of distribution sampling from arbitrary continuous probability densities.

II. Conditional Bit Sampling

We present our method by example. Suppose a normal number is to be computed with $\mu = 0$ and $\sigma = 1$. Further, the range will be restricted to 8σ. Let $SX_1X_2 \cdot X_3X_4 \ldots$ be the binary expansion ($S = $ sign) of the normal number to be generated. The bits of the normal number are generated from high to low order. For the normal $P(X_1 = 1)$ equals the probability that the entire pseudo-normal number lies between $10.000 \ldots_2 = 2_{10}$ and $11.111 \ldots_2 = 4_{10}$. Thus $P(X_1 = 1) = 2[\phi(4) - \phi(2)] = 2(0.5000 - 0.4773) = 0.0454$. Care must be taken. If another high-order bit was appended to the left of X_1, then $X_1 = 1$ could also occur when the normal number was between $110.000 \ldots_2 = 6_{10}$ and $111.111 \ldots_2 = 8_{10}$, and so forth; however, for this example $\phi(8) - \phi(6) \simeq 0$. Of course, $P(X_1 = 0) = 1 - P(X_1 = 1) = 0.9546$.

The next step is to examine the conditional probability $P(X_2 = 1 \mid X_1)$. Specifically,

$$P(X_2 = 1 \mid X_1 = 0) = \frac{P(X_1 = 0 \text{ and } X_2 = 1)}{P(X_1 = 0)} \tag{2}$$

thus

$$P(X_2 = 1 \mid X_1 = 0) = \frac{\phi(01.111 \ldots = 10.000 \ldots) - \phi(01.000 \ldots)}{\phi(01.111 \ldots = 10.000 \ldots) - \phi(00.000 \ldots)}$$

$$= \frac{2[\phi(2) - \phi(1)]}{2[\phi(2) - \phi(0)]}$$

$$= \frac{2(0.4773 - 0.3412)}{0.9546}$$

$$= 0.285 \tag{3}$$

Since the twos cancel from the numerator and denominator of the conditional probability statement, it is only necessary to consider one side of the distribution. In a similar manner

$$P(X_2 = 1 \mid X_1 = 1) = \frac{P(X_1 = 1 \quad \text{and} \quad X_2 = 1)}{P(X_1 = 1)}$$

$$= \frac{\phi(11.111 \ldots = 100.000 \ldots) - \phi(11.000 \ldots)}{\phi(11.111 \ldots = 100.000 \ldots) - \phi(10.000 \ldots)}$$

$$= \frac{\phi(4) - \phi(3)}{\phi(4) - \phi(2)}$$

$$= 0.0573 \tag{4}$$

Suppose the first j bits of the number are selected and this number is written $\bar{X}_j = X_1 X_2 \cdot X_3 X_4 \ldots X_j$. It is inductively apparent from previous computations that the probability of selecting $X_{j+1} = 1$ is

$$P(X_{j+1} = 1 \mid \bar{X}_j) = \frac{\phi(\bar{X}_j + 2^{2-j}) - \phi(\bar{X}_j + 2^{1-j})}{\phi(\bar{X}_j + 2^{2-j}) - \phi(\bar{X}_j)}$$

$$(j = 0, 1, 2, \ldots) \tag{5}$$

and define $P(X_1 = 1 \mid \bar{X}_0) = P(X_1 = 1)$. Substitute $Z_j = \bar{X}_j + 2^{2-j}$ and divide the numerator and denominator by 2^{2-j} and obtain

$$P(X_{j+1} = 1 \mid \bar{X}_j) = \frac{1}{2} \frac{[\phi(Z_j) - \phi(Z_j - 2^{1-j})]/2^{1-j}}{[\phi(Z_j) - \phi(Z_j - 2^{2-j})]/2^{2-j}} \tag{6}$$

The limit as $j \to \infty$ of the above expression can be taken (noting the familiar expression for the derivative)

$$\lim_{j \to \infty} P(X_{j+1} = 1 \mid \bar{X}_j) = \frac{1}{2} \frac{\Phi'(Z_\infty)}{\Phi'(Z_\infty)} = \frac{1}{2} \tag{7}$$

Theorem *If F is a differentiable cumulative probability distribution and a number X is sampled from F, then the probability of occurrence of the jth digit of X is uniformly distributed in the limit as $j \to \infty$.*

Proof The proof is immediate from the above limit consideration. Graphically (see Fig. 1), the area under dF/dX between $\bar{X}_j + 2^{1-j}$ and $\bar{X}_j + 2^{2-j}$ is divided by the area under the curve between $\bar{X}_j$ to $\bar{X}_j + 2^{2-j}$. If dF/dX is only piecewise continuous, the theorem will not hold at the points of discontinuity. Consider the random variable $X = 0.b_1 b_2 b_3 \ldots$, where the b's are independently distributed with $P(b_i = 0) = \frac{1}{3}$ and $P(b_i = 1) = \frac{2}{3}$. Although we are presently unable to explicitly write down the distribution function of the random variable, the above theorem clearly does not apply.

As with most limit theorems, no information is given in the theorem on how large j must be before the distribution of the jth digit is close to uniformity. Fortunately, the number is usually small (particularly for the normal).

A normal conditional bit generator, for clarity, will be developed. Extension to any differentiable cdf F is immediate. The aim in construction of the generator is to select as few high-order bits as possible with the use of tables and then attach a pseudorandom number to form the low-order digits. The idea of representing a random variable Y as $Y = X + V$, where X is discrete and V is uniform on some small interval, is not new [Marsaglia (1962, 1963)]. The technique present here is quite different from these previous endeavors. First, however, tables of conditional probabilities must be computed.

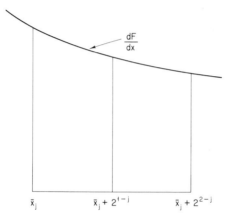

FIG. 1. Areas under typical *cdf* used to compute conditional probability.

III. Computation of Conditional Probabilities

Numerical integration of $dF(X)$ can be accomplished to compile tables of $P(X_{j+1} = 1 \mid \bar{X}_j)$ of Eq. (5). For the normal example define $R_j = 2^{2-j}$ or, recursively,

$$R_j = R_{j-1}/2, \qquad R_0 = 4 \tag{8}$$

and

$$U_j = \bar{X}_j + R_j \tag{9}$$

$$V_j = \bar{X}_j + R_{j+1} \tag{10}$$

The numerator of Eq. (5) is

$$\phi(U_j) - \phi(V_j) = \int_{V_j}^{U_j} d\Phi(X) \tag{11}$$

Similarly substitute

$$U_j' = U_j, \qquad V_j' = \bar{X}_j \tag{12}$$

in Eq. (11) to obtain the denominator. Integration over the largest interval occurs first, and as j increases integration is computed over smaller intervals, thus maintaining accuracy.

The next set of probabilities, $P(X_{j+2} = 1 \mid \bar{X}_{j+1})$, can be computed by replacement:

$$R_{j+1} \leftarrow R_j/2 \tag{13}$$

$$U_{j+1} \leftarrow \bar{X}_{j+1} + R_{j+1} \tag{14}$$

```
      SUBROUTINE PTABLE(L,K,TABLE,BIGERR,ERR,EPS,N,MAXE)
C RCMBERG QUADRATURE, COMM. ACM, 12, 6, (JUNE 1969), 324.
C L=TCTAL NUMBER CF BITS SAMPLED USING TABLE.
C K=NUMBER CF BITS TO THE LEFT OF BINARY POINT.
C BIGERR=-1.=ERRCR IN ROMINT.
C ERR,EPS,N,MAXE=RCMINT PARAMETERS.
C R=2**(K-J) IN ITERATIVE FORM.
C J=BIT CCUNTER, CCUNTS TO L.
C J2=2**J IN ITERATIVE FORM.
      DCUBLE PRECISION TOP,BOT,ERR,EPS,UJ,VJ,XBAR,P,B
      DIMENSICN TABLE(1)
      BIGERR=0
      R=2**(K+1)
      J2=1
      DC 1 JJ=1,L
      J=JJ-1
      R=R/2
C II=PERMUTATICN CCUNTER.
      DC 2 II=1,J2
C CBTAIN BIT CCMEINATICN IN INTEGER FORM.
      I=II-1
C NCRMALIZE BIT COMBINATION TO GET XBAR(J).
      XBAR=I*R
C CALCULATE UPPER AND LCWER LIMITS.
      UJ=XBAR+R
      VJ=XBAR+R/2
C INTEGRATE NUMERATOR, THEN CENOMINATOR.
      MAXE1=MAXE
      CALL RCMINT(TOP,ERR,EPS,VJ,UJ,N,MAXE1)
      IF(MAXE1.EC.0.OR.TOP.LT.C) GO TO 3
      MAXE2=MAXE
      CALL RCMINT(BOT,ERR,EPS,XBAR,UJ,N,MAXE2)
      IF(MAXE2.EC.0.OR.BOT.LT.C) GO TO 3
C CALCULATE TABLE ENTRY.
      TABLE(I+J2)=TOP/BCT
      GC TC 2
    3 TABLE(I+J2)=-1.
      BIGERR=-1.
    2 CCNTINUE
      J2=2*J2
    1 CCNTINUE
      RETURN
      ENC
C RCMINT IS ALGCRITHM 351, FAIRWEATHER, G. MCDIFIED
```

FIG. 2. Fortran subprogram PTABLE to compute conditional probabilities.

and

$$V_{j+1} \leftarrow \bar{X}_{j+1} + R_{j+2} \tag{15}$$

All possible permutations of the bits of $\bar{X}_j$ must be enumerated in order while maintaining the correct binary point. Let there be k bits to the left of the binary point. All possible 2^j combinations of ones and zeros of $\bar{X}_j$ can be computed by evaluation of

$$i/2^{j-k}, \qquad i = 0, 1, ..., 2^{j-1} \tag{16}$$

in floating point. The table of $P(X_{j+1} = 1 \mid \bar{X}_j)$, $j = 0, ..., L - 1$, consists of $2^{L+1} - 1$ conditional probabilities. A Fortran subprogram which generates this table is presented (see Fig. 2) and calls the numerical integration routine ROMINT. The quadrature scheme we used was Algorithm 351, an improved Romberg technique [Fairweather (1969)]. The variable names in this subprogram closely resemble the notation used in the text and Algorithm 351. Complete tables of $P(X_{j+1} = 1 \mid \bar{X}_j)$ are given in Table I up to $j = 4$, and selected values are given for $j = 5, 6, 7$, and 8.

IV. Pseudonormal Number Generator

A pseudorandom number U and a compare is used to sample each bit. If

$$U \leq P(X_{j+1} = 1 \mid \bar{X}_j) \tag{17}$$

is true, a one bit is generated for X_{j+1} and a zero bit if the inequality is false. The generated bits are used to form table index m. The mapping function to retrieve $P(X_{j+1} = 1 \mid \bar{X}_j)$ can be written in terms of $\bar{X}_j$ as

$$m = \bar{X}_j + 2^{j-1}, \qquad j = 1, ..., L \tag{18}$$

that is, m is the address of $P(X_{j+1} = 1 \mid \bar{X}_j)$ in the table. When as many bits as necessary have been sampled using the conditional probability table, a uniform number W is appended to the low-order positions to form a full-length floating point number C. This results when

$$C = \frac{\bar{X}_L + W}{2^{L-k}} \tag{19}$$

is evaluated in floating point (L bits generated from table; k integer bits). With use of a quality pseudorandom number generator [Lewis et al. (1969), Payne et al. (1969), Payne (1970), Whittlesey (1968)] called RAND, a generator used to produce pseudonormal numbers can be easily coded in Fortran. We present a possible coding in Fig. 3. We found that high-quality pseudonormal numbers could be obtained by sampling six bits using the conditional probability table (63 entries) and then appending a pseudorandom number to form the low-order bits.

TABLE I

TABLE VALUES FOR K= 2AND L= 9

XBAR(0)	P(X(1)=1|XBAR(0))
0.0	0.04543980
XBAR(1)	P(X(2)=1|XBAR(1))
0.0	0.28476858
2.00000	0.05797400
XBAR(2)	P(X(3)=1|XBAR(2))
0.0	0.43907970
1.00000	0.32420421
2.00000	0.22708505
3.00000	0.15228540
XBAR(3)	P(X(4)=1|XBAR(3))
0.0	0.48445958
0.50000	0.45350313
1.00000	0.42290229
1.50000	0.39288020
2.00000	0.36364293
2.50000	0.33537388
3.00000	0.30823034
3.50000	0.28234071
XBAR(4)	P(X(5)=1|XBAR(4))
0.0	0.49609876
0.25000	0.48829854
0.50000	0.48050398
0.75000	0.47271895
1.00000	0.46494716
1.25000	0.45719236
1.50000	0.44945818
1.75000	0.44174832
2.00000	0.43406636
2.25000	0.42641592
2.50000	0.41880035
2.75000	0.41122317
3.00000	0.40368766
3.25000	0.39619714
3.50000	0.38875473
3.75000	0.38136363
XBAR(5)	P(X(6)=1|XBAR(5))
0.0	0.49902374
0.12500	0.49707127
0.25000	0.49511892
0.37500	0.49316669
0.50000	0.49121469
0.62500	0.48926294
0.75000	0.48731154
0.87500	0.48536050
1.00000	0.48340988
1.12500	0.48145986
1.25000	0.47951031
•	•
•	•
•	•
•	•

3.50000	0.44458240
3.62500	0.44265473
3.75000	0.44072872
3.87500	0.43880457
XBAR(6)	P(X(7)=1|XBAR(6))
0.0	0.49975586
0.06250	0.49926764
0.12500	0.49877936
0.18750	0.49829113
0.25000	0.49780291
0.31250	0.49731469
0.37500	0.49682647
•	•
•	•
•	•
•	•
3.75000	0.47049564
3.81250	0.47000915
3.87500	0.46952266
3.93750	0.46903628
XBAR(7)	P(X(8)=1|XBAR(7))
0.0	0.49993896
0.03125	0.49981689
0.06250	0.49969482
0.09375	0.49957275
0.12500	0.49945068
•	•
•	•
•	•
•	•
3.87500	0.48480719
3.90625	0.48468524
3.93750	0.48456329
3.96875	0.48444134
XBAR(8)	P(X(9)=1|XBAR(8))
0.0	0.49998474
0.01563	0.49995422
0.03125	0.49992371
0.04688	0.49989319
0.06250	0.49986267
0.07813	0.49983215
0.09375	0.49980164
0.10938	0.49977112
0.12500	0.49974060
•	•
•	•
•	•
•	•
3.92188	0.49232543
3.93750	0.49229491
3.95313	0.49226445
3.96875	0.49223393
3.98438	0.49220341

```
      FUNCTICN CCNBIT(L,K,TABLE)
      CIMENSICN TABLE(1)
      EQUIVALENCE (IW,W)
C FCT FCCNBT RETURNS A PSEUDORANDOM VARIATE BY BIT SAMPLING
C USING TABLE(M) CF CONCITIONAL PROBABILITIES.
C L=TCTAL NUMBER CF BITS SAMPLED USING TABLE.
C K=NUMBER CF BITS TO THE LEFT OF BINARY POINT.
C RAND=REAL PSEUCORANDOM NUMBER ON (0,1).
C J=BIT CCUNTER, CCUNTS TO L.
C I=INTEGER FCRM OF PCCNBT.
C M=MAFPING CF TABLE ACDRESS FCR P(X(J+1)=1=XBAR(J)).
C W=IW=UNIFORM PSEUDORANCOM NUMBER APPENDED TO TABLE
C GENEFATED EITS.  LOW CRDER BIT OF W USED FOR SIGN.
C J2=2**(J-1) IN ITERATIVE FORM.
      J2=1
      I=0
      DC 1 J=1,L
      M=I+J2
      I=2*I
      IF(RANC(NC).LE.TABLE(M)) I=I+1
C TEST FCR 1 CR 0 IN J-TH IIT POSITION.
      J2=2*J2
1     CCNTINUE
      W=RAND(NC)
      SIGN=2*(IW-2*(IW/2))-1.0
      CCNBIT=SIGN*(I+ )/(2.0**(L-K))
      RETURN
      END


      FUNCTION CONBIT(L,K,TABLE)
      DIMENSION TABLE(1)
      EQUIVALENCE (IW,W)
C     FCT CONBIT RETURNS A NORMAL DEVIATE BY BIT SAMPLING
C     USING TABLE(M) OF CONDITIONAL PRCBABILITIES, (SEE M BELOW)
C     L=TOTAL NUMBER OF BITS SAMPLED USING TABLE.
C     K=NUMBER OF BITS TO THE LEFT OF BINARY POINT.
C     RAND=REAL PSEUDORANDOM NUMBER ON (0,1).
C     J=BIT COUNTER.
C     I=INTEGER FORM OF CONBIT.
C     M=MAPPING OF TABLE ADDRESS FOR P(X(J+1)=1|XBAR(J)).
C     W=IW=UNIFORM PSEUDORANDOM NUMBER APPENDED TO TABLE
C     GENERATED BITS.  LOW ORDER BIT OF W USED FOR SIGN.
      I=0
      DO 10 J=1,L
      M=I+2**(J-1)
      I=2*I
      IF(RAND(NO).LE.TABLE(M)) I=I+1
C     TEST FOR 1 OR 0 IN JTH BIT POSITION COMPLETED.
10    CONTINUE
      W=RAND(NO)
      SIGN=2*(IW-2*(IW/2))-1.0
      CONBIT=SIGN*(I+W)/(2.0**(L-K))
      RETURN
      END
```

FIG. 3. Fortran fuction CONBIT to compute pseudonormal numbers.

V. Discussion

Effectively conditional bit sampling is a binary-directed search on the abcissa governed by conditional probabilities. Since one pseudorandom number is needed to generate each bit of the number sampled from the arbitrary continuous distribution, the algorithm is relatively slow. Arbitrary accuracy in the tails of the distribution is easily determined by the number of high-order bits selected by table compares. Accuracy of the fraction is determined from the departure from $\frac{1}{2}$ of the conditional probabilities of the jth bit. Thus fractional accuracy, measured by convergence in probability, can be arbitrarily set.

A sampling method, somewhat similar to conditional bit sampling, has been used extensively at the Los Alamos Scientific Laboratory [Cashwell and Everett (1959), Zelen and Severo (1964)]. The abscissa is divided into an equal number of parts with boundaries x_i and $F(x_i)$ is computed; a *single* pseudorandom number is selected; a binary-directed search on the *ordinate* probabilities determines an interval on the abscissa (see Fig. 4); and then either a linear or quadratic interpolation determines the sampled number within the selected interval.

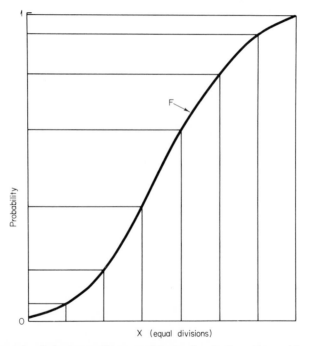

FIG. 4. Example of abscissa–ordinate partitioning for the Los Alamos binary directed table search and conditional bit distribution sampling algorithms.

By making the interval divisions on the abscissa some power of two, the table search part of the Los Alamos algorithm is equivalent to the table search part of the conditional bit algorithm, but uses only one pseudorandom number. We make the following recommendations: compute conditional probabilities to determine the accuracy (number of bits) needed in both the high order and fraction of the binary expansion of the number to be sampled; then use a combination of conditional bit sampling of the Los Alamos binary-directed table search technique. We give an example sampling from the normal; the probabilities are computed in subroutine PROBS (see Fig. 5) and the sampling algorithm is SAMPLE (see Fig. 6).

A general sampling algorithm in close competition with the Los Alamos conditional bit method is the table lookup method. The ordinate is split

```
        SUBROUTINE PROBS(L,K,TABLE,BIGERR,ERR,EPS,N,MAXE)
C RCMINT IS ALGORITHM 351, FAIRWEATHER, G.  MODIFIED
C RCMBERG QUADRATURE, COMM. ACM, 12, 6, (JUNE 1969), 324.
C L=TOTAL NUMBER CF BITS SAMPLED USING TABLE.
C K=TOTAL NUMBER CF BITS TO THE LEFT OF BINARY POINT.
C BIGERR=-1.=ERROR IN ROMINT.
C ERR,EPS,N,MAXE=ROMINT PARAMETERS.
        DOUBLE PRECISION TOP,BOT,ERR,EPS
        DIMENSION TABLE(1)
        BIGERR=0
        FCT=-2**K
        LTCP=2**L
        DC 1 I=1,LTCP
        TCF=-2**K+I/2.DO**(L-K)
        MAXEO=MAXE
        CALL RCMINT(TABLE(I),ERR,EPS,BOT,TOP,N,MAXEC)
        IF(MAXEO.EC.0)GC TO 2
        GC TO 3
2       TABLE(I)=-1.0
        BIGERR=-1.C
3       CCNTINUE
1       TABLE(I)=TABLE(I)+TABLE(1)
        RETURN
        END
```

FIG. 5. Fortran subroutine PROBS for Los Alamos probability table.

```
        FUNCTION SAMPLE(L,K,TABLE)
C FCT SAMPLE RETURNS A NORMAL CEVIATE BY A COMBINATION
C CF THE LOS ALAMOS BINARY DIRECTED SEARCH TECHNIQUE
C AND CCNDITICNAL BIT SAMPLING USING A TABLE OF 2**L
C PROBABILITIES.
C L=TOTAL NUMBER CF BITS SAMPLED USING TABLE.
C K=NUMBER CF BITS TO THE LEFT OF BINARY POINT.
C RAND=REAL PSEUDORANDOM NUMBER ON (C,1).
C W=IW=UNIFORM PSEUDORANDOM NUMBER APPENDED TO TABLE
C GENERATED BITS. LOW-CRDER BIT OF W USED FOR SIGN.
        DIMENSION TABLE(1)
        EQUIVALENCE (IW,W)
        R=RAND(NC)
        INDX=2**(L-1)
        INCRMT=INDX
        DC 1 I=2,L
        IF(R.GT.TABLE(INDX))INDX=INDX+INCRMT
        INCRMT=INCRMT/2
1       INDX=INDX-INCRMT
        VAR=2.0**K-INDX*2.0**(K-L)
        W=RAND(NC)
        SIGN=2*(IW-2*(IW/2))-1.0
        SAMPLE=SIGN*(VAR+W/2.0**(L-K))
        RETURN
        END
```

FIG. 6. Fortran subroutine SAMPLE for computing pseudonormal numbers from the Los Alamos tables.

into (2^n)th equal parts (see Fig. 7); a fixed point pseudorandom number between 1 and 2^n is chosen and used as an index to select the interval. Either an inverse interpolation [Muller (1958)] developed a Chebychev polynomial interpolation method, where the maximum pointwise error was less than 4×10^{-4} in the range $(-5, 5)$, (i.e., accurate to about 11 or 12 bits to the

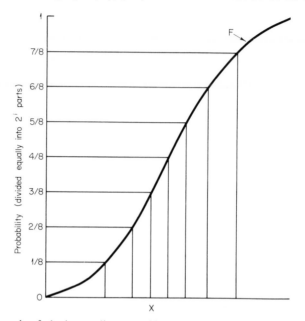

FIG. 7. Example of abscissa–ordinate partitioning for the Mullers and the GPSS table look-up sampling algorithm.

right of the binary point, which is unnecessarily excessive, considering the results presented here), or a modification of the GPSS sampling algorithm using the rejection technique [Butler (1970)] is used to compute the low-order digits. The trouble with the algorithm is obvious: accuracy in the tails is hard to achieve [cf. Muller (1957) and Fig. 7] since the interval over which interpolation accuracy must hold is very large. Further, the often difficult equation $x_i = F^{-1}(i/2^n)$ must also be evaluated. Neither of these problems occurs in the Los Alamos conditional bit sampling method.

Marsaglia (1962, 1963) developed a procedure which essentially converts the table search technique into a modified table look-up method. Unlike Los Alamos or conditional bit sampling, Marsaglia's technique is not very general and requires extensive modification whenever the distribution to be sampled is changed.

As of spring 1970, the Los Alamos conditional bit sampling method is the most attractive general distribution sampling method. Although conditional bit sampling is most useful for accuracy analysis of the Los Alamos technique with software computer sampling, it is of prime importance in hardware sampling. We are presently developing a conditional bit sampling hard-wired device which will produce pseudonoise with arbitrary distribution. The flow chart in Fig. 8 makes clear the simplicity of the conditional bit algorithm which is essential for a hard-wired device.

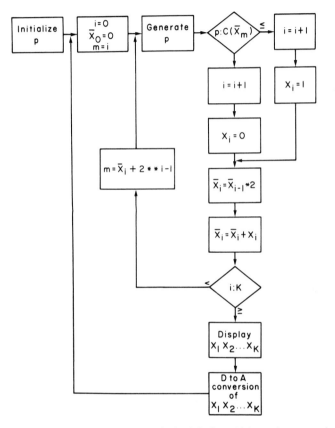

FIG. 8. Flow chart for conditional bit hard-wired device which produces pseudonoise for arbitrary distributions. p is a shift register sequence generated pseudorandom number.

None of the distribution sampling methods mentioned in this paper has the generality and simplicity of conditional bit sampling (the algorithm works unchanged for the example of a "bad" distribution given in Sect. II): perhaps microprogramable computers will make the algorithm more attractive with respect to speed in the future.

ACKNOWLEDGMENTS

Appreciation is expressed to O. W. Rechard for pointing out to me the similarity of conditional bit sampling to the method he and his associates developed at the Los Alamos Scientific Laboratory, and to Herman Rubin for his general comments on this paper and an anonymous ACM referee for the example of a "bad" distribution given in Sect. II.

REFERENCES

Bell, J. R. (1968). Algorithm 334-Normal random deviates. *Comm. ACM* **11**, 498.

Box, G., and Muller, M. (1958). A Note on the Generation of Normal Deviates. *Ann. Math. Stat.* **28**, 610.

Butler, E. L. (1970). Algorithm 370: General random number generator. *Comm. ACM* **13**, 49.

Cashwell, E. D., and Everett, C. J. (1959). "Monte Carlo Method for Random Walk Problems." Pergamon Press, Oxford.

Donnelly, T. (1969). Some techniques for using pseudorandom numbers in computer simulation. *Comm. ACM* **12**, 392–394.

Fairweather, G. (1969). Algorithm 351—Modified Romberg quadrature. *Comm. ACM* **12**, 324–325.

Hastings, C. (1955). "Approximations for Digital Computations," p. 192. Princeton Univ. Press, Princeton, New Jersey.

Kronmal, R. (1964). Evaluation of a pseudo-random normal number generator. *J. ACM* **11**, 357–363.

Kuck, D. J. (1968). Illiac IV software and applications programming. *IEEE Trans. Comput.* **8**, C-17, 758–770.

Lewis, P. A. W., Goodman, A. S., and Miller, J. M. (1969). A pseudo-random number generator for the System/360. *IBM Systems J.* **8**, 136.

Marsaglia, G. (1962). Random variables and computers. *In Trans. Prague Conf.*, 3rd, pp. 449–512. Czechoslovak Academy of Sciences, Prague.

Marsaglia, G. (1963). Generating discrete random variables in a computer. *Comm. ACM* **6**, 38.

Marsaglia, G., and Bray, T. A. (1964). A convenient method for generating normal variables. *Comm. ACM* **6**, 260–264.

Marsaglia, G., MacLaren, M. P., and Bray, T. A. (1964). A fast procedure for generating normal random variables. *Comm. ACM* **7**, 4–10.

Muller, M. E. (1957). Generation of Normal Deviates, Tech. Rep. No. 13. Statistical Techniques Research Group, Dept. of Math., Princeton Univ.

Muller, M. E. (1958). An inverse method for the generation of random normal deviates on large-scale computers. *Math. Tables Other Aids Comp.* **2**, 167.

Muller, M. E. (1959). A comparison of methods for generating normal deviates on digital computers. *J. ACM* **6**, 376–383.

Naylor, T. H., Balintfy, J. L., Burdick, D. S., and Chu, K. (1966). "Computer Simulation Techniques." Wiley, New York.

Payne, W. H., Rabung, J. R., and Bogyo, T. (1969). Coding the Lehmer pseudo-random number generator. *Comm. ACM* **12**, 85–86.

Payne, W. H. (1970). FORTRAN Tausworthe pseudorandom number generator. *Comm. ACM* **13**, 57.

Teichroew, D. (1953). Distribution sampling with high speed computers. Ph.D. Thesis, North Carolina State College.

Tocher, K. D. (1963). "The Art of Simulation," p. 27. English Univ. Press, London.
von Neumann, J. (1959). Various techniques used in connection with random digits. *Nat. Bur. Standards Appl. Math. Ser.* **36.**
Whittlesey, J. R. B. (1968). A comparison of correlational behavior of random number generators for the IBM 360. *Comm. ACM* **11,** 641–655.
Winje, G. L. (1969). Random number generators for Illiac IV. Dept. of Computer Science, Univ. of Illinois at Urbana-Champaign, Urbana, Illinois.
Zelen, M., and Severo, N. C. (1964). Probability Functions. *In* "Handbook of Mathematical Functions" (M. Abramowitz and L. Stegun, eds.), Vol. 55, pp. 927–995. U.S. Government Printing Office, Washington D.C.

5.20 Applications of Singular Value Analysis*

C. L. Lawson

JET PROPULSION LABORATORY

PASADENA, CALIFORNIA

I. Introduction

The singular value decomposition of a matrix provides quantitative information about a matrix which is very pertinent in a number of situations which arise in scientific computation.

I believe it is fair to say that at the present time singular value analysis is not widely used in scientific computing. Most scientific computing facilities probably do not have a singular value decomposition subroutine on their library, and most persons specifying or writing programs for scientific or engineering applications are probably not aware of the situations in which singular value analysis would be useful to them.

About two years ago, Dr. Richard Hanson and I, inspired primarily by Golub and Kahan (1965) and Golub and Businger (1967), and some previous experience with eigenvalue analysis of least-squares problems, started recommending the use of singular value analysis to scientists and eingineers at JPL when their computational problems involved a matrix whose condition number was *large with respect to the precision of the input data.* Our experience in consulting on these applications has led us to regard singular value analysis as an extremely important tool in numerical linear algebra. Correspondingly we regard a carefully designed singular value analysis subroutine as a very important component in a computer system's mathematical subroutine library.

My purpose in giving this talk at this symposium is to suggest to those of you who are responsible for the mathematical services of computing facilities that singular value analysis merits a place in your user education program and in your subroutine library. I would also like to suggest that the singular value decomposition should appear in the academic curriculum wherever linear transformations are discussed.

* Research carried out under NASA Contract NAS7–100.

I will briefly review the mathematical concept of the singular value decomposition and give an example of one way in which it can be used to analyze a problem of solving a system of linear equations. I will mention some features of the algorithms and subroutines we have used and indicate some of the applications which have been made of this type of analysis at JPL.

II. Definition of the Singular Value Decomposition

Let A be a real m by n matrix. There exist matrices U, S, and V such that

$$A = USV^T$$

where U and V are square orthonormal matrices of orders m and n, respectively, and S is an m by n matrix whose only nonzero elements are on the principal diagonal. The diagonal elements of S are the *singular values* of A and may be required to satisfy

$$s_1 \geqq s_2 \geqq \cdots \geqq s_k > s_{k+1} = \cdots = s_t = 0$$

where k is the rank of A and $t = \min(m, n)$.

One may interpret the column vectors $u^{(j)}$ of U as an orthogonal basis for m-space, and the column vectors $v^{(j)}$ of V as an orthogonal basis for n-space, having the property that relative to these bases the linear transformation $\mathfrak{A}$ associated with the matrix A is represented by the diagonal m by n matrix S. Thus the linear transformation

$$Ax = b$$

is equivalent to the uncoupled system

$$Sp = g$$

if one makes the orthogonal changes of variables

$$x = Vp$$

and

$$b = Ug$$

III. Singular Value Analysis of Systems of Linear Equations

Consider the (possibly inconsistent) system of equations

$$Ax \cong b \tag{1}$$

For any vector x we define the residual vector r by

$$r = b - Ax \tag{2}$$

We propose to use the euclidean norm of r (denoted by $\| r \|$ and defined by $\| r \| = [\sum_{i=1}^{m} r_i^2]^{1/2}$) as a measure of how well a vector x satisfies the system (1).

Knowing the singular value decomposition of A ($A = USV^T$) permits (2) to be transformed to an equivalent expression whose structure is so simple that information helpful in assessing the effects of data uncertainties and in choosing a *useful* solution of (1) is directly available.

The transformed problem is obtained by replacing x in (2) by Vp and then left multiplying (2) by U^T, obtaining

$$\tilde{r} = g - Sp \qquad (3)$$

where

$$\tilde{r} = U^T r$$

and

$$g = U^T b$$

Due to the orthogonality of U and V it follows that

$$\| p \| = \| x \|$$
$$\| g \| = \| b \|$$
$$\| \tilde{r} \| = \| r \|$$

and for orthogonally invariant matrix norms such as the spectral norm or the Frobenius norm

$$\| S \| = \| A \|$$

Furthermore, if, as is usually the case, the vector b is only known to some limited accuracy, then the uncertainty in b maps directly into uncertainty in g of the same magnitude, in the following sense: If the uncertainty in b is represented by an unknown vector db about which it is known that

$$\| db \| \leq \beta$$

then it follows that g has an uncertainty of

$$dg = U^T db$$

and thus

$$\| dg \| = \| db \| \leq \beta$$

Similar considerations apply to the mapping of uncertainty in A into uncertainty in S and the mapping of uncertainty in p into uncertainty in x.

It should be noted that the use of the (uniformly weighted) euclidean norm to describe uncertainty in a vector is appropriate only if uncertainties are assumed to be of about the same size in all components of the vector.

If this is not the case then it can and should be achieved by appropriate preliminary scaling. Thus the relative size of uncertainties in components of b can be balanced by appropriate row scaling of the augmented matrix $[A:b]$ or, more generally, by left multiplication of $[A:b]$ by an appropriate matrix.

Similarly if there is a priori information on the relative size of components of the solution vector x, one can balance these by column scaling of A or, more generally, by right multiplication of A by an appropriate matrix. Column scaling or right multiplication of A requires a compensating un-scaling of the solution vector x.

To summarize this section we assert that

(a) system (3) is equivalent to system (2) with respect to a number of properties which have relevance in the practical solution of linear equations, and

(b) the simple uncoupled structure of system (3) makes information directly available, which is very helpful in choosing a *useful* solution.

This latter point will be illustrated by an example in the next section.

IV. An Example of Singular Value Analysis

Consider the system of equations

$$\begin{bmatrix} 780 & 563 \\ 913 & 659 \end{bmatrix} \cdot x = \begin{bmatrix} 480 \\ 561 \end{bmatrix} \tag{4}$$

We will assume that numbers α and β are known such that the uncertainty in the data can be described by $\| dA \| \leq \alpha$ and $\| db \| \leq \beta$. Here the matrix norm is either the spectral or the Frobineous norm. These norms can differ by a factor of $n^{1/2}$, but that will not be significant for our discussion.

This matrix has previously been used by Moler (1969) and by Forsythe (1970) in expository talks on numerical linear algebra. The right-side vector is my own, however.

Inputting this problem to the nonsingular square system solver on the JPL mathematical subroutine library (UNIVAC 1108, single precision) one obtains the solution

$$x = \begin{bmatrix} 476.52809 \\ -659.34621 \end{bmatrix} \tag{5}$$

along with an indication that the system is near-singular.

Using the singular value decomposition subroutine on our library we find the decomposition

$$USV^T = \begin{bmatrix} -0.650 & 0.760 \\ -0.760 & -0.650 \end{bmatrix} \cdot \begin{bmatrix} 1480.9520 & 0.0 \\ 0.0 & 0.00067 \end{bmatrix} \cdot \begin{bmatrix} -0.811 & -0.585 \\ 0.585 & -0.811 \end{bmatrix}$$

(6)

and

$$g = U^T b = \begin{bmatrix} -738.323 \\ 0.550 \end{bmatrix}$$

Thus, for this example, (3) appears as

$$\tilde{r} = \begin{bmatrix} -738.323 \\ 0.550 \end{bmatrix} - \begin{bmatrix} 1480.9520 & 0.0 \\ 0.0 & 0.00067 \end{bmatrix} \cdot p$$

(7)

or

$$\tilde{r}_1 = -738.323 - 1480.9520 p_1$$

$$\tilde{r}_2 = \qquad 0.550 - \qquad 0.00067 p_2$$

We note that a large part of the mass of the g vector is contained in its first component. Specifically the norm of $\tilde{r}$ can be reduced to 0.55 by solving only for p_1 and setting $p_2 = 0$. If this is done we obtain

$$p^{(1)} = \begin{bmatrix} -0.49855 \\ 0.0 \end{bmatrix}$$

$$x^{(1)} = V p^{(1)} = \begin{bmatrix} 0.40424281 \\ 0.29178072 \end{bmatrix}$$

and

$$\| r^{(1)} \| = \| \tilde{r}^{(1)} \| = 0.55$$

The superscripts on p, x, r, and $\tilde{r}$ have been affixed to identify this as the "first-candidate solution."

Solving for both p_1 and p_2 would give a second-candidate solution

$$p^{(2)} = \begin{bmatrix} -0.49855 \\ 818.92 \end{bmatrix}$$

and

$$x^{(2)} = V p^{(2)} = \begin{bmatrix} 479.68759 \\ -663.72347 \end{bmatrix}$$

(8)

Using infinite precision arithmetic this second-candidate solution would be the unique exact solution to (4) as would (5), and thus the residuals would be zero. Using 8-decimal floating point arithmetic one obtains

$$\| \tilde{r}^{(2)} \| \cong 0.7E - 5$$

and

$$\| r^{(2)} \| \cong 0.72E - 2$$

(9)

Note that the norm of the residual computed using (5) is $0.86E - 2$. I do not attach any significance to the fact that this happens to be larger than (9).

I regard $x^{(2)}$ and (5) as equally acceptable representations of the solution to the full rank problem, even though they have a relative difference of about 0.0066, since this is the order of magnitude of the expected relative uncertainty in the solution due to uncertainty of about 10^{-8} in the elements of A. If I were assuming that A is known to more precision than this, then I would use higher-precision arithmetic.

Under what circumstances might one prefer solution $x^{(1)}$ to solution $x^{(2)}$? This choice should depend upon the use to which the solution is to be put, and the size of the initial error estimates, $\| dA \| \leq \alpha$ and $\| db \| \leq \beta$. If this problem arose as a linearized approximation to a nonlinear problem, as in Newton's method, for example, and the solution vector x is actually to be used as an increment by which some nominal parameter values will be changed, then it may be advantageous to use the smaller solution vector $x^{(1)}$. In such a case the residual in the linear problem is only an estimate of the residual which will be obtained in the nonlinear problem when the nominal parameter values are incremented by x. In general the validity of this estimate decreases with larger x due to the nonlinearity of the true problem. Thus it may very well happen that $x^{(1)}$ is a better increment to use in the nonlinear problem than $x^{(2)}$.

If β is about 0.55 or larger, then the smaller residual produced by $x^{(2)}$ is of no real significance and one might prefer the smaller solution $x^{(1)}$. In many cases a smaller solution vector represents in some sense a simpler, more economical, explanation of the real-world phenomenon being modeled, and this in itself could be grounds for preferring it if the difference in residuals is not significant relative to β.

If α is about 0.00067 or larger then we cannot be sure that the matrix of our problem is nonsingular. If it is singular, say it is the matrix $\tilde{A}$ obtained by replacing 0.00067 by zero in (6) and then multiplying the resulting matrices together, then the residual norm cannot be reduced below 0.55 and there is a one-parameter family of solutions, expressible as

$$x_\lambda = x^{(1)} + \lambda v^{(2)}$$

$$= \begin{bmatrix} 0.404 \\ 0.292 \end{bmatrix} + \lambda \begin{bmatrix} 0.585 \\ -0.811 \end{bmatrix}$$

all members of which achieve the minimal residual norm of 0.55. Since $v^{(2)}$ is orthogonal to $x^{(1)}$, the solution of minimum norm is this family is obtained by setting $\lambda = 0$; namely, it is $x^{(1)}$.

We summarize the interpretation of α and β by saying that if $\alpha \geq 0.00067$ or $\beta \geq 0.55$, then there may be no significant digits in the full rank solution

$x^{(2)}$ and it may be preferable to use the solution $x^{(1)}$. If $\alpha \leq 0.000067$ and $\beta \leq 0.055$, then the solution $x^{(2)}$ has at least one significant decimal digit and one may prefer to use $x^{(2)}$. There is an interval of indecision between these two situations, but this simply reflects the fact that these suggested interpretations are only given as guidelines and not as rigid rules.

V. Algorithms and Subroutines

A review of the singular value decomposition of a matrix was given in Golub and Kahan (1965). This included a bibliography dealing with applications and algorithms, and some work toward a new algorithm. This work toward a new algorithm was completed in Golub and Businger (1967), where a very satisfactory computational method is given for computing U, S, and V.

I translated the Algol code given in Golub and Businger (1967) to Fortran (originally IBM 7094 Fortran IV, later UNIVAC 1108 Fortran V). We subsequently reorganized the subroutine with a view to

(a) conserving storage,
(b) computing quantities particularly relevant to the singular value analysis of a system of linear equations,
(c) printing these quantities in a convenient format.

Reinsch and Golub (1970) have also developed a reorganized code covering points (a) and (b) above, as well as introducing some changes in the iterative portion of the algorithm.

The main observation that permits a substantial saving of storage is that for the analysis of a system of equations one does not need to produce the U matrix explicitly but only the vector $g = U^T b$. It transpires that, for the case of $m \geq n$, the storage requirement can be limited to four arrays as follows:

(1) an m by n array initially containing A; A can be overwritten by V, which in turn can be overwritten by a set of n candidate solutions $x^{(1)}, \ldots, x^{(n)}$;
(2) an m-array initially containing b and to be overwritten by g;
(3) an n-array to contain $s_1, \ldots, s_n$;
(4) an n-array of temporary working space.

The case of $m < n$ can be handled in a similar way [see Reinsch and Golub (1970)]; however, I think the context in which such problems arise is sufficiently different from the context of problems having $m \geq n$ that they require separate discussion, and I do not propose to discuss the case of $m < n$ in this talk.

If it is inconvenient or impossible to accommodate an m by n array in main storage but an n by n array can be accommodated, then methods can be used to transform the m by n problem to an n by n problem having the same singular values, the same V-matrix, and the same first n components of the g-vector. This permits singular value analysis to be accomplished. One method of performing such a transformation is through the use of sequential Householder orthonormal transformations, as described in Hanson and Lawson (1969).

VI. Experience in Using Singular Value Analysis

We have set the tolerance in our library subroutine for solving square nonsingular systems somewhat higher than is commonly done, so that it is biased toward calling systems near-singular. For example, the problem discussed above in Sect. IV caused this subroutine to take its alternate return, indicating near-singularity. We hope that this motivates a user to give fresh thought to the validity of his mathematical model and his computational approach.

When a user informs us that he has encountered this near-singularity warning (this has occurred about six times in the past twelve months), our first suggestion is that he use the singular value analysis subroutine, which is also on the library, to obtain a printed listing of the various quantities mentioned above in Sect. IV.

In one case the user was doing cubic polynomial least-squares curve fitting for $35 \le t \le 45$ with no transformation of the independent variable, and forming normal equations. The matrix involved had a condition number of about 10^7. A translation of the independent variable gave him a stable formulation of his problem.

Another case was a 15 by 15 system being solved at every step during the numerical solution of an O.D.E. system. The two smallest singular values of this matrix were of the order of 10^{-8} times the largest singular value. Columns 14 and 15 of the V-matrix each had three dominant components, thus giving specific identification of the columns of A that were nearly linearly dependent. With this information in hand the user is re-examining his problem formulation. At this writing an explanation has not yet been found.

We have developed a Newton method subroutine (for solving m nonlinear equations in n unknowns) which uses singular value decomposition to generate shorter steps when the full length, full rank, step fails. Our success with this has been mixed, but it has been used effectively to obtain convergence in "shooting method" solutions of some two-point and multipoint boundary problems for systems of ordinary differential equations.

Singular value analysis has been used as a smoothing technique in the numerical solution of linear integral equations and also in spline surface fitting to data. In both of these applications the u-vectors associated with the smaller singular values tend to be highly oscillatory and setting the associated p-components to zero tends to give smoother solutions.

Singular value analysis has also been used in a conventional regression analysis context, in which case it is mathematically equivalent to eigenvalue analysis of the matrix of a quadratic form.

VII. Conclusions

In deference to readers who are acquainted with concepts such as condition number, pseudoinverse, and statistical interpretations of (generally over-determined) systems of linear equations, I would like to acknowledge that there are very close and well-known connections between those concepts and the ideas discussed in this paper. I have not stressed these connections because I wished to present the properties given in Sects. II and III with a minimum of supplementary definitions.

The singular value decomposition of a matrix can be used

(a)   in the analysis of a system of linear equations, and

(b)   in the synthesis of a solution to such a system.

In my opinion the singular value decomposition is particularly valuable for (a), as it provides insight into the effect of data uncertainties on candidate solutions and specific information such as the identification of sets of columns of A which are nearly dependent. I believe that a problem analyst would find it cost-effective to use singular value analysis any time he is faced with questions as to the reliability of a solution to a system of linear equations (assuming the storage requirements mentioned in Sect. V can be met). With regard to (b), we note that there are many alternative techniques for stabiliz-ing the solution of a system of linear equations which is ill conditioned with respect to the data uncertainties, and these other techniques must be con-sidered on their merits. Even if the singular value decomposition is used for (b), there are a variety of strategies available. Generally, if the spectral norm of the uncertainty dA is bounded by α, then one wishes to change each singular value by less than α so that each modified singular value is either zero or greater than α. There are, however, an unlimited numbers of ways of doing this.

We conclude that the singular value decomposition is especially valuable for (a), and is one of a number of mathematical concepts upon which algorithms for (b) can be based.

REFERENCES

Forsythe, G. (1970). Pitfalls in Computation or Why a Math Book Isn't Enough, *Amer. Math. Monthly* **77,** 931–956.

Golub, G., and Businger, P. (1967). Least Squares, Singular Values and Matrix Approximations, An ALGOL Procedure for Computing the Singular Value Decomposition, Tech. Rep. No. CS73, Stanford Univ.

Golub, G., and Kahan, W. (1965). Calculating the Singular Values and Pseudo-inverse of a Matrix, *J. SIAM Numer. Anal.* Ser. B, **2,** 205–224.

Hanson, R., and Lawson, C. (1969). Extensions and Applications of the Householder Algorithm for Solving Linear Least Squares Problems, *Math. Comput.* **23,** 787–812.

Moler, C. B. (1969). Numerical Solution of Matrix Problems. *Proc. Joint Conf. Math. Comput. Aids Design.* Conference sponsored by ACM, SIAM, and IEEE.

Reinsch, C., and Golub, G. (1970). Singular Value Decomposition and Least Squares Solutions. *Numer. Math.* **14,** 403–420.

5.21 Numerical Implementation of Variational Methods for Eigenvalue Problems*

B. N. Parlett

UNIVERSITY OF CALIFORNIA

BERKELEY, CALIFORNIA

O. G. Johnson

INTERNATIONAL MATHEMATICAL AND STATISTICAL LIBRARIES, INC.

HOUSTON, TEXAS

I. Introduction

There has been a surge of activity recently in the area of using piecewise polynomial spaces in conjunction with variational methods for the solution of a variety of problems. A large portion of this activity has been devoted to the space of cubic splines and piecewise cubic Hermite functions. Many researchers include these methods in the general category of finite element methods. See (Babuska (1970), Bosarge and Johnson (1969), and Johnson (1968), for instance. Has this effort resulted in the dissemination of superior programs for solving these problems? To judge from the literature the answer is no.

Two extreme attitudes on this phenomenon may be distinguished. The first claims that the translation of the methods presented in the literature into viable programs is a pedestrian task which can safely be delegated to a graduate student. There are no difficulties which warrant public presentation. The second attitude, equally extreme, regards the theoretical analysis as the simplest aspect of a very difficult engineering task, namely the creation of automatic programs. This attitude suggests that we do not understand computing well enough to be able to set values to the decision parameters which appear in our programs.

What we do here is to take a simple problem, the one-dimensional Sturm-Liouville eigenvalue problem, and discuss the issues which arise in making use of the new found error bounds [see Birkhoff *et al.* (1966), Ciarlet *et al.*

* Research supported by ONR Contract N00014–A–0200–1017.

(1969), or Johnson (1969)] to produce a good automatic program. The issues in this case are typical of those which arise in the implementation of more complicated problems.

The mathematical impulse here is in complete contrast to the current fashion for generality. We wish to capitalize on all that is special to this problem. In particular, only recently have there appeared methods for finding selected eigenvalues and eigenvectors which take advantage of all the properties of the approximating matrices.

II. A Sequence of Related Problems

In order to write an automatic program for the Sturm–Liouville problem we identify a sequence of problems, each an approximation to the one above and each important in the problem of programming the variational method.

P1. Find the first few stationary values and functions of the functional

$$N[\phi] \equiv \int_b^a [p(\tau)(\phi'(\tau))^2 + q(\tau)^2(\tau)]\, d\tau + \xi_1 p(a)\, \phi^2(a) + \xi_2 p(b)\, \phi^2(b) \quad (1)$$

on the surface

$$D[\phi] \equiv \int_a^b r(x)\, \phi^2(x)\, dx = 1 \quad (2)$$

in the space $\mathscr{A}$ (absolutely continuous functions which satisfy the corresponding boundary conditions, say). This problem is of course the same as that of finding the smallest eigenvalues γ_i and eigenvectors of the Sturm–Liouville problem.

P2. Find the first few stationary values of $N[\phi]$ on the surface $D[\phi] = 1$ in the finite-dimensional space $S \subset \mathscr{A}$. For instance, S may be the space of piecewise cubic Hermite functions on a mesh

$$\tau: \quad a = \tau_0 < \tau_1 < \cdots > \tau_m = b \quad (3)$$

Problem 2 is equivalent to the problem of finding the eigenvalues of

$$Ax - \lambda Bx = 0 \quad (4)$$

where

$$A_{ij} = \int_a^b [p(\tau)\, S_i'(\tau)\, S_j'(\tau) + q(\tau)\, S_i(\tau)\, S_j(\tau)]\, d\tau$$
$$+ \xi_1 p(a)\, S_i(a)\, S_j(a) + \xi_2 p(b)\, S_i(b)\, S_j(b) \quad (5)$$

$$B_{ij} = \int_a^b r(\tau)\, S_i(\tau)\, S_j(\tau)\, d\tau \quad (6)$$

and the $\{S_i(\tau)\}$, $i = 1, \ldots, n$, form a basis for S. If the Hermite patch basis is used, for instance, A and B will be band, seven diagonal matrices [Johnson (1968)].

P3. Find the eigenvalues of

$$\bar{A}y - \mu\bar{B}y = 0 \tag{7}$$

where the elements of $\bar{A}$ and $\bar{B}$ are numerical quadrature approximations to (5) and (6). Thus $\bar{A}$ and $\bar{B}$ are machine representable matrices which differ from A and B by the truncation error in the quadrature scheme and the round-off in its evaluation.

P4. Construct a method and a corresponding computer program for the solution of *P3* which utilizes the symmetry, definiteness, and band structure of the matrices.

P5. Determine bounds on an equivalent perturbation

$$\hat{A}z - \lambda\hat{B}\hat{z} = 0$$

for the computed solutions. That is, find $\hat{A}$ close to $\bar{A}$ and $\hat{B}$ close to $\bar{B}$ such that the computed quantities λ and z satisfy the above equation.

III. Problems in Writing the Matrices Generation Routine

Throughout the rest of this paper we will assume, to be definite, that S is the space of cubic Hermite functions on π. For this space it has been shown that the solutions of *P2* are $O(h^6)$ approximations to the solutions of *P1*. where h is the norm of π. In fact, convergence may be even higher with appropriate assumptions on the coefficient functions [Birkoff and Varga (1968)].

But what about the approximation of *P3* to *P2*? The integration formula in the matrix generation routine may create truncation error out of proportion to the high-order approximation of *P2* to *P1*. Here we settle the question: What order quadrature scheme is compatible? Since a three-point Gaussian scheme gives $O(h^6)$ accuracy to the integral of any given function, one might expect such a scheme to be quite appropriate for use in the matrices generation routine. In fact, it is quite inadequate and may allow perturbations of much lower order than $O(h^6)$. In order to understand why this is so we must look at the perturbation problem for *P2*.

In Section XI we shall show, somewhat surprisingly, that there is at most negligible extra cost in solving the algebraic problem to an accuracy much greater than the approximation of *P2* to *P1*.

IV. Perturbations in (Ax − λBx)

We note that perturbations in the eigenvalues of *P2* depend on the eigenvalues of the B matrix as well as the perturbations in the elements of A and B.

Thus one can bound the perturbations by the norms of the perturbation matrices and, as it turns out, the inverse of the smallest eigenvalue of B. To see this, simply let

$$\bar{A} = A - E, \qquad \bar{B} = B - F$$

Also let μ_i and y_i, $i = 1, \dots, n$, be eigenvalues and eigenvectors of (7). Let $\mu_1 + \Delta u_1 = \lambda_1$ and $y_1 + \Delta y_1 = x_1$ be the smallest eigenvalue and corresponding eigenvector of (4). We can assume that $\bar{A}$ and $\bar{B}$ are also symmetric and positive definite. For details see Johnson (1968). Thus

$$y_i{}^t \bar{A} y_i = \mu_i y_i{}^t \bar{B} y_i = \mu_i \delta_{ij}$$

where δ_{ij} is the Kroneker delta. Hence, in particular, the y_i are linearly independent and we can write

$$x_1 = \alpha \left[y_i + \sum_{i=2}^{n} \tau_i y_i \right]$$

For h small enough, it can be shown that the τ_i are the same magnitude as the maximum error in E and F [Wilkerson (1965)]. Inserting this form of x_1 into the perturbed equation, we have

$$(\bar{A} + E)\left(y_1 + \sum_{i=1}^{n} \tau_i y_i \right) = \tau_i (\bar{B} + F)\left(y_i + \sum_{i=1}^{n} \tau_i y_i \right)$$

Hence, multiplying on the left by $y_1{}^t$ we get

$$y_1{}^t E y_1 + y_1{}^t E \sum_{i=1}^{n} \tau_i y_i$$

$$= (\lambda_1 - \mu_1) y_1{}^t \bar{B} y_1 + \lambda_1 y_1{}^t F y_1 + \lambda_1 y_1{}^t F \sum_{i=1}^{n} \tau_i x_i$$

Ignoring terms which contain both τ_i and E (or F) we have

$$\frac{\lambda_1 - \mu_1}{\lambda_1} y_1{}^t B y_1 \doteq y_1{}^t (\lambda_1^{-1} E - F) y_1$$

or

$$\frac{\lambda_1 - \mu_1}{\lambda_1} \doteq \left(\frac{y_1{}^t y_1}{y_1{}^t B y_1} \right) \left[\frac{y_1{}^t (\lambda_1^{-1} E - F) y_1}{y_1{}^t y_1} \right]$$

This analysis is valid for any eigenvalue. Hence, dropping subscripts, we have

$$\frac{\lambda - \mu}{\lambda} \doteq \frac{y^t y}{y^t \bar{B} y} \left[\frac{y^t (\lambda^{-1} E - F) y}{y^t y} \right] \qquad (8)$$

where μ, λ, and y are any of the respective solutions to (8) and (7). But λ^{-1} is $O(1)$ since λ converges to γ. Hence if E and F are $O(h^\sigma)$,

$$\frac{\lambda - \mu}{\lambda} \doteq \frac{y^t[O(h^\sigma)]y}{y^t \bar{B} y} \leqq \frac{O(h^\sigma)}{\bar{\beta}_{\min}} \tag{9}$$

where $\bar{\beta}_{\min}$ is the smallest eigenvalue of $\bar{B}$. This bound can be made rigorous. Those interested should see Johnson (1968).

V. A Lower Bound on the Smallest Eigenvalue of B

We now assume, again for definiteness, that our basis is the fundamental patch basis for the piecewise cubic Hermite space. This basis $\{S_i(x)\}$ $i = 1, \ldots, n$, has the property (for appropriate boundary conditions) that if

$$\phi(x) = \sum_{i=1}^{n} v_i S_i(x) \tag{10}$$

then

$$\phi(x_i) = v_{2i+1}, \qquad i = 0, \ldots, m - 1$$

and

$$\phi'(x_i) = h^{-1} v_{2i}, \qquad i = 1, \ldots, m$$

(Note: $n = 2m$.)

We will show that $\beta_{\min}$, the smallest eigenvalue of B, satisfies

$$\beta_{\min} \geqq K_1 h + K_2 h^2 \tag{11}$$

for some constants K_1, K_2.

Provided $\| F \|$ is $O(h^2)$, that is $\sigma \geqq 2$, standard perturbation theory combined with (11) shows

$$\bar{\beta}_{\min} \geqq Kh \tag{12}$$

for sufficiently small h.

To obtain (11) we note that

$$\int_{\tau_j}^{\tau_{j+1}} r\phi^2 \geqq r_{\min} \int_j^{\tau_{j+1}} \phi^2$$

Now ϕ in the piecewise cubic Hermite space can be written

$$\phi(x) = \sum_{v=0}^{3} \frac{1}{v!} (x - x_j)^v u_j^{(v)}$$

and so ϕ^2 is a quadratic form in the $u_j^{(v)}$, $v = 0, 1, 2, 3$. We have to minimize this form subject to

$$\| V \|^2 = \sum_{j=0}^{m-1} u_j^2 + \sum_{j=1}^{m} (u_j')^2 = 1$$

Let us first find the minimum in terms of the u_j. To this end let

$$u = \begin{pmatrix} hu_j' \\ h^2 u'' \\ h^1 u''' \end{pmatrix}, \qquad z = \begin{pmatrix} 1/2 \\ 1/3 \\ 1/24 \end{pmatrix} u_j, \qquad J = \begin{pmatrix} 1/3 & 1/16 & 1/30 \\ 1/16 & 1/20 & 1/72 \\ 1/30 & 1/72 & 1/42 \end{pmatrix}$$

Then

$$\int_{\tau_j}^{\tau_{j+1}} \phi^2 = u^T J u + u^T z + h u_j{}^2$$

Since J is positive definite the minimum of the right side is

$$\tfrac{1}{2} \hat{u}^T z + h u_j{}^2$$

where $\hat{u}$ satisfies $2J\hat{u} + z = 0$. Solving this equation gives

$$\int_{\tau_j}^{\tau_{j+1}} \phi^2 \geqq K_3 h u_j{}^2$$

where $K_3 \doteq \tfrac{1}{4}$. A similar argument yields

$$\int_{\tau_j}^{\tau_{j+1}} \phi^2 \geqq K_4 h^3 (u_j')^2$$

Summing over all the intervals,

$$2 \int_a^b \phi^2 \geqq \min(K_3, K_4) h (\Sigma u_j{}^2 + \Sigma h^2 (u_j')^2) = \min(K_3, K_4) h$$

A similar argument establishes (12) for the other admissible boundary conditions.

VI. The Compatible Quadrature Order

The results of sections four and five show that the elements of A and B must be computed with $O(h^7)$ accuracy if we are to maintain an $O(h^6)$ approximation to PI. This does not mean, however, that four-point Gaussian quadrature is sufficient. The elements of the basis $\{S_1\}$ vary with h, and, in particular, their higher derivatives grow as powers of h^{-1}. These derivatives enter into the quadrature error. Hence we must pick a quadrature order high enough to overcome these negative contributions. Fortunately this is possible.

Suppose we compute the elements of A and B with a Gaussian v-point quadrature scheme. We know that

$$\int_{-1}^{+1} f(x)\, dx = \sum_{i=1}^{v} w_i^{(v)} f(x_i^{(v)}) + R_{(v)}(f)$$

where

$$R^{(v)} = C_v\, d^{2v} f(n)/dx^2, \qquad -1 \leqq n \leqq 1$$

$$C_v = \frac{2^{2v+1}}{(2v+1)!} \left[\frac{(v!)^2}{(2v)!} \right]^2 \qquad \text{[Krylov (1962)]}$$

Thus

$$\int_{x}^{x_{j+1}} f(x) = \frac{h}{2} \int_{-1}^{+1} f\left(\frac{h}{2}\bar{x} + \frac{x_{j+1} + x_j}{2}\right) d\bar{x}$$

$$= \frac{h}{2} \sum_{i=1}^{v} w_i^{(v)} f(\bar{x}_i^{(v)}) + C_v\left(\frac{h}{2}\right)^{2v+1} f^{(2v)}(x)$$

It would thus seem that $O(h^7)$ convergence is possible for $v = {}^-3$, as remarked previously.

However, we recall that f changes with h. In Sect. II, we saw that the elements of B are of the form

$$\int_a^b rs_i s_j$$

Thus

$$\frac{d^{2v}}{dx^{2v}}[rs_i s_j] = \left(\frac{2v}{6}\right) r^{(2v-6)} \frac{6!}{4h^6} + O(h^{-5})$$

In order for the error term to be $O(h^7)$, we thus need $2v + 1 - 6 = 7$ or $v = 6$.

For $v = 6$ we have

$$\left| C_v\left(\frac{h}{2}\right)^{2v+1} \frac{d^{2v}}{dx^{2v}}[rs_i s_j] \right|$$

$$\leqq \frac{11}{117,600} \max_{x\varepsilon[a,\, b]} |r^{(6)}(x)|\, h^7 + O(h^8)$$

That is, the coefficient of the leading term in the error is about 10^{-4} times $\max|r^{(6)}|$.

Thus, for $\max_{x\varepsilon[a,\, b]} |r^{(6)}(x)| \leqq 10^4$, our error will be less than single-precision machine accuracy for $h \leqq h_0$ in the following cases:

Machine	Working precision accuracy	h_0
IBM 7094	2^{-28}	2^{-4}
CDC 6400	2^{-49}	2^{-7}
IBM 360	2^{-53}	$2^{-7.5}$

VII. Invariance and Optimality of the Compatible Order

It is possible to choose a basis for our space S such that the first few derivatives are bounded independently of h. It might seem that a lower-order quadrature scheme would be acceptable for such a basis. This of course is not true since the eigenvalues of B are adversely affected by such a choice of basis. In fact, it is easily seen that the compatible order is independent of such a choice.

Secondly, we note that the largest eigenvalue of B is bounded above by an $O(h)$ function. This is easily seen from a Gersgorin argument since the elements of B are $O(h)$. Hence $\beta_{max}/\beta_1 = O(1)$. Thus the replacement of $(y^t By)$ by μ_{min} in inequality (9) preserves the order of $(y^t By)$. Hence the compatible order is, in this sense, best possible.

VIII. Observations on the Sharpness of the Perturbation Bounds

We have seen that both the smallest and largest eigenvalues of B are $O(h)$, and so our bound on the perturbation of an eigenvalue is sharp in terms of h. However, in problems where the mass density r varies sharply, the ratio of the largest eigenvalue to the smallest will be large. This ratio is called the condition number of $\bar{B}$ with respect to inversion.

It is of great practical significance that in these cases we can replace the smallest eigenvalue of $\bar{B}$ by the largest in estimating the perturbation of the *smallest* eigenvalue of $\bar{A} - \lambda\bar{B}$. This is another example where we can take advantage of the fact that we are only interested in the smallest few roots.

Let μ_1 be the smallest root of $\det(\bar{A} - \mu\bar{B}) = 0$ and let y_1 be its unit-length eigenvector. Then

$$y_1{}^T \bar{B} y_1 = y_1{}^T \bar{A} y_1 / \mu_1$$

$$\geqq \alpha_1 / \mu_1$$

where α_1 is the minimum eigenvalue of $\bar{A}$. This is often a very much better bound than β_1, the smallest eigenvalue of $\bar{B}$.

IX. Quadrature Schemes with Basis Elements as Weights

Basis functions and derivatives of basis functions for some piecewise polynomial spaces do not change signs within any subinterval. Such is the case, for instance, if one uses the cubic spline patch basis. Hence products of the form $u_{ij}(x) = S_i(x) S_j(x)$ and $v_{ij}(x) = S_i{}'(x) S_j{}'(x)$ are always positive or always negative for any given interval. We may thus use $u_{ij}(x)$ and $v_{ij}(x)$ as weight functions in the development of quadrature schemes for (5) and (6).

This approach has the advantage that fewer points are needed to achieve a given accuracy, since the error depends only on the derivatives of the coefficient functions p, q, and r. These derivatives, unlike those of $u_{ij}p$, for example, are independent of h. Further, there is essentially no difference in the computation time for Gaussian quadrature with weights and without.

The disadvantage of this approach is that there will be eight distinct quadrature schemes for each subinterval $[x_j, x_{j+1}]$; those for u_{ij}, $i = j, \ldots,$ $j + 3$, and v_{ij}, $i = j, \ldots, j + 3$. If each scheme is a four-point formula there will be sixteen functional evaluations for each of the coefficient functions p, q, and r. With the previous six-point Gaussian scheme there would, of course, be only six functional evaluations per point.

We could, of course, write

$$u_{ij}(x) = u_{ii}(x) \, S_j(x)/S_i(x)$$

and use the scheme with weight u_{ii} throughout the ith row. This device would give us higher precision in the diagonal element than the off-diagonal elements. If the eigenvalues of $\bar{A}x - \lambda\bar{B}x = 0$ are all distinct, further perturbation analysis shows that this is permissible [see Johnson (1968)]. We have no guarantee, however, that the eigenvalues are distinct, even though those of the original Sturm-Liouville problem are. Of course, if h is small enough, the pertinent eigenvalues will be distinct. But how small need h be? Our feeling is that this problem requires further investigation. It should not be difficult to establish a value of h for which theoretical separation is a certainty.

X. Methods Which Utilize the Structure of the Algebraic Problem

A variety of methods have been used in the past to solve systems of the form $(\bar{A} - \lambda\bar{B})x = 0$. Until recently there was no method which took advantage of our three properties:

(1) $\bar{A}, \bar{B}$ are both symmetric and positive definite,

(2) $\bar{A}, \bar{B}$ both have narrow bandwidth,

(3) Only the smallest few eigenvalues and their eigenvectors are required.

We describe here, very briefly, the method outlined in Peters and Wilkerson (1969) and developed at the National Physical Laboratory, England. Let $C = \bar{A} - \lambda\bar{B}$. The leading principal minors of C form a Sturm sequence of polynomials in λ. The number of sign agreements in the sequence of values for a given λ is the number of eigenvalues exceeding λ. By use of this property, any desired eigenvalue may be isolated and then the value of the last minor, which is the determinant of $\bar{A} - \lambda\bar{B}$, may be used to achieve rapid convergence to that eigenvalue.

The difficult task is to determine the minors in a stable manner. Triangular decomposition of C without interchanges is unstable, and the usual partial-pivoting strategy does not preserve the minors. However a more complicated use of row interchanges yields a satisfactory stable method. Let us describe the typical step.

Imagine that the first $k - 1$ rows of C have been reduced to upper triangular form U by use of elimination and interchanges. The kth row is now considered for the first time. If $|c_{k1}| \leq |u_{11}|$, then c_{k1} is eliminated by subtracting a multiple of row 1 from row k. Otherwise rows 1 and k are interchanged and the new $(k, 1)$ element is eliminated. The interchanges are recorded. Next row 2 and row k are examined and the $(k, 2)$ element is eliminated, again using an interchange if necessary. The process is continued until the first $(k - 1)$ elements of row k have been eliminated and another row of U has been found. However if C is a band matrix of width $2w + 1$ then only the last w of these $(k - 1)$ steps need be performed. The bandwidth of U will be $2w + 1$.

The product of the first k diagonal elements of U is the kth principal minor except for sign which is $(-1)^N$ where N is the number of interchanges used. The number of multiplications required to determine U is approximately $w(2w + 1)n$.

When an eigenvalue has been accepted, the associated eigenvector is determined using inverse iteration on U, $Ux^{(1)} = x^{(0)}\sigma_0$, where σ_0 is a normalizing factor. For $x^{(0)}$ we recommend the vector $(1, 1, \ldots, 1)^T$ with a sign pattern corresponding to the corresponding eigenvector of the model problem. Thus for the jth eigenvector we use $x_k^{(0)} = \sin kj\pi/(n + 1)$, $k = 1, \ldots, n$.

There is an interesting detail of mathematical software here. It is normal to compute eigenvalues as close to working accuracy as possible. In this case even two cycles of inverse iteration are seldom necessary. On the other hand, if a customer requests a very modest number of figures in the eigenvalues, then more cycles will be required. However, for $i > 1$ we must solve $Cx^{(i)} = x^{(i-1)}\sigma_{i-1}$, and this requires that the right-hand side receive the same elimination treatment which reduced C to U. This entails the saving of the multipliers (wn locations) and the interchanges. To solve $Ux = y$ requires $(2w + 1)n$ multiplications. To reprocess the right-hand side requires wn multiplications. Thus in our case each cycle of inverse iteration requires only $10n$ multiplications.

XI. The Selection of Program Arguments Governing Error

An important principle in the writing of subroutines is not to burden the user with the unnecessary choice of parameters. We wish to point out that recent advances have reduced the importance of certain variables in our program.

Let us illustrate this. Let $h = (b - a)/m$, where $[a, b]$ is the given interval. Our variational methods produce matrices of order αm ($\alpha = 2$ for the Hermite subspace) and, the techniques employed to solve the algebraic problem for several eigenvalues were essentially $O(m^3)$, despite the band structure. This computation time was sensitive to changes in h, and many customers would want control of it. The procedures described in the previous section require only $O(m)$ arithmetic operations. Thus with a fixed bandwidth and a fixed number of desired eigenvalues the just determination of a suitable h is much less crucial. The determination of the bandwidth is discussed in the next section.

Using certain a priori estimates from the model problem, our program can determine as large an h as seems reasonable to yield a modest number of significant figures in the eigenvalues. This choice can be overriden by the user if he wishes. The program then calculates its solution and delivers it together with the a posteriori error bounds and the value of m actually used.

In the light of this information the user may wish to call the subroutine again with a different value of m.

For example, with the Hermite space discussed in this paper the bandwidth of A and B is 7. It follows that approximately $21ksm$ operations are required to find k eigenvalues, where s is the average number of triangular decompositions required in the inverse interpolation search for an eigenvalue. A reasonable estimate for s is 20. As pointed out in the last section the cost of computing the eigenvector increases as the accuracy of the eigenvalue decreases. It is this sort of fact which renders so difficult the appropriate tolerances. It is our view that the extra cost required to calculate the eigenvalues of $\bar{A} - \lambda\bar{B}$ to 6 decimals rather than 3 is negligible. Of course this accuracy is not to be confused with the accuracy with which the numbers approximate the eigenvalues of the continuous problem.

XII. Cost and Accuracy

Thus far we have looked at cubic Hermite functions. What about piecewise polynomials of other orders? If the coefficient functions are sufficiently differentiable, might our program be improved by using quantic Hermites or piecewise linear functions? Such questions cannot be answered independently of the methods used in the program. For each mesh size h and each polynomial order w there is an associated error $E(h, w)$ given by the error bounds. Also there is an associated cost $C(h, w)$ in operation counts say. Ideally, we should try to minimize $E(h, w)$ subject to $C(h, w) = c$, where c is given. The usual error bounds relating the kth eigenvalues of $P1$ and $P2$ have the form

$$E(h, w) = K_w h^{2w} \sum_{i=1}^{k} \mu_i^{(w+1)}$$

The cost, using triangular factorization with multiple interchanges, can be written

$$C(h, w) = Kw(2w + 1)/h$$

where $K/(b - a)k$ is the average number of determinant evaluations per eigenvalue. A preliminary analysis of these functions indicates that one would use as large a value of w as possible. This analysis ignored the variation of K_w with w as well as the effects due the μ_i.

A computer program embodying the methods described will be published as a Technical Report of the Computer Center, University of California, Berkeley. Full documentation and experiments with various w will be included.

Current programs, which use the cubic Hermite space and the techniques described in this paper, are from four to eight times faster than programs which reduce the band-structured general eigenvalue problem to full standard form. Of more importance is the fact that the savings in storage permit the use of finer meshes. Such meshes are essential for approximating eigenvalues greater than the smallest two in difficult cases, such as the Mathien equation with parameter values exceeding 1000.

The error bounds accurately predict the rate at which accuracy is lost in the higher eigenvalues.

REFERENCES

Babuska, I. (1970). Error-Bounds for the Finite Element Method. Techn. Note BN-630. The Inst. for Fluid Dynamics and Appl. Math., Univ. of Maryland.

Birkhoff, G., and Varga, R. (1968). Hermite Interpolation in One and Two Variables with Application to Partial Differential Equations. *Numer. Math.* **11**, 232–256.

Birkhoff, G., de Boor, C., Swartz, B., and Wendroff, B. (1966). Rayleigh–Ritz Approximations by piecewise cubic polynomials. *J. SIAM Numer. Anal.* **3**, 188–203.

Bosarge, E., and Johnson, O. (1969). High order approximations to the state regulator problem via piecewise polynomial subspaces. *J. SIAM Control* (to appear Feb. 1971).

Ciarlet, P., Schultz, M., and Varga, R. (1969). Numerical methods of high order accuracy for nonlinear boundary value problems. III: Eigenvalue Problems. *Numer. Math.* **12**, 120–133.

Johnson, O. (1968). Convergence, Error Bounds, Sensitivity, and Numerical Comparisons of Certain Absolutely Continuous Rayleigh–Ritz Methods for Sturm–Liouville Eigenvalue Problems (thesis), Tech. Rep. 23. Univ. of California Computer Center, Berkeley, California.

Johnson, O. (1969). Error Bounds for Sturm–Liouville Eigenvalue Approximations by Several Piecewise Cubic Rayleigh Ritz Methods. *J. SIAM Numer. Anal.* **6**, 317–333.

Krylov, V. (1962). "Approximate Calculation of Integrals." MacMillan, New York.

Peters, G., and Wilkinson, J. (1969). Eigenvalues of $Ax = \lambda Bx$ with band symmetric A and B. *Comput. J.* **12**, No. 4, 398–404.

Wilkinson, J. (1965). "The Algebraic Eigenvalue Problem." Oxford Univ Press (Clarendon), London and New York.

5.22 Taylor Series Methods for Ordinary Differential Equations—An Evaluation

D. Barton, I. M. Willers, and R. V. M. Zahar

CAMBRIDGE UNIVERSITY

CAMBRIDGE, ENGLAND

I. Introduction

The Taylor series method for the solution of systems of ordinary differential equations has often been referred to in the literature of integration
procedures. Indeed, it has been stated [for example, in Collatz (1960)]
that for certain problems the method is the most efficient of all known
procedures. For problems in astronomy it has been shown in the articles of
Steffensen (1956) and of Deprit and Price (1965) that the method is effective
in achieving a high degree of accuracy. It was indicated by Deprit and
Zahar (1967) that in some cases it can do so with small values of computing
time.

In the usual formulations, the Taylor series method for the solution of
differential equations is not presented as a general purpose algorithm. In the
first step of the procedure, the right-hand members of the differential system
are manipulated algebraically, and these manipulations are performed
either by hand or on a computer by the use of an advanced algebra system.
For this reason very few theoretical treatises on the numerical solutions of
differential equations consider the method of Taylor series extensively. In
particular, it has been omitted from most of the recent surveys which compare
the efficiencies of integration procedures.

By using recurrence schemes, it is possible to devise an algorithm which
enables the coefficients of the Taylor series to be calculated numerically. To
our knowledge, the first step-by-step Taylor series procedure which partially
employed recurrences was described by J. R. Airey for the calculation of
Emden Functions in the British Association Mathematical Tables (1932).
A full recurrence scheme in which all series coefficients were calculated
numerically was used extensively by J. C. P. Miller in computing the Airey
Integral for the British Association Mathematical Tables (1946). It wasn't
until 1954, however, that Miller developed a general recurrent power series

369

method, applicable to a class of differential equations. It was this full recurrence algorithm that was used for Weber's equation in the National Physical Laboratory Tables (1955).

It is the purpose of this paper first to indicate how the Taylor series method can be programmed as a general purpose algorithm. Then, we will show for a reasonably wide class of problems that the method can be used in practice to generate accurate numerical solutions to differential equations and that, on the basis of computing time, it can be considerably more efficient than other modern integration procedures.

Hull (1967) has outlined a general and worthwhile philosophy to be used when searching for optimum integration procedures. Using his terminology, we are in this initial paper testing the Taylor series method on a class of problems, P, large enough to include most problems usually presented to a computing facility, but not so large as to include problems such as those involving stiff equations which at present require specialized techniques. Fortunately, a large compendium of problems of this type has already been presented by Crane and Fox (1969). Secondly, the class of methods, M, that we shall use for comparison will not be extensive, for we feel that, in the case of our P, a suitable M has already been narrowed considerably by other researchers such as Hull and Creemer (1963), Hull and Johnston (1964), and Clark (1968). Lastly, as a measure of goodness G, we shall use computing time not only because it is unreasonable to apply other criteria such as the number of function evaluations to the Taylor series method but also because we feel that, for a given accuracy, computing time is the criterion of greatest interest in the final analysis.

Despite the nature of our results, we emphasize that we are not at this time attempting to prove that the Taylor series method m is best in the sense that $G(m, P) \leq G(m', P)$ for all $m' \in M$. We do not feel that P has been described precisely enough or that the numerical aspects of m have been investigated thoroughly enough for such a statement to be made. Nevertheless, we are endeavoring to present a very strong case for the inclusion of m, not only in future comparisons of integration techniques but also as a candidate procedure in polyalgorithms for the integration of systems of ordinary differential equations.

II. The Taylor Series Method

Briefly, the method can be described as an application of the process of analytic continuation. Consider the system of differential equations given by

$$dy_i/dt = f_i(t, y_1, \ldots, y_n)$$

$$y_i(t_0) = a_i$$

(1)

($i = 1, 2, \ldots, N$). Theoretically, it is desired to obtain Taylor series expansions about $t = t_0$ for each of the dependent variables:

$$y_i(t) = \sum_{j=0}^{\infty} \frac{y_i^{(j)}(t_0)}{j!} (t - t_0)^j \tag{2}$$

One then evaluates these series at time $t = t_1$, expands the y_i in a new series about t_1, and continues. Proceeding in this manner, one generates a sequence of Taylor expansions which are valid successively in overlapping intervals, the totality of these intervals completely covering the desired domain of integration time.

We shall now describe the technical aspects of the procedure in some detail. A more complete description can be found in Moore (1966). Obviously, any differential system which can eventually be reduced to a form whose right-hand members are rational functions of the dependent variables and of the auxiliary variables introduced in the reduction can be treated by this technique.

If the functions f_i contain arbitrary combinations of nonrational functions (such as nonintegral powers, exp, log, sin, cos), we assume that these functions are obtainable from a known rational differential system. Thus, the function is replaced by an auxiliary dependent variable and the appropriate equations are added to system (1). For example, if y^q occurs for an arbitrary expression y, we could denote $z = y^q$ and add

$$z' = qzy'/y$$

with the appropriate initial condition. In Sect. IV we will describe an alternative procedure for dealing with such functions.

We may now assume that the differential system is in rational form. It is then reduced to a *canonical form* as follows. We introduce auxiliary variables $T_s(t)$, $s = 1, 2, \ldots, p$, until each f_i can be expressed as a result $T_{k_i}(t)$, $1 \leq k_i \leq p$. During this procedure each T_s is defined by a relation of the form

$$T_s = T_l * T_m, \qquad l, m < s$$

or

$$T_s = \odot T_m, \qquad m < s$$

where, for each s, $*$ is one of the diadic operations $+$, $-$, $\cdot$, and $/$, and $\odot$ is a monadic operation such as negate and, as we shall see, integrate.

Now, following Miller (1966), we assume each dependent variable y_i and each auxiliary variable T_s to be expanded in a power series about the generic point $t = t_r$,

$$y_i(t) = \sum_{j=0}^{\infty} y_i^{(j)}(t - t_r)^j$$

$$T_s(t) = \sum_{j=0}^{\infty} T_s^{(j)}(t - t_r)^j$$

using reduced derivatives $y_i^{(j)}$ and $T_s^{(j)}$ to avoid the division by $j!$. (For this reason, the method is often referred to as the method of power series.) We then apply the appropriate rule of calculation, for each j in turn, of the sequence of reduced derivatives $T_s^{(j)}, j = 1, 2, \ldots, p$. For example, in the case of division, $T_s = T_l/T_m$, we have the recurrence relation

$$T_s^{(j)} = \frac{1}{T_m^{(0)}}\left[T_l^{(j)} - \sum_{k=1}^{j} T_m^{(k)}T_s^{(j-k)}\right]$$

which may be used to obtain $T_s^{(j)}$ when $T_l^{(r)}$ and $T_m^{(r)}$ are known for $r = 0, 1, \ldots, j$ and $T_s^{(r)}$ for $r = 0, 1, \ldots, j - 1$.

As a result of the above procedure, we obtain an ordered sequence of recurrence relations for calculating $T_s^{(j)}$ when all T_s are known up to order $j - 1$ and all dependent variables are known up to order j. Next, we add to this sequence the relations obtained by substituting series into the differential equations themselves, which yields recurrences for increasing the order of the dependent variables. And, in so doing, we interpret the integral sign $\int$ to be a monadic operation. Thus, we obtain a complete set of ordered recurrences for generating all series coefficients to any order from the initial conditions.

As an example, consider the equation

$$y_1' = y_1 + \sin(y_1^{3/2})$$

which, by the assignment,

$$y_2 = y_1^{3/2}, \qquad y_3 = \sin y_2, \qquad y_4 = \cos y_2$$

is first reduced to the rational form

$$y_1' = y_1 + y_3 \qquad y_3' = y_4 y_2'$$
$$y_2' = (3/2)y_2 y_1'/y_1 \qquad y_4' = -y_3 y_2'$$

It further reduces to the canonical form

$$
\begin{aligned}
T_1 &= y_1 + y_3 & T_2 &= y_2 T_1 \\
T_3 &= T_2/y_1 & T_4 &= (3/2)T_3 \\
T_5 &= y_4 T_4 & T_6 &= y_3 T_4 \\
T_7 &= -T_6 & y_1 &= \int T_1 \\
y_2 &= \int T_4 & y_3 &= \int T_5 \\
y_4 &= \int T_6
\end{aligned}
\tag{3}
$$

To obtain the ordered sequence of recurrence relations, it is only necessary now to translate each equation of (3) into its corresponding recurrence.

In comparison with discrete procedures such as the Runge-Kutta or predictor–corrector techniques, the automatic Taylor series method has the disadvantage of not being easily adaptable to equations whose right-hand sides contain functions which are not obtainable from rational differential equations. Also, it is expected that for certain equations adverse cancellation can occur when the series obtained are evaluated for a given t_r, and this type of error should be carefully monitored during the running of the program.

However, the Taylor series method has the advantage that the series can be evaluated easily at nonstep points. In addition, it can be programmed so that at each step the order of the method and the step length can be modified without added difficulty. A priori, it has been found that the step lengths taken with the Taylor series method can be considerably larger—often thousands of times larger—than those taken by other methods of comparable accuracy, without fear of instability occurring in the calculation.

III. Numerical Techniques

For the method described in the previous section to become a viable numerical procedure, it is necessary to incorporate into it techniques for predicting the step-length h to be taken at time t_r for measuring the local error committed per step and for choosing the number of terms n in the series.

We considered various techniques for choosing step length. It is immediately obvious that, for a prediction technique of this type to be practicable, it must be constructed to depend on the rate of convergence of the series or on the error arising from the truncation of the series to n terms. A formula based on the error term for a given geometric series would not even be accurate enough for all geometric series, let alone for either logarithmic or exponential series. We investigated briefly methods for categorizing the series obtained and then predicting step length according to series type. Besides the mathematical problems involved in this categorization, the programming can become quite tedious and the amount of computing time can become considerable if the tests are to be performed on the series for all the dependent variables, so these techniques were put aside.

We finally decided to make the choice of step length dependent on the measurement of local error in a manner motivated by the general error term. At step t_r, the error involved in the truncation to n terms of a Taylor series for y is usually expressed as

$$E(t) = \frac{y^{(n+1)}(\xi)}{(n+1)!}(t - t_r)^{n+1}, \qquad t_r \leqq \xi \leqq t$$

or, for a step-length $h = t - t_r$,

$$E(h) = R(h) h^{n+1}$$

If it is desired that the local error be kept at least as small as ε, we wish to choose h so that $E(h) = \varepsilon$. Suppose that $R(h)$ may be estimated by a constant R. Then h may be estimated initially by

$$h_0 = c_r(\varepsilon/R)^{1/(n+1)} \qquad (4)$$

where it is assumed that c_r has been calculated from the information at the previous step. In addition, suppose that the error term $E(h_i)$ can be accurately estimated after the new series have been generated at $t_r + h_i$. We calculate a sequence of h_i from the formulas

$$\alpha_i = (\varepsilon\delta/E(h_i))^{1/(n+1)} \qquad (5)$$
$$h_{i+1} = \alpha_i h_i$$

where δ is a fixed constant less than 1 if $E(h_i) > \varepsilon$ and $\delta = 1$ if $E(h_i) \leq \varepsilon$. The iteration is terminated if $E(h_k) \leq \varepsilon$, where we choose the step length to be $h = h_k$ and the next point $t_{r+1} = t_r + h$. If we define

$$\alpha^{(r)} = \alpha_0 \alpha_1 \alpha_2 \cdots \alpha_k$$

then the c_r are defined by $c_0 = 1$ and

$$c_{r+1} = \alpha^{(r)} c_r$$

    This procedure is simply a generalization of the popular method of halving or doubling a step length used for discrete methods. It has been introduced merely because it would be too drastic to halve or double the step length in a high-order method. To completely define the above procedure, we must describe the method for calculating R and for approximating the local error $E(h_i)$. Because Eq. (4) contains the scaling factor c_r, it is only important that R be proportional to $R(h)$ for each t_r, but it is not necessary that R be a good approximation to $R(h)$. We calculate R by the formula

$$R = \max_{\substack{1 \leq i \leq N \\ n-2 \leq j \leq n}} |y_i^{(j)}|$$

(where N is the number of independent variables) which has been used successfully in the past.

    On the other hand, if our program is to be reliable in the sense that it maintains a local error less than the prescribed maximum, it is essential that our estimate of $E(h)$ be an accurate one. We shall derive an estimate based on a classical formula of Darboux. Consider the points t_r and t_{r+1} and a single dependent variable $y(t)$. Suppose that we have obtained the coefficients c_j and d_j in the two expansions for y about t_r and t_{r+1}, respectively:

$$y(t) = \sum_{j=0}^{n} c_j(t - t_r)^j + \frac{y^{(n+1)}(\xi_1)}{(n + 1)!} (t - t_r)^{n+1} \tag{6}$$

$$y(t) = \sum_{j=0}^{n} d_j(t - t_{r+1})^j + \frac{y^{(n+1)}(\xi_2)}{(n + 1)!} (t - t_{r+1})^{n+1}$$

As a special case of a formula due to Darboux [see Whittaker and Watson (1927, p. 125)] we have

$$y(t_{r+1}) - y(t_r) = \sum_{j=1}^{n} (c_j + (-1)^{j-1}d_j) \left(\frac{t_{r+1} - t_r}{2}\right)^j + s_{n+1} \tag{7}$$

where

$$|s_{n+1}| \leq \left| \frac{2f^{(n+1)}(\xi_3)}{(n + 1)!} \left(\frac{h}{2}\right)^{n+1} \right|, \qquad t_r \leq \xi_3 \leq t_{r+1}$$

Thus, putting $h = t_{r+1} - t_r$, the error term $E(h)$ in (6) can be approximated by the expression

$$\sum_{j=1}^{n} \left[(2^j - 1)c_j + (-1)^j d_j \right] \left(\frac{h}{2}\right)^j \tag{8}$$

which is obtained by subtracting (7) from (6) evaluated at $t = t_{r+1}$, noting that $c_0 = y(t_r)$ and ignoring the error term in (7). It should be mentioned that, even if we assume the coefficients c_j to be calculated exactly, the d_j will be in error because they were calculated using the truncated expansion obtained from expression (6). However, if the errors in the d_j behave like $E(h)$, as one would expect, those errors can be neglected from (8). But if the d_j contain errors due to adverse cancellation in (6), expression (8) should yield a large estimate of $E(h)$, which is desirable. In our program, we calculate $E_i(h)$ from (8) for each of the dependent variables y_i, and in expression (5) we use

$$E(h) = \max_{1 \leq i \leq N} |E_i(h)|$$

which has proven to be an extremely sensitive check on the local error.

Since the procedure of decreasing h for a given time t_r involves the generation of new series for all the variables, it is in our interest to keep k, the number of decreases in h, as small as possible. In particular, the average k for the entire span of integration should be less than 1 if our method is to compare favorably with the method of Richardson's extrapolation. In all the problems we considered, including problems in which t was allowed to approach a singularity, the average k was very near 0 when δ was chosen to be 0.2. This result is a reflection of the fact that the series expansions provide a great deal more information than can be obtained by discrete methods.

For the above method of choosing step length and for a fixed required accuracy, we found that an efficient procedure was obtained by keeping the number of terms n in the series fixed for the entire integration time. The best number n was not particularly sensitive to the equations integrated and was only slightly dependent on the required accuracy. A rule of thumb stated by Moore (1966), which is consistent with past results [Deprit and Price (1965) and Deprit and Zahar (1966)], is to choose n to be approximately equal to the number of significant decimal digits that can be carried by the computer.

IV. Implementation

The method described in Sect. II has been partially automated by some researchers. Gibbons (1960) and Moore (1966) have described programs that accept a differential system in canonical form and from this form produce the series solutions. Our program can be used as a general purpose algorithm that automatically reduces the input differential system to canonical form, calculates the appropriate series, and then prints the solution at specified points. It is not necessary for the user to perform any previous manipulations whatsoever on the differential system when our algorithm is used.

The program will accept as input a differential system of the form

$$y_i^{(n_i)} = f_i(t, y_1, \dots, y_1^{(n_1-1)}, \dots, y_n^{(n_N-1)}, r_1, \dots, r_M)$$
$$(i = 1, \dots, N)$$

$$r_i = g_i(t, y_1, \dots, y_1^{(n_1-1)}, \dots, y_N^{(n_N-1)}, r_1, \dots, r_{i-1})$$
$$(i = 1, \dots, M)$$

(9)

with initial conditions

$$y_i^{(k)}(t_0) = a_i^{(k)}, \qquad k = 0, 1, \dots, (n_i - 1)$$
$$(i = 1, \dots, N)$$

where $y_i^{(k)}$ is the kth derivative of y_i with respect to time. The functions f_i occurring in (9) may be arbitrarily complicated combinations of the variables t, y_i, and their derivatives, and elementary functions such as sin, cos, etc. The program performs the reduction to canonical form in an optimal manner.

A detailed description of the program can be found in Barton et al. (1970b). Briefly, Eqs. (9) are initially treated in much the same way as an arithmetic expression is compiled by a conventional compiler. The compilation of the differential equations proceeds in three stages. The first stage involves the reduction to canonical form and this is represented internally by a tree structure. At the second stage the canonical form is optimized by the recognition of operations between series and numerical constants and also by the removal of common subexpressions. Hence, at this stage, it is recognized that in expressions such as

$$(y + \sin y) + \cos(\sin y + y)$$

the element $y + \sin y$ need only be evaluated once. Finally, each operation on the variables and auxiliary variables introduced by the compiler is replaced by a sequence of machine code instructions that perform the desired recurrences. This final ordered sequence of machine instructions forms a subroutine that is driven by the overall numerical program.

In Sect. II it was mentioned that the elementary functions ϕ can be treated by augmenting the differential system with appropriate defining equations. Alternatively, if an algorithm is available that enables the reduced derivatives of ϕ to be calculated when those of its arguments are known, then the function ϕ may be treated as an operator for which a rule of calculation is available. Thus, instead of altering the differential system, the corresponding machine code instructions need only be incorporated for use in the last stage of compilation. The compiler is able to treat nonrational functions either by automatically augmenting the differential system or alternatively by using a known recurrence procedure. It is an interesting commentary on machine-generated instructions to note that, when both techniques for dealing with these functions were incorporated in our system, the machine-generated code was actually more efficient than the initial code generated by hand.

Compilation of the equations of a differential system is only possible if those equations are presented to our program in their original textual form. In this manner the Taylor series system differs fundamentally from a system for the integration of equations by discrete methods. The reduction of a differential system to canonical form and the derivation of recurrence relations from that form is a formal mathematical procedure that must be carried out exactly and, in our case, is performed by compilation techniques. The compilation section of our system is quite independent of the program for numerical control of the integration, the compiler being a substantial program using the logical and nonnumerical facilities of a computer. Such a program is most conveniently written in a language that provides extensive facilities for nonnumerical work, and for this reason our system is written in a programing language that is under development for use with the Camal Algebra System which is described by Barton *et al.* (1970a). The compiler could, of course, be written in Fortran and the machine orders generated from the canonical form of the differential system stored as data. Provided that this data could be obeyed as a subroutine, the entire system could be duplicated in that language. While the ability to obey data is absent from the standard specification of the Fortran language, it is nevertheless possible in most Fortran facilities.

It is true that the utility of the system must be judged on numerical grounds in comparison with other integration techniques; however, its implementation is dependent on the nonnumerical techniques of compilation and algebraic manipulation. Since the formal reduction to canonical form is performed by the machine, it becomes practicable to use the Taylor series method as easily as discrete techniques. That the method is also a practical numerical procedure we hope to show in the next section. In Fig. 1, an example of the respective input and output for our system is presented.

```
INTEGRATE
Y" =Y+SIN(2Y)
WITH INITIAL CONDITIONS
Y =1
Y' =0
FOR T=0.5: 4.5 PRINT T,Y
```

```
       T                        Y
   +0                      +1.CCCCCCCCCC
   +5.0C00000000*-1        +1.1329972590
   +1.CC000CCCCC           +1.6334842646
   +1.5C0000CCCC           +2.751887C128
   +2.CC000CCCCC           +4.5327844644
   +2.5C00000C000          +7.48C5341286
   +3.CC0000CCCC           +1.2335469C48*1
   +3.5C0000CC0C           +2.C235823864*1
   +4.CC000CCC0C           +3.3528692730*1
   +4.5C00000C000          +5.5279686187*1
```

FIG. 1. Example of the system's input and output.

V. Results

In testing our procedure, we were motivated to choose a variety of situations which would help us determine the validity of our numerical techniques, the subclass of problems for which the Taylor series method was effective, and the cost of the method in comparison with other procedures. This general approach could be considered slightly too ambitious, but in our case we felt it necessary to proceed in this manner because of the lack of pertinent literature on the Taylor series method.

All computer tests were performed on the Atlas computer in Cambridge. The Atlas uses a floating point word length of 48 bits, with a 39-bit fraction and an 8-bit exponent. Because the exponent is interpreted as a power of 8, the fraction is normalized so that the first 3 bits contain a nonzero representation. The arithmetic performed is *statistically rounded* in the sense that the 39th bit is forced to be 1 if the least significant half of the accumulator is nonzero, and otherwise the 39th bit is left unchanged. The machine can obey approximately 250,000 orders per second.

Originally, the parameter denoting maximum tolerable local error ε, the number of terms in the series n, and the δ occurring in Eq. (5) were left variable and were input to the program. After performing a few trial tests, it was decided that $\delta = 0.2$ and $n = 11$ were suitable, so these values were used for all the results presented in this section. For the purpose of the following analysis, we varied ε between 10^{-3} and 10^{-11} so that the results for a wide range of accuracy could be obtained. It is our belief, however, that in future practice the user should not be required to provide ε, for the same reasons that he is not required to specify the accuracy of his square roots or his cosines. We feel that ε should be set beforehand so that the program will return accuracy above a certain minimum level in the majority of cases and that the user should be warned about the exceptional cases.

Because of the wide range and the nature of the problems we intended to consider, it was decided that only programs capable of changing step length would be included in our comparisons. In our search for suitable procedures, we were aided by the previous results of Crane and Fox (1969). First, we obtained a program of the Bulirsch–Stoer rational extrapolation method, which has produced highly favorable results in other comparisons. From the IBM 360 Subroutine Package, which was easily accessible, we chose HPCG, a program of the fourth-order Hamming predictor–corrector, and RKGS, a program of the fourth-order Runge-Kutta–Gill method. In addition, we used the popular fourth-order Runge-Kutta variable step procedure [see Henrici (1963, p. 122)] which we programmed ourselves. All four programs were coded in Fortran, and all subroutine calls except the one for function evaluation were eliminated.

Contrary to some of the results appearing in Crane and Fox (1969), the HPCG program with appropriate initial step lengths did not perform particularly well on any of the problems considered. Except in Problem 4, HPCG returned larger values of computing time than RKGS throughout the entire range of accuracy. In the high-accuracy range, HPCG was often worse by a factor of five, and for this reason it has been omitted from our graphs. The performance of our Runge-Kutta program and RKGS were almost identical, our program returning only marginally better computing times in the medium-accuracy range. As standards for comparison from these four programs, therefore, we decided to include only the results obtained from the Bulirsch–Stoer (BS) and the Runge-Kutta–Gill (RK) procedure on our graphs. The Taylor series plot is denoted by TS.

Each differential system was integrated twice—the first time to print out values of the solutions and their errors at specified intervals, and the second time with printing suppressed. In all problems the maximum absolute errors occurred near the end of the interval of integration, as expected. The absolute error η was defined by the L_1 norm

$$\eta = \| z \| = \sum_{i=1}^{N} | z_i |$$

where z_i represents the difference between the calculated and the true solutions. On each graph, a horizontal broken line is drawn to denote the limit of expected maximum accuracy caused by using a fixed, finite word length—that is, a final relative error of 2^{-40}. In determining computing time, we counted only the amount of time taken during execution, with the printing off. This time is measured by the local clock, which always rounds up to the nearest 0.01 sec. The initial step lengths for the methods other than Taylor series were chosen so that the step length did not need to be altered near the beginning of the integration. In fact, since the timing of the Bulirsch–Stoer program was very sensitive to the initial step length, the program was run for various appropriate initial step lengths and the lowest computing time was plotted. As previously mentioned, the Taylor series program determines its own initial step length so that no such experimentation was needed. In the graphs that follow, we plotted the absolute error versus the computing time in seconds, each on a log scale with base 10.

We considered five different categories of problems and chose one or two problems for each category. Most of the problems are discussed briefly in Crane and Fox (1969). The first category contains two straightforward problems which are often used in comparisons of this type. Problem 1 is defined by

$$\begin{aligned} y_1' &= y_2, & y_1(0) &= 1 \\ y_2' &= -y_1/(y_1{}^2 + y_2{}^2)^{3/2}, & y_2(0) &= 0 \end{aligned}$$

which has the solution

$$y_1 = \cos t, \qquad y_2 = -\sin t$$

The integration interval was taken to be [0, 5]. For problem 2 we took

$$\begin{aligned} y_1' &= y_2, & y_1(0) &= 1 \\ y_2' &= y_1, & y_2(0) &= -1 \end{aligned}$$

with the solution

$$y_1 = e^{-t}, \qquad y_2 = -e^{-t}$$

and integrated over the interval [0, 4]. The error plots for these problems are shown in Figs. 2 and 3, respectively.

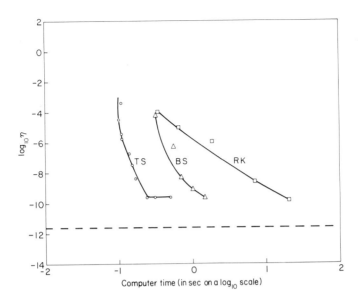

FIG. 2. Problem 1.

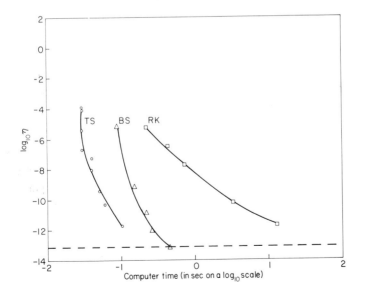

FIG. 3. Problem 2.

The behavior of the curves for both problems are similar qualitatively. The maximum accuracies attainable by all three methods are slightly lower and the computing times longer for Problem 1 than for Problem 2, as expected from the nature of the respective functions f. In Problem 1, the maximum accuracy attained by each method is the same, whereas in Problem 2, BS achieved a higher maximum accuracy than TS and RK, mainly because of fortunate rounding. Nevertheless, throughout the ranges of accuracy, TS is faster than BS by a factor of 3 or 4 and TS is faster than RK by factors between 10 and 50.

In the second category are two problems whose solutions are, in some sense, ill-behaved. Problem 3 is

$$y_1' = y_1^2 y_2, \qquad y_1(0) = 1$$
$$y_2' = -1/y_1, \qquad y_2(0) = 1$$

having the solution

$$y_1 = \frac{y_1(0)}{\cosh(t) - y_1(0)\, y_2(0) \sinh(t)}$$

$$y_2 = y_2(0) \cosh(t) - \sinh(t)/y_1(0)$$

the interval of integration being $[0, 5]$. Because y_1' becomes very large and y_2' very small, the integration procedures have difficulty in choosing the appropriate step length; for large step lengths will result in poor accuracy and small step lengths will introduce excessive rounding errors. The difficulty is more pronounced in the RK plot than in the plots for TS and BS because the latter methods employ considerably larger effective step lengths. As seen in Fig. 4, the maximum accuracies attainable by both TS and BS are significantly greater than that of RK, even at much lower computing times. Again, TS is faster than the other methods throughout almost the entire accuracy range.

For Problem 4, we chose the system

$$y_1' = y_2, \qquad\qquad y_1(6) = 1.20194993061 \times 10^{-6}$$

$$y_2' = \left(\frac{256}{t^2} - 1\right) y_1 - \frac{y_2}{t}, \qquad y_2(6) = \frac{8 y_1(6)}{3} - 2.18720051176 \times 10^{-7}$$

and the interval of integration $[6, 24]$, in which the problem behaves quite well. This system for the Bessel function $y_1 = J_{16}$ has been considered by Nordsieck (1962) and others. Figure 5 shows that the times for TS and BS are considerably smaller than those for RK, and that TS was better than BS by a factor of approximately 3. In addition, we integrated the problem by Taylor series for the interval $[6, 6006]$, and achieved a maximum accuracy greater than 10^{-8} at the end of the interval.

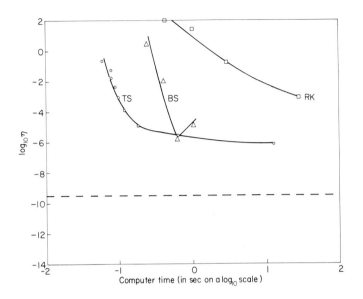

FIG. 4. Problem 3.

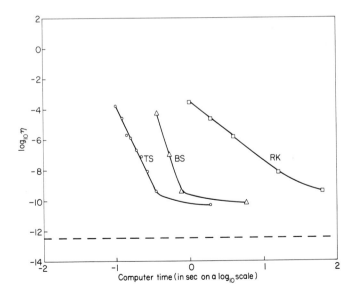

FIG. 5. Problem 4.

Next, we considered an equation containing a singularity. Problem 5 is

$$y' = 1/(1 - t), \qquad y(0) = 1$$

with the solution

$$y = 1 - \ln(1 - t)$$

and interval [0, 0.995]. All three methods performed well. At low accuracies, RK was faster than BS, a fact often noted previously. TS achieved a slightly better accuracy than both of the other methods, with approximately one fifth the computing time of BS. (See Fig. 6.)

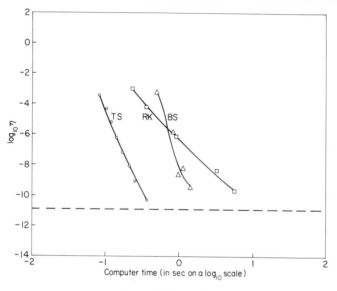

FIG. 6. Problem 5.

For the next category of problems, we considered two sets of equations which were prone to numerical instability because of the presence of dominant, but unwanted, solutions. Problem 6 [see Hull and Creemer (1963)] is

$$y' = y - 2t/y$$

with general solution

$$y = \pm(2t + 1 + ce^{2t})^{1/2}$$

For the choice of $y(0) = 1$, one obtains $c = 0$ and the positive solution. This problem is particularly appropriate because the terms in the right-hand side of the equation become nearly equal as t increases, introducing rounding errors and thus the exponential into the numerical solution at quite low values of t. Figure 7 indicates the results obtained by straightforward integration in [0, 5]. TS was only slightly better than BS in this example and, again the integration times for RK were exceedingly long.

Problem 7 is defined by

$$y' = 10(y - t^2)$$

which has the general solution

$$y = 0.02 + 0.2t + t^2 + ce^{10t}$$

The exponential is eliminated by the choice $y(0) = 0.02$. Forward integration in [0, 1] by the two methods produces the plot in Fig. 8. In this case, the Taylor series can achieve significantly higher maximum accuracy than both BS and RK, in extensively smaller computing times, TS being faster than BS by a factor greater than 15 in the high-accuracy range. The points in the TS plot seem slightly erratic because the effect of discretized computing times is manifest in this region of low values.

For the eighth and final problem, we chose to integrate an orbit of a satellite in the earth–moon system described by the planar restricted problem of three bodies. We considered this problem partly because it is longer and more complicated than the previous examples. The equations are

$$y_1'' = y_1 + 2y_2' - (1 - \mu)\frac{(y_1 + \mu)}{r_1{}^3} - \mu\frac{(y_1 + \mu - 1)}{r_2{}^3}$$

$$y_2'' = y_2 - 2y_1' - (1 - \mu)\frac{y_2}{r_1{}^3} - \mu\frac{y_2}{r_2{}^3}$$

$$r_1 = ((y_1 + \mu)^2 + y_2{}^2)^{1/2}$$
$$r_2 = ((y_1 + \mu - 1)^2 + y_2{}^2)^{1/2}$$

with initial conditions

$$y_1(0) = 0.994, \qquad y_1'(0) = 0$$
$$y_2(0) = 0, \qquad y_2'(0) = -2.03173262956$$

and $\mu = 0.012277471$. Other results for this problem are described extensively in Clark (1968) and also in Crane and Fox (1969). We chose the measure of absolute error to be the L_1 norm, as before, but calculated at the end of the orbit which has period $T = 11.1243403373$. The "true solution" was found from a double-precision program. Figure 9 shows that the plots for the three methods are not as spread out as they are in our previous problems. However, TS is still better than BS in terms of computing time and maximum accuracy achieved. The TS and RK plots are quite like those appearing in Deprit and Zahar (1967), where a similar problem is treated.

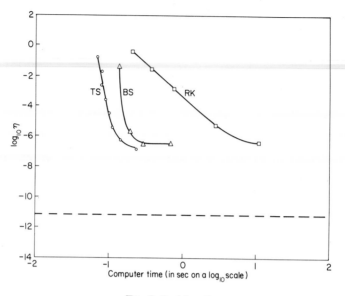

FIG. 7. Problem 6.

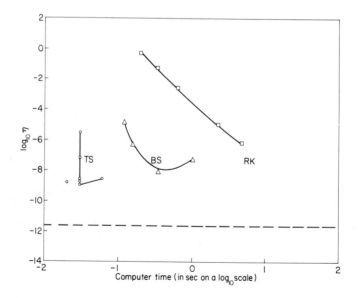

FIG. 8. Problem 7.

We shall conclude this section with a general discussion of some of the features occurring in the graphs. In Figs. 2, 4, and 5, it may be noted that, in the region of maximum accuracy, the Taylor series plot can move sharply to the right, representing a small increase in accuracy for a large increase in computing time. This behavior exists because we have not limited the smallness of the step length calculated in the method. If desired, simple tests such as those that exist in the Runge-Kutta–Gill and other programs could be incorporated to curtail this loss of efficiency. It may be noted, however, that the Bulirsch–Stoer program suffers from a similar difficulty, in some problems, the plot moving sharply upwards to the right of the maximum accuracy point.

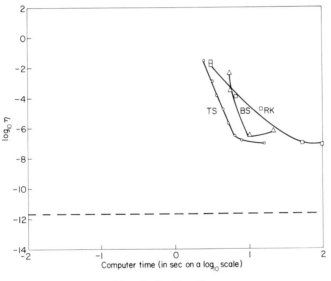

Fig. 9.   Problem 8

It is possible that our graphs may be criticized for three reasons. First, the methods other than Taylor series were coded in Fortran and the Taylor series was not. Second, some of the more efficient integration methods, such as the high-order predictor–corrector procedures, were omitted in our comparisons. Finally, computing time as a criterion is too machine dependent.

It is possible that the Taylor series method could be programmed in non-standard Fortran. However, as mentioned previously, we feel that the efficiency of our compiler could be considerably decreased in the process even though it could produce the same run-time code. It is possible also that some saving in computing time—realistically, a factor of 2—could be gained by coding the other methods in machine code.

We realize that other methods can be more efficient in some problems than both the Bulirsch–Stoer method and, of course, the fourth-order Runge-Kutta methods. Previous comparisons have shown, however, that, for a given accuracy, the gain in computing time over the Bulirsch–Stoer method rarely, if ever, exceeds a factor of 2, especially if one considers only general purpose procedures.

We have attempted to minimize the machine dependence of our results by considering only run-times and by suppressing print-outs. (The compile time for the Taylor series program was less than that for the Fortran programs, partly because our compiler is load and go, whereas the Fortran compiler is not.) Inevitably, discrepancies between machines will exist, particularly because of the different relative times of arithmetic operations. In spite of the fact that machine dependence can cause substantially different results in terms of absolute computing time, we hope that it will be unimportant in relative comparisons such as ours. Nevertheless, even if criteria such as the number of function evaluations could be adequately applied to the Taylor series method, we feel that such criteria are not easily transformed into time factors and are often misleading.

VI. Conclusions

The method of Taylor series is conceptually straightforward, yet mathematically elegant. Its use has been restricted and its numerical theory neglected merely because adequate software in the form of automatic programs for the method has been nonexistent. Because it has usually been formulated as an *ad hoc* procedure, it has generally been considered too difficult to program, and for this reason has tended to be unpopular.

The method is one of the oldest and most reliable procedures for integrating systems of differential equations. It can be formulated in general for differential systems which are reducible to rational form. In addition, it can be programmed as a general purpose algorithm which requires no previous manipulations on the differential system, accepting only the defining system as input.

Experience over a wide range of problems and required accuracies has indicated that it does not suffer from numerical instability. It is a flexible, variable-step method which produces a piecewise polynomial solution, valid throughout the entire domain of integration. In many problems, it can attain a significantly greater maximum accuracy than the Runge-Kutta–Gill method, and it often proceeds with unusually large step lengths. Throughout a wide range of accuracy, the Bulrisch–Stoer method often requires five times the amount of computing time taken by the Taylor series method. In comparison with fourth-order predictor–corrector and Runge-Kutta methods, the Taylor series method can achieve an appreciable saving in computing time, often by a factor of 100.

ACKNOWLEDGMENTS

We are grateful to J. C. P. Miller, who was responsible originally for much of the Taylor series formulation in terms of recurrent power series, and to him and D. J. Wheeler for their helpful suggestions, many of which have been incorporated in our program.

REFERENCES

Barton, D., Bourne, S. R., and Horton, J. R. (1970a). Structure of the Cambridge Algebra System, *Comput. J.* **13**, 243–247.

Barton, D., Willers, I. M., and Zahar, R. V. M. (1970b). An Implementation of the Taylor series method for ordinary differential equations. *Comput. J.* **14.** (In press).

British Association for the Advancement of Science Mathematical Tables (1932). "Emden Functions," Vol. II. B.A., London.

British Association for the Advancement of Science Mathematical Tables (1946). "The Airy Integral" (prepared by J. C. P. Miller), Part-Vol. B. Cambridge Univ. Press, London and New York.

Clark, N. W. (1968). A study of some numerical methods for the integration of systems of first-order ordinary differential equations. Rep. ANL-7428, Argonne National Laboratory.

Collatz, L. (1960). "The Numerical Treatment of Differential Equations," 3rd ed. Springer, Berlin.

Crane, P. C., and Fox, P. A. (1969). A comparative study of computer programs for integrating differential equations, Numerical Mathematics Program Library Project, Bell Telephone Lab., Vol. 2, Issue 1.

Deprit, A., and Price, J. (1965). The computation of characteristic exponents in the planar Restricted Problem of Three Bodies, *Astronom. J.* **70**, 836–842.

Deprit, A., and Zahar, R. V. M. (1966). Numerical integration of an orbit and its concomitant variations by recurrent power series, *ZAMP* **17**, 425–430.

Gibbons, A. (1960). A program for the automatic integration of differential equations using the method of Taylor series, *Comput. J.* **3**, 108–111.

Henrici, P. (1963). "Discrete Variable Methods in Ordinary Differential Equations." Wiley, New York.

Hull, T. E. (1967). A search for optimum methods for the numerical integration of differential equations. *SIAM Rev.* **9**, 647–654.

Hull, T. E., and Creemer, A. L. (1963). Efficiency of predictor–corrector procedures, *JACM* **10**, 291–301.

Hull, T. E., and Johnston, R. L. (1964). Optimum Runge-Kutta methods. *Math. Comput.* **18**, 306–310.

Miller, J. C. P. (1966). The Numerical Solution of Ordinary Differential Equations. "Numerical Analysis, An Introduction" (J. Walsh, ed.), Chapter 4. Academic Press, New York and London.

Moore, R. A. (1966). "Interval Analysis." Prentice-Hall, Englewood Cliffs, New Jersey.

National Physical Laboratory (1955). Tables of Weber Parabolic Cylinder Functions (Introduction by J. C. P. Miller).

Nordsieck, Arnold (1962). On numerical integration of ordinary differential equations. *Math. Comput.* **16,** 22–49.

Steffensen, J. F. (1956). On the Restricted problem of Three Bodies, *Kong. Danske. Videnskab. Selskab. Mat-Fys. Medd.* **30.**

Whittaker, E. T., and Watson, G. N. (1927). "Modern Analysis," 4th ed. Cambridge Univ. Press, London and New York.

5.23   A New Algorithm for Nonlinear Least-Squares Curve Fitting*

K. M. Brown† and J. E. Dennis

CORNELL UNIVERSITY

ITHACA, NEW YORK

I. Introduction and Description of the Method

In this paper we present a new algorithm for the problem of fitting a given set of tabular data points with a curve in the nonlinear least-squares sense. We give convergence theorems for the method and report the results of computational investigations in which the algorithm was tested against currently used minimization techniques.

Let (p_i, y_i), $i = 1, \ldots, M$, be given data, let $x \in E^N$, and suppose we wish to fit a function of the form $g(x; p)$ to the data in such a way that

$$\phi(x) = \| g(x; p) - y \|_2^2 = \sum_{i=1}^{M} (g(x; p_i) - y_i)^2$$

is a minimum; i.e., we seek a point $x^* \in E^N$ which minimizes the slacar function ϕ. Now every such relative minimum will be found among the zeros of $\phi'(x)$, the gradient of the ϕ. If we set

$$f(x) \equiv g(x; p) - y$$

and

$$G(x) \equiv J_f^T(x) f(x)$$

where $J_f(x)$ denotes the $M \times N$ Jacobian matrix of $f = (f_1, \ldots, f_M)^T$, then our task is to find the zeros of $G(x)$ and hence of the gradient $\phi'(x)$.

Let $H_i(x)$ denote the Hessian matrix of f_i at x. By direct calculation we have that

$$J_G(x) = \sum_{k=1}^{M} f_k(x) H_k(x) + J_f(x)^T J_f(x) \tag{1}$$

* Supported in part by NSF Grant GJ-844.
† Present address: Yale University, New Haven, Connecticut.

so that Newton's method applied to

$$G(x) = 0 \qquad (2)$$

is given by

$$x_{n+1} = x_n - \left[\sum_{k=1}^{M} f_k(x_n)\, H_k(x_n) + J_f(x_n)^T J_f(x_n) \right]^{-1} J_f^{T}(x_n)\, f(x_n) \qquad (3)$$

The latter formula requires (assuming continuous second partial derivatives of ϕ) the calculation of $M \cdot N \cdot (N + 1)/2$ second partial derivatives per iterative step. The Gauss–Newton and Levenberg–Marquardt algorithms are two frequently used attempts to circumvent this difficulty. The former simply drops the term $\sum_{k=1}^{M} f_k(x_n)\, H_k(x_n)$, and the latter approximates it with a diagonal matrix $\mu_n I$. Both methods work well locally when $\| f(x_n) \|$ is very small at a zero of G; for example, in Brown and Dennis (1970), we have shown that the Levenberg–Marquardt iteration converges quadratically to x^*, a zero of f, if $\mu_n = O(\| f(x_n)\|)$ and $J_f(x^*)$ has full rank. Obviously the Gauss–Newton method behaves likewise. When the stationary points have large residuals, the Levenberg–Marquardt algorithm can degenerate into an awkward descent method, for then the H_k in (3) are no longer damped out.

In order to approximate the H_k without requiring additional derivative or function evaluations, we propose the following algorithm.

Algorithm 1.1   Let x_n, $J_f(x_n)$, and $f(x_n)$ be given along with M matrices $B_{1,n}, \ldots, B_{M,n}$, each of size $N \times N$. [Initially the $B_{i,0}$ may be chosen to approximate the $H_i(x_0)$ by, say, using first differences on the entries of $J_f(x)$.] Obtain

$$
\begin{aligned}
x_{n+1} &= x_n - \left[\sum_{k=1}^{M} f_k(x_n)\, B_{k,n} + J_f(x_n)^T J_f(x_n) \right]^{-1} J_f(x_n)^T\, f(x_n) \\
&\equiv x_n - A_n^{-1} J_f(x_n)^T\, f(x_n)
\end{aligned} \qquad (4)
$$

and compute $J_f(x_{n+1})$ and $f(x_{n+1})$.

Now update the B_i by means of

$$
\begin{aligned}
B_{i,n+1} = B_{i,n} &+ [\nabla f_i(x_{n+1})^T - \nabla f_i(x_n)^T \\
&- B_{i,n}(x_{n+1} - x_n)] \frac{(x_{n+1} - x_n)^T}{\| x_{n+1} - x_n \|_2^2}
\end{aligned} \qquad (5)
$$

for each $i = 1, \ldots, M$. Continue the process until termination criteria are met.

Remark 1.1 $\nabla f_i(x)$ is just the ith row of $J_f(x)$.

Remark 1.2   Equation (5) is the appropriate generalization of Broyden's "single-rank" approximation to $H_i(x^n)$ [Broyden (1970)].

Remark 1.3 The algorithm requires no more function or derivative evaluations than do the Gauss–Newton or Levenberg–Marquardt algorithms; however, more storage space is needed. The additional storage requirement is offset on the one hand by superior local behavior (stability near a root) and on the other hand by a gain in speed of convergence.

II. Convergence Results

The purpose of this section is to present theorems which characterize the local convergence properties of the algorithm. (The proofs will be given elsewhere.)

The following lemma bounds the error in the Hessian approximations given by (5).

Lemma 1 Let Ω be an open convex neighborhood of x^*, and let $K \geq 0$ be a constant and P be a Frechet differentiable function mapping Ω into E^N such that, for every $x \in \Omega$,

$$\| J_P(x^*) - J_P(x) \| \leq K \| x^* - x \| \qquad (L1)$$

Let B be a real $N \times N$ matrix and let $x, x' \in \Omega$. Define B' by

$$B' \equiv B + [P(x') - P(x) - B(x' - x)] \frac{(x' - x)^T}{\| x' - x \|^2} \qquad (5')$$

Under these hypotheses

$$\| B' - J_P(x') \| \leq \| B - J_P(x) \| + 2K(\| x - x^* \| + \| x' - x^* \|)$$

Lemma 2 If, for each $i = 1, \ldots, M$, H_i satisfies (L1) with constants K_i on a compact convex subset C of Ω, then there is a constant γ_1 such that J_G satisfies (L1) with $K = \gamma_1$ on C.

Theorem 1 Let x^* be a zero of $J_f^T(\cdot) f(\cdot)$ and let $\varepsilon > 0$, $K_i \geq 0$, $i = 1, \ldots, M$, be constants such that, for every $x \in \Omega$, $\| H_i(x) - H_i(x^*) \| \leq K_i \| x - x^* \|$. Then, if $J_G(x^*)$ is nonsingular, there exist constants $\delta > 0$, $\varepsilon > 0$, such that, if $\| x_0 - x^* \| < \varepsilon$ and $\| B_{i,0} - H_i(x_0) \| \leq \delta$, $i = 1, \ldots, M$, Algorithm 1.1 converges to x^* from x_0.

Theorem 2 If the hypotheses of Theorem 1 hold and $\| f(x^*) \| = 0$, then the iteration defined by Algorithm 1.1 converges at least quadratically.

If we assume the stronger continuity condition

$$\| J_P(x) - J_P(y) \| \leq K \| x - y \| \qquad (L2)$$

then it is not necessary to assume the existence of a zero, x^*, of G; that is, by making assumptions about the behavior of the function and its derivatives in an open convex subset of E^N we are able to prove a "Kantorovich theorem" for the iteration defined by Algorithm 1.1 in which the existence of x^* is deduced as a part of the proof.

III. Numerical Results

Example 3.1   In order to test the method against a variety of algorithms in current use (see Table I), we referred to the very fine survey paper by Box (1966). The test function used was

$$\phi(x_1, x_2, x_3) = \sum_p [e^{-x_1 p} - e^{-x_2 p}) - x_3(e^{-p} - e^{-10p})]^2$$

where the summation is over the values $p = 0.1(0.1)1.0$. This problem has a zero residual at (1, 10, 1) and whenever $x_1 = x_2$ with $x_3 = 0$. We used those starting points for which $\phi(x^0)$ was large:

I. $x_1 = 0$, $x_2 = 10$, $x_3 = 20$, $\phi = 1031.154$
II. $x_1 = 0$, $x_2 = 20$, $x_3 = 20$, $\phi = 1021.655$

TABLE I

NUMBER OF FUNCTION EVALUATIONS REQUIRED TO REDUCE ϕ
TO LESS THAN 10^{-5}

Method	Starting point I	Starting point II
Swann (1964)	Failed	Failed
Rosenbrock (1960)	350	246
Nelder and Mead (1965), Spendley et al. (1962)	307	315
Powell (1964)	Failed	Failed
Fletcher and Reeves (1964)	92	188
Davidon (1959), Fletcher and Powell (1963)	140	140
Powell (1965)	28	33
Barnes (1965)	37	59
Algorithm 1.1, including evaluations done to approximate $H_i(x^0)$	37	33
Algorithm 1.1, when $H_i(x^0)$ was given approximately	24	20

Remark 3.1   Many of the methods above behave linearly and could not be expected to rapidly reduce ϕ from 10^{-5} to 10^{-10}; however, Algorithm 1.1 showed quadratic convergence in this range.

Example 3.2 This example is given by Nielsen (1968) and is an illustration of how quadrature weights and nodes may be calculated by nonlinear least-squares techniques (see Table II).

Data Vectors:

p		y	
0.0	5.0	2.0	0.0
1.0	6.0	0.0	2/7
2.0	7.0	2/3	0.0
3.0	8.0	0.0	2/9
4.0	9.0	2/5	0.0

Functional relationship: $g(x; p) = x_1 x_3{}^p + x_2 x_4{}^p$.
Initial approximation: $x_1 = 1.0$, $x_2 = 1.0$, $x_3 = -0.75$, $x_4 = 0.75$.

TABLE II

n	Gauss–Newton[a]	$\phi(x^n)$	Algorithm 1.1	$\phi(x^n)$
1	0.95493	8.93674(−02)	0.96373	10.117(−02)
	0.95493		0.96301	
	−0.68949		−0.69911	
	0.68949		0.69947	
4	0.97719	7.46872(−02)	0.97596	0.74687(−02)
	0.97719		0.97183	
	−0.65194		−0.65219	
	0.65194		0.65047	
6	Not reported		0.97754	0.746847(−02)
			0.97754	
			−0.65140	
			0.65140	
8	0.97754	7.46847(−02)	Ten significant digits of accuracy	
	0.97754			
	−0.65140			
	0.65140			

[a] Nielson (1968, p. 41).

Remark 3.2 The above example contrasts the slower convergence rate of the Gauss–Newton method with that of Algorithm 1.1, even in the presence of a small residual at the root.

REFERENCES

Barnes, J. G. P. (1965). An algorithm for solving non-linear equations based on the secant method. *Comput. J.* **8**, 66.

Box, M. J. (1966). A comparison of several current optimization methods, and the use of transformations in constrained problems. *Comput. J.* **9**, 67.

Brown, K. M., and Dennis, J. E., Jr. (1970). Derivative Free Analogues of the Levenberg–Marquardt and Gauss Algorithms for Nonlinear Least Squares Approximation. IBM Philadelphia Scientific Center Technical Report No. 320-2994.

Broyden, C. G. (1970). The convergence of single-rank quasi-Newton methods. *Math. Comput.* **24**, 365.

Davidon, W. C. (1959). Variable Metric Method for Minimization, ANL-5990 (Rev.). A.E.C. Res. and Develop. Rep.

Fletcher, R., and Powell, M. J. D. (1963). A rapidly convergent descent method for minimization. *Comput. J.* **6**, 163.

Fletcher, R., and Reeves, C. M. (1964). Function minimization by conjugate gradients. *Comput. J.* **7**, 149.

Nelder, J. A., and Mead, R. (1965). A simplex method for function minimization. *Comput. J.* **7**, 308.

Nielson, G. M. (1968). Nonlinear Approximations in the l_2 Norm. (Master's Thesis.) Univ. of Utah, Salt Lake City, Utah.

Powell, M. J. D. (1964). An efficient method of finding the minimum of a function of several variables without calculating derivatives. *Comput. J.* **7**, 155.

Powell, M. J. D. (1965). A method for minimizing a sum of squares of non-linear functions without calculating derivatives, *Comput. J.* **7**, 303.

Rosenbrock, H. H. (1960). An automatic method for finding the greatest or least value of a function. *Comput. J.* **3**, 175.

Spendley, W., Hext, G. R., and Himsworth, F. R. (1962). Sequential applications of simplex designs in optimization and evolutionary operation. *Technometrics* **4**, 441.

Swann, W. H. (1964). Report on the development of a new direct searching method of optimisation, Central Instrument Laboratory Research Note 64/3. I.C.I. Ltd.

Part Three

SELECTED MATHEMATICAL SOFTWARE

The final part of this book contains four chapters, each of which describes and presents an algorithm of mathematical software. It is hoped that these algorithms will serve as examples of quality mathematical software. Yet each author is acutely aware that there always seems to be another improvement possible in any program. In addition to being examples, each of these algorithms is for a very basic mathematical problem and thus of considerable interest in itself.

CHAPTER 6

SELF-CONTAINED POWER

ROUTINES

*N. W. Clark, W. J. Cody**

ARGONNE NATIONAL LABORATORY

ARGONNE, ILLINOIS

H. Kuki

THE UNIVERSITY OF CHICAGO

CHICAGO, ILLINOIS

I. Introduction

The standard approach to the computation corresponding to the Fortran statement

$$Z = X ** Y$$

is to use existing library subroutines for the exponential and logarithm functions to compute

$$Z = \exp(Y * \ln(X))$$

For sound theoretical reasons, the accuracy of this computation is poor, independent of the quality of the exponential and logarithm routines, whenever

$$W = Y * \ln(X)$$

is significantly larger than 1 [Clark and Cody (1969)]. The crux of the problem is that the relative error in Z, neglecting the round-off error intro-

* Work performed under the auspices of the United States Atomic Energy Commission.

duced during the exponential computation, is proportional to the absolute error in W. The finite word length of the computer assures that this absolute error is, in turn, proportional to the magnitude of W. To combat this problem, the effective word length of the computer can be extended in various ways. For single-precision floating point routines, the computation of ln(X), W, and Z can be carried out in the power routine in fixed point arithmetic, without calling upon existing library programs. This approach was taken at the University of Toronto and at Argonne National Laboratory in the early 1960's and later at the University of Chicago. For double-precision floating point, more subtle techniques are needed if the resulting program is to be competitive in speed with the traditional program.

Section II of this paper presents a Fortran version of the program described by Clark and Cody (1969), to illustrate the principles involved in this self-contained approach to the computation. Section III presents a polished assembly language program for the IBM 360, extending the general principles so as to further improve performance. This latter version also incorporates entries for various logarithm and exponential computations.

II. A Fortran Program (Fig. I)

We slightly modify the formulation of the exponential computation as follows. Let

$$U = 2^W$$

where

$$W = Y * \log_2(X)$$

Then

$$U = X^Y$$

As pointed out above, the key to an accurate floating point computation of U is the computation of W to higher than working precision in the machine. The technique reported in Clark and Cody (1969) uses only standard floating point arithmetic to accomplish this.

Essentially, the technique involves the representation of appropriate floating point numbers in reduced form. Let Z be a floating point number and let

$$Z_1 = [16Z + 0.5]/16$$

where the brackets denote "the integer part of." Then Z is expressed in reduced form as

$$Z = Z_1 + Z_2$$

where Z_1 and Z_2 are standard floating point numbers. If we decompose X into

$$X = 16^P * 2^{-(Q+B/16)} * (M * 2^{B/16})$$

where P is an integer, Q is an integer between 0 and 3, B is an odd positive integer less than 16, and

$$2^{-1/16} \leq M * 2^{B/16} \leq 2^{1/16}$$

then

$$\log_2(X) = (4P - Q - B/16) + \log_2(M * 2^{B/16})$$

is naturally representable in reduced form. In particular,

$$| \log_2(M * 2^{B/16}) | < 1/16$$

By working entirely with the reduced forms of Y and $\log_2(X)$, the quantity W can also be obtained in reduced form,

$$W = W_1 + W_2$$

Finally, we can decompose W_1 into

$$W_1 = 4 * P1 + Q1 - B1/16$$

and compute

$$U = 16^{P1} * 2^{Q1 - B1/16} * 2^{W_2}$$

where 2^{W_2} is found by the use of an appropriate approximation. Note that the reduced form of $\log_2(X)$ assures us that ΔW, the absolute error in W, is 2^{-4} times the normal relative error in floating point notation. This means that Y must be greater than 16 before ΔW begins to affect the computation U. In practice, Y is about 32 before the effects of ΔW can be detected over the error generated during the computation of 2^{W_2}.

The Fortran code for the IBM System/360 presented at the end of this section (see Fig. 1) is not intended to be implemented as a standard subroutine on any particular machine, for it cannot compete in speed or storage with a tight assembly language program, such as that presented in the next section. It is intended to illustrate the technique detailed in Clark and Cody (1969) and described briefly above. The major difference in the Fortran code presented here is the use of a truncated Taylor series to evaluate the logarithm. The tests reported in Clark and Cody (1969) have been repeated for this code, and the results included in the comments to the program. Comparison against the results in the reference indicate only a slight deterioration in accuracy over the assembly language version. (Essentially the same deterioration was obtained when a minimax approximation was used for the logarithm.)

The program has been converted for the CDC 3600 by suitably modifying the various approximations used and the sections devoted to determining P, Q, P1, Q1, etc. Tests similar to those mentioned above show that the expected accuracy was achieved.

```
      SUBROUTINE FDXPD(XX,YY,POW)
C
C     SELF CONTAINED ROUTINE TO COMPUTE POW = XX**YY.  THIS FORTRAN
C     PROGRAM IS INTENDED TO DEMONSTRATE A PRINCIPLE RATHER THAN TO
C     BE A POLISHED PROGRAM.  THE CODE IN ITS PRESENT FORM IS INTENDED
C     FOR AN IBM/360.  THUS, CERTAIN MACHINE DEPENDENT CONSTANTS
C     MUST BE CHANGED BEFORE THE CODE CAN BE USED ON OTHER MACHINES.
C     IN ADDITION, CERTAIN SEQUENCES OF ARITHMETIC STATEMENTS, DESIGNED
C     TO MAINTAIN PRECISION ON A HEXADECIMAL COMPUTER, CAN BE SIMPLIFIED
C     FOR A BINARY COMPUTER.
C
C     THE FOLLOWING TABLE GIVES THE RESULTS OF RANDOM ARGUMENT
C     ACCURACY TESTS FOR THIS ROUTINE ON AN IBM/360.
C
```

ARGUMENT RANGE	FREQUENCY OF BIT ERRORS NO. OF BITS IN ERROR						MAX. REL. ERROR	RMS REL. ERROR
	0	1	2	3	4	5		
X (2**- 4.,2**+ 4.) Y (- 4., 4.)	1250	726	24	0	0	0	0.222E-15	0.626E-16
X (2**-16.,2**+16.) Y (-16.,16.)	1067	809	85	34	5	0	0.243E-15	0.651E-16
X (2**-32.,2**+32.) Y (- 8., 8.)	1019	737	74	134	35	1	0.435E-15	0.865E-16
X (2**-64.,2**+64.) Y (- 4., 4.)	832	607	157	296	103	5	0.435E-15	0.113E-15
X (2**- 8.,2**+ 8.) Y (-32.,32.)	870	812	159	132	27	0	0.442E-15	0.782E-16
X (2**- 4.,2**+ 4.) Y (-64.,64.)	613	701	260	341	85	0	0.433E-15	0.110E-15

```
C*******
      DOUBLE PRECISION A,B,D1,D2,D3,D4,D5,D6,INF,LOG2E,O625,SIXTEN,
     1TEMP,TWOMQ,W1,W2,X,XMAX,XMIN,XPON,XX,Y,YN,YY,Z,ZN,ZSQ,ZZ,Z2,POW
      INTEGER I,N,NA,NB,NBIAS,NP,NQ,NSHIFT,NT,NX,NZ
      EQUIVALENCE (NX,X),(NT,TEMP)
      DIMENSION A(8),B(17),NT(2),NX(2),TWOMQ(5)
C**********
C
C     THE FOLLOWING MACHINE-DEPENDENT CONSTANTS ARE NEEDED.  IT IS
C     ESSENTIAL THAT THESE CONSTANTS BE IN OCTAL OR HEXADECIMAL
C     FORM TO AVOID ERRORS IN THE LEAST SIGNIFICANT BIT INTRODUCED
C     BY THE COMPILER DURING CONVERSION FROM DECIMAL.
C
C     B(I) = (2**(I/16))/2**(I/16)  I=1,...,17,  ROUNDED TO
C     WORKING PRECISION.
C
C     A(I) = CORRECTION TERMS TO BE USED WITH APPROPRIATE B(I)
C     SO THAT  2**((2I-1)/16) = B(2I) + A(I),  I=1,...,8  TO
C     BEYOND WORKING PRECISION.
C
C     NBIAS = INTEGER REPRESENTATION OF BIAS ON FLOATING POINT
C     EXPONENT.
C
C     NSHIFT = INTEGER REPRESENTATION OF UNBIASED FLOATING POINT
C     EXPONENT OF 1.
C
C     LOG2E = LOG-BASE-2(E) - 1
C           = .44269...
C     TO WORKING PRECISION
C
C     INF = LARGEST MACHINE ACCEPTABLE FLOATING POINT NUMBER.
C
C     XMAX = LARGEST UNBIASED FLOATING POINT EXPONENT (BASE 2)
C     ACCEPTABLE TO MACHINE WITHOUT CAUSING OVERFLOW.
C
C     XMIN = SMALLEST UNBIASED FLOATING POINT EXPONENT (BASE 2)
C     ACCEPTABLE TO MACHINE WITHOUT CAUSING UNDERFLOW.
C
C**********
```

Fig. 1.

```
      DATA NSHIFT,NBIAS/Z01000000,Z40000000/
      DATA SIXTEN/16.0D0/,0625/.0625D0/,XMAX,XMIN/252.0D0,-260.0D0/
      DATA B/1.0D0,Z40F5257D152486CC,Z40EAC0C6E7DD2439,
     1    Z40CE0CCDEEC2A94E1,Z40D744FCCAD69D6B,Z40CE248C151F8481,
     2    Z40C5672A115506DB,Z40BD08A39F58DC37,Z40B504F333F9DE65,
     3    Z40AD583EEA42A14B,Z40A5FED6A9B15139,Z409EF5326091A112,
     4    Z409837F0518DB8A9,Z4091C3D373AB11C3,Z408B95C1E3EA8BD7,
     5    Z4085AAC367CC487B,0.5D0/
      DATA A/Z322C7B9C00000000,Z3211065900000000,ZB21C1DCA00000000,
     1       ZB241577E00000000,ZB239B67F00000000,ZB2525E6F00000000,
     2       Z32360FD700000000,Z3214C5CC00000000/
      DATA LOG2E/Z4071547652B82FE2/
C******
C
C     BEST FIT COEFFICIENTS TO COMPUTE 2**(-W2), WHERE ABS(W2)
C     .LE. 1/16.  THIS SIXTH DEGREE POLYNOMIAL WITH CONSTANT
C     TERM EQUAL 1 APPROXIMATES WITH A MAXIMUM RELATIVE ERROR
C     .LE. 6.9D-18.
C
C******
      DATA D6,D5,D4,D3,D2,D1/.1507371568055265CD-3,.1333073476631756 0D-2
     1,.96181170994625860D-2,.55504108402448D-1,.24022650695637260D0
     2,.69314718055993460D0/
      DATA TWOMQ/1.0D0,.5D0,.25D0,.125D0,.0625D0/,INF/Z7FFFFFFFFFFFFFFFF/
C******
C
C     OBTAIN ARGUMENTS
C
C******
      X=XX
      Y=YY
C******
C
C     TEST FOR ILLEGAL ARGUMENTS
C
C******
      IF(X) 5, 10, 20
    5 PRINT 1000
 1000 FORMAT(26H ILLEGAL ARGUMENT IN FDXPD)
      RETURN
   10 IF(Y) 15, 5, 15
   15 POW=0.0D0
      RETURN
C******
C
C     AN ARGUMENT Z WILL BE IN REDUCED FORM WHEN IT IS EXPRESSED
C     AS A SUM Z1 + Z2 WHERE Z1 IS THE INTEGER PART OF 16*Z
C     AND Z2 IS THE REMAINDER
C
C     PUT Y INTO REDUCED FORM
C
C******
   20 N=SIXTEN*Y
      YN=N
      YN=YN*0625
      Y=Y-YN
C******
C
C     LET X=(16**P)*(2**-Q)*M  WHERE  .5 .LE. M .LT. 1.0,
C               AND Q IS AN INTEGER BETWEEN 0 AND 3.
C
C     DETERMINE P
C
C******
      NP=(NX(1)-NBIAS)/NSHIFT
      IF(X .LT. 0625) NP=NP-1
C******
C
C     SET FLOATING POINT EXPONENT ON X TO 0.
C
C******
```

FIG. 1. (cont.)

```
        NX(1)=NX(1)-NP*NSHIFT
C*******
C
C       DETERMINE Q
C
C*******
        NQ=0
        DO 35 I=1,3
        IF(X-.5D0) 30, 40, 40
   30 X=X+X
   35 NQ=NQ+1
C*******
C
C       SCALE M BY 1/A, WHERE A=2**(-B/16), B AN ODD INTEGER BETWEEN 1
C       AND 15, SO THAT IT LIES BETWEEN 2**(-1/16) AND 2**(1/16).
C
C       DETERMINE A, B AND XPON=(4P-Q-B/16)*16
C
C*******
   40 NB=1
      IF(X-B(9)) 45, 45, 50
   45 NB=9
   50 IF(X-B(NB+4)) 55, 55, 60
   55 NB=NB+4
   60 IF(X-B(NB+2)) 65, 65, 70
   65 NB=NB+2
   70 XPON=(4*NP-NQ)*16-NB
C*******
C
C       COMPUTE 2Z=(M-A)/(M/2+A/2).
C
C       SUBTRACT A IN TWO STEPS TO PRESERVE ACCURACY.  ARRAY B CONTAINS
C       MOST SIGNIFICANT PART OF ELEMENTS OF A.  ARRAY A ARE CORRECTIONS.
C       HANDLING OF M/2 AND A/2 DICTATED BY HEX FLOATING POINT FOIBLES.
C
C*******
      NA=NB/2
      Z2=((X-B(NB+1))-A(NA+1))/(X*.5D0+B(NB+1)/2.0D0)
C*******
C
C       COMPUTE LOGBASE2((1+Z)/(1-Z)) USING TAYLOR-S SERIES. NOTE THAT
C       ABS(Z) .LE. .022.  INCREASED EFFICIENCY IS POSSIBLE IF ONE
C       USES A BEST-FIT POLYNOMIAL, INSTEAD.  HERE WE USE
C          2Z*LOG2(E)*SUM(I=0,4)((2Z)**(2I)/((2I+1)*2**(2I))).
C       THE TERM FOR I=0 IS TREATED SEPARATELY TO PRESERVE
C       SIGNIFICANCE.
C
C       PUT RESULT IN REDUCED FORM
C*******
      ZSQ=Z2*Z2
      Z=(((ZSQ/2304.0D0+1.0D0/448.0D0)*ZSQ+1.0D0/ 80.0D0)*ZSQ
     1  +1.0D0/12.0D0)*ZSQ*Z2
      Z=Z+Z*LOG2E
      NZ=SIXTEN*Z
      TEMP=NZ
      TEMP=TEMP*0625
      Z=Z-TEMP
      XPON=XPON*0625+TEMP
C*******
C
C       FORM 2Z*LOG2(E) AS 2Z(LOG2(E)-1)+2Z
C       TO PRESERVE SIGNIFICANCE.
C
C*******
      ZZ=Z2*LOG2E
      NZ=SIXTEN*ZZ
      TEMP=NZ
      TEMP=TEMP*0625
      ZZ=ZZ-TEMP
      XPON=XPON+TEMP
```

FIG. 1. (cont.)

```
      Z=Z+ZZ
      NZ=Z2*SIXTEN
      TEMP=NZ
      TEMP=TEMP*0625
      Z2=Z2-TEMP
      Z=Z+Z2
      XPON=XPON+TEMP
C******
C
C     COMPUTE W=Y*LOG2(X) IN REDUCED FORM
C
C******
      TEMP=Z*YY+XPON*Y
      NZ=TEMP*SIXTEN
      ZN=NZ
      ZN=ZN*0625
      W1=ZN
      W2=TEMP-ZN
      TEMP=XPON*YN
      NZ=TEMP*SIXTEN
      ZN=NZ
      ZN=ZN*0625
      W1=W1+ZN
      W2=(TEMP-ZN)+W2
      NZ=W2*SIXTEN
      ZN=NZ
      ZN=ZN*0625
      XPON=W1+ZN
      W2=W2-ZN
C******
C
C     CHECK TO SEE IF EXPONENT IS IN BOUNDS
C
C******
      IF(XPON-XMAX) 105,100,100
  100 POW=INF
      RETURN
  105 IF(XPON-XMIN) 110,115,115
  110 POW=0.0D0
      RETURN
C******
C
C     ARGUMENT FOR EXPONENTIAL BASE 2 IS NOW IN FORM
C         4P1+Q1+B1/16+W2 = XPON + W2
C     WHERE ABS(W2) .LE. 1/16.  FORCE W2 TO BE NEGATIVE AND ADD
C     1 TO B1, IF NECESSARY, SINCE COMPUTATION OF 2**(-W2) IS MORE
C     ACCURATE THAN COMPUTATION OF 2**(W2) IN HEX FLOATING POINT.
C
C******
  115 IF(W2) 120,120, 117
  117 W2=W2-0625
      XPON=XPON+0625
C******
C
C     DETERMINE P1,Q1,B1
C
C******
  120 NZ=XPON*SIXTEN
      I=1
      IF(NZ .LT. 0) I=0
      N=NZ/16+I
      NB=16*N-NZ
      NP=N/4+I
      NQ=4*NP-N
C******
C
C     COMPUTE 2**(-W2) AS POLYNOMIAL IN W2 MINUS THE CONSTANT TERM 1.
C
C******
      TEMP=((((((D6*W2+D5)*W2+D4)*W2+D3)*W2+D2)*W2+D1)*W2
```

FIG. 1. (cont.)

```
C ••••••
C
C       MULTIPLY BY 2**(-B1/16) AND BY 2**(-Q1).   FIX UP CONSTANT TERM 1.
C
C ••••••
        TEMP=TEMP*B(NB+1)+B(NB+1)
        TEMP=TEMP*TWOMQ(NQ+1)
C ••••••
C
C       ADD P1 TO BIASED EXPONENT OF RESULT AND EXIT.
C
C ••••••
        NT(1)=NT(1)+NP*NSHIFT
        POW=TEMP
        RETURN
        END
```

FIG. 1. (cont.)

III. An Assembler Language Program (Fig. 2)

This program attempts to carry out the principle described in Sect. I with two hexadecimal guard digits beyond the double precision of the S/360 computers. In order to economize storage, this program combines the following five functions:

(1) real-to-real exponentiation,
(2) base-e exponential function,
(3) base-2 exponential function,
(4) base-e logarithm function,
(5) base-10 logarithm function.

Besides the use of guard digits, final rounding is attempted for exponentiation functions. Therefore, virtually full accuracy is obtained for computation of the three exponentiation functions except for marginal cases where the base for the real-to-real exponentiation is very close to 1.0. On the other hand, final multiplications for the two logarithm functions are carried out in straight double precision. Therefore, the last digit of the result for these functions is in doubt.

This consolidation of five functions is not without penalty. The real-to-real exponentiation is slightly faster than the much less accurate version of the standard library. However, the exponential function and the two logarithmic functions are 15 to 25% slower than the stand-alone versions of the standard library, which are more or less of the same accuracy as this version.

The choice of two guard digits is made so that we obtain an accurate result a^b whenever a, b, and a^b are all integral quantities which can be exactly represented in the double precision. Our test shows that if a is an integer less than 1000, and if b is an integer such that a^b is less than 2^{56}, then a^b is computed exactly without any round-off whatsoever. It is expected that this statement should hold true for any integral value of a. In other words, this routine is at least as accurate as the A $**$ I routine in all cases, and is far more accurate than the A $**$ I for many cases.

The main strategy is as follows. Extended accuracy requires two things: a more accurate formula and a more accurate arithmetic. Of these the former is quite easy to implement. An increase of one degree in the approximation formula goes quite far in increasing the accuracy. On the other hand, an increase in precision, in the absence of a backup hardware precision, requires a simulated arithmetic which is in general quite expensive. Therefore, we should limit use of the simulated arithmetic. This can be done if we reduce the given argument to a very narrow basic range so that the result can be written as $b_n + g(r)$, where b_n is a constant in an extra precision which depends only on the reduction index n, and $g(r)$ is a small perturbation part in the working precision which is a function of the reduced argument r.

Write $x = 16^p \cdot 2^{-q} \cdot m$, where $0.5 \leq m < 1$. Choose an appropriately distributed set of base points $\{\alpha_n\}$ in $[0.5, 1]$, and for the given m choose the closest α_n. Then letting $z = (m - \alpha_n)/(m + \alpha_n)$, we have

$$\log_2 (x) = 4p - q + \log_2 \alpha_n + \log_2((1 + z)/(1 - z))$$

Here $4p - q + \log_2 \alpha_n$ is the dominant part, and $\log_2((1 + z)/(1 - z))$ is the perturbation part. Since z is small, it is sufficient to compute the latter in the working precision. $\log_2 \alpha_n$ must be supplied as a table of extra-precision constants.

We now face a problem. In order to improve accuracy, z must be very small. This means we must have many base points α_n. In order to limit the perturbation part to $\pm 16^{-2}$, one needs at least 129 base points α_n and 129 extra-precision constants $\log_2 \alpha_n$. Moreover, search for the closest base point becomes time consuming. This problem is solved as follows.

Estimate $\log_2 m$ very crudely, and obtain three indices $0 \leq j \leq 8$, $0 \leq k \leq 3$, and $0 \leq l \leq 4$ so that $20j + 5k + l$ is the nearest integer to $-160 \log_2 m$. Using these indices, select three constants β_j, γ_k, and δ_l, where $\beta_j = [2^{-j/8}]$, $\gamma_k = [2^{-k/32}]$, and $\delta_l = [2^{-l/160}]$. Here the brackets denote rounding to their closest 17 bits quantity. These 18 constants β_j, γ_k, δ_l and their logarithms (in 20 hexadecimal digits) are encoded in the program. Then the economically computable product $\alpha_{jkl} = \beta_j \gamma_k \delta_l$ is taken to be the base point for m. This is sufficiently close to m. In fact, $\log_2((1 + z)/(1 - z))$ is not much bigger than $1/320$ in magnitude. The 18 constants β_j, γ_k, and δ_l thus are used to generate 161 base points α_{jkl}. Extra precision is maintained as follows: The integer $4p - q$ and the high-order part of $\log_2 \alpha_{jkl}$ $(= \log_2 \beta_j + \log_2 \gamma_k + \log_2 \delta_l)$ comprise the dominant part of $\log_2 x$ and are kept in single precision. The sum of the low-order part of

$\log_2 \alpha_{jkl}$ and $\log_2((1 + z)/(1 - z))$ comprises the perturbation part and is kept in double precision.

For computation of x^y, the multiplication of $\log_2(x)$ by y must be carried out in a simulated extra precision. Having retained extra accuracy in the product $w = y \log_2(x)$, the answer $x^y = 2^w$ is computed with the same care as above. In this second half of the computation, the constants $\{\beta_j\}$ $\{\gamma_k\}$ and their logarithms are utilized again. Aided by these, the fraction part of the answer is developed beyond the working precision, so that an effective final rounding operation can be given.

Accuracy tests involving random samples of 2000 cases each gave the following results:

	(0.5, 2.0)	(0.1, 10.0)	(0.01, 100.0)
$x \in$			
$y \in$	$(-250.0, 250.0)$	$(-70.0, 70.0)$	$(-35.0, 35.0)$
Max. rel. error	1.22 E–16	1.10 E–16	1.05 E–16
RMS rel. error	2.97 E–17	2.83 E–17	2.71 E–17
2 bits short	1	0	0
1 bit short	152	73	35
Exactly rounded	1647	1806	1882
1 bit excess	199	121	83
2 bits excess	1	0	0

Execution time is estimated to be as follows:

	X ** Y	DLOG	DEXP
360/50	1643 μsec	898 μsec	905 μsec
360/65	323	176	175
360/75	175	98	97

```
IHCFCXFC CSECT
         SPACE 2
*        CCMPREHENSIVE EXPCNENTIAL/LOGARITHM ROUTINE (LONG FORM)
*        1.  L(GE(X)   = LOGE(2)*LOG2(X)
*            LCG10(X)  = LOG1C(2)*LOG2(X)
*            EXP(Y)    = 2**(Y*LOC2(E))
*            2**(Y)    = 2**Y
*            X**Y      = 2**(Y*LOC2(X))
*        2.  THUS THESE FUNCTIONS CAN BE OBTAINED BY COMPLTATION CF
*            LCG2(X) AND 2**Z.    IN ORDER TO ATTAIN HIGH ACCLRACY FCR
*            X**Y, LCG2(X) AND Y*LOG2(X) ARE OBTAINED IN HIGHER THAN
*            THE LCNG PRECISION.   THE RESULT OF EXPCNENTIATICN IS
*            RCUNCED.
*        3.  LCG2(X) IS CBTAINED AS FOLLOWS
*            WRITE X = (16**A)*(2**-B)*M   WHER M IS IN (0.5,1)
*            LCG2(X) = 4A-B-LOC2(M)
*            USING APPROPRIATELY CHOSEN TRIPLE CONSTANTS ALPHA(J),
*              BETA(K), AND CAMMA(L), CONSTRLCT A BASE VALLE
*              PHI(J,K,L) WHICH IS LOGARITHMICALLY CLOSEST TO M
*              AMCNG THE 160 PCSSIBLE VALUES OF PHI.
*            LCG2(M) = LOG2(PHI(J,K,L))+LOG2((1+L)/(1-L))
*                        WHERE  U = (M-PHI(J,K,L))/(M+PHI(J,K,L))
*            THE SECONC TERM IS LESS THAN 16**-2
*            THE CCMPUTEC LOC2(X) CONSISTS OF 2 PARTS, 6 HEX DIGITS CF
*            HIGH PART ANC 14 HEX DIGITS OF LOW PART
*        4.  2**Z IS COMPUTEC AS FOLLOWS
*            WRITE Z = 4*A-B-R,  B = 0,1,2,OR 3, R IS IN (C,1)
*            USING SAME POOL OF CONSTANTS AS IN 3., FIND PHI(J,K)
*              WHOSE BASE 2 LOGARITHM IS CLOSEST TC -R, AMCNG THE
*              32 POSSIBLE SUCH VALUES.
*            LET S = -R-LOC2(PHI(J,K)).   /S/ IS LESS THAN 1/64
*            THEN 2**Z = (16**A)*(2**-B)*PHI(J,K)*(2**S)
         SPACE
GRA      ECU    1               ARGUMENT POINTER
GRS      EQU    13              SAVE AREA POINTER
GRR      EQU    14              RETURN REGISTER
GRL      EQU    15              LINK REGISTER
GRX      EQU    11              SET NON C FOR X**Y ENTRY
GRO      EQU    0               SCRATCH REGISTERS
GR1      ECU    1
GR2      EQU    2
GR3      EQU    3
GR4      ECU    4
GR5      EQU    5
GR6      EQU    6
GR7      ECU    7
GR8      EQU    8
GR9      ECU    9
GR10     EQU    1C
GR12     EQU    12
FRO      EQU    0               ANSWER REGISTER
FR2      EQU    2               SCRATCH FLOAT POINT REGISTERS
FR4      EQU    4
FR6      ECU    6
         SPACE 2
         ENTRY CEXP             EXPONENTIAL BASE E
         ENTRY FCXP2=           EXPONENTIAL BASE 2
         ENTRY FCXPC=           POWER ROUTINE  X**Y
         ENTRY CLCC1O           LOGARITHM BASE 1C
         ENTRY CLCC             LOGARITHM BASE E
```

FIG. 2. S/360 assembler code for comprehensive exponential/logarithm.

```
          SPACE
          USING *,GRL
DEXP      B      PRLOG1             ENTRY FOR E**Y
          DC     AL1(4)
          DC     CL5'DEXP'
PRCLCG1   STM    GRR,GR12,12(GRS)   SAVE REGISTERS AND
          L      GR2,0(GRA)             LET GR2 POINT TC ARGLMENT
          SR     GRX,GRX
          LA     GRL,BASE           ADJUST BASE REGISTER TO COMMON BASE
          USING  EASE,GRL
          LD     FR4,0(GR2)         ARGUMENT Y IN FR4
          LD     FR0,LOG2EH         HIGH ORDER 6 DIGITS CF LCG(E) BASE 2
          LD     FR2,LOG2EL         LOW ORDER 14 DIGITS CF LOG(E) BASE 2
          B      EXP1               MERGE WITH MAIN EXPONENTIAL SECTICN
          SPACE
          USING  *,GRL
FDXP2=    B      PRCLOG2            ENTRY FOR 2**Y
          DC     AL1(6)
          DC     CL7'FDXP2='
PRCLCG2   STM    GRR,GR12,12(GRS)   SAVE REGISTERS AND
          L      GR2,0(GRA)             LET GR2 PCINT TC ARGLMENT
          SR     GRX,GRX
          LA     GRL,BASE           ADJUST BASE REGISTER TO CCMMCN BASE
          USING  EASE,GRL
          LD     FR0,0(GR2)         ARGUMENT Z IN FRC
          SDR    FR2,FR2            CLEAR FR2 WHICH IS EXTENSICN CF FR0
          B      EXP2               MERGE WITH MAIN EXPONENTIAL SECTICN
          SPACE
          USING  *,CRL
FDXPD=    B      PRCLOG3            ENTRY FOR X**Y
          DC     AL1(6)
          DC     CL7'FDXPD='
PRCLCG3   STM    GRR,GR12,12(GRS)   SAVE REGISTERS AND LET
          LM     GR2,GR3,0(GRA)        GR2, GR3 POINT TO ARGUMENTS
          LA     GRL,BASE           ADJUST BASE REGISTER TO CCMMCN BASE
          USING  EASE,GRL
          BALR   GRX,0              SET GRX TO 0 FOR THIS CASE
          B      LCG1               MERGE WITH MAIN LOGARITHMIC SECTICN
          SPACE
          USING  *,GRL
DLCG10    B      PRCLOG4            ENTRY FOR COMMCN LOGARITHM
          DC     AL1(6)
          DC     CL7'DLOC10'
PRCLCG4   STM    GRR,GR12,12(GRS)   SAVE REGISTERS AND LET
          LA     GR10,LOCT2            GR1C POINT TO LCG(2) BASE 10
          BAL    GRL,LOGO           ADJUST BASE REGISTER AND MERGE WITH LCG
          SPACE
          USING  *,GRL
BASE      EQU    *
DLCG      B      PRCLOG5            ENTRY FOR NATURAL LOGARITHM
          DC     AL1(4)
          DC     CL5'DLOC'
PRCLCG5   STM    GRR,GR12,12(GRS)   SAVE REGISTERS AND LET
          LA     GR10,LOCE2            GR1C POINT TO LCG(2) BASE E
          SPACE
LOGO      L      GR2,0(GRA)         COMMON PROLOG FOR DLCG, DLOG1C.
          SR     GRX,GRX
          SPACE  2
LCG1      LM     GR4,GR5,0(GR2)     MAIN LOG SECTION.  PICK UP X IN GR4-GR5
          LTR    GR0,GR4            HIGH ORDER X TO GRC
          BC     12,LOGER           IF X 0 OR NEGATIVE, CC TO LOG ERRCR
```

FIG. 2. (cont.)

```
SPACE
SRDL    GR0,24          DECOMPOSE X = (16**A)*(2**-B)*M
SLL     GR0,14          4*(A+64)*16**3 IN GR0
SRL     GR1,29          1ST 3 BITS OF HEX MANTISSA IN GR1
IC      GR1,LEADO(GR1)  B = = OF LEAD C OF HEX MANTISSA IN GR1
SLDL    GR4,0(GR1)      NORMALIZE X IN BINARY FASHION
STM     GR4,GR5,BUFF    TO OBTAIN BINARY MANTISSA M
MVI     BUFF,X'40'      AND FLOAT IT IN BUFF
SPACE
LM      GR5,GR9,K5      PRELOAD REGS WITH CONSTANTS, AND
LA      GR4,0(GR4)      COMPUTE -32*LOG2(M)+0.1 ROUGHLY AS
AR      GR7,GR4         D0*(D1-M+D2/(M+D3))
DR      GR6,GR7         GR4 HAS M IN B7, GR7 HAS D3 IN B7
AR      GR8,GR7         GR6 HAS D2 IN B15, GR8 HAS D1 IN B7
SR      GR8,GR4         GR9 HAS D0 IN B23
MR      GR8,GR8         RESULT IS IN GR8-GR9 IN B31
LR      GR6,GR8         DECOMPOSE THIS AS 4*J+K+L/5 TO GET
SRDL    GR6,2           INDICES J,K, L SO THAT LOG2(M) IS
SRL     GR7,27          WITHIN 1/3CC OF -J/8-K/32-L/16C
SLL     GR6,3           8*K IN GR7, 8*J IN GR6
SRL     GR9,1
MR      GR8,GR5         GR5 HAS 5 IN B3C
SLL     GR8,3           8*L IN GR8
SPACE
SDR     FR0,FR0         OBTAIN THE ORIGIN PHI(J,K,L) BY
LD      FR6,BUFF        MULTIPLYING 3 17 BITS QUANTITIES
LE      FR2,ALPHA(GR6)  ALPHA(J) APPROX 2**-J/8
ME      FR2,BETA(GR7)   BETA(K) APPROX 2**-K/32 AND
LE      FR0,GAMMA(GR8)  GAMMA(L) APPROX 2**-L/160
MDR     FR2,FR0         PHI(J,K,L) IN FR2, M IN FR6
SPACE
HDR     FR4,FR6         COMPUTE REDUCED ARG W = 2*U/LOGE(2)
SDR     FR6,FR2         WHERE U = (M-PHI)/(M+PHI)
HDR     FR2,FR2
ADR     FR4,FR2         NAMELY
MD      FR4,LOGE2       W = (M-PHI)/((C.5M+C.5PHI)*LOGE(2))
DDR     FR6,FR4
SPACE
LDR     FR4,FR6         COMPUTE LOG2((1+L)/(1-L)) = LCG2(M/PHI)
MDR     FR4,FR4         AS,
LER     FR2,FR4
ME      FR2,E2          W + E1*W**3 + E2*W**5
AD      FR2,F1
MDR     FR2,FR4         THE SUM IS LESS THAN 1/3CC
MDR     FR2,FR6         DEFER ADDITION OF W FOR BETTER ACCUR.
SPACE
LE      FR0,AH(GR6)     COMBINE HIGH PART OF LOG2(ALPHA(J)),
AE      FR0,BH(GR7)     LOG2(BETA(K)), LOG2(GAMMA(L)) TO
AE      FR0,CH(GR8)     GET HIGH PART OF LOG2(PHI(J,K,L))
AD      FR2,AL(GR6)     COMBINE LOW PARTS OF LOG2(PHI(J,K,L))
AD      FR2,BL(GR7)     WITH LOG2((1+U)/(1-L)) TO FORM THE
AD      FR2,CL(GR8)     LOW PART OF LOG2(M)
ADR     FR2,FR6         NOW ADD IN W
SPACE
LA      GR1,256(GR1)    ADD IN 4*A-B TO HIGH PART LOG2(M)
SLL     GR1,12          (B+4*64)*16**3 IN GR0
ST      GR1,BUFF
MVI     BUFF,X'43'      FLOAT B+4*64
SE      FR0,BUFF        SUBTRACT IT FROM HIGH PART LOG2(M)
ST      GR0,BUFF        (4*A+4*64)*16**3 IN GR1
MVI     BUFF,X'43'      FLOAT 4*A+4*64
```

FIG. 2. (cont.)

```
        AE      FR0,BUFF            ADD IT TO HIGH PART LOG2(M)
        SPACE
        LTR     GRX,GRX            IF X**Y ENTRY CONTINUE ON
        BNZ     EXP0               TO EXPONENTIAL SECTION
        SPACE
        ADR     FR0,FR2            IF DLOG OR DLOG1C, COMBINE TWO PARTS
        MD      FR0,0(GR10)        OF LOG2(X), AND MULTIPLY BY LOGE(2)
        B       RETURN             OR LOG1C(2).
        SPACE 2
EXP0    LD      FR4,0(GR3)         FOR X**Y, LOAD Y IN FR4
        SPACE
EXP1    LPER    FR6,FR4            E**Y AND X**Y MERGE HERE
        CE      FR6,MAX1           IF /Y/ GREATER THAN 16**17, THERE IS
        BH      EXPER              A DANGER OF OVERFLOW/UNDERFLOW
        CE      FR6,MIN1           IF /Y/ LESS THAN 16**-17, GIVE ANS = 1
        BL      ANSONE             TO AVOID INTERMEDIATE UNDERFLOW
        MDR     FR2,FR4            OBTAIN PRODUCT Z=Y*LOG2(X) OR LOG2(E)
        SDR     FR6,FR6            FR2 HAS Y*(LOW PART LOG)
        LER     FR6,FR4            HIGH PART Y IN FR6
        SDR     FR4,FR6            LOW PART Y IN FR4
        MDR     FR4,FR0            (LOW Y) * (HIGH LOG)
        MER     FR0,FR6            (HIGH Y) * (HIGH LOG) = HIGH Z IN FR0
        ADR     FR2,FR4            LOW PART OF Z IN FR2
        SPACE
EXP2    SR      GP8,GR8            COMMON EXPONENTIAL SECTION STARTS
        SDR     FR6,FR6            ADD TWO PARTS OF Z TENTATIVELY.
        LDR     FR4,FR0             IF Z IS NON POSITIVE,
        ADR     FR4,FR2             SET GRE=C, AND FR6=C
        BNH     SKIP1              IF Z IS POSITIVE,
        BCTR    GR8,0               SET GRE=ALL F, AND FR6 = 1
        LE      FR6,ONE
SKIP1   CE      FR4,MAX            IF Z IS GREATER THAN OR EQUAL
        BH      OVFLO              TO 252, THE RESULT IS OVERFLOW
        CE      FR4,MIN            IF Z IS LESS THAN -260,
        BL      UNDFLO             THE RESULT UNDERFLOWS
        SPACE
        AW      FR4,SCALER         DECOMPOSE Z AS 4A-B-R, A,B INTEGERS
        STD     FR4,BUFF           SAVE Z IN FIXED POINT B31
        AER     FR6,FR6            SAVE INTEGER PART OF Z (+1 IF Z POS)
        LM      GR6,GR7,BUFF       PICK UP Z IN GR6, GR7 FIXED POINT
        XR      GR6,GR8            IF Z IS POSITIVE, TAKE 1'S COMPLE-
        XR      GR7,GR8             MENT OF BOTH PARTS OF FIX POINT Z
        SRL     GR7,23             NOW GR6 HAS -4A+B, GR7 HAS R APPROX
        LA      GR4,8(GR7)         16*(32*R+C.5) IN GR4
        SRDL    GR6,2
        SRL     GR7,30             B IN GR7, VALUE = C,1,2,OR 3
        SLL     GR6,24             -A IN GR6, B7 AS CHAR ADJUSTER
        SRL     GR4,4              32*R ROUNDED
        LR      GR5,GR7            CLEAR GR5
        SRDL    GR4,2              DECOMPOSE ROUNDED 32*R AS 4J+K
        SRL     GR5,27             8*K IN GR5   R IS APPROXIMATELY
        SLL     GR4,3              8*J IN GR4     J/8 + K/32
        SPACE
        AE      FR6,AH(GR4)        REDUCE Z AS 4A-B+LOG2(PHI(J,K))+S
        AE      FR6,BH(GR5)        PHI(J,K)=ALPHA(J)*BETA(K), AND LOG2
        SDR     FR0,FR6            OF THIS IS APPROXIMATELY -J/8-K/32
        ADR     FR0,FR2            ADD HIGH PART OF THIS TO 4A-B,
        SD      FR0,AL(GR4)         SUBTRACT IT FROM HIGH Z, EASE IN
        SD      FR0,BL(GR5)         LOW Z, AND SUBTRACT LOW LOG2(PHI)
        SPACE
        LDR     FR2,FR0            COMPUTE T=2**S-1.   /S/ IS LE 1/64
```

FIG. 2. (cont.)

```
          MDR    FR2,FR2          S*S
          LD     FR4,D1
          MDR    FR4,FR2
          AD     FR4,D0
          MDR    FR0,FR4          S*(D0+D1*S*S)     DENOTE THIS BY P
          AD     FR2,D2
          SDR    FR2,FR0          D2+S*S-S*(D0+D1*S*S) DENOTE THIS BY C
          ADR    FR0,FR0
          DDR    FR0,FR2          T = 2*P/Q
          SPACE
          LE     FR2,ALPHA(GR4)   COMPUTE PHI(J,K) = ALPHA(J)*BETA(K)
          ME     FR2,BETA(GR5)     EACH COMPONENT IS EXACTLY 17 BITS
          LTR    GR7,GR7            LONG
          BZ     SKIP2
HALVE     HDR    FR2,FR2          HALVE PHI(J,K) B TIMES
          BCT    GR7,HALVE
          SPACE
SKIP2     MDR    FR0,FR2          ASSEMBLE 2**(-B-R) AND ROUND AS
          AD     FR0,ROUND         (2**-B)*PHI + ((2**-B)*PHI*T+2**-57)
          HDR    FR2,FR2
          ADR    FR0,FR2          ADD (2**-B)*PHI IN 2 STEPS
          ADR    FR0,FR2           TO EFFECT GOOD ROUNDING
          SPACE
          STE    FR0,BUFF         ADD A TO CHAR OF THE RESULT
          LCR    GR6,GR6           TO OBTAIN THE ANSWER
          A      GR6,BUFF
          ST     GR6,BUFF
          LE     FR0,BUFF
          SPACE
RETURN    LM     GR8,GR12,12(GR8)    RESTORE REGISTERS AND
          MVI    12(GR8),X'FF'       RETURN TO CALLING PROGRAM
          BR     GR8
          SPACE 2
ANSONE    LD     FR0,ONE          CASE WHEN /Z/ IS VERY SMALL
          B      RETURN            GIVE 1.0 AS Z**Z
          SPACE
UNDFLO    LE     FR0,BOMB         CASE WHEN Z IS NEGATIVE AND BIG
          MER    FR0,FR0           GIVE 0 AS ANSWER VIA UNDERFLOW
          B      RETURN             WARNING
          SPACE
ANSZERO   SDR    FR0,FR0          CASE WHEN POSITIVE POWER OF C
          B      RETURN            WAS ASKED.  GIVE C QUIETLY
          SPACE 2
LOGER     LTR    GRX,GRX          X WAS C OR NEGATIVE FOR LOG OR X**Y
          BZ     ERROR            IF LOG(X) OR LOG10(X), ERROR CASE
          LD     FR0,0(GR2)       X**Y CASE.  LOAD X
          LD     FR2,0(GR3)                   LOAD Y
          LTER   FR0,FR0          TEST X
          BM     PWRER1
          LTER   FR2,FR2          CASE X=0
          BP     ANSZERO          IF Y POSITIVE, ANSWER =C
          B      ERROR            IF Y 0 OR NEGATIVE, ERROR CASE
PWRER1    LTER   FR2,FR2          CASE X NEGATIVE
          BZ     ANSONE           IF Y IS C, ANSWER = 1
          B      ERROR            IF Y IS NOT C, ERROR CASE
          SPACE
EXPER     AER    FR0,FR2          /Y/ WAS VERY LARGE FOR EXP(Y) OR X**Y
          BZ     ANSONE           IF X**Y AND LOG2(X)=C, X**Y = 1.0
          BM     EXPER1
          LTER   FR4,FR4          IF EXP(Y) OR X**Y AND LOG2(X) POSITIVE
          BP     OVFLO            AND IF Y IS POSITIVE, OVERFLOW
```

Fig. 2. (cont.)

414   N. W. CLARK, W. J. CODY, AND H. KUKI

```
        B     UNCFLO         ANC IF Y IS NEGATIVE, UNDERFLOW
EXPER1  LTER  FR4,FR4        IF X**Y ANC LOG2(X) NEGATIVE
        BP    UNCFLO         AND IF Y IS POSITIVE, UNDERFLCW
        SPACE
CVFLC   EQU   *              EXP(Y), 2**Y OR X**Y CVERFLCW CASES
        SPACE
ERRCR   ABEND 99
        SPACE 2
K5      DC    F'10'          5   B 3C       THESE 5 CONSTANTS
        DC    X'00037038'    3.43835  B15   MUST BE LCCATED
        DC    X'005A2616'    .352144  B7    CCNSECUTIVE
        DC    X'FE7693F1'    -1.53681 B7
        DC    X'00001010'    16.C638  B23
MAX1    DC    X'52100CC0'    16**17
MIN1    DC    X'30100CC0'    16**-17
MAX     DC    X'42FBFFFE'    A LITTLE SHCRT CF 252
MIN     DC    X'C3103FFF'    USED AS TEST FOR -26C
E2      DC    X'3EBD19B8'    E2=+.2885444CE-2
        DS    0C
CNE     EQU   *
ALPHA   DC    X'41100C00'    2**-0/8  ROUNDED TO 17 BITS
AH      DC    X'0C000CC0'    HIGH PART OF LCG2(ALPHA(C))
        DC    X'40EAC1C0'    2**-1/8  ROUNDED TC 17 BITS
        DC    X'C0200CC0'    HIGH PART OF LCG2(ALPHA(1))
        DC    X'40D745C0'    2**-2/8  ROUNDED TO 17 BITS
        DC    X'C0400C00'    HIGH PART OF LCG2(ALPHA(2))
        DC    X'40C567C0'    2**-3/8  ROUNDED TO 17 BITS
        DC    X'CC600000'    HIGH PART OF LCG2(ALPHA(3))
        DC    X'40B50500'    2**-4/8  ROUNDED TO 17 BITS
        DC    X'C0800000'    HIGH PART OF LCG2(ALPHA(4))
        DC    X'40A5FFCC'    2**-5/8  ROUNDED TC 17 BITS
        DC    X'CCA00000'    HIGH PART OF LCG2(ALPHA(5))
        DC    X'40983800'    2**-6/8  ROUNDED TO 17 BITS
        DC    X'C0C00C00'    HIGH PART OF LCG2(ALPHA(6))
        DC    X'408B9600'    2**-7/8  ROUNDED TC 17 BITS
        DC    X'C0E00000'    HIGH PART OF LCG2(ALPHA(7))
        DC    X'40800000'    2**-8/8  ROUNDED TO 17 BITS
        DC    X'C1100CC0'    HIGH PART OF LCG2(ALPHA(8))
BETA    DC    X'41100CC0'    2**-0/32  RCUNDED TC 17 BITS
BH      DC    X'0C000CC0'    HIGH PART OF LCG2(BETA(0))
        DC    X'40FA8380'    2**-1/32  RCUNDED TO 17 BITS
        DC    X'BF800C00'    HIGH PART CF LCG2(BETA(1))
        DC    X'40F52580'    2**-2/32  RCUNDED TC 17 BITS
        DC    X'C0100000'    HIGH PART OF LCG2(BETA(2))
        DC    X'40EFE480'    2**-3/32  ROUNDED TO  7 BITS
        DC    X'C0180C00'    HIGH PART OF LCG2(BETA(3))
GAMMA   DC    X'41100C00'    2**-0/16C  ROUNDED TO 17 BITS
CH      DC    X'0C000000'    HIGH PART OF LCG2(GAMMA(0))
        DC    X'40FEE480'    2**-1/16C  ROUNDED TO 17 BITS
        DC    X'BF1A0000'    HIGH PART OF LCG2(GAMMA(1))
        DC    X'40FCCA80'    2**-2/16C  RCUNDED TC 17 BITS
        DC    X'BF330C00'    HIGH PART OF LCG2(GAMMA(2))
        DC    X'40FCB200'    2**-3/16C  ROUNCED TO 17 BITS
        DC    X'BF4D0CC0'    HIGH PART OF LCG2(GAMMA(3))
        DC    X'40FB9AC0'    2**-4/16C  ROUNCED TC 17 BITS
        DC    X'BF660000'    HIGH PART OF LCG2(GAMMA(4))
AL      DC    X'0000000000000C00'  LOW PART CF LCG2(ALPHA(C))
        DC    X'3C59C31659E93B12'  LOW PART CF LCG2(ALPHA(1))
        DC    X'3B580C63130ABC3C'  LOW PART CF LCG2(ALPHA(2))
        DC    X'BC4EB4C99C573C16'  LOW PART CF LCG2(ALPHA(3))
        DC    X'3C1A1BFFB6CA5B85'  LOW PART CF LCG2(ALPHA(4))
```

FIG 2. (cont.)

```
         DC    X'3C5BF8F4075CC1A6'    LOW PART CF LCG2(ALPHA(5))
         DC    X'3C260C4CF15E2FBC'    LOW PART CF LCG2(ALPHA(6))
         DC    X'3CA4561F0F541E9C'    LOW PART CF LCG2(ALPHA(7))
BL       DC    X'00000C0000000C00'    LOW PART CF LCG2(BETA(C)), AL(8)
         DC    X'BC4AFA6A7C398CFC'    LOW PART CF LCG2(BETA(1))
         DC    X'3B46531122CDCEC1C'   LOW PART CF LCG2(BETA(2))
         DC    X'BC58B13BEA649752'    LOW PART CF LCG2(BETA(3))
CL       DC    X'0C000C00000000C00'   LOW PART CF LCG2(GAMMA(C))
         DC    X'3C61BC6E675593B2'    LOW PART CF LCG2(GAMMA(1))
         DC    X'BC36262EC1B1EA70'    LOW PART CF LCG2(GAMMA(2))
         DC    X'3C388A277E052468'    LOW PART CF LCG2(GAMMA(3))
         DC    X'BC697885A2A41SC2'    LOW PART CF LCG2(GAMMA(4))
LCG2EH   DC    X'411715470C0C0C00'    HIGH PART LCG2(E)
LCG2FL   DC    X'3B652B82FE1777C1'    LOW PART  LCG2(E)
LCGT2    DC    X'404D104C427CE7FC'    LOG10(2)
LCGE2    DC    X'40B17217F7C1CF7A'    LOGE(2)
LEADO    DC    X'030201010C0CCC00'    TABLE OF LEADING ZERC'S
E1       DC    X'3FA3FE9FFC63ACCE'    E1=+.4CC3775115982529E-1
DO       DC    X'41736A66C3098257'    DC=+.7213476S672C4S02E+1
D1       DC    X'3FEC980C031C3342'    D1=+.5776215447C12637E-1
D2       DC    X'4214CC4E4F088F09'    D2=+.2CE13ES4ES63C655E+2
SCALER   DC    X'46000C000C0C0C00'
RCUNC    DC    X'3280000000000C00'
BUFF     DS    2C
BCME     EQU   C1+4
         END
```

FIG. 2. (cont.)

REFERENCE

Clark, N. W., and Cody, W. J. (1969). Self-Contained Exponentiation. *AFIPS Conf. Proc.* **35,** 701–706.

CHAPTER 7

CADRE: AN ALGORITHM
FOR NUMERICAL QUADRATURE

Carl de Boor

PURDUE UNIVERSITY

LAFAYETTE, INDIANA

I. Introduction

The program CADRE attempts to solve the following problem: Given the name F of a real function subprogram, two real numbers A and B, and two nonnegative numbers AERR and RERR, find a number CADRE such that

$$\left| \int_A^B F(x)\, dx - \text{CADRE} \right| \leq \max\left(\text{AERR}, \text{RERR} * \left| \int_A^B F(x)\, dx \right| \right)$$

For this, the program employs an adaptive scheme, whereby CADRE is found as the sum of estimates for the integral of $F(x)$ over suitably small subintervals of the given interval of integration. Starting with the interval of integration itself as the first such subinterval, the program attempts to find an acceptable estimate on a given subinterval by cautious Romberg extrapolation (see Sect. II). If this attempt fails, the subinterval is divided into two subintervals of equal length, each of which is now considered separately. For the sake of economy, values of $F(x)$, once calculated, are saved until they are successfully used in estimating the integral over some subinterval to which they belong.

II. Mathematical Analysis

Let BEG and END be the two end points of a subinterval of the interval of integration, of length

$$\text{STEP} = \text{END} - \text{BEG}$$

417

We assume, for the convenience of the discussion only, that

$$BEG < END$$

In order to estimate

$$I = \int_{BEG}^{END} F(x)\,dx$$

the program uses the composite trapezoid sum. For certain classes of integrands, the composite trapezoid sum exhibits a known and characteristic convergence behavior which the algorithm attempts to detect and then to exploit through cautious extrapolation.

A. CAUTIOUS EXTRAPOLATION

Let $S(h)$ be a function defined for positive h which satisfies

$$I = S(h) + A_1 h^{\gamma_1} + \cdots + A_k h^{\gamma_k} + o(h^{\gamma_k}) \qquad (1)$$

where $A_1, \ldots, A_k$ are fixed constants and

$$0 < \gamma_1 < \cdots < \gamma_k$$

In order to estimate

$$I = S(0+) = \lim_{h \to 0} S(h)$$

one customarily evaluates $S(h)$ for various positive values of h and then *extrapolates* from these values to $h = 0$. This amounts to combining the calculated values of $S(h)$ in such a way that the first few of the error terms $A_i h^{\gamma_i}$ in (1) drop out. Since, for example,

$$0 = S(h) - S(2h) + \sum_{i=1}^{k} A_i h^{\gamma_i}(1 - 2^{\gamma_i}) + o(h^{\gamma_k})$$

one has, for any $r \neq 1$,

$$I = S(h) + (S(h) - S(2h))/(r-1) + \sum_{i=1}^{k} A_i h^{\gamma_i}(r - 2^{\gamma_i})/(r-1) + o(h^{\gamma_k})$$

Hence, with the particular choice $r = 2^{\gamma_1}$, the function

$$S(h; r) \underset{d}{=} S(h) + (S(h) - S(2h))/(r-1)$$

satisfies

$$I = S(h; r) + \sum_{i=2}^{k} A_i' h^{\gamma_i} + o(h^{\gamma_k}),$$

with

$$A_i' = A_i(r - 2^{\gamma_i})/(r-1)$$

Since $\gamma_1 < \gamma_2 < \cdots$, the estimate $S(h; r)$ is then clearly of *higher order* than the estimate $S(h)$. The well-known principle that "higher order is better" therefore implies that $S(h;r)$ is a better estimate for I than is $S(h)$.

Once this is accepted, one goes, of course, all the way. Choosing

$$r_i = 2^{\gamma_i}, \qquad i = 1, \ldots, k$$

one calculates

$$S(h;r_1, \ldots, r_j) \underset{d}{=} S(h;r_1, \ldots, r_{j-1}) + \frac{S(h;r_1, \ldots, r_{j-1}) - S(2h;r_1, \ldots, r_{j-1})}{r_j - 1}$$

and finds to one's satisfaction that

$$I = S(h;r_1, \ldots, r_j) + \sum_{i=j+1}^{k} A_i^{(j)} h^{\gamma_i} + o(h^{\gamma_k})$$

with

$$A_i^{(j)} = A_i \prod_{n=1}^{j} \frac{r_n - 2^{\gamma_i}}{r_n - 1}$$

Unfortunately, this satisfaction lasts only until one uses extrapolation in actual calculation. For, it then becomes painfully clear that $S(h;r_1)$ is a better estimate for I than is $S(h)$ only if the eliminated error term $A_1 h^{\gamma_1}$ accounts for most of the discrepancy between $S(h)$ and I. If this is not so, then the increase in the coefficients of the higher-order terms may actually make $S(h;r_1)$ a worse estimate for I than is $S(h)$. Also, amazing things may go on behind the "o."

An insurance of sorts against such unpleasantries has been proposed by Lynch (1967): If the "dominant" error term $A_j^{(j-1)} h^{\gamma_j}$ for $S(h;r_1, \ldots, r_{j-1})$ is in fact dominant, i.e., if

$$I \approx S(h;r_1, \ldots, r_{j-1}) + A_j^{(j-1)} h^{\gamma_j}, \qquad \text{for} \quad h \leq 4h_0 \tag{2}$$

then, for $h \leq 2h_0$,

$$S(h;r_1, \ldots, r_{j-1}) - S(2h;r_1, \ldots, r_{j-1}) \approx A_j^{(j-1)} h^{\gamma_j}(2^{\gamma_j} - 1) \tag{3}$$

therefore, for $h \leq h_0$,

$$R_{j-1}(h) \underset{d}{=} \frac{S(2h;r_1, \ldots, r_{j-1}) - S(4h;r_1, \ldots, r_{j-1})}{S(h;r_1, \ldots, r_{j-1}) - S(2h;r_1, \ldots, r_{j-1})} \approx 2^{\gamma_j} \tag{4}$$

The cautious man will accept $S(h;r_1, \ldots, r_j)$ as an estimate for I, which is better than $S(h;r_1, \ldots, r_{j-1})$, only if

$$R_{j-1}(h) \approx 2^{\gamma_j}$$

Further, under this condition, he will put his faith into (2) and therefore believe that

$$|I - S(h;r_1, \ldots, r_j)| \leq |S(h;r_1, \ldots, r_j) - S(h;r_1, \ldots, r_{j-1})|$$
$$= |S(h;r_1, \ldots, r_{j-1}) - S(2h;r_1, \ldots, r_{j-1})|/(r_j - 1)$$

This is the essence of cautious extrapolation.

We have so far assumed that the exponents $\gamma_1, \ldots, \gamma_k$ are known in advance. In such a case, computational use of cautious extrapolation amounts to taking smaller and smaller h until finally

$$R_0(h) = \frac{S(2h) - S(4h)}{S(h) - S(2h)} \approx 2^{\gamma_1}$$

Typically, one might test for

$$| R_0(h) - 2^{\gamma_1} | \leq 2^{\gamma_1}/10$$

and, when this finally occurs, one believes that

$$| I - S(h; 2^{\gamma_1}) | \leq | S(h) - S(2h) |/(2^{\gamma_1} - 1)$$

If this bound on the error in the estimate $S(h; 2^{\gamma_1})$ for I is not small enough, one continues taking smaller h until

$$| R_1(h) - 2^{\gamma_2} | \leq 2^{\gamma_2}/10$$

say, and, when this finally happens, one believes the estimate

$$| I - S(h; 2^{\gamma_1}, 2^{\gamma_2}) | \leq | S(h; 2^{\gamma_1}) - S(2h; 2^{\gamma_1}) |/(2^{\gamma_2} - 1)$$

etc.

If the exponents $\gamma_1, \ldots, \gamma_k$ are not known in advance, one might be tempted to calculate them from various computed values of $S(h)$. This can be done quite efficiently using Peter Wynn's ε-algorithm (1963) if one is willing to *assume* that the o-term in (1) is, in fact, zero. More naively, one might take smaller and smaller h until $R_0(h)$ seems to have converged to one or two places, and then use $r = R_0(h)$ for extrapolation. One verifies quite easily that the resulting extrapolated value

$$S(h; r)$$

coincides with the result of applying Aitken's Δ^2-process to the three numbers $S(4h)$, $S(2h)$, and $S(h)$. Unfortunately, the validity of the bound

$$| I - S(h; r) | \leq | S(h) - S(2h) |/(r - 1) \tag{5}$$

is now more in doubt, as the following lemma shows.

Lemma   Suppose that $S(h)$ satisfies

$$I = S(h) + Ah^{\alpha} + Bh^{\beta}, \qquad \text{all} \quad h > 0$$

where $0 < \alpha, \beta$. If

$$| R_0(h) - 2^{\alpha} | = C$$

then

$$| I - S(h; 2^{\alpha}) | = \frac{| S(h) - S(2h) |}{2^{\alpha} - 1} \frac{C}{2^{\beta} - 1}$$

The difficulty arises when

$$\alpha \gg \beta \quad \text{and} \quad |A| \gg |B|$$

so that the higher-order term is (initially) dominant. In such a case, $R_0(h)$ will be (initially) relatively close to 2^α, so that extrapolation with $r \simeq 2^\alpha$ seems indicated; yet the existence of the (as yet unnoticed) lower-order term spoils the estimate (5). To give a simple example, let

$$I = S(h) + 10^3 h^6 + h^{1/2}$$

and take $h = \frac{1}{4}$. Then

$$R_0(\tfrac{1}{4}) - 2^6 = 0.8 = C$$

so that $R_0(\tfrac{1}{4})$ is relatively close to $2^6 = 64$. But

$$C/(2^\beta - 1) = 2$$

hence the error in $S(\tfrac{1}{4}; 2^6)$ predicted by (5) (with $r = 2^6$) is only half the actual error.

To summarize: When using values of $S(h)$ to determine the unknown exponents $\gamma_1, \dots, \gamma_k$ (as in Aitken's Δ^2-process), it is important to use all available information about the γ_i in order to avoid extrapolation with $R_0(h) \simeq 2^\alpha$ while the error expansion for $S(h)$ still contains h^β terms with $\beta \ll \alpha$.

B. CAUTIOUS EXTRAPOLATION FOR THE COMPOSITE TRAPEZOID SUM

With

$$I = \int_{\text{BEG}}^{\text{END}} F(x)\, dx, \qquad \text{STEP} = \text{END} - \text{BEG}$$

the composite trapezoid sum approximation to I based on subintervals of length $h = \text{STEP}/2^L$ is

$$S(h) = h\left[\tfrac{1}{2}F(\text{BEG}) + \tfrac{1}{2}F(\text{END}) + \sum_{i=1}^{2^L - 1} F(\text{BEG} + ih)\right]$$

Note that $S(h)$ can be computed as

$$S(h) = S(2h)/2 + h \sum_{i=1}^{2^{L-1}} F(\text{BEG} + (2i - 1)h)$$

once $S(2h)$ is known.

1. Smooth Integrand

If $F(x)$ is $(2k + 1)$ times continuously differentiable, then the Euler–MacLaurin sum formula [see, for example, Davis and Rabinowitz (1967, p. 92)] gives

$$I = S(h) + A_1 h^2 + A_2 h^4 + \cdots + A_k h^{2k} + o(h^{2k})$$

so that cautious extrapolation applies with

$$\gamma_i = 2i, \qquad i = 1, \ldots, k$$

The program does not deal directly with the numbers $S(h; 2^2, \ldots, 2^{2j})$. Rather, the numbers

$$T(1, 1) = \tfrac{1}{2}[F(\text{BEG}) + F(\text{END})]$$

$$T(L, 1) = T(L - 1, 1)/2 + 2^{1-L} \sum_{i=1}^{2^{L-2}} F(\text{BEG} + (2i - 1) * \text{STPLM1})$$

$$T(L, J + 1) = T(L, J) + (T(L, J) - T(L - 1, J))/(4^J - 1)$$

$$J = 1, \ldots, L - 1; \quad L = 2, 3, \ldots$$

are calculated, with

$$\text{STPLM1} = \text{STEP}/2^{L-1}$$

But it is clear that

$$\text{STEP} * T(L, J) = S(\text{STPLM1}; 4, 16, \ldots, 2^{2J})$$

Hence, the numbers

$$\text{STEP} * T(L, J), \qquad J = 1, \ldots, L; \quad L = 1, 2, \ldots$$

are the entries of the customary triangular T-Table of Romberg integration. One would now look initially for a ratio

$$R_0(h) = [T(L - 1, 1) - T(L - 2, 1)]/[T(L, 1) - T(L - 1, 1)]$$

of about $4 = 2^2$ in the first column, the column of trapezoid sums. Once such a ratio is found, one then looks at the further columns of the T-table in the cautious manner described under Sect. II,A.

2. Integrand with Jump Discontinuity

Assume that $F(x)$ has a jump discontinuity at the point

$$\xi = \text{BEG} + \zeta * \text{STEP}$$

in the interval [BEG, END]. For sufficiently small STEP, $F(x)$ is then essentially a step function. Since the trapezoid sum approximation is exact for constants, we can assume that $F(x)$ is of the form

$$F(x) = \begin{cases} s, & x < \xi \\ 0, & x > \xi \end{cases}$$

i.e., $F(x)$ has a jump of (signed) size s at $x = \xi$. With $S(h)$ the trapezoid sum approximation on $2^L + 1$ points, we have $h = \text{STEP}/2^L$, and the only difference between I and $S(h)$ occurs in some interval

$$[BEG + jh, \ BEG + (j + 1)h]$$

which contains the jump point ξ. If we write ζ as a binary fraction,

$$\zeta = c_1 2^{-1} + c_2 2^{-2} + \cdots$$

then

$$j = c_1 2^{L-1} + c_2 2^{L-2} + \cdots + c_L 2^0$$

Integrating $F(x)$ over this subinterval gives the number

$$sh(\zeta 2^L - j) = sh(c_{L+1} 2^{-1} + c_{L+2} 2^{-2} + \cdots)$$

while the trapezoid rule for this subinterval gives

$$\tfrac{1}{2} sh$$

Hence,

$$E(h) = \int_{BEG}^{END} F(x)\, dx - S(h) = \tfrac{1}{2} sh[(c_{L+1} + c_{L+2} 2^{-1} + \cdots) - 1]$$

It follows that

$$S(h) - S(2h) = E(2h) - E(h)$$
$$= \tfrac{1}{2} sh[2(c_L + c_{L+1} 2^{-1} + \cdots) - 2 - (c_{L+1} + c_{L+2} 2^{-1}$$
$$\qquad + \cdots) + 1]$$
$$= \tfrac{1}{2} sh(2c_L - 1)$$
$$= \begin{cases} \tfrac{1}{2} sh, & c_L = 1 \\ -\tfrac{1}{2} sh, & c_L = 0 \end{cases}$$

Consequently,

$$R_0(h) = \pm 2 \tag{6}$$

and, with $h = STEP/2^L$,

$$|s| = |S(h) - S(2h)| * 2^{L+1}/STEP \tag{7}$$

while

$$\left| \int_{BEG}^{END} F(x)\, dx - S(h) \right| = \tfrac{1}{2} |sh| |(c_{L+1} + c_{L+2} 2^{-1} + \ldots) - 1|$$
$$\leqq \tfrac{1}{2} |sh| = |S(h) - S(2h)| \tag{8}$$

is the best possible error bound to be obtained.

Hence, if STEP is small enough, then a jump discontinuity can be recognized by having

$$|R_0(STEP/2^L)| \approx 2$$

Since no extrapolation from the trapezoid sums is possible in this case, there is little merit in calculating further trapezoid sums for the whole interval [BEG, END] if the error bound (8) is unsatisfactory. It seems then better to subdivide the interval further.

3. *Integrand with Algebraic End Point Singularity*

Suppose that $F(x)$ is of the form

$$F(x) = (x - \text{BEG})^\alpha g(x), \qquad (\text{or,} \quad F(x) = (\text{END} - x)^\alpha g(x)) \qquad (9)$$

where $\alpha \in (-1, 1)$, $\alpha \neq 0$, and $g(x)$ is $(2k + 1)$ times continuously differentiable. For negative α, assume further that

$$F(\text{BEG}) = 0 \qquad (\text{or,} \quad F(\text{END}) = 0)$$

Under these assumptions, Lyness and Ninham (1967) have shown that the composite trapezoid sum approximation $S(h)$ satisfies

$$I = S(h) + \sum_{1 \leq i \leq 2k+1-\alpha} A_i h^{i+\alpha} + \sum_{i=1}^{k} B_i h^{2i} + O(h^{2k+1}) \qquad (10)$$

where the A_i and B_i do not depend on h.

If the exponent α were known to the algorithm, cautious extrapolation could be applied much as in the case of a smooth integrand. The only minor complication is provided by the necessity to eliminate the h^{γ_i} terms in order of increasing exponents. If, for example, $\alpha = \frac{1}{2}$, then

$$\gamma_1 = \tfrac{3}{2}, \quad \gamma_2 = 2, \quad \gamma_3 = \tfrac{5}{2}, \quad \gamma_4 = \tfrac{7}{2}, \quad \gamma_5 = 4, \quad \dots$$

while, for $\alpha = -0.8$,

$$\gamma_1 = 0.2, \quad \gamma_2 = 1.2, \quad \gamma_3 = 2, \quad \gamma_4 = 2.2, \quad \gamma_5 = 3.2, \quad \dots$$

But our assumptions about the input to this program do not allow for such knowledge. Hence, the algorithm must somehow recognize that the integrand is of the form (9) and, further, estimate reasonably accurately the exponent before cautious extrapolation can be used. This is done in the program as follows. According to (10), one has, for an integrand of the form (9), that

$$R_0(h) \underset{h \to 0}{\to} 2^{1+\alpha}$$

Hence, if $R_0(h)$ seems to converge to some number between 1 and 4 (other than 2), the integrand is suspected of having the form (9). This is further tested while estimating α (or, rather, the number 2^α): The usual Romberg T-Table is constructed, extrapolating with $\gamma_i = 2i$, $i = 1, 2, \dots$. This should eliminate the h^{2i} terms in (10), leaving the term $A_1 h^{1+\alpha}$ as the dominant term (if h is small enough). Hence, if the term $A_1 h^{1+\alpha}$ is present in the error expansion at all, then the ratios of the various columns of the T-Table should more and more approach $2^{1+\alpha}$. The program takes the ratio in the highest column with at least three entries as the approximation to $2^{1+\alpha}$, provided this ratio is reasonably close to $R_0(h)$.

4. *Integrand with Other Troubles*

The program is not equipped to recognize jump discontinuities in derivatives. Neither can it recognize algebraic singularities with exponent greater than 1. Both of these features could be added—within limits. Thus it is not really necessary to recognize a jump discontinuity in the fourth derivative of the integrand merely to calculate the integral correctly.

Further, if the integrand has an algebraic singularity at $A + \zeta(B - A)$, where $0 < \zeta < 1$, then the algorithm, as presently written, will recognize this singularity only if the binary fraction for ζ has a string of zeros or a string of ones. This is the same as saying that the location of the singularity be very close to an end point of one of the subintervals considered.

Finally, the algorithm is not very well equipped to deal with integrands which are noisy relative to the given error requirements. It will often give up on these.

III. Numerical Procedures

A. INTERVAL STACKING

In the course of the program, various subintervals of the interval of integration are examined. If, for a given subinterval, no satisfactory estimate of the integral can be found, the subinterval is cut into two subintervals of equal size, each of which is examined separately, information about the other half being temporarily stacked.

Starting with the given interval of integration as the first entry into the stack, the algorithm may generate a stack of subintervals yet to be examined. The flow chart in Table I may help to explain how subintervals are added and deleted from this stack. The logical variable RIGHT and the integer ISTAGE are essential to this process.

To put a subinterval onto the stack means, of course, that certain information about the subinterval is being stacked. How this is done is commented upon in the program. We remark here only on the saving of function values belonging to a subinterval.

The current subinterval [BEG, END] has certain function values available from earlier calculations, stored in TS(I), I = IBEG, ..., IEND. Specifically,

$$\text{TS(I)} = F(\text{BEG} + (\text{I} - \text{BEG})/(\text{IEND} - \text{IBEG}) * \text{STEP})$$

$$= F(\text{END} - (\text{IEND} - \text{I})/(\text{IEND} - \text{IBEG}) * \text{STEP})$$

where, as before, STEP = END − BEG. [Actually, TS(IBEG) = $\frac{1}{2}F$(BEG), TS(IEND) = $\frac{1}{2}F$(END); but, having mentioned it this once, we will not bother about it anymore.] Hence, if

$$\text{IEND} - \text{IBEG} = 2^{\text{L}}$$

then the first $L + 1$ trapezoid sums for this subinterval can be computed without any call to the function program F. If it becomes necessary to calculate $T(L + 2, 1)$ also, then IEND is increased by 2^L and the stored values are spaced out,

$$TS(IBEG + J) \to TS(IBEG + 2 * J), \qquad J = 1, \ldots, 2^L$$

and, for $J = 1, \ldots, 2^L$,

$$F(BEG + (2 * J - 1)/2 ** (L + 1) * STEP)$$

is calculated and put into $TS(IBEG + 2 * J - 1)$.

TABLE I

FLOW CHART FOR INTERVAL STACKING

What happens when a subinterval is done with (successfully or unsuccessfully) is best explained by the following, admittedly odd, flow chart, in Fig. 1, which parallels the earlier one. The strokes on the lines represent consecutive entries in the stack TS. The values of IBEG and IEND upon entering and upon leaving a station in the flow chart are indicated above the line and below the line, respectively. For simplicity of the flow chart, it is assumed that the integration attempt on the current subinterval always leaves the subinterval with five function values. In reading this flow chart (and the earlier one), it might be helpful to keep in mind that the logical variable RIGHT changes value after each integratioñ attempt, whether successful or not. Thus if RIGHT is now TRUE (FALSE), then it must have been FALSE (TRUE) for the previous subinterval.

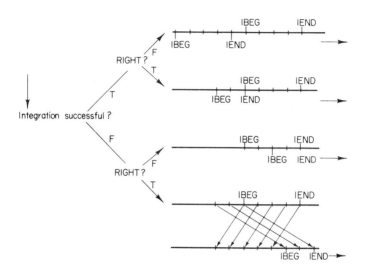

FIG. 1. Flow chart for the use of the stack TS.

B. Round-off Error Control (or the Lack of It)

The program was developed for the CDC 6500 which carries a 48 binary mantissa in floating point arithmetic. Hence, round-off error was not considered (nor has it shown itself to be) a problem. On shorter-word-length machines, one might carry out both the summation in the calculation of trapezoid sums and the accumulation of results from the various subintervals in double precision.

A less obvious source of round-off error annoyance is the saving of function values. Any subinterval (other than the interval of integration) uses function values calculated earlier and, possibly, values calculated during consideration of this subinterval. The abscissas for these function values are determined from the end points of the first subinterval which needs the values. If the calculation of end points involves round-off, then a saved function value may eventually fail to be the value of the integrand at the point it is imagined to be. Tests have shown this to be no problem on the CDC 6500. Should this be of trouble on a shorter-word-length machine, one might consider the following remedy: With

$$\text{LENGTH} = B - A$$

replace the integrand internally by

$$G(X) = F(A + \text{LENGTH} * X)$$

using as interval of integration the interval $[0, 1]$, and multiplying the final result by LENGTH. [The absolute error requirement must, of course, also be adjusted.]

C. PROGRAM PARAMETERS

The specific numerical values given to the nine parameters in the program have been chosen after a limited amount of experimentation. There is no reason to believe them best, mainly because the absence of an acceptable definition of the "desired domain" for the algorithm makes it difficult to ascertain whether some values are better than others.

But it can be ascertained that program performance is more sensitive to changes in some of the parameters than in others. In this sense, the most important parameters are TOLMCH, H2TOL, AITTOL, and JUMPTL. TOLMCH should reflect the precision of the arithmetic used. The other three parameters are used in recognizing the behavior of the integrand. $R_0(h)$ is believed to converge to 4 if

$$| R_0(h) - 4 | \leq \text{H2TOL} \qquad (= 0.15, \quad \text{at present})$$

$R_0(h)$ is believed to be, at least, converging if

$$| R_0(h) - R_0(2h) | \leq \text{AITTOL} | R_0(h) |$$

At present, AITTOL $= 0.1$. Finally, the integrand is believed to have a jump discontinuity if

$$| | R_0(h) | - 2 | \leq \text{JUMPTL} \qquad (= 0.01, \quad \text{at present})$$

Additional parameters are:

AITLOW ($= 1.1$, at present), giving the smallest ratio acceptable for use in extrapolation.

MAXTS $= 2049$, the length of the stack TS.

MAXTBL $= 10$, the upper bound on the number of rows in any one Romberg T-table.

MXSTGE $= 30$, the upper bound on the number of subintervals in the stack waiting to be dealt with.

During execution, additional parameters are calculated from these and from information about the quadrature problem. Chief among these is STEPMN, a lower bound on the absolute length of subintervals considered. An attempt to stack more than MXSTGE subintervals or to integrate over a subinterval of length less than STEPMN results in termination of the program.

IV. Fortran Listing of CADRE (Fig. 2)

```
      FUNCTION CADRE(F,A,B,AERR,RERR,LEVEL,ERRCR,IFLAG)
C THIS FUNCTION RETURNS AN ESTIMATE *CADRE* FCR THE NUMBER
C    INT = INTEGRAL CF *F*(X) FROM *A* TC *B*
C WHICH HOPEFULLY SATISFIES
C    ABS(INT - *CADRE*)  .LE.  AMAX1(*AERR*, *RERR* TIMES ABS(INT)).
C    THE PROGRAM USES CAUTIOUS ADAPTIVE RCMBERG EXTRAPCLATICN.
C IN THIS SCHEME, THE INTEGRAL IS CALCULATED AS THE SUM CF INTEGRALS
C OVER SUITABLY SMALL SUBINTERVALS. ON EACH SUBINTERVAL, AN ESTIMATE
C *VINT*, WITH ESTIMATED ABSOLUTE ERROR *ERROR*, IS FOUND BY CAUTIOUS
C RCMBERG EXTRAPOLATION. IF *ERROR* IS SMALL ENOUGH, *VINT* IS ACCEPT
C EC AND ADDED TO *CADRE*, AND *ERROR* IS ADDED TC *ERRCR*. OTHERWISE
C THE SUBINTERVAL IS HALVED, AND EACH HALF IS CONSIDERED SEPARATELY,
C INFORMATION ABOUT THE OTHER HALF BEING TEMPORARILY STACKED.
C
C                      *****   INPUT   *****
C
C F          THE NAME OF A SINGLE-ARGUMENT REAL FUNCTION SUBPROGRAM
C            THIS NAME MUST APPEAR IN THE CALLING PROGRAM IN AN
C            EXTERNAL STATEMENT.
C A,B        THE TWO ENDPOINTS OF THE INTERVAL OF INTEGRATION
C AERR
C RERR       DESIRED ABSOLUTE AND RELATIVE ERROR IN THE ANSWER
C LEVEL      AN INTEGER INDICATING DESIRED LEVEL OF PRINTOUT
C            .LE. 1,  NO PRINTOUT,
C               = 2,  SUCCESS OR FAILURE MESSAGE, AND LIST CF SINGULAR-
C                     ITIES ENCOUNTERED (IF ANY),
C               = 3,  IN ADDITION, ALL SUBINTERVALS CONSIDERED ARE LISTED
C                     TOGETHER WITH THE KIND OF REGULAR BEHAVICUR FCUND
C                     (IF ANY),
C               = 4,  IN ADDITION, ALL RATIOS CONSIDERED ARE LISTED AS IS
C                     INFO ON WHICH DECISION PROCEDURE IS BASED,
C            .GE. 5,  IN ADDITION, ALL T-TABLES ARE LISTED.
C
C                      *****   OUTPUT   *****
C
C CADRE      ESTIMATE OF THE INTEGRAL, RETURNED VIA THE FUNCTICN CALL,
C ERRCR      ESTIMATED BOUND ON THE ABSOLUTE ERROR OF THE NUMBER *CADRE*
C IFLAG      AN INTEGER BETWEEN 1 AND 5 INDICATING WHAT DIFFICULTIES
C            WERE MET WITH  SPECIFICALLY
C            = 1, ALL IS WELL,
C            = 2, ONE ORE MORE SINGULARITIES WERE SUCCESSFULLY HANDLED,
C            = 3, IN SOME SUBINTERVAL(S), THE ESTIMATE *VINT* WAS ACCEPT
C                 EC MERELY BECAUSE *ERROR* WAS SMALL, EVEN THOUGH NC
C                 REGULAR BEHAVIOUR COULD BE RECOGNIZED,
C            = 4, FAILURE, OVERFLOW OF STACK *TS* (THIS HAS NEVER HAPPEN
C                 EC,  - SO FAR),
C            = 5, FAILURE, TOO SMALL A SUBINTERVAL IS REQUIRED  THIS MAY
C                 BE CUE TO TOO MUCH NOISE IN THE FUNCTION (RELATIVE TC
C                 THE GIVEN ERROR REQUIREMENTS) CR DLE TC A PLAIN CRNERY
C                 INTEGRAND.
C A VERY CAUTIOUS MAN WOULD ACCEPT *CADRE* ONLY IF IFLAG IS 1 CR 2
C THE MERELY REASONABLE MAN WOULD KEEP THE FAITH EVEN IF IFLAG IS 3.
C THE ADVENTUROUS MAN IS QUITE OFTEN RIGHT IN ACCEPTING *CADRE*
C EVEN IF IFLAG IS 4 OR 5.
C
C      *****   LIST OF MAJOR VARIABLES   *****
C
C CUREST     BEST ESTIMATE SO FAR FOR
C            INT - (INTEGRAL OVER CURRENTLY CCNSIDERED SUBINTERVAL).
C FNSIZE     MAXIMUM AVERAGE FUNCTION SIZE SO FAR ENCOUNTERED,
C ERRR       RELATIVE ERROR REQUIREMENT USED  DERIVED FROM INPUT *RERR*
C            AND CHOSEN TO LIE BETWEEN .1 AND 1C TIMES *TOLMCH*,
C ERRA       = ABS(*AERR*)
C STAGE      (MORE OR LESS) EQUAL TO 2 TO THE -(*ISTAGE*)
C THESE FIVE QUANTITIES ARE USED IN THE DETERMINATICN CF THE LCCAL
C ERRCR REQUIREMENT.
```

FIG. 2.

```
C
C   STEPMN   MINIMUM SUBINTERVAL LENGTH PERMITTED,
C   TS       STACK OF VALUES OF F(X) SO FAR COMPUTED BUT NOT YET
C            SUCCESSFULLY USED,
C   ISTAGE   AN INTEGER INDICATING THE HEIGHT OF THE STACK CF INTERVALS
C            YET TO BE PROCESSED.
C
C       *****   LIST OF PARAMETERS   *****
C
C   TOLMCH   DEPENDS ON THE LENGTH OF FLOATING POINT MANTISSA   SHOULD BE
C            ABOUT 1.E-7 FOR 27 BINARY BIT MANTISSA AND
C            ABOUT 1.E-13 FOR 48 BINARY BIT MANTISSA.
C   AITLOW   SHOULD BE SOMEWHAT GREATER THAN 1.
C   H2TOL,
C   AITTOL,
C   JUMPTL   TOLERANCES USED IN THE DECISION PROCESS TO RECOGNIZE
C            H**2 CONVERGENCE, X**ALPHA TYPE CONVERGENCE, CR
C            JUMP-TYPE CONVERGENCE OF THE TRAPEZOID SUMS.
C   MAXTS,
C   MAXTBL,
C   MXSTGE   ARE THE THREE DIFFERENT UPPER LIMITS FOR THE DIMENSICN OF
C            THE VARIOUS ARRAYS.
C
C       *****   PROGRAM LAYOUT   *****
C
C            INITIALIZATION
C   5,6      BEGIN WORK ON NEXT SUBINTERVAL
C   9-14     GET NEXT TRAPEZOID SUM
C   15-19    GET RATIOS. PRELIMINARY DECISION PROCEDURE.
C   20-      ESTIMATE *VINT* ASSUMING SMOOTH INTEGRAND
C   30-      ESTIMATE *VINT* ASSUMING INTEGRAND HAS X**ALPHA TYPE
C            SINGULARITY
C   40-      NO LUCK WITH THIS TRAPEZOID SUM. GET NEXT ONE CR GET OUT.
C   50-      ESTIMATE *VINT* ASSUMING INTEGRAND HAS JUMP
C   60-      ESTIMATE *VINT* ASSUMING INTEGRAND IS STRAIGHT LINE
C   70-      ESTIMATE *VINT* ASSUMING VARIATION IN INTEGRAND
C            IS MOSTLY NOISE.
C   80-      INTEGRATION OVER CURRENT SUBINTERVAL SUCCESSFUL.
C            SET UP NEXT SUBINTERVAL, IF ANY, OR RETURN.
C   90-      INTEGRATION OVER CURRENT SUBINTERVAL NOT SUCCESSFUL.
C            MARK CURRENT SUBINTERVAL FOR SUBDIVISION AND SET UP
C            NEXT SUBINTERVAL.
C   900-     FAILURE.
C
      DIMENSION T(10,10),R(10),AIT(10),DIF(10),RN(4),
     *          TS(2049),IBEGS(30),BEGIN(30),FINIS(30),EST(30)
      REAL LENGTH, JUMPTL
      LOGICAL H2CONV,AITKEN,RIGHT,REGLAR,REGLSV(30)
      DATA TOLMCH,AITLOW,H2TOL,AITTOL,JUMPTL,MAXTS,MAXTBL,MXSTGE
     *  / 2.E-13, 1.1 , .15 , .1 , .01 , 2049, 10 , 30/
      DATA RN/.7142005,.34662815,.843751,.1263306/
      DATA ALG402 /.30102999566239795/
      CADRE = 0.
      ERROR = 0.
      IFLAG = 1
      LENGTH = ABS(B-A)
      IF (LENGTH .EQ. 0.)                RETURN
      ERRR = AMIN1(.1,AMAX1(ABS(RERR), 10.*TOLMCH))
      ERRA = ABS(AERR)
      STEPMN = AMAX1(LENGTH/FLOAT(2**MXSTGE),
     *         AMAX1(LENGTH,ABS(A),ABS(B))*TOLMCH)
      STAGE = .5
      ISTAGE = 1
      CUREST = 0.
      FNSIZE = 0.
      PREVER = 0.
      REGLAR = .FALSE.
```

FIG. 2. (cont.)

```
C
C  THE GIVEN INTERVAL OF INTEGRATION IS THE FIRST INTERVAL CONSIDERED.
      BEG = A
      FBEG = F(BEG)/2.
      TS(1) = FBEG
      IBEG = 1
      END = B
      FEND = F(END)/2.
      TS(2) = FEND
      IEND = 2
C
    5 RIGHT = .FALSE.
C
C  INVESTIGATION OF A PARTICULAR SUBINTERVAL BEGINS AT THIS POINT.
C         *****   MAJOR VARIABLES   *****
C  BEG,
C  END      ENDPOINTS OF THE CURRENT INTERVAL
C  FBEG,
C  FEND     ONE HALF THE VALUE OF F(X) AT THE ENDPOINTS
C  STEP     SIGNED LENGTH OF CURRENT SUBINTERVAL
C  ISTAGE   HEIGHT OF CURRENT SUBINTERVAL IN STACK OF SUBINTERVALS
C           YET TO BE DONE
C  RIGHT    A LOGICAL VARIABLE INDICATING WHETHER CURRENT SUBINTERVAL
C           IS RIGHT HALF OF PREVIOUS SUBINTERVAL. NEEDED IN 80FF AND
C           90FF TO DECIDE WHAT INTERVAL TO LOOK AT NEXT.
C  TS(I), I=IBEG,...,IEND, CONTAINS THE FUNCTION VALUES FOR THIS
C           SUBINTERVAL SO FAR COMPUTED. SPECIFICALLY,
C           TS(I) = F(BEG + (I-IBEG)/(IEND-IBEG)*STEP), ALL I
C           EXCEPT THAT TS(IBEG) = FBEG, TS(IEND) = FEND
C  REGLAR   A LOGICAL VARIABLE INDICATING WHETHER OR NOT THE CURRENT
C           SUBINTERVAL IS REGULAR (SEE NOTES)
C  H2CONV   A LOGICAL VARIABLE INDICATING WHETHER H**2 CONVERGENCE OF
C           THE TRAPEZOID SUMS FOR THIS INTERVAL IS RECOGNIZED,
C  AITKEN   A LOGICAL VARIABLE INDICATING WHETHER CONVERGENCE OF RATIOS
C           FOR THIS SUBINTERVAL IS RECOGNIZED
C  T        CONTAINS THE FIRST *L* ROWS OF THE ROMBERG T-TABLE FOR THIS
C           SUBINTERVAL IN ITS LOWER TRIANGULAR PART. SPECIFICALLY,
C           T(I,1) = TRAPEZOID SUM (WITHOUT THE FACTOR *STEP*)
C                 ON 2**(I-1) + 1 EQUISPACED POINTS, I=1,...,L,
C           T(I,J+1) = T(I,J) + (T(I,J)-T(I-1,J))/(4**J - 1),
C                 J=2,...,I-1, I=2,...,L.
C           FURTHER, THE STRICTLY UPPER TRIANGULAR PART OF T CONTAINS
C           THE RATIOS FOR THE VARIOUS COLUMNS OF THE T-TABLE.
C           SPECIFICALLY,
C           T(J,I) = (T(I,J)-T(I-1,J))/(T(I+1,J)-T(I,J)),
C                 I=J+1,...,L-1,  J=1,...,L-2.
C           FINALLY, THE LAST OR L-TH COLUMN CONTAINS
C           T(J,L) = T(L,J) - T(L-1,J), J=1,...,L-1.
    6 STEP = END - BEG
      ASTEP = ABS(STEP)
      IF (ASTEP .LT. STEPMN)              GO TO 950
      IF (LEVEL .GE. 3) WRITE(6,609) BEG,STEP,ISTAGE
  609 FORMAT(10H DEC,STEP ,2E16.8,I5)
      T(1,1) = FBEG + FEND
      TABS = ABS(FBEG) + ABS(FEND)
      L = 1
      N = 1
      H2CONV = .FALSE.
      AITKEN = .FALSE.
                                          GO TO 10
C
    9 IF (LEVEL .GE. 4)  WRITE(6,692) L,T(1,LM1)
   10 LM1 = L
      L = L + 1
```

Fig. 2. (cont.)

```
C
CALCULATE THE NEXT TRAPEZOID SUM, T(L,1), WHICH IS BASED CN
C   *N2*  + 1 EQUISPACED POINTS. HERE,  N2 = N*2 = 2**(L-1) .
      N2 = N*2
      FN = N2
      ISTEP = (IEND - IBEG)/N
      IF (ISTEP .GT. 1)                 GO TO 12
      II = IEND
      IEND = IEND + N
      IF (IEND .GT. MAXTS)              GO TO 9CO
      HCVN = STEP/FN
      III = IEND
      CC 11 I=1,N2,2
      TS(III) = TS(II)
      TS(III-1) = F(END - FLOAT(I)*HOVN)
      III = III-2
   11 II = II-1
      ISTEP = 2
   12 ISTEP2 = IBEG + ISTEP/2
      SUM = 0.
      SUMABS = 0.
      CC 13 I=ISTEP2,IEND,ISTEP
      SUM = SUM + TS(I)
   13 SUMABS = SUMABS + ABS(TS(I))
      T(L,1) = T(L-1,1)/2. + SUM/FN
      TABS = TABS/2. + SUMABS/FN
      ABSI = ASTEP*TABS
      N = N2
C
C   GET PRELIMINARY VALUE FOR *VINT* FROM LAST TRAPEZOID SUM AND
C   UPDATE THE ERROR REQUIREMENT *ERGOAL* FCR THIS SUBINTERVAL.
C   THE ERROR REQUIREMENT IS NOT PRORATED ACCORDING TO THE LENGTH CF
C   THE CURRENT SUBINTERVAL RELATIVE TO THE INTERVAL OF INTEGRATION,
C   BUT ACCORDING TO THE HEIGHT *ISTAGE* OF THE CURRENT SUBINTERVAL
C   IN THE STACK OF SUBINTERVALS YET TO BE DONE.
C   THIS PROCEDURE IS NOT BACKED BY ANY RIGOROUS ARGUMENT, BUT
C   SEEMS TO WORK.
      IT = 1
      VINT = STEP*T(L,1)
      TABTLM = TABS*TOLMCH
      FNSIZE = AMAX1(FNSIZE,ABS(T(L,1)))
      ERGCAL = AMAX1(ASTEP*TOLMCH*FNSIZE,
     *           STAGE*AMAX1(ERRA,ERRR*ABS(CLREST+VINT)))
C
COMPLETE ROW L AND COLUMN L OF *T* ARRAY.
      FEXTRP = 1.
      CC 14 I=1,LM1
      FEXTRP = FEXTRP*4.
      T(I,L) = T(L,I) - T(L-1,I)
   14 T(L,I+1) = T(L,I) + T(I,L)/(FEXTRP-1.)
      ERRER = ASTEP*ABS(T(1,L))
C-------------------------------------------------------------------
C--- PRELIMINARY DECISION PROCEDURE -------------------------------
C
C   IF L = 2 AND T(2,1) = T(1,1), GO TO 6C TO FOLLOW UP THE IMPRESSION
C   THAT INTEGRAND IS STRAIGHT LINE.
      IF (L .GT. 2)                     GO TO 15
      IF (ABS(T(1,2)) .LE. TABTLM)      GO TO 6C
                                        GO TO 1C
C
CALCULATE NEXT RATIOS FOR COLUMNS 1,...,L-2 OF T-TABLE
C   RATIO IS SET TO ZERO IF DIFFERENCE IN LAST TWO ENTRIES CF
C   CCLUMN IS ABOUT ZERO.
   15 CC 16 I=2,LM1
      CIFF = 0.
```

FIG. 2. (cont.)

```
         IF (ABS(T(I-1,L)) .GT. TABTLM) DIFF = T(I-1,LM1)/T(I-1,L)
      16 T(I-1,LM1) = DIFF
C
C     T(1,LM1) IS THE RATIO DERIVED FROM LAST THREE TRAPEZOID SUMS, I.E.,
C          T(1,LM1) = (T(L-1,1)-T(L-2,1))/(T(L,1)-T(L-1,1)) .
C     IF THIS RATIO IS ABOUT 4, GO TO 20 FOR ROMBERG EXTRAPOLATION.
C     IF THIS RATIO IS ZERO, I.E., IF LAST TWO TRAPEZOID SUMS ARE ABOUT
C          EQUAL, GO TO 18 FOR POSSIBLE NOISE CHECK.
C     IF THIS RATIO IS ABOUT 2 IN ABSOLUTE VALUE, GO TO 50 WITH THE
C          BELIEF THAT INTEGRAND HAS JUMP DISCONTINUITY.
C     IF THIS RATIO IS, AT LEAST, ABOUT EQUAL TO THE PREVIOUS RATIO, THEN
C          THE INTEGRAND MAY WELL HAVE A NICE INTEGRABLE SINGULARITY.
C          GO TO 30 TO FOLLOW UP THIS HUNCH.
         IF (ABS(4.-T(1,LM1)) .LE. H2TOL) GO TO 20
         IF (T(1,LM1) .EQ. 0.)               GO TO 18
         IF (ABS(2.-ABS(T(1,LM1))) .LT. JUMPTL) GO TO 50
         IF (L .EQ. 3)                       GO TO 9
         H2CONV = .FALSE.
         IF (ABS((T(1,LM1)-T(1,L-2))/T(1,LM1)) .LE. AITTOL)
     *                                       GO TO 30
C     AT THIS POINT, NO REGULAR BEHAVIOUR WAS DETECTED.
C     IF CURRENT SUBINTERVAL IS NOT REGULAR AND ONLY FOUR TRAPEZOID SUMS
C     WERE COMPUTED SO FAR, TRY ONE MORE TRAPEZOID SUM.
C     IF , AT LEAST, LAST TWO TRAPEZOID SUMS ARE ABOUT EQUAL, THEN
C     FAILURE TO RECOGNIZE REGULAR BEHAVIOUR MAY WELL BE DUE TO NOISE.
C     GO TO 70 TO CHECK THIS OUT.
C     OTHERWISE, GO TO 90 FOR FURTHER SUBDIVISION.
C
      17 IF (REGLAR)                         GO TO 18
         IF (L .EQ. 4)                       GO TO 9
      18 IF (ERRER .LE. ERGOAL)              GO TO 70
         IF (LEVEL .GE. 4) WRITE (6,652) L,T(1,LM1)
                                             GO TO 91
C------------------------------------------------------------------------
CAUTIOUS ROMBERG EXTRAPOLATION -------------------------------------------
C
C     THE CURRENT, OR L-TH, ROW OF THE ROMBERG T-TABLE HAS L ENTRIES.
C     FOR J=1,...,L-2, THE ESTIMATE
C                 STEP*T(L,J+1)
C     IS BELIEVED TO HAVE ITS ERROR BOUNDED BY
C          ABS(STEP*(T(L,J)-T(L-1,J))/(4**J - 1))
C     IF THE LAST RATIO
C          T(J,LM1) = (T(L-1,J)-T(L-2,J))/(T(L,J)-T(L-1,J))
C     FOR COLUMN J OF THE T-TABLE IS ABOUT 4**J.
C     THE FOLLOWING IS A SLIGHTLY RELAXED EXPRESSION OF THIS BELIEF.
      20 IF (LEVEL .GE. 4) WRITE (6,619) L,T(1,LM1)
     619 FORMAT(I5,E16.8,5X6HH2CONV)
         IF (H2CONV)                         GO TO 21
         AITKEN = .FALSE.
         H2CONV = .TRUE.
         IF (LEVEL .GE. 3) WRITE (6,620) L
     620 FORMAT(22H H2 CONVERGENCE AT ROW,I3)
      21 FEXTRP = 4.
      22 IT = IT + 1
         VINT = STEP*T(L,IT)
         ERRER = ABS(STEP/(FEXTRP-1.)*T(IT-1,L))
         IF (ERRER .LE. ERGOAL)              GO TO 80
         IF (IT .EQ. LM1)                    GO TO 40
         IF (T(IT,LM1) .EQ. 0.)              GO TO 22
         IF (T(IT,LM1) .LE. FEXTRP )         GO TO 40
         IF (ABS(T(IT,LM1)/4.-FEXTRP)/FEXTRP .LT. AITTOL)
     *          FEXTRP = FEXTRP*4.
                                             GO TO 22
C------------------------------------------------------------------------
C--- INTEGRAND MAY HAVE X**ALPHA TYPE SINGULARITY------------------------
C     RESULTING IN A RATIO OF *SING* = 2**(ALPHA + 1)
      30 IF (LEVEL .GE. 4) WRITE (6,629) L,T(1,LM1)
```

FIG. 2. (cont.)

```
629 FCRMAT( I5,E16.8,5X6HAITKEN)
      IF (T(1,LM1) .LT. AITLOW)           GO TO 91
      IF (AITKEN)                         GO TO 31
      F2CCNV = .FALSE.
      AITKEN = .TRUE.
      IF (LEVEL .GE. 3) WRITE (6,63C) L
630 FCRMAT(14H AITKEN AT ROW,I3)
 31 FEXTRP = T(L-2,LM1)
      IF (FEXTRP .GT. 4.5)                GO TO 21
      IF (FEXTRP .LT. AITLOW)             GO TO 91
      IF (ABS(FEXTRP-T(L-3,LM1))/T(1,LM1) .GT. H2TCL)
     *                                    GO TO 91
      IF (LEVEL .GE. 3) WRITE (6,631) FEXTRP
631 FCRMAT(6H RATIO,F12.8)
      SING = FEXTRP
      FEXTM1 = FEXTRP - 1.
      CC 32 I=2,L
      AIT(I) = T(I,1) + (T(I,1)-T(I-1,1))/FEXTM1
      R(I) = T(1,I-1)
 32 CIF(I) = AIT(I) - AIT(I-1)
      IT = 2
 33 VINT = STEP*AIT(L)
      IF (LEVEL .LT. 5) GO TO 333
      WRITE (6,632) (R(I+1),I=IT,LM1)
      WRITE (6,632) (AIT(I),I=IT,L)
      WRITE (6,632) (CIF(I+1),I=IT,LM1)
632 FCRMAT(1X,8E15.8)
333 ERRER = ERRER/FEXTM1
      IF (ERRER .GT. ERGOAL)              GO TO 34
      ALPHA = ALOG10(SING)/ALG402 - 1.
      IF (LEVEL .GE. 2) WRITE (6,633) ALPHA,BEG,END
633 FCRMAT(11X42HINTEGRAND SHOWS SINGULAR BEHAVICR OF TYPE
     *      4HX**(F4.2,9H) BETWEENE15.8,4H ANDE15.8)
      IFLAG = MAXO(IFLAG,2)
                                          GO TO 8C
 34 IT = IT + 1
      IF (IT .EQ. LM1)                    GO TO 4C
      IF (IT .GT. 3)                      GO TO 35
      H2NEXT = 4.
      SINGNX = 2.*SING
 35 IF (H2NEXT .LT. SINGNX)              GO TO 36
      FEXTRP = SINGNX
      SINGNX = 2.*SINGNX
                                          GO TO 37
 36 FEXTRP = H2NEXT
      H2NEXT = 4.*H2NEXT
 37 CC 38 I=IT,LM1
      R(I+1) = 0.
 38 IF (ABS(CIF(I+1)) .GT. TABTLM) R(I+1) = DIF(I)/DIF(I+1)
      IF (LEVEL .CE. 4) WRITE (6,638) FEXTRP,R(L-1),R(L)
638 FCRMAT(16H FEXTRP + RATIOS,3E15.8)
      H2TFEX = -H2TOL*FEXTRP
      IF (R(L) - FEXTRP .LT. H2TFEX)      GO TO 4C
      IF (R(L-1)-FEXTRP .LT. H2TFEX)      GO TO 4C
      ERRER = ASTEP*ABS(DIF(L))
      FEXTM1 = FEXTRP - 1.
      CC 39 I=IT,L
      AIT(I) = AIT(I) + DIF(I)/FEXTM1
 39 CIF(I) = AIT(I) - AIT(I-1)
                                          GO TO 33
C-------------------------------------------------------------------
C    CURRENT TRAPEZOIC SUM AND RESULTING EXTRAPOLATED VALUES DIC NCT GIVE
C    A SMALL ENCUGH *ERRER*.
C    IF LESS THAN FIVE TRAPEZOID SUMS WERE CCMPLTED SO FAR, TRY NEXT
C       TRAPEZCIC SUM.
C    CTHERWISE, CECIDE WHETHER TO GO CN OR TC SUBDIVIDE AS FCLLCWS.
```

FIG. 2. (cont.)

```
C     WITH T(L,IT) GIVING THE CURRENTLY BEST ESTIMATE, GIVE UP ON DEVELOP
C     ING THE T-TABLE FURTHER IF  L .GT. IT+2, I.E., IF EXTRAPOLATION
C     DID NOT GO VERY FAR INTO THE T-TABLE.
C     FURTHER, GIVE UP IF REDUCTION IN *ERRER* AT THE CURRENT RATE
C     DOES NOT PREDICT AN *ERRER* LESS THAN *ERGOAL* BY THE TIME
C     *MAXTBL* TRAPEZOID SUMS HAVE BEEN COMPLTED.
C     ---NOTE---
C     HAVING PREVER .LT. ERRER  IS AN ALMOST CERTAIN SIGN OF BEGINNING
C     TROUBLE WITH NOISE IN THE FUNCTION VALUES. HENCE,
C     A WATCH FOR, AND CONTROL OF, NOISE SHOULD BEGIN HERE.
   40 FEXTRP = AMAX1(PREVER/ERRER,AITLOW)
      PREVER = ERRER
      IF (L .LT. 5)                          GO TO 10
      IF (LEVEL .GE. 3) WRITE (6,641) ERRER,ERGOAL,FEXTRP,IT
  641 FORMAT(23H ERRER,ERGOAL,FEXTRP,IT,2E15.8,E14.5,I3)
      IF (L-IT .GT. 2 .AND. ISTAGE .LT. MXSTGE) GO TO 90
      IF (ERRER/FEXTRP**(MAXTBL-L) .LT. ERGOAL) GO TO 10
                                             GO TO 90
C----------------------------------------------------------------------
C----    INTEGRAND HAS JUMP (SEE NOTES)  ------------------------------
   50 IF (LEVEL .GE. 4) WRITE (6,649) L,T(1,LM1)
  649 FORMAT(I5,E16.8,5X4HJUMP)
      IF (ERRER .GT. ERGOAL)                 GO TO 90
C     NOTE THAT  2*FN = 2**L
      DIFF = ABS(T(1,L))*2.*FN
      IF (LEVEL .GE. 2) WRITE (6,650) DIFF,BEG,END
  650 FORMAT(13X36HINTEGRAND SEEMS TO HAVE JUMP OF SIZEE13.6,
     *       8H BETWEENE15.8,4H ANDE15.8)
                                             GO TO 80
C----------------------------------------------------------------------
C----  INTEGRAND IS STRAIGHT LINE  ------------------------------------
C     TEST THIS ASSUMPTION BY COMPARING THE VALUE OF THE INTEGRAND AT
C     FOUR 'RANDOMLY CHOSEN' POINTS WITH THE VALUE OF THE STRAIGHT LINE
C     INTERPOLATING THE INTEGRAND AT THE TWO ENDPOINTS OF THE SUB-
C     INTERVAL. IF TEST IS PASSED, ACCEPT *VINT*
   60 IF (LEVEL .GE. 4) WRITE (6,660) L
  660 FORMAT(I5,21X13HSTRAIGHT LINE)
      SLOPE = (FEND-FBEG)*2.
      FBEG2 = FBEG*2.
      DO 61 I=1,4
      DIFF = ABS(F(BEG+RN(I)*STEP) - FBEG2-RN(I)*SLOPE)
      IF (DIFF .GT. TABTLM)                  GO TO 72
   61 CONTINUE
      IF (LEVEL .GE. 2) WRITE (6,667) BEG, END
  667 FORMAT(27X43HINTEGRAND SEEMS TO BE STRAIGHT LINE BETWEEN
     *        E15.8,4H ANDE15.8)
                                             GO TO 80
C----------------------------------------------------------------------
C------  NOISE MAY BE DOMINANT FEATURE  -------------------------------
C     ESTIMATE NOISE LEVEL BY COMPARING THE VALUE OF THE INTEGRAND AT
C     FOUR 'RANDOMLY CHOSEN' POINTS WITH THE VALUE OF THE STRAIGHT LINE
C     INTERPOLATING THE INTEGRAND AT THE TWO ENDPOINTS. IF SMALL
C     ENOUGH, ACCEPT *VINT*.
   70 IF (LEVEL .GE. 4) WRITE (6,670) L,T(1,LM1)
  670 FORMAT(I5,E16.8,5X5HNOISE)
      SLOPE = (FEND-FBEG)*2.
      FBEG2 = FBEG*2.
      I = 1
   71 DIFF = ABS(F(BEG+RN(I)*STEP) - FBEG2-RN(I)*SLOPE)
   72 ERRER = AMAX1(ERRER,ASTEP*DIFF)
      IF (ERRER .GT. ERGOAL)                 GO TO 91
      I = I+1
      IF (I .LE. 4)                          GO TO 71
      IF (LEVEL .GE. 2) WRITE (6,671) BEG,END
  671 FORMAT(15H NOISE BETWEEN ,E15.8,4H AND,E15.8)
      IFLAG = 3
C----------------------------------------------------------------------
C---  INTEGRATION OVER CURRENT SUBINTERVAL SUCCESSFUL -----------------
C     ADD *VINT* TO *CADRE* AND *ERRER* TO *ERROR*, THEN SET UP NEXT
```

FIG. 2. (cont.)

```
C   SUBINTERVAL, IF ANY.
    80 CADRE = CADRE + VINT
       ERROR = ERROR + ERRER
       IF (LEVEL .LT. 3)                    GO TO 83
       IF (LEVEL .LT. 5)                    GO TO 82
       DO 81 I=1,L
    81 WRITE (6,692) I,(T(I,J),J=1,L)
    82 WRITE (6,682) VINT,ERRER,L,IT
   682 FORMAT(12H INTEGRAL IS,E16.8,7H, ERROR,E15.8,9H FROM T(,
      *       I1,1H,,I1,1H))
C
    83 IF (RIGHT)                           GO TO 85
       ISTAGE = ISTAGE - 1
       IF (ISTAGE .EQ. 0)                   RETURN
C
       REGULAR = REGLSV(ISTAGE)
       BEG = BEGIN(ISTAGE)
       END = FINIS(ISTAGE)
       CUREST = CUREST - EST(ISTAGE+1) + VINT
       IEND = IBEG - 1
       FEND = TS(IEND)
       IBEG = IBEGS(ISTAGE)
                                            GO TO 94
    85 CUREST = CUREST + VINT
       STAGE = STAGE*2.
       IEND = IBEG
       IBEG = IBEGS(ISTAGE)
       END = BEG
       BEG = BEGIN(ISTAGE)
       FEND = FBEG
       FBEG = TS(IBEG)
                                            GO TO 5
C--------------------------------------------------------------------
C--- INTEGRATION OVER CURRENT SUBINTERVAL IS UNSUCCESSFUL------------
C   MARK SUBINTERVAL FOR FURTHER SUBDIVISION. SET UP NEXT SUBINTERVAL.
    90 REGULAR = .TRUE.
    91 IF (ISTAGE .EQ. MXSTGE)              GO TO 950
       IF (LEVEL .LT. 5)                    GO TO 93
       DO 92 I=1,L
    92 WRITE (6,692) I,(T(I,J),J=1,L)
   692 FORMAT(I5,7E16.8/3E16.8)
    93 IF (RIGHT)                           GO TO 95
       REGLSV(ISTAGE+1) = REGULAR
       BEGIN(ISTAGE) = BEG
       IBEGS(ISTAGE) = IBEG
       STAGE = STAGE/2.
    94 RIGHT = .TRUE.
       BEG = (BEG+END)/2.
       IBEG = (IBEG+IEND)/2
       TS(IBEG) = TS(IBEG)/2.
       FBEG = TS(IBEG)
                                            GO TO 6
    95 NNLEFT = IBEG - IBEGS(ISTAGE)
       IF (IEND+NNLEFT .GE. MAXTS)          GO TO 900
       III = IBEGS(ISTAGE)
       II = IEND
       DO 96 I=III,IBEG
       II = II + 1
    96 TS(II) = TS(I)
       DO 97 I=IBEG,II
       TS(III) = TS(I)
    97 III = III + 1
       IEND = IEND + 1
       IBEG = IEND - NNLEFT
       FEND = FBEG
       FBEG = TS(IBEG)
       FINIS(ISTAGE) = END
```

FIG. 2. (cont.)

```
      ENC = BEG
      BEG = BEGIN(ISTAGE)
      BEGIN(ISTACE) = END
      REGLSV(ISTACE) = REGLAR
      ISTAGE = ISTAGE + 1
      REGLAR = REGLSV(ISTACE)
      EST(ISTAGE) = VINT
      CURFST = CUREST + EST(ISTAGE)
                                           GO TO 5
C----------------------------------------------------------------
C--- FAILURE TC HANCLE GIVEN INTEGRATICN PROBLEM-----------------
  900 IF (LEVEL .GE. 2) WRITE (6,6SCC) BEG, END
 6900 FCRMAT(37H TOO MANY FUNCTION EVALUATIONS AROUND/
     *         10X,E15.8,4H AND,E15.8)
      IFLAG = 4
                                           GO TO SS9
  950 IFLAG = 5
      IF (LEVEL .LT. 2)                    GO TO SS9
      IF (LEVEL .LT. 5)                    GO TO SS9
      CC 958 I=1,L
  958 WRITE (6,692) I,(T(I,J),J=1,L)
  959 WRITE (6,6959) BEG, END
 6959 FCRMAT(12X38HINTEGRAND SHOWS SINGULAR BEHAVICUR CF
     *        20HUNKNOWN TYPE BETWEENE15.8,4H ANDE15.8)
  909 CADRE = CUREST + VINT
                                           RETLRN
      ENC
```

FIG. 2. (cont.)

V. Testing and Examples

The testing of quadrature algorithms is largely unexplored territory. Supposedly, such testing should consider a wide variety of quadrature problems in an attempt to delineate the domain of the algorithm. This is usually done by picking problems from published reports on quadrature routines, or by constructing problems which look difficult, or even by asking someone down the hall what kind of function he has integrated lately.

James Lyness has questioned the utility of taking, in this manner, "pot shots" at the domain, and proposed the following alternative. Pick a one-parameter family $f(x; \alpha)$ of functions which exhibit, as α increases, say, a particular disagreeable property to an ever more marked degree. Attempt to calculate $\int f(x; \alpha) \, dx$ using various error requirements, and plot the actual error achieved and the number of function evaluations needed against α. In this way, precise information about the domain of the algorithm can be obtained, at least along certain (hopefully) important rays or lines in the "space" of all quadrature problems.

Below, samples of both approaches are given. The twenty-one sample problems are those used by Kahaner in the comparison of quadrature algorithms, as reported earlier in these proceedings. As simple disagreeable properties, I picked:

(i) peaking as exhibited by

$$f(x; \alpha) = 2^{-\alpha}/(2^{-2\alpha} + x^2) \qquad \text{on} \quad [-1, 1]$$

(ii) an algebraic end-point singularity,

$$f(x; \alpha) = \begin{cases} 0, & x = 0 \\ x^\alpha, & x > 0 \end{cases} \quad \text{on} \quad [0, 1]$$

(iii) oscillation, as shown by

$$f(x; \alpha) = \cos(\alpha \pi x) + 1 \quad \text{on} \quad [0, 1]$$

All calculations were run on a CDC 6500, with CADRE precisely as listed under Sect. IV.

Some comments are in order. The case of the peaking function (see Fig. 3),

$$f(x; \alpha) = 2^{-\alpha}/(4^{-\alpha} + x^2) \quad \text{on} \quad [-1, 1]$$

is relatively uninteresting since CADRE performs well. Note that the height of the peak, $f(0; \alpha) = 2^\alpha$, and the curvature at the peak increase exponentially, while the number of function evaluations increases only like $\sqrt{\alpha}$. This is, of course, due to the adaptive behavior of CADRE. There seems to be a slight shift of tactics around $\alpha = 24$, after which the error goes up to pass the error requirement just before CADRE breaks down, with IFLAG = 5, at $\alpha = 31$. At this point, the peak has height $\approx 10^{10}$, at $x = 0$, while already, at $x = 10^{-5}$, $f(x; \alpha)$ is less than 1 and, at the end of the interval, $f(\pm 1; \alpha) \approx 10^{-10}$.

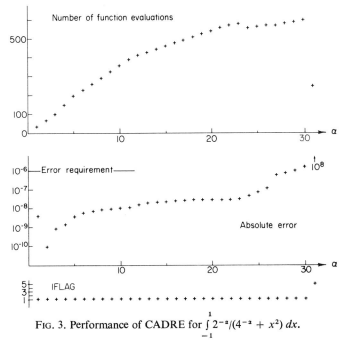

FIG. 3. Performance of CADRE for $\int_{-1}^{1} 2^{-\alpha}/(4^{-\alpha} + x^2)\, dx$.

440 CARL DE BOOR

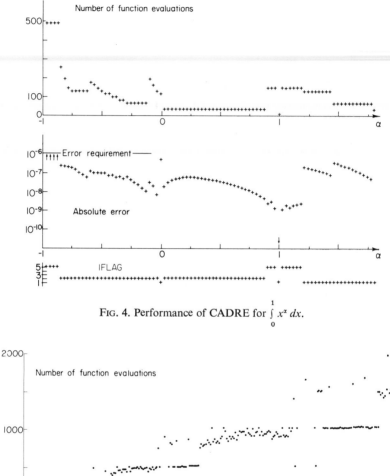

FIG. 4. Performance of CADRE for $\int_0^1 x^x\, dx$.

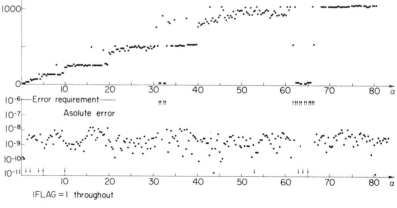

FIG. 5. Performance of CADRE for $\int_0^1 [\cos(\alpha\pi x) + 1]\, dx$.

The case of an algebraic end point singularity (see Fig. 4),

$$f(x; \alpha) = x^\alpha \qquad \text{on} \quad [0, 1]$$

is more interesting. Since CADRE is not trained to recognize such a singularity for $\alpha > 1$, one can get a feeling for what such training would buy in efficiency by comparing function evaluations for $\alpha > 1$ with those for $\alpha < 1$. Near $\alpha = 1$, CADRE produces very accurate results but returns IFLAG = 5. This happens as follows: In any subinterval having 0 as end point, the ratio $R_0(h)$ converges rather slowly to $2^{1+\alpha}$. Hence, although regularity of ratios is noted right away, the ratio for extrapolation found from some column of the ordinary Romberg T-table (see Sect. II,B,3) is too far from $R_0(h)$ to be accepted (ironically, because it is too accurate an estimate for $2^{1+\alpha}$). In such a case, it is the program's policy to subdivide the subinterval further, although for this particular integrand it would be better to calculate further rows of the Romberg table for the subinterval. The IFLAG = 5 return for α near -1 is, by contrast, genuine. The arrows in the error plot indicate an absolute error in the answer well above the plotting range. This breakdown occurs (for $\alpha \leq -0.875$ in the plot) since the program refuses to accept ratios $R_0(h) = 2^{1+\alpha}$ which are less than the constant AITLOW = 1.1 = $2^{1-0.845}$.

The third and final graph (see Fig. 5) shows how CADRE copes (or fails to cope) with the oscillatory function

$$f(x; \alpha) = \cos(\alpha \pi x) + 1 \qquad \text{on} \quad [0, 1]$$

The most interesting feature here arises from the common difficulty with oscillatory integrands: When only a few function values are considered (as happens at the beginning of CADRE), the integrand looks at times like quite a different function. This is clearly indicated near $\alpha = 32$, and again near $\alpha = 64$, where after 9, or 17, or 33 evaluations the algorithm is certain that it has a very simple integrand at hand and thus comes up with the wrong answer. Note that, for $\alpha = 32$ or 64, CADRE gets the right answer. For, in this case, the integrand looks initially *too* simple, it seems to be a straight line, and CADRE checks such supposed behavior by a few "random" evaluations before believing it. In these cases (and indeed for every α divisible by 4) the integrand is subjected to, and fails, this test.

The graph also indicates a difficulty of such parameter studies. Although quite a few values of α have been considered [$\alpha = \frac{1}{3}$ ($\frac{1}{3}$) 83 $\frac{1}{3}$], quite a few breakdowns have not been caught. CADRE comes up with the wrong answer near (but not at) every multiple of 8. This failure is due to "resonance" between the oscillation frequency and the regular choice of points at which the integrand is examined. The rate of failure can, therefore, be decreased dramatically at essentially no extra cost by calculating, in cases of oscillatory integrands,

$$\int_A^B F(x)\,dx \qquad \text{as} \qquad \int_A^C F(x)\,dx + \int_C^B F(x)\,dx$$

where C is a randomly chosen number between A and B.

There follows (Fig. 6) an output listing from a test run with Kahaner's twenty-one sample problems.

FUNCTION NUMBER 1, WITH A = 0.00000 AND B = 1.00000   F(X) = EXP(X)
THE CORRECT VALUE OF THE INTEGRAL IS   1.71828182E+00

TOLERANCE	*CADRE*	ABSOLUTE ERROR	RELATIVE ERROR	*ERROR*	EVALUATIONS	*IFLAG*
1.0E-03	1.718281842E+00	1.38184E-08	8.04200E-09	2.31200E-06	9	1
1.0E-06	1.718281829E+00	2.75235E-10	1.60239E-10	1.45377E-07	17	1
1.0E-09	1.718281828E+00	6.03666E-11	3.51285E-11	2.14970E-10	17	1

FUNCTION NUMBER 2, WITH A = 0.00000 AND B = 1.00000   F(X) = STEP FUNCTION WITH ONE JUMP AT .3
THE CORRECT VALUE OF THE INTEGRAL IS   7.00000000E-01

TOLERANCE	*CADRE*	ABSOLUTE ERROR	RELATIVE ERROR	*ERROR*	EVALUATIONS	*IFLAG*
INTEGRAND SEEMS TO HAVE JUMP OF SIZE 1.00000E+00 BETWEEN 2.96875000E-01 AND 3.00781250E-01						
1.0E-03	6.99707031E-01	2.92969E-04	4.18527E-04	4.88281E-04	53	1
INTEGRAND SEEMS TO HAVE JUMP OF SIZE 1.00000E+00 BETWEEN 2.99999237E-01 AND 3.00001144E-01						
1.0E-06	7.00000477E-01	4.76837E-08	6.81196E-08	2.38419E-07	119	1
INTEGRAND SEEMS TO HAVE JUMP OF SIZE 1.00000E+00 BETWEEN 2.99999997E-01 AND 3.00000001E-01						
1.0E-09	6.99999997E-01	2.75366E-10	3.99137E-10	4.65661E-10	173	1

FUNCTION NUMBER 3, WITH A = 0.00000 AND B = 1.00000   F(X) = SQRT(X)
THE CORRECT VALUE OF THE INTEGRAL IS   6.66666667E-01

TOLERANCE	*CADRE*	ABSOLUTE ERROR	RELATIVE ERROR	*ERROR*	EVALUATIONS	*IFLAG*
INTEGRAND SHOWS SINGULAR BEHAVIOR OF TYPE X**(.50) BETWEEN 0. AND 1.00000000E+00						
1.0E-03	6.66667581E-01	9.14606E-08	1.37191E-07	1.03580E-04	17	2
INTEGRAND SHOWS SINGULAR BEHAVIOR OF TYPE X**(.50) BETWEEN 0. AND 1.00000000E+00						
1.0E-06	6.66666990E-01	3.23422E-08	4.85146E-08	4.43388E-08	33	2
INTEGRAND SHOWS SINGULAR BEHAVIOR OF TYPE X**(.50) BETWEEN 0. AND 1.00000000E+00						
1.0E-09	6.66666668E-01	1.20566E-10	1.80879E-10	1.79859E-10	129	2

FIG. 6.

443

FUNCTION NUMBER 4, WITH A = -1.00000 AND B = 1.00000   F(X) = .46 (EXP(X) + EXP(-X)) - COSX
THE CORRECT VALUE OF THE INTEGRAL IS   4.794282267E-01

TOLERANCE	*CACRE*	ABSOLUTE ERROR	RELATIVE ERROR	*ERROR*	EVALUATIONS	*IFLAG*
1.0E-03	4.794282577E-01	3.10146E-08	6.46908E-08	3.04776E-06	17	1
1.0E-06	4.794282272E-01	4.84736E-10	1.01107E-09	4.00051E-08	33	1
1.0E-09	4.794282267E-01	1.33227E-13	2.77887E-13	4.84601E-10	33	1

FUNCTION NUMBER 5, WITH A = -1.00000 AND B = 1.00000   F(X) = 1 / (X**4 + X**2 + .9)
THE CORRECT VALUE OF THE INTEGRAL IS   1.582232564E+00

TOLERANCE	*CACRE*	ABSOLUTE ERROR	RELATIVE ERROR	*ERROR*	EVALUATIONS	*IFLAG*
1.0E-03	1.582232871E+00	9.28417E-08	5.86776E-08	4.64362E-04	33	1
1.0E-06	1.582232974E+00	1.01078E-08	6.38831E-09	2.96449E-07	49	1
1.0E-09	1.582232964E+00	4.24154E-12	2.68098E-12	3.45406E-10	129	1

FUNCTION NUMBER 6, WITH A = 0.00000 AND B = 1.00000   F(X) = X**1.5
THE CORRECT VALUE OF THE INTEGRAL IS   4.00000000E-01

TOLERANCE	*CACRE*	ABSOLUTE ERROR	RELATIVE ERROR	*ERROR*	EVALUATIONS	*IFLAG*
1.0E-03	4.000536050E-01	5.36050E-05	1.34013E-04	1.18222E-04	9	1
1.0E-06	4.000027743E-01	2.74251E-07	6.85728E-07	4.59762E-07	65	1
1.0E-09	4.000000003E-01	2.70158E-10	6.75495E-10	5.55244E-10	529	1

FUNCTION NUMBER 7, WITH A = 0.00000 AND B = 1.00000   F(X) = 1 / SQRT(X) , F(0) = 0
THE CORRECT VALUE OF THE INTEGRAL IS   2.00000000E+00

TOLERANCE	*CACRE*	ABSOLUTE ERROR	RELATIVE ERROR	*ERROR*	EVALUATIONS	*IFLAG*
1.0E-03	INTEGRAND SHOWS SINGULAR BEHAVIOR OF TYPE X**(-.50) BETWEEN 0. AND 1.00000000E+00					
	1.99837495E+00	1.62505E-04	8.12524E-05	4.15952E-04	33	2
1.0E-06	INTEGRAND SHOWS SINGULAR BEHAVIOR OF TYPE X**(-.50) BETWEEN 0. AND 6.25000000E-02					
	2.00000089E+00	8.9C754E-08	4.45377E-08	3.76451E-07	129	2
1.0E-09	INTEGRAND SHOWS SINGULAR BEHAVIOR OF TYPE X**(-.50) BETWEEN 0. AND 5.96046448E-08					
	2.00000000E+00	1.29469E-10	6.47447E-11	1.99954E-09	625	2

Fig. 6. (cont.)

444

FUNCTION NUMBER 8, WITH A = 0.0000 AND B = 1.0000   F(X) = 1 / (1 + X**4)
THE CORRECT VALUE OF THE INTEGRAL IS   8.6697259739E-01

TOLERANCE	*CADRE*	ABSOLUTE ERROR	RELATIVE ERROR	*ERROR*	EVALUATIONS	*IFLAG*
1.0E-03	8.6697220056E-01	7.8176E-07	9.0174E-07	4.42172E-05	9	1
1.0E-06	8.6697300466E-01	1.72957E-08	1.99513E-08	5.24399E-08	17	1
1.0E-09	8.6697298730E-01	2.05610E-13	2.41772E-13	5.11213E-11	65	1

FUNCTION NUMBER 9, WITH A = 0.0000 AND B = 1.0000   F(X) = 2 / (2 + SIN(31.4159 X))
THE CORRECT VALUE OF THE INTEGRAL IS   1.1547006695E+00

TOLERANCE	*CADRE*	ABSOLUTE ERROR	RELATIVE ERROR	*ERROR*	EVALUATIONS	*IFLAG*
1.0E-03	1.1547008410E+00	1.72451E-07	1.49347E-07	9.03810E-04	183	1
1.0E-06	1.1547006660E+00	3.32596E-09	2.88C36E-09	4.10028E-07	409	1
1.0E-09	1.1547006690E+00	8.3172CE-12	7.25495E-12	9.45265E-10	785	1

FUNCTION NUMBER 10, WITH A = 0.0000 AND B = 1.0000   F(X) = 1 / (1 + X)
THE CORRECT VALUE OF THE INTEGRAL IS   6.9314718066E-01

TOLERANCE	*CADRE*	ABSOLUTE ERROR	RELATIVE ERROR	*ERROR*	EVALUATIONS	*IFLAG*
1.0E-03	6.9314790150E-01	7.20521E-07	1.04CC7E-06	3.31459E-05	9	1
1.0E-06	6.9314719430E-01	1.37371E-08	1.98184E-08	4.58522E-07	17	1
1.0E-09	6.9314718060E-01	3.59856E-12	5.19211E-12	1.66568E-10	33	1

FUNCTION NUMBER 11, WITH A = 0.0000 AND B = 1.0000   F(X) = 1 / (1 + EXP(X))
THE CORRECT VALUE OF THE INTEGRAL IS   3.7985493CE-01

TOLERANCE	*CADRE*	ABSOLUTE ERROR	RELATIVE ERROR	*ERROR*	EVALUATIONS	*IFLAG*
1.0E-03	3.7988349660E-01	1.99648E-06	5.25548E-06	2.80548E-04	5	1
1.0E-06	3.7988549560E-01	2.544C8E-C9	6.69696E-C9	1.24939E-07	9	1
1.0E-09	3.7988549300E-01	1.35714E-12	3.5724SE-12	1.67044E-10	17	1

Fig. 6. (cont.)

445

FUNCTION NUMBER 12, WITH A = 0.00000 AND B = 1.00000   F(X) = X / (EXP(X) - 1) , F(0) = 1
THE CORRECT VALUE OF THE INTEGRAL IS   7.77504634341E-01

TOLERANCE	*CARE*	ABSOLUTE ERROR	RELATIVE ERROR	*ERROR*	EVALUATIONS	*IFLAG*
1.0E-03	7.775045939E-01	4.0231E-08	5.1744E-08	2.10080E-04	9	1
1.0E-06	7.775046343E-01	1.5584E-10	2.0044E-10	4.03874E-08	9	1
1.0E-09	7.775046341E-01	2.2204E-12	2.8586E-12	2.40039E-12	17	1

FUNCTION NUMBER 13, WITH A = .10000 AND B = 1.00000   F(X) = SIN(314.159 X)/(3.14159 X)
THE CORRECT VALUE OF THE INTEGRAL IS   9.09645256E-03

TOLERANCE	*CARE*	ABSOLUTE ERROR	RELATIVE ERROR	*ERROR*	EVALUATIONS	*IFLAG*
1.0E-03	9.098526099E-03	1.1915E-07	1.3096E-05	6.87399E-04	1028	1
1.0E-06	9.098644839E-03	4.1696E-10	4.5828E-08	2.18079E-06	1449	1
1.0E-09	9.098645256E-03	3.7255E-13	4.1243E-11	1.78358E-09	3505	1

FUNCTION NUMBER 14, WITH A = 0.00000 AND B = 10.00000   F(X) = SQRT(50)*EXP(-50*3.14159 X**2)
THE CORRECT VALUE OF THE INTEGRAL IS   5.00002112E-01

TOLERANCE	*CARE*	ABSOLUTE ERROR	RELATIVE ERROR	*ERROR*	EVALUATIONS	*IFLAG*
1.0E-03	INTEGRAND SEEMS TO HAVE JUMP OF SIZE 1.805974-106 BETWEEN 1.25000000E+00 AND 2.50000000E+00					
	INTEGRAND SEEMS TO HAVE JUMP OF SIZE 1.59C493E-26 BETWEEN 6.25000000E-01 AND 1.25000000E+00					
	INTEGRAND SEEMS TO HAVE JUMP OF SIZE 1.535364E-06 BETWEEN 3.12500000E-01 AND 6.25000000E-01					
	5.000015285E-01	1.31720E-06	2.63461E-06	1.24951E-04	62	3
	INTEGRAND SEEMS TO HAVE JUMP OF SIZE 1.805974-106 BETWEEN 1.25000000E+00 AND 2.50000000E+00					
1.0E-06	INTEGRAND SEEMS TO HAVE JUMP OF SIZE 1.59C493E-26 BETWEEN 6.25000000E-01 AND 1.25000000E+00					
	INTEGRAND SEEMS TO HAVE JUMP OF SIZE 1.539364E-06 BETWEEN 3.12500000E-01 AND 6.25000000E-01					
	5.000002574E-01	4.62227E-08	9.24453E-08	4.54250E-07	89	1
	INTEGRAND SEEMS TO HAVE JUMP OF SIZE 1.805974-106 BETWEEN 1.25000000E+00 AND 2.50000000E+00					
1.0E-09	INTEGRAND SEEMS TO HAVE JUMP OF SIZE 1.59C493E-26 BETWEEN 6.25000000E-01 AND 1.25000000E+00					
	INTEGRAND SHOWS SINGULAR BEHAVIOR OF TYPE X**(-.30) BETWEEN 4.68750000E-01 AND 6.25000000E-01					
	5.00002112E-01	1.16174E-12	2.32347E-12	1.86840E-10	202	3

FUNCTION NUMBER 15, WITH A = 0.00000 AND B = 10.00000   F(X) = 25 EXP(-25 X)
THE CORRECT VALUE OF THE INTEGRAL IS   1.00000000E+00

FIG. 6. (cont.)

446

TOLERANCE | *CADRE* | ABSOLUTE ERROR | RELATIVE ERROR | *ERRCR* | EVALUATIONS | *IFLAG*

TOLERANCE	*CADRE*	ABSOLUTE ERROR	RELATIVE ERRCR	*ERRCR*	EVALUATIONS	*IFLAG*
		INTEGRAND SEEMS TO HAVE JUMP OF SIZE 1.2916CE-53 BETWEEN		5.00000000E+00 ANC 1.00000000E+C1		
		INTEGRAND SEEMS TO HAVE JUMP OF SIZE 1.796945E-26 BETWEEN		2.50000000E+00 ANC 5.00000000E+C0		
		INTEGRAND SEEMS TO HAVE JUMP OF SIZE 6.697C88E-13 BETWEEN		1.25000000E+00 ANC 2.50000000E+C0		
1.0E-03	1.000001027E+C0	1.026E8E-C6	1.02668E-C6	1.32827E-04	88	3
		INTEGRAND SEEMS TO HAVE JUMP OF SIZE 1.2916CE-53 BETWEEN		5.00000000E+00 ANC 1.00000000E+C1		
		INTEGRAND SEEMS TO HAVE JUMP OF SIZE 1.796945E-26 BETWEEN		2.50000000E+00 ANC 5.00000000E+C0		
		INTEGRAND SEEMS TO HAVE JUMP OF SIZE 6.697C88E-13 BETWEEN		1.25000000E+00 ANC 2.50000000E+C0		
1.0E-06	1.000000005E+C0	4.915C7E-C9	4.91507E-C9	3.84416E-07	140	3
		INTEGRAND SEEMS TO HAVE JUMP OF SIZE 1.2916CE-53 BETWEEN		5.00000000E+00 ANC 1.00000000E+C1		
		INTEGRAND SEEMS TO HAVE JUMP OF SIZE 1.796945E-26 BETWEEN		2.50000000E+00 ANC 5.00000000E+C0		
		INTEGRAND SEEMS TO HAVE JUMP OF SIZE 6.69708E-13 BETWEEN		1.25000000E+00 ANC 2.50000000E+C0		
1.0E-09	1.000000000E+C0	8.5975E-12	8.59757E-12	4.88338E-10	215	3

FUNCTICN NUMBER 16, WITH A = 0.00000 AND B = 1C.CCCCC $F(X) = 50 / (3.14159 (2500 X^2 + 1))$
THE CCRRECT VALUE OF THE INTEGRAL IS 4.95263EC29E-01

TOLERANCE	*CADRE*	ABSOLUTE ERROR	RELATIVE ERROR	*ERRCR*	EVALUATIONS	*IFLAG*
1.0E-03	4.993917E1E-C1	5.3722SE-C6	1.07602E-C5	8.91151E-04	81	1
1.0E-06	4.9936390.48E-C1	1.01S234E-C7	2.04128E-07	1.12156E-06	145	1
1.0E-09	4.9936380.29E-C1	1.29958E-11	2.60248E-11	1.29052E-09	337	1

FUNCTICN NUMBER 17, WITH A = .01000 AND B = 1.CCCCC $F(X) = SIN(50*3.14159 X)^2 / 50(3.14159 X)^2$
THE CCRRECT VALUE OF THE INTEGRAL IS 1.12135669E-C1

TOLERANCE	*CADRE*	ABSOLUTE ERROR	RELATIVE ERROR	*ERRCR*	EVALUATIONS	*IFLAG*
1.0E-03	1.124982671E-C1	3.58657E-C4	3.19867E-C3	9.19022E-04	512	3
1.0E-06	1.121395710E-01	1.38E25E-C9	1.23796E-08	2.12313E-06	1237	1
1.0E-09	1.121395696E-C1	2.88081E-12	2.56895E-11	1.99374E-09	2329	1

FUNCTICN NUMBER 18, WITH A = 0.00000 AND B = 3.14159 $F(X) = CCS(CCSX + 3SINX + 2CCS2X + SIN2X + 3COS3X + 2SIN2X)$
THE CCRRECT VALUE OF THE INTEGRAL IS 8.3E676223.4E-01

TOLERANCE	*CADRE*	ABSOLUTE ERROR	RELATIVE ERROR	*ERRCR*	EVALUATIONS	*IFLAG*

Fig. 6. (cont.)

447

1.0E-03	8.38676459SE-01	1.36164E-C7	1.62356E-07	7.92000E-05	1C7	1
1.0E-06	8.38676327TE-01	4.31918E-C9	5.15000E-C9	4.40650E-07	177	1
1.0E-09	8.38676323 4E-01	1.53477E-12	1.82999E-12	4.32018E-10	417	1

FUNCTION NUMBER 19, WITH A = 0.00000 AND B = 1.00000   F(X) = NATURAL LOG OF X ,   F(0) = 0
THE CORRECT VALUE OF THE INTEGRAL IS   -1.0000000CE+C0

TOLERANCE	*CADRE*	ABSOLUTE ERROR	RELATIVE ERROR	*ERROR*	EVALUATIONS	*IFLAG*
	INTEGRAND SHOWS SINGULAR BEHAVIOR OF TYPE X**(-.09) BETWEEN 0.			AND 1.95312500E-C3		
1.0E-03	-1.00000141E+00	4.14C98E-C6	4.14098E-C6	6.83516E-04	137	2
	INTEGRAND SHOWS SINGULAR BEHAVIOR OF TYPE X**(-.C6) BETWEEN 0.			AND 1.9073486 3E-C6		
1.0E-06	-1.0000C013E+C0	1.2965 7E-C8	1.29657E-C8	7.88460E-07	233	2
	INTEGRAND SHOWS SINGULAR BEHAVIOR OF TYPE X**(-.C6) BETWEEN 0.			AND 1.9073486 3E-C6		
1.0E-09	-1.0000C00005E+C0	4.5CC3E-C9	4.5CCC3E-09	1.06760E-09	369	2

FUNCTION NUMBER 20, WITH A = -1.00000 AND B = 1.0CCCC   F(X) = 1 / (X**2 + 1.005)
THE CORRECT VALUE OF THE INTEGRAL IS   1.56439 6443E+CC

TOLERANCE	*CADRE*	ABSOLUTE ERROR	RELATIVE ERROR	*ERROR*	EVALUATIONS	*IFLAG*
1.0E-03	1.56439 7146E+C0	7.032C5E-C7	4.49508E-C7	3.78417E-06	17	1
1.0E-06	1.56439 6448E+C0	4.8C524E-C9	3.07424E-C9	1.82610E-08	33	1
1.0E-09	1.56439 6444E+C0	1.069CCE-C9	6.83333E-10	3.51042E-12	129	1

FUNCTION NUMBER 21, WITH A = 0.00CC0 AND B = 1.0CCCC   F(X) = SUM OF POWERS CF HYPERBOLIC SECANT, SEE KAHANER
THE CORRECT VALUE OF THE INTEGRAL IS   2.10602 7354E-01

TOLERANCE	*CADRE*	ABSOLUTE ERROR	RELATIVE ERROR	*ERROR*	EVALUATIONS	*IFLAG*
1.0E-03	2.09736358 1E-C1	1.C662 8E-C3	5.05865E-C3	1.15653E-04	1C8	1
1.0E-06	2.09736073 2E-C1	1.C6666E-C3	5.C6000E-C3	4.72801E-07	189	1
1.0E-09	2.10802735 5E-C1	1.12262E-1C	5.32639E-1C	1.27031E-09	661	1

Fig. 6. (cont.)

REFERENCES

Davis, P. J., and Rabinowitz, P. (1967). "Numerical Integration." Ginn (Blaisdell), Boston, Massachusetts.

Lynch, R. E. (1967). Generalized trapezoid formulas and errors in Romberg quadrature, Blanch anniversary volume, pp. 215–229. Aerospace Research Laboratories, Office of Aerospace Research, United States Air Force.

Lyness, J. N., and Ninham, B. W. (1967). Numerical quadrature and asymptotic expansions, Mathematics of Computations, XXI, pp. 162–178.

Wynn, P. (1963). Acceleration techniques in numerical analysis, with particular reference to problems in one independent variable, "Information Processing 1962" (C. M. Popplewell, ed.). North-Holland Publ., Amsterdam.

CHAPTER 8

SQUARS: AN ALGORITHM FOR
LEAST-SQUARES APPROXIMATION*

John R. Rice

PURDUE UNIVERSITY

LAFAYETTE, INDIANA

I. Introduction

A. THE LEAST-SQUARES PROBLEM

The simplest case of this problem occurs when one is given a function $f(x)$ defined on the interval $[a, b]$ and one desires the polynomial $P_n(x)$ of specified degree n which is the best approximation to $f(x)$ in the sense that

$$\int_a^b (f(x) - P_n(x))^2 \, dx \tag{1}$$

is minimized among all polynomials of degree n. The algorithm described here attempts to solve this problem and a number of closely related ones.

B. MAIN ALGORITHM FEATURES

The main features of the algorithm are as follows:

1. Exact integration is replaced by approximate integration using the trapezoidal rule or the rectangular rule. Note that minimization of the sum of the squares of the errors (as in most statistical regression algorithms) is just the rectangular rule to approximate the integral.

* This work was supported in part by NSF Grant GP-05850.

2. The polynomials may be replaced by an arbitrary set of basis functions supplied by the user. The use of polynomials is the default condition.

3. A weight function may be specified in the integral (but not with the rectangular rule), and, in particular, a weight function providing relative error approximation is supplied upon request.

4. A trend function may be inserted to multiply the approximating function (but not with the rectangular rule).

5. A desired error of approximation (in the least-squares sense) may be specified and the degree (or number of supplied basis functions) is increased until that error is achieved (if possible).

In summary, the algorithm attempts to solve the problem

$$\int_a^b \left[f(x) - \text{TREND}(X) \sum_{I=1}^{MB} \text{COEF}(I)\, \text{BASE}(I, X) \right]^2 \text{WEIGHT}(X)\, dx$$
$$= \text{MINIMUM} \tag{2}$$

or

$$\int_a^b \left[f(x) - \text{TREND}(X) \sum_{I=1}^{MB} \text{COEF}(I)\, \text{BASE}(I, X) \right]^2 \text{WEIGHT}(X)\, dx$$
$$\leq \text{EPS}^2 \tag{3}$$

C. EXAMPLE PROGRAM USES

The algorithm may be used by any one of four names. The following examples (Fig. 1) indicate the typical uses. More detailed definitions are given later.

```
          REAL X(200),CATA(200)
     C    ----------
          PDEG = 6
          NPTS = 150
          ERR = PCLCEG(X,CATA,NPTS,MDEG)
     C         POLCEG CIVES POLYNOMIAL APPROX WITH DEGREE MDEG
     C    ----------
          EPS = .00001
          ERR = FCLEPS(X,CATA,NPTS,EPS)
     C         POLEPS CETERMINES THE DEGREE TO GET AN ERROR OF EPS OR LESS
     C    ----------
          IB = 1
          IT = 1
          IW = 1
          LV = 1
          ERR = SQUARS(X,CATA,NPTS,EPS,IB,IT,IW,MDEG,LV)
     C         SQUARS IS THE PRIMARY ALGORITHM WITH ARGUMENTS
     C         IB = SWITCH FCR NCN-PCLYNOMIAL BASIS SUPPLIED
     C         IW = SWITCH FCR WEIGHT FUNCTION USES (3 CHOICES)
     C         IT = SWITCH FCR TREND FUNCTION SUPPLIED
     C         LV = LEVEL OF OUTPUT DESIRED (4 CHOICES)
     C    ----------
          ERR = SUMSCR(X,CATA,NPTS,EPS,IB,MDEG,LV)
     C         SUMSCR IS A VERSION WHICH USES THE SUM OF SQUARES
     C         TO APPROXIMATE THE INTEGRAL (RECTANGULAR RLLE)
     C         FURTHER EXAMPLES USES ARE GIVEN AFTER THE PROGRAM
```

FIG. 1. Typical uses of the versions of SQUARS.

D. ALGORITHM OUTPUT AND RESULTS

The algorithm returns information to the user in three distinct ways:

1. *Printed Output*

There are four levels:

-1 level = no printed output at all
0 level = success/failure message
1 level = some general information
2 level = some details of the computation

2. *The Best Approximation*

There is a function POLFIT(X) which is the best approximation obtained in the most recent execution of SQUARS.

3. *Variables in the COMMON block SQUARA*

A variety of information is available to the user in a labeled COMMON block SQUARA. This information is specified in detail in the comments for SQUARS, and it includes almost all variables which are meaningful to the user of the algorithm.

II. Mathematical Analysis

This section briefly outlines the mathematical procedures used in the algorithm. More detailed discussions of orthogonal polynomials and their three-term recurrence relation are given in Rice (1964, Chap. 2) and in Hart *et al.* (1968, Sect. 4.5). For more details on Gram–Schmidt orthogonalizations see Rice (1964, Chap. 2), Rice (1966), and Rice (1969, Sect. 12–10).

Let $P_k(x)$ denote a polynomial of degree k, then we say that $P_k(x)$ and $P_m(x)$ are orthogonal on the interval $[a, b]$ if

$$\int_a^b P_k(x) P_m(x) \, dx = 0 \tag{4}$$

A set of polynomials $\{P_k(x), k = 0, 1, 2, \dots, M\}$, is said to be an orthonormal set of polynomials if

$$\int_a^b P_k(x) P_m(x) \, dx = \begin{cases} 0, & k \neq m \\ 1, & k = m \end{cases} \tag{5}$$

454   JOHN R. RICE

A set of orthonormal (or even only orthogonal) polynomials satisfy a three-term recurrence relationship

$$P_{k+1}(x) = (A_k x + B_k)\,P_k(x) - C_k\,P_{k-1}(x) \qquad (6)$$

This three-term relationship is used in SQUARS to define the orthogonal polynomials and to evaluate them. It is clear that Eq. (6) may be used to evaluate $P_k(x)$ (as is done in the function POL). Less clear, but very significant, is the fact that the polynomial

$$Q(x) = \sum_{I=0}^{M} \text{COEF}(I)\,P_I(x)$$

may be evaluated (with I descending) by the scheme

$$V_M = \text{COEF}(M), \qquad V_{M-1} = (A_{M-1}x + B_{M-1})V_M + \text{COEF}(M-1)$$
$$V_K = (A_K x + B_K)V_{K+1} - C_{K+1}V_{K+2} + \text{COEF}(K), \qquad K = M-2,\dots,0,$$
$$Q(x) = V_0 \qquad (7)$$

as is done in the function POLFIT.

One may compute the three-term recurrence coefficients A_k, B_k, and C_k as follows [Forsythe (1957)]: Start with

$$P_0(x) = \left(\int_a^b dx\right)^{-1/2}, \qquad P_1(x) = A_1 x + B_1$$

where

$$\beta = \int_a^b x\,dx, \qquad \alpha = \int_a^b (1 - \beta x)^2\,dx$$
$$A_1 = 1/\alpha, \qquad B_1 = \beta/\alpha$$

This construction simply makes $P_0(x)$ and $P_1(x)$ orthonormal on $[a, b]$. The remaining recurrence coefficients are obtained by noting that normalization and Eq. (6) imply

$$B_k/A_k = -\int_a^b x(P_k(x))^2\,dx \qquad (8)$$

$$C_k/A_k = \int_a^b x\,P_k(x)\,P_{k-1}(x)\,dx \qquad (9)$$

The final condition is that $P_{k+1}(x)$ is normalized; i.e.,

$$\int_a^b (P_{k+1}(x))^2\,dx = 1 \qquad (10)$$

Note that SQUARS indexes these coefficients differently because Fortran does not allow nonpositive subscripts. Thus SQUARS uses

$$P_0(x) = A0$$
$$P_{k+1}(x) = (AAA_{k+1}x + BBB_{k+1})P_k(x) - CCC_{k+1}P_{k-1}(x) \qquad (11)$$

The evidence available at this time indicates that for the computation of least-squares polynomial approximations this approach is one of the best conditioned, and, further, it is significantly more efficient for higher-degree polynomials (say degree ≥ 5).

When a set of basis functions other than polynomials is specified one cannot use the three-term recurrence relationship, and, in SQUARS, it is replaced by the Gram–Schmidt orthogonalization process. Let $\{\phi_i(x) \mid i = 1, 2, ..., M\}$ be a given set of functions (one may identify them as the basis functions supplied by a user of SQUARS) and we desire to obtain an equivalent set $\psi_i(x)$. The process is started as follows:

$$\psi_1(x) = \phi_1(x) \Big/ \left(\int_a^b (\phi_1(x))^2 \, dx \right)^{1/2}$$

$$\phi_j^2(x) = \phi_j(x) - \psi_1(x) \int_a^b \psi_1(x) \, \phi_j(x) \, dx, \qquad j = 2, 3, ..., M \tag{12}$$

Then we have

$$\psi_2(x) = \phi_2^2(x) \Big/ \left(\int_a^b (\phi_2^2(x))^2 \, dx \right)^{1/2}$$

$$\phi_j^3(x) = \phi_j^2(x) - \psi_2(x) \int_a^b \psi_2(x) \, \phi_j^2(x) \, dx, \qquad j = 3, 4, ..., M$$

and the general kth step is

$$\psi_k(x) = \phi_k^k(x) \Big/ \left(\int_a^b (\phi_k^k(x))^2 \, dx \right)^{1/2}$$

$$\phi_j^{k+1}(x) = \phi_j^k(x) - \psi_k(x) \int_a^b \psi_k(x) \, \phi_j^k(x) \, dx, \qquad j = k + 1, ..., M \tag{13}$$

The primary alternative to this approach is the use of Householder transformations [Businger and Golub (1965), Hanson and Lawson (1969), and Golub and Reinsch (1970)]. The use of Householder transformations is occasionally significantly better conditioned than the modified Gram–Schmidt process, but it is likely to be slower.

The primary work in least-squares approximations is obtaining the orthonormal functions or polynomials. Let $\psi_i(x)$ be such a set; then the best least-squares approximation to $f(x)$ is

$$Q(x) = \sum_{I=1}^{M} \text{COEF}(I) \, \psi_I(x)$$

where

$$\text{COEF}(I) = \int_a^b f(x) \, \psi_I(x) \, dx \tag{14}$$

III. Numerical Procedures

There are six points in the algorithms where the numerical procedures used merit some comment. These are discussed one by one below.

A. NUMERICAL INTEGRATION

The function subprogram TRAPD defines the inner product of the least-squares approximations. It can be replaced by a completely different inner product. This function has, in fact, two inner products in it selected according to the setting of ISUMSQ. With the normal setting ISUMSQ = 0 the trapezoidal rule is used to approximate the integral. This approximate integral is a linear function of the integrand, and the appropriate coefficients (weights) are computed and stored in the arrays TTW and TW. These coefficients include the step size, and the weight and trend functions (if any). The advantage is a gain in speed, but the cost is the need for two fairly large arrays (currently with 500 elements) to be allocated whether needed or not. One may easily modify TRAPD to compute these quantities each time and thereby save this memory space.

B. SUM OF SQUARES

The second part of TRAPD uses an inner product based simply on a sum of squares. The primary reason for such an inner product is that one does not need to have data of a one-dimensional nature. One can adapt almost any problem into this form with a little ingenuity. Consider, for a simple example, data defined on a rectangular grid in the (x, y) plane with increments in the x and y variables of XSTEP and YSTEP, and suppose these are XPTS points in the x direction. In order to use polynomials of two variables with maximum x degree of NDEG, one can use the following function (Fig. 2) to define the basis (NDEG1 = NDEG + 1).

```
       FUNCTICN BASE(J,T)
   C   THIS PRCVICES THE VALUE CF THE J-TH BASIS AT THE T-TH POINT
       CCMMON/ INFC / NCEG1,XSTEP,YSTEP,XPTS
       BASE = 1.
       IF( J .EC. 1 )            RETURN
       JY = (J-1)/NCEG1
       JX = J-JY*NCEG1-1
       K = (T-1.)/XPTS
       Y = FLCAT(K)*YSTEP
       X = (T-1.-FLCAT(K))*XSTEP
       BASE = X**JX*Y**JY
       RETURN
       END
```

FIG. 2.　A Fortran subroutine to generate a set of basis functions for a two-dimensional approximation problem.

A secondary reason for providing this feature is that a large number of people are in the habit of using it even when an integral is more appropriate.

C. Linear Dependence Test for Basis Functions

As each new basis function is brought into the Gram–Schmidt process, a test is made of its dependence on the previous basis functions. This test is made on the size of the orthogonal function just before it is normalized. If the size of this new orthogonal function is less than 100 times a one-bit unit in the machine word, then this basis function is discarded. This test is made using the machine-dependent constant BIT. If the basis functions supplied by the users are, in fact, powers of x, one can expect some basis functions to be discarded soon after the Mth one, where M is the number of decimal digits in a floating point machine word. The three-term recurrence approach does not have this difficulty, and thus the user should not supply the powers of x as basis functions except for numerical experiments.

D. Attainment of Specified Accuracy

When an accuracy of approximation is specified, one increases the polynomial degree (or the number of basis functions) until it is achieved. However, there is a reasonable probability that the user will specify an accuracy that is unattainable. In order to protect the user against this possibility, a check is made in the subprogram ERRCHK. If the least-squares error does not decrease by 10% for three consecutive increases in the degree, the computation is terminated. Furthermore, if it fails to decrease by 10% for two consecutive times and the accuracy is within twice that requested, the computation is terminated. This process is controlled by three numerical control constants marked in POLAPX and ERRCHK.

E. Iterative Refinement of the Solution

Many least-squares problems lead to substantial errors due to round-off effects. Thus one might not have completely orthogonal functions or might not have accurate coefficients due to these effects. Iterative refinement is one of the two numerical procedures in this algorithm which alleviate this difficulty somewhat.

In iterative refinement one recomputes the error curve in double precision and a correction to the coefficients is obtained by approximating this new error curve. Note that the algorithm is organized so that it always approximates the error curve (UERR). The sizes of the corrections are used to decide when the refinement is to stop. The current algorithm is also limited to three refinements. Note that the orthogonal functions (or polynomials) are *not* recomputed during this process, and thus it is relatively cheap. This technique is effective in obtaining extremely accurate solutions when the error (or least-squares residuals) are relatively small. For example, SQUARS obtains full accuracy on the test problems Y1 (with zero residuals) and Y1∗ (with residuals the order of 2.0) of Wampler (1969). This is with computation on a CDC 6500.

Refinement may be bypassed by specifying the number (NPTS) of points in the data as minus the actual number.

F. REORTHOGONALIZATION FOR THE GRAM–SCHMIDT PROCESS

The basis functions that arise naturally in applications have a good probability of being nearly linearly dependent. Extreme linear dependence is detected by the algorithm and the corresponding basis functions ignored. However, much before this occurs the modified Gram–Schmidt orthogonalization breaks down. A spot check is made in GRMCHK for significant lack of orthogonality and, if it is detected, a reorthogonalization is performed. In fact the entire computation is redone starting with the basis that resulted from the first attempt at orthogonalization. The experiments of Rice (1966) indicate that one reorthogonalization (or reinforcement) of the modified Gram–Schmidt process is normally sufficient and only one is done in SQUARS.

G. EVALUATION OF THE BASIS FUNCTIONS

A very substantial improvement in the speed of SQUARS can be achieved at the expense of a very substantial increase in the memory required. This is a typical instance of the trade-off between these two aspects of an algorithm. The choice of slow speed and smaller memory for SQUARS is arbitrary. Reasons for it include: (a) the multiprograming environment in which SQUARS was developed has excess central processor time relative to memory capacity; (b) typical problems involving a low degree (say ≤ 5) and a small number of points (say ≤ 20) execute very rapidly in any case. There certainly are circumstances where SQUARS should be modified so that the values of the basis functions (whether orthogonal polynomials, user specified or orthogonalized sets of the user-specified basis) are kept in arrays.

Approximately 80–90% of the execution time of SQUARS is spent in the evaluation of the basis functions. Thus one can expect a modified version of SQUARS to run from four to eight times faster than this version. The current dimensions of SQUARS allow 500 data points and 30 basis functions. Thus the auxiliary array required to save all the evaluations of the basis functions has 15,000 elements.

IV. The Algorithm SQUARS

The function subprogram SQUARS is preceded by a BLOCK DATA subprogram (see Fig. 3). This subprogram pertains to machine, system, and compiler dependencies and would normally not be present in any specific library version of SQUARS.

```
       BLCCK DATA
       CCMMON/ FCATAQ / LX,BIT,PRIOR,EPSMIN,SIZE,
      *          MAXBAS,MAXLNC,JUERR,JSIZE,JORTHO,TW(5CC),TTW(5CC)
C      ****** MACHINE, SYSTEM ANC COMPILER DEPENDENCIES ******
C
C MACHINE - THE MAIN VARIABLE EXPLICITLY MACHINE DEPENDENT IS BIT
C            THE ACCURACIES OBTAINABLE DEPEND, CF CCLRSE, ON THE
C            FLCATING POINT WORC LENGTH AND THE CONCITICNING
C          - THE CCNSTANT .2E+38 IS USED AS THE FAILLRE VALLE CF SCLARS
C            IT APPEARS NEAR 55-POLAPX, 120-SQLARS,2C-WWWW
C
C SYSTEM  - THE STANDARC SYSTEMS OUTPUT TAPE OR FILE IS TAKEN TC BE 6
C          - THE SYSTEM IS ASSUMED TO PUT A PROGRAM INTO EXECUTICN
C            EVEN WITH UNSATISFIED EXTERNALS. OTHERWISE, FLNCTICN
C            SUBRCUTINES TRENC AND WEIGHT MUST BE SLPPLIED BY THE LSER
C
C CCMPILER- LABELLED CCMMON IS USEC BUT MAY REPLACED BY LSING ARGLMENTS
C            BLANK COMMCN SHOULC NOT BE USED FOR VARIABLES CF SQLARS
C          - ARRAYS PASSEC AS ARGUMENTS ARE DIMENSICNED AT 1. CCMPILERS
C            WITH SUBSCRIPT CHECKING, ETC. WILL PRODLCE BAD CCDE HERE.
C          - MIXEC COUBLE PRECISION AND SINGLE PRECISION ARITHMETIC IS
C            USED IN TRAPC. IT IS ASSUMED THAT SINGLE PRECISION VARIABLES
C            ARE CCNVERTEC TO DCUBLE PRECISION IN MIXED EXPRESSICNS
C            ****** MACHINE CEPENDENT CONSTANT ******
C      BIT = 100 TIMES RELATIVE PRECISION OF MACHINE FLOATING POINT
C            IF FLOATING PT. WORD HAS K BITS THEN BIT = 1CC*2**(-K)
       CATA BIT / 1.E-12 /
C
C      ****** DIMENSION CEPENCENCIES ******
C      TWC BASIC VARIABLES CETERMINE THE SIZE OF PRCBLEMS FOR SQLARS
C            MAXLNG = MAXIMUM LENGTH OF THE DATA
C            MAXBAS = MAXIMUM NUMBER OF BASIS FLNCTIONS
C      THE FCLLCWING ARRAYS HAVE DIMENSION - MAXLNG - ALL ARE IN CEMMCN
C            UERR - TW - TTW
C      THE FCLLCWING ARRAYS HAVE DIMENSION - MAXBAS -
C            CCEF-PCOE-AAA-BBB-CCC-TR-ERRSAV            THESE ARE IN CCMMCN
C            CSAVE-PSAVE                                THESE ARE IN PCLAPX
C      THE FCLLCWING VARIABLES ARE DEPENDENT ON MAXBAS AND SET IN SCLARS
C            JUERR = MAXBAS + 1 = PSEUDO INCEX CF THE ERROR CLRVE
C            JSIZE = MAXBAS + 2 = PSEUDO INDEX OF ERRCR CLRVE SQUARED
C            JORTHC= MAXBAS +10 = PSEUDO INDEX FOR 1-ST ORTHOG. BASIS
       CATA  MAXLNG,MAXBAS,JUERR,JSIZE,JORTHO/5CC,3C,31,32,4C/
       ENC
       FUNCTICN SCUARS(X,Y,NPTS,EPS,IB,IT,IW,MP,LV)
C      THE VALUE CF SQUARS IS THE LEAST SQUARES ERRCR OBTAINED
C            EXCEPT IN THE CASE OF FAILURE, THEN IT IS .2E+38
C-INPUT = X,Y,NPTS,EPS,IB,IT,IW,MP,LV (ARGUMENTS) ISLMSQ,MAXBAS,MAXLNG
C-CUTPUT= NREINF,LX,EPSMIN,IBASE,ITREND,IWEGHT,MBPCL,LERR
C         MPRNT1,MPRNT2,MPRNT3,MPRNT4,MDEG,SQUARS,ISUMSQ,TW,TTW
       REAL X(1),Y(1)
C
C                 ARGUMENTS ARE
C      X    = ARRAY OF ABSCISSA - CALLED  XX  AT MANY POINTS
C      Y    = ARRAY OF CRCINATES- CALLED  U   AT MANY POINTS
C      NPTS = NUMBER OF CATA POINTS - ABSOLUTE VALLE TAKEN
C             ALSC USEC AS SWITCH TO SIGNAL NO REINFCRCEMENT WHEN NEGATIVE
C      EPS  = CESIREC LEAST SQUARES ERROR OF APPROXIMATICN (REQLIRES MP=-1)
C      IB   = 1 IF A SET CF BASIS FUNCTIONS ARE PROVIDED
C           = 0 IF PCLYNCMIALS ARE TO BE USED
C      IT   = 1 IF A TRENC FUNCTION IS PROVIDED
C             IF IT = 1 THEN A FUNCTION SUBPROGRAM NAMED TREND(X)
C                       MUST BE PROVIDED
C           = 0 IF 1. IS TO BE USED
C      IW   = 0 IF WEIGHT = 1. IS TO BE USED
C           = 1 IF WEIGHT CIVING RELATIVE ERROR FIT IS TO BE LSED
C           = 2 IF WEIGHT IS PROVICED
C             IF IW = 2 THEN A FUNCTION SUBPROGRAM NAMED WEIGHT(X)
C                       MUST BE PROVIDED
C      MP   = -1 IF EPS ACCURACY IS TO BE MET
C           = CEGREE (OR NUMBER CF BASIS FUNCTIONS ) CTHERWISE
C      LV   = -1 IF NO OUTPUT AT ALL IS DESIREC
C           = 0 IF SUCESS/FAILURE MESSAGE ONLY IS DESIRED
```

FIG. 3.   The algorithm SQUARS.

```
C                = 1 IF GENERAL INFORMATION IS DESIRED
C                = 2 IF DETAILED INFORMATION IS DESIRED
C
C                    POLFIT(X)
C       IS A FUNCTION AUTOMATICALLY PROVIDED WHICH IS THE APPROXIMATION
C       COMPUTED BY SQUARS. THUS ONE CALLS ON SQLARS AND THEN ONE CAN
C       EVALUATE THE FIT OBTAINED AT ANY POINT DESIRED BY MEANS OF POLFIT
C
C       MANY RESULTS ARE RETURNED VIA THE COMMON BLOCK / SQUARA /
C       THESE VARIABLES ARE DEFINED BELOW
C
C---------- EVALUATION SCHEME FOR POLY APPROX OBTAINED -------
C---------- USE THE FORTRAN CODE OF FUNCTION POLFIT
C
C                SUBROUTINE LIST
C       SQUARS   = HOUSEKEEPPING DRIVER, SETS PRINTS, ETC.
C       POLAPX   = MAIN CONTROL PROGRAM FOR LEAST SQUARES APPROXIMATION
C       PAFFRX   = CONTAINS VARIOUS ORTHOGONALIZATION CODES
C                  COMPUTES COEFFICIENTS( ONE PER CALL) AND ERROR CURVE
C       ERRCHK   = COMPUTES VARIOUS ERRORS, TEST VARIABLES FOR TERMINATION
C       GRMCHK   = SIMPLE TEST OF THE SUCCESS OF GRAM-SCHMIDT ORTHOGONALIZATION
C       TRAFO    = TRAPEZOICAL RULE FOR DOT PRODUCTS
C       POL      = FUNCTION - THE ORTHONORMAL POLYNOMIALS GENERATED
C       POLFIT   = FUNCTION - THE APPROXIMATION OBTAINED
C                  OR OF THE ORIGINAL GIVEN BASIS FUNCTIONS
C       QBASE    = FUNCTION - THE BASIS FUNCTIONS PLUS THE ORTHONORMAL ONES
C       hhhh     = WEIGHT FUNCTION
C       TTTT     = TREND FUNCTION
C       POWERS   = CONVERTS ORTHOG POLYNOMIAL COEFS. TO POWERS OF X COEFS.
C       POLEPS   = SIMPLE VERSION OF SQUARS (FEWER ARGUMENTS)
C       POLDEG   = SIMPLE VERSION OF SQUARS (FEWER ARGUMENTS)
C       SUMSQR   = VERSION OF SQUARS WITH TRAPEZOIDAL RULE REPLACED BY RECT.
C                  RULE - THIS GIVES SUM OF SQUARES MINIMIZATION
C                  NO TREND OR WEIGHT FUNCTION ALLOWED
C                  - ENTRY POINT GRMCHK IS USED IN ERRCHK, GRMCHK CAN BE MADE
C                  INTO AN INDEPENDENT FUNCTION WITHOUT LOSS OF EFFICIENCY
C       NOTE ******* THE USER MIGHT ALSO SUPPLY
C       TREND    = HIS OWN TREND FUNCTION
C       WEIGHT   = HIS OWN WEIGHT FUNCTION
C
C                ***** COMMON BLOCK   SQUARA *****
C                ERRL2    = LEAST SQUARES ERROR
C                ERRMAX   = MAXIMUM ERROR - OCCURS AT THE POINT XX(LOCMAX)
C                ERRAVE   = AVERAGE ERROR
C                MB       = NO. OF BASIS FUNCTIONS( OR POLY. DEGREE +1 )
C                MDEG     = FINAL POLYNOMIAL DEGREE = MB - 1
C                ERRSAV   = THE VALUES OF ERRL2 FOR MB = 1,2,...
C                PCOE     = BEST APPROXIMATION COEFFICIENTS OF ORTHOGONAL POLYS
C                COEF     = BEST APPROXIMATION COEFFICIENTS OF POWERS OF X
C                           OR OF THE ORTHOGONALIZED BASIS FUNCTIONS
C                AO       = 1-ST ORTHOG POLYNOMIAL VALUE = CONSTANT
C                AAA      = 3-TERM RECURRECE COEFFICIENTS FOR ORTHOGONAL POLYNOMIALS
C                BBB          P(N,X) =
C                CCC      (AAA(N)*X + BBB(N))*P(N-1,X) - CCC(N)*P(N-2,X)
C                TR       = MATRIX TO TRANSFORM ORIG BASIS INTO ORTHOG BASIS
C                UERR     = CURRENT ERROR CURVE - CHANGES WITH EACH CALL OF PAFFRX
C       COMMON/ SQUARA / ERRL2,ERRMAX,LOCMAX,ERRAVE,MB,MDEG,ERRSAV(30),
C     *    CCEF(30),PCOE(30),AO,AAA(30),BBB(30),CCC(30),TR(30,30),UERR(500)
C
C                ***** COMMON BLOCK  SWTCHZ *****
C                IBASE   = IB OF SQUARS ARGUMENT
C                ITREND  = IT OF SQUARS ARGUMENT
C                IWEGHT  = IW+1 OF SQUARS ARGUMENT
C                MBPOL   = MP OR MP+1 OF SQUARS( DEPENDS ON IB = 1 OR 0 )
C                ISUMSQ  = 0 NORMALLY
C                        = 1 IF SQUARS ENTERED VIA SUMSQR - USED IN TRAPD
C                IWEFT   = 1 NORMALLY
C                        = 2 TO INSERT A FACTOR OF X IN TRAPD FOR FORSYTHE RLLE
C                NP      = COUNTER ON DEGREE INCREASES WITHOUT SUFF ERROR DECREAS
C                IREINF  = SWITCH FOR REFINEMENT MODE - LOGICAL VARIABLE
C                JPASS   = SWITCH FOR REORTHOGONALIZATION OF GIVEN BASIS FUNCTIONS
C                MPRNT1  = PRINT CONTROL FOR LEVEL   C
```

FIG. 3. (cont.)

```
C        MPRNT2 = PRINT CONTROL FOR LEVEL   1
C        MPRNT3 = PRINT CONTROL FOR LEVEL   2
C        MPRNT4 = PRINT CONTROL FOR SUCESS-FAILURE RESULT
C        NREINF = NUMBER OF ITERATIVE REFINEMENTS MADE - ALSO A SWITCH
C
      COMMON/ SWTCHZ / IBASE,ITREND,IWEGHT,MBPOL,ISUMSQ,IWEHT,NP,IREINF,
     *          JPASS,MPRNT1,MPRNT2,MPRNT3,MPRNT4,NREINF
C        ***** COMMON BLOCK   FDATAQ *****
C        LX     = NUMBER OF DATA POINTS
C        BIT    = MACHINE DEPENDENT CONSTANT
C        PRIOR  = PRECEEDING VALUE OF ERRL2
C        EPSMIN = DESIRED ERROR IN APPROXIMATION
C        SIZE   = ESTIMATE OF THE SIZE OF THE DATA
C        MAXBAS,MAXLNG = DIMENSION DEPENDEND CONSTANTS
C          JUERR = MAXBAS+1, JSIZE = MAXBAS+2, JORTHO = MAXBAS+10
C        TW,TTW = COMBINATIONS OF TRENDS, WEIGHTS AND X-VALUES USED TO
C                 SPEED THE COMPUTATION OF THE DOT PRODUCT IN TRAPD
C                 TO SAVE STORAGE AT THE EXPENSE OF SPEED ONE CAN
C                 REPLACE THESE BY APPROPRIATE FUNCTION CALLS
C
      COMMON/ FDATAQ / LX,BIT,PRIOR,EPSMIN,SIZE,
     *      MAXBAS,MAXLNG,JUERR,JSIZE,JORTHO,TW(500),TTW(500)
C
      ISUMSQ = 0
C        CHECK FOR CALL VIA SUMSQR - HAS IT = -1070207
      IF( IT .NE. -1070207 )            GO TO 1
      IT = 0
      ISUMSQ  = 1
C              INITIALIZE REFINEMENT CONTROLS
    1 NREINF = 0
C        NEGATIVE NPTS SIGNALS NO REFINEMENT OR REORTHOGONALIZATION
      IF(NPTS .LT. 0 ) NREINF = -1
C                 CHECK ARGUMENTS
C        CHANGE SOME NAMES TO PUT THINGS IN COMMON
      LX = IABS(NPTS)
      IF( LX .GT. MAXLNG .OR. LX .EQ. 0)        GO TO 105
      EPSMIN = ABS(EPS)
      IF(IABS(IB*(IB-1))+IABS(IT*(IT-1))+IABS(IW*(IW-1)*(IW-2)) .NE. 0)
     *                    GO TO 100
      IBASE = IB
      ITREND = IT
      IWEGHT = IW + 1
      MBPOL = MP
      IF( EPS.GT.0.0)                   GO TO 2
      IF( MP .LE. -1)                   GO TO 100
      IF( IABS(IB-1)+IABS(MP) .EQ. 0)   GO TO 100
                                        GO TO 3
    2 IF( MP .EQ. 0 )  MBPOL = -1
    3 IF( IB .EQ. 0 )  MBPOL = MBPOL + 1
      IF( MBPOL.GT.MAXBAS .OR. MBPOL.GT.LX)     GO TO 105
      IF( ISUMSQ .EQ. 1 )              GO TO 7
C        INITIALIZE TW, TTW ARRAYS, CHECK ON ORDER OF THE X(I)
C        THESE ARRAYS ARE THE COEF. OF THE TRAPEZOIDAL RULE AS A LINEAR
C        FUNCTION OF THE INTEGRAND
      DO 5 K = 1,LX
      XK = X(K)
      DX = X(K+1)-X(K-1)
      IF( K .EQ. 1) DX = X(2) -XK
      IF( K .EQ.LX) DX = XK - X(LX-1)
      IF( DX .LT. 0.)                  GO TO 110
      TK = 1.
      IF( ITREND .EQ. 1 ) TK = TREND(XK)
      TW(K) = .5*TK*DX
      IF( IWEGHT .NE. 1 ) TW(K)= TW(K)*WWWW(K,XK,Y)
      TTW(K) = TW(K)*TK
C        PROTECT AGAINST DIVIDING BY ZERO IN TRAPD WHEN TREND = 0
    5 IF( TTW(K) .EQ. 0. ) TTW(K) = 1.E-30
C        PUT THE Y-ARRAY INTO UERR TO WORK ON IT
    7 DO 8 I = 1,LX
    8 UERR(I) = Y(I)
C                 SET PRINT CONTROLS
      MPRNT1 = 0
      MPRNT2 = 0
```

FIG. 3. (cont.)

```
          MPRNT3 = 0
          MPRNT4 = 0
          IF( LV .EQ.-1 )              GO TO 1C
          MPRNT1 = 101
          IF( LV .EQ. 0 )              GO TO 1C
          MPRNT2 = 101
          IF( LV .EQ. 1 )              GO TO 1C
          MPRNT3 = 101
   10 CALL PCLAPX(X,Y)
C       SQUARS IS THE L2-ERROR UNLESS A FAILURE OCCLRS- THEN .GT. 1C**37
   50 SQUARS = PRIOR
          MDEG = MB-1
          RETURN
C
C          OUTPUT FOR ERRORS/INCONSISTANCIES IN THE ARGLMENTS
  100 IF( LV .NE. -1 ) WRITE(6,101) EPS,IB,IT,IW,MP
  101 FCRMAT(5X41H*** ILLEGAL/INCOMPATIBLE SQUARS ARGLMENTS /8X 5HEPS =
     *  E14.5,18H IB,IT,IW ANC MP = 4I5 )
                                      GO TO 12C
  105 IF( LV .NE. -1 ) WRITE(6,106) MAXLNG,MAXBAS,LX,MBPOL
  106 FCRMAT(5X45H***SCUARS CIMENSION INCONSISTANCIES, BCLNDS = 214,
     *          10H, ACTUAL = 214 )
                                      GO TO 12C
  110 IF( LV.NE. -1 ) WRITE(6,111) K, X(K)
  111 FCRMAT(5X40H*** SCUARS AESCISSA NOT ORCERED NEAR PT. I4,12H WITH V
     *ALUE E14.5)
  120 SCUARS = .2E+38
          RETURN
          END
          FUNCTION PCLEPS(X,Y,NPTS,EPS)
C       THIS IS A SIMPLE VERSION OF SQUARS FOR PCLYNCMIAL APPRCXIMATICN
C       TC THE SPECIFIEC ACCURACY EPS
          REAL X(1),Y(1)
          FCLEPS = SCUARS(X,Y,NPTS,EPS,C,O,C,-1,C)
          RETURN
          END
          FUNCTION PCLCEG(X,Y,NPTS,NCEG)
C       THIS IS A SIMPLE VERSION OF SQUARS FOR PCLYNCMIAL APPRCXIMATICN
C          OF SPECIFIEC CEGREE  NDEG
          REAL X(1),Y(1)
          FCLCEG = SCUARS(X,Y,NPTS,O.,C,C,C,NDEG,C)
          RETURN
          END
          FUNCTION SUMSQR(X,Y,NPTS,EPS,IB,NDEG,LV)
C       THIS PRCGRAM SETS THE SWITCH ISUMSQ = 1 FOR LSE IN TRAPD
          REAL X(1),Y(1)
C          USE CF SUMSCR SIGNALED BY SETTING IT = -1C7C2C7
          IT = -1070207
          SUMSCR = SCUARS(X,Y,NPTS,EPS,IB,IT,C,NCEG,LV)
          RETURN
          END
          SUBROUTINE PCLAPX(XX,U)
C
C       THIS IS THE MAIN CONTROL PROGRAM WHICH COMPLIES THE APPRCXIMATICNS
C       (USING PAPERX), TEST THEM USING ERRCHK AND GRMCHK  AND DOES THE
C       ITERATIVE REFINEMENT
C
C-INPUT = IBASE,MBPOL,NP,MPRNT1,NREINF,LX,MAXBAS,XX,CCEF,BIT
C-CUTPUT= NREINF,IWEHT,TR,COEF,ERRL2,MB,PRIOR,IREINF,PCCE
          REAL XX(1),U(1),PSAVE(3C),CSAVE(30)
          DCUBLE PRECISION APPX,XJC
          LCGICAL IREINF,ERRCHK,GRMCHK
          CCMMON/ SWTCHZ / IBASE,ITREND,IWEGHT,MBPCL,ISLMSC,IWEHT,NP,IREINF,
     *          JPASS,MPRNT1,MPRNT2,MPRNT3,MPRNT4,NREINF
          CCMMON/ SCUARA / ERRL2,ERRMAX,LOCMAX,ERRAVE,MB,MDEG,ERRSAV(3C),
     *  CCEF(30),PCOE(30),AO,AAA(30),BBB(3C),CCC(3C),TR(3C,30),LERR(5CO)
          CCMMON/ FDATAQ / LX,BIT,PRIOR,EPSMIN,SIZE,
     *          MAXEAS,MAXLNC,JUERR,JSIZE,JORTHO,TW(5CC),TTW(5CC)
          REINF = 0.
          IREINF = .FALSE.
          JPASS = 0
          IWEHT = 1
```

Fig. 3. (cont.)

```
    8 IF( IBASE .EC. 0 )                         GO TO 15
C          INITIALIZE TR, COEF ARRAYS
      CC 10 I = 1,MAXBAS
      CC 10 J = 1,I
   10 TR(I,J) = 0.
   15 CC 20 I = 1,MAXBAS
      CCEF(I) = 0.
   20 TR(I,I) = 1.
   25 IF(MBPCL.GT.0)                             GO TO 6C
C
C       CEGREE (CR NC. OF BASIS FUNCTIONS) NOT SPECIFIED
C       WANT TC INCREASE NUMBER UNTIL EPSMIN ERRCR IS ACHIEVED
   29 ERRL2 = 0.
      MB = 0
C                 LCCP TC INCREASE DEGREE
   30 MB = MB+1
      FRICR = ERRL2
   33 CALL PAPPRX(MB,XX,U)
C
C       CHECK ERRCR FOR TERMINATION
   36 IF( MB.EC.MAXBAS .CR. MB.EQ.LX ) GO TC 5C
      IF((NP .EC. 0).ANC.(MB .CT. 4))       GO TO 5C
C          .9 IS A NUMERICAL CONTROL CONSTANT
C          INCICATES AVERACE CECREASE IN ERRL2 EXPECTED PER CHANGE IN MB
      IF( .NCT. ERRCHK(.9,XX,U) )            GO TC 3C
C       FAVE CBTAINEC ACCEPTABLE APPROXIMATION
      FRICR = ERRL2
      IF( IBASE .EC. 0 ) CALL POWERS
C
C          DC THE REINFCRCEMENT AND ACCURACY CHECK
      IF( MB .GT. 4 )                        GO TC 1CC
C
C                  FINAL OUTPUT
   41 IF( MPRNT1 .EQ. 101 )    CALL SQPRNT
                                             RETURN
C
C       UNABLE TC CBTAIN ACCEPTABLE APPROXIMATION
C          CHECK FCR NEEC OF RECRTHOGONALIZATION
   50 IF( .NCT. CRMCHK(XX,U) )               GO TO 2S
      FRICR = .2E+38
      MPRNT4 = 101
      IF( MPRNT1 .EQ. 101 )    CALL SQPRNT
                                             RETURN
C
C       CEGREE (CR NC. OF BASIS FUNCTIONS) SPECIFIED
   60 CC 65 MB = 1,MBPCL
   65 CALL PAPPRX(MB,XX,U)
      MB = MBPCL
C          CHECK FCR NEEC OF RECRTHOGONALIZATION
      IF( .NCT. CRMCHK(XX,U) )               GO TC 6C
      IF( IBASE .EC. 0 ) CALL POWERS
C
C          DC THE REINFCRCEMENT AND ACCURACY CHECK
      IF( MB .GT. 4 )                        GO TC 1CC
C          USE ERRCHK CNLY TO CET ERROR INFORMATION
   66 IREINF = ERRCHK(.9,XX,U)
      IF( MPRNT1 .EQ. 101 )    CALL SQPRNT
                                             RETURN
C
C       ****** ITERATIVE REFINEMENT STARTS HERE *****
C       TEST FCR NC REINFORCEMENT - NREINF IS NEGATIVE
  100 IF( NREINF .LT. 0 )                    GO TO 22C
      IREINF = .TRUE.
C          DIGIT = APPRCXIMATE NUMBER OF CECIMAL DIGITS IN FLOATING FCINT
C          ESTIMATE THE SIZE OF THE CATA COEFFICIENTS
      CIGIT = 2. - ALOC10(BIT)
      YSIZE = 0.
      CC 110 I= 1,MB
  110 YSIZE = YSIZE + COEF(I)**2
      YSIZE = SCRT(YSIZE)
C          SAVE THE COEFFICIENTS
      CC 115 I= 1,MB
```

FIG. 3. (cont.)

```
      CSAVE(I) = COEF(I)
      CCEF (I) = 0.
  115 FSAVE(I) = PCOE(I)
C                     REFINEMENT LOOP
  120 REINF = REINF + 1.
C                     RECOMPUTE ERROR CURVE IN DCLBLE PRECISION
      CC 130 J= 1,LX
      XJ = XX(J)
      IF( IBASE .EC. 0 )                GO TC 125
      APPX = 0.
      CC 118 I= 1,MB
  118 APPX = CBLE(CSAVE(I))*CBLE(QBASE(I,XJ)) +APPX
                                        GO TC 13C
  125 APPX = CSAVE(MB)
      IF( MB .EC. 1 )                   GO TO 13C
      XJD = XJ
      CC 122 I= 2,MB
      KI = MB+1-I
  122 APPX = APPX*XJD + CBLE(CSAVE(KI))
  130 UERR(J) = CBLE(U(J)) - APPX*DBLE(TTTT(XJ))
      MBSAVE = ME
      CC 135 MB = 1,MBSAVE
  135 CALL PAPPRX(MB,XX,U)
      MB = MBSAVE
C       CCMPUTE CCEFS OF POWERS OF X FROM ORTHCG POLYS AND THEIR CCEFS
      IF( IBASE .EC. 0 ) CALL POWERS
C
C                     CHECK ACCURACY
C       ESTIMATE THE SIZE OF THE COEFFICIENTS CHANGES FOLND
      YREINF = 0.
      CC 140 I = 1,MB
  140 YREINF = YREINF + COEF (I)**2
      YREINF = SCRT(YREINF)
C
C       DIGNCW ESTIMATES CURRENT ACCURACY( IN DECIMAL DIGITS )
      CIGNCW = ALOG10(YSIZE/YREINF)
C       ACC THE COEFFICIENT CORRECTIONS ON
      CC 160 I =1,MB
      CSAVE(I) = CSAVE(I) + COEF(I)
      CCEF(I) = 0.
  160 FSAVE(I) = PSAVE(I) + PCCE(I)
C       TEST TC STOP REFINEMENT - COMPARE DIGNOW AND DIGIT
      IF(DIGNCW + CIGNCW/REINF .GE. DIGIT )    GO TO 2CC
C       LIMIT THE REFINEMENTS TO 3
      IF( REINF .LT. 2.5 )                GC TC 12C
C       WARNING OF UNRELIABLILITY
      IF( MPRNT1 .EQ. 101 )  WRITE(6,119C)
 1190 FCRMAT(3X46H***WARNING - POORLY CONDITIONED CCMPLTATICN*** )
C       RESTCRE THE IMPROVEC COEFFICIENTS
  200 CC 210 I = 1,MB
      FCCE(I) = PSAVE(I)
  210 CCEF(I) = CSAVE(I)
  220 MREINF = REINF
      IF( MBFCL)                          41,41,66
      ENC
      LCGICAL FUNCTION ERRCHK(RATEST,XX,U)
C       ERRCHK = TRUE  MEANS QUIT
C       ERRCHK = FALSE MEANS REPEAT - INCREASE DEGREE BY CNE
C-INPUT = ERRL2,MBPOL,EPSMIN,PRIOR,RATEST,XX,U,LERR,SIZE,BIT
C-CUTPUT= ERRAVE,ERRMAX,LOCMAX,NP
C       THE ERROR TESTS ARE AS FOLLOWS
C          1) IF L2-ERRCR IS LESS THAN EPSMIN, QUIT. OTHERWISE
C          2) CHECK TC SEE IF FIRST TIME THRU 'ERRCHK'.
C             IF SO (PRIOR=0), SET COUNTER NP AND REPEAT. CTHERWISE
C          3) CHECK TC SEE IF ERROR HAS DECREASED SLFFICIENTLY
C             SINCE LAST APPROXIMATION ATTEMPI.
C             IF SO, RESET COUNTER NP AND REPEAT. CTHERWISE,
C          4) IF ERROR HAS NOT DCREASED SUFFICIENILY, CHECK TO
C             SEE IF ERRL2 IS CLOSE TO EPSMIN AND NOT DECREASING MLCH
C             IF SO, CUIT. OTHERWISE,
C          5) ACC 1 TC CCUNTER NP AND REPEAT (NP IS CHECKED IN
```

Fig. 3. (cont.)

```
C                EACH OF THE RESPECTIVE CALLING ROUTINES---MAXIMUM
C                NUMBER OF REPETITIONS IS USUALLY 3 IE IF NP=C)
C                NF IS INITIALIZED AT -3
      REAL XX(1),U(1)
      CCMMON/ SCUARA / ERRL2,ERRMAX,LOCMAX,ERRAVE,MB,MDEG,ERRSAV(3C),
     *  CCEF(30),PCOE(30),AO,AAA(30),BBB(3C),CCC(3C),TR(3C,30),LERR(5CO)
      CCMMON/ FDATAQ / LX,BIT,PRIOR,EPSMIN,SIZE,
     *          MAXEAS,MAXLNC,JUERR,JSIZE,JORTHO,TW(5CC),TTW(5CC)
      CCMMON/ SWTCHZ / IBASE,ITREND,IWEGHT,MBPCL,ISLMSC,IWEHT,NP,IREINF,
     *          JPASS,MPRNT1,MPRNT2,MPRNT3,MPRNT4,NREINF
C
C        **ERROR TESTS FOR TERMINATION
      IF( MBPCL .GT. 0 )                  GO TO 4C
      IF( EPSMIN .GE. ERRL2 )             GO TO 4C
      IF( PRICR .EQ. 0. )                 GO TO 5C
      IF( RATEST .GE. ERRL2/PRIOR )       GO TO 5C
C             .5 IS A NUMERICAL CONTROL CONSTANT
C        DEFINES WHEN ERRL2 IS CLOSE TO EPSMIN
      IF( EPSMIN .GE. .5*ERRL2 .AND. NP .GE. -1) GO TO 40
      NP = NF+1
      IF( NP .EC. 0 )                     GO TO 4C
   30 ERRCHK = .FALSE.
      RETURN
   40 ERRCHK = .TRUE.
C
C        **COMPUTE L2, L1, MAX ERRORS
      ERRAVE = 0.
      ERRMAX = 0.
      CC 10 I=1,LX
      CIF = ABS(UERR(I))
      IF( IWEGHT - 2 )                   E,7,7
    7 CIF = CIF*WWW(I,XX,U)
    8 IF( ERRMAX .GT. CIF)               GO TO 2C
      LCCMAX = I
      ERRMAX = CIF
   20 ERRAVE = ERRAVE + CIF
   10 CCNTINUE
      ERRAVE = ERRAVE/FLOAT(LX)
      RETURN
C             INITIALIZE NP
C                -3 IS A NUMERICAL CONTROL CONSTANT
   50 NP = -3
      CC TO 30
      ENC
      LCGICAL FUNCTION CRMCHK(XX,U)
      REAL XX(1),U(1)
      EXTERNAL CEASE
C
C     GRMCHK MAKES TWC SIMPLE TESTS TO SEE IF THE GRAM-SCHMIDT
C     CRTHOGCNALIZATION HAS PRCDUCEC NEARLY CRTHOGCNAL FUNCTICNS
      CCMMON/ SCUARA / ERRL2,ERRMAX,LOCMAX,ERRAVE,MB,MDEG,ERRSAV(3C),
     *  CCEF(30),PCOE(30),AO,AAA(3C),BBB(3C),CCC(3C),TR(3C,30),LERR(5CO)
      CCMMON/ FDATAQ / LX,BIT,PRIOR,EPSMIN,SIZE,
     *          MAXEAS,MAXLNC,JUERR,JSIZE,JORTHO,TW(5CC),TTW(5CC)
      CCMMON/ SWTCHZ / IBASE,ITREND,IWEGHT,MBPCL,ISLMSC,IWEHT,NP,IREINF,
     *          JPASS,MPRNT1,MPRNT2,MPRNT3,MPRNT4,NREINF
      CRMCHK = .TRUE.
C        TEST RELEVANCE OF CRMCHK - JPASS=JCRTHO AFTER CNE TEST
      IF((MB-4)*(JCRTHC-JPASS)*IBASE*(NREINF+1) ) 1CC,1CO,6C
   60 JPASS = JCRTHO
      ERRCRT = AES(TRAPC(QBASE,MB+JORTHO,MB+JORTHO-2,XX,U))
     *          +AES(TRAPC(QBASE,MB+JORTHO,2,XX,U))
C        ERRCRT = 0 FCR PERFECT ORTHCGCNALIZATION
      CRMCHK = .FALSE.
      IF( MPRNT1 .EQ. 1C1 ) WRITE(6,1C2O)
 1020 FCRMAT(4X47H*** WARNINC NEARLY LINEARLY CEPENDENT FUNCTICNS /
     *8X42HHAC TC RECO CRAM-SCHMIDT ORTHOGONALIZATICN )
  100 RETURN
      ENC
      SUBROUTINE SCPRNT
      CCMMON/ SWTCHZ / IBASE,ITREND,IWEGHT,MBPCL,ISLMSC,IWEHT,NP,IREINF,
     *          JPASS,MPRNT1,MPRNT2,MPRNT3,MPRNT4,NREINF
```

FIG. 3. (cont.)

```
      CCMMON/ SCUARA / ERRL2,ERRMAX,LOCMAX,ERRAVE,MB,MDEG,ERRSAV(3C),
     *  CCEF(30),PCOE(30),AO,AAA(30),BBB(30),CCC(3C),TR(3C,30),LERR(5CO)
      CCMMON/ FDATAC / LX,BIT,PRIOR,EPSMIN,SIZE,
     *        MAXEAS,MAXLNC,JUERR,JSIZE,JORTHO,TW(5CC),TTW(5CC)
C     THIS SUBRCUTINE CCES ALL OUTPUT EXCEPT WARNINGS
C
      IF( MPRNT2 .NE. 101 )                 GO TC 1CC
C                  GENERAL INFORMATION OUTPUT
      MD = ME-1
      IF( IBASE .EC. 0 ) WRITE(6,1010) MD
 1010 FCRMAT(3X56H*** LEAST SQUARES APPROXIMATION BY POLYNOMIALS CF CEGR
     *EE I3 )
      IF( IBASE .NE. 0 ) WRITE(6,1C2C) MB
 1020 FCRMAT(3X36H*** LEAST SQUARES APPROXIMATICN WITH I3,16H BASIS FLNC
     *TICNS )
      IF( ITRENC .EQ.1 ) WRITE(6,1C45)
 1045 FCRMAT(7X27HTIMES THE FUNCTION TREND(X) )
      IF( IWEGHT .EQ.2 ) WRITE(6,1050)
 1050 FCRMAT(7X29HRELATIVE ERRCR CRITERION USEC )
      IF( IWEGHT .EQ.3 ) WRITE(6,1C55)
 1055 FCRMAT(7X41HERROR CRITERION CEFINEC BY WEIGHT(X) LSED )
      IF( MPRNT3 .NE. 101 )                 GO TC 1CC
      WRITE(6,1060)
 1060 FCRMAT(4X3CHPEST APPROX COEF.(ORTHOG ONES),6X13HLEAST SCLARES /
     *8X1HK 36X5HERROR )
      WRITE(6,1065) (J,PCOE(J),ERRSAV(J),J=1,MB)
 1065 FCRMAT(4X,3HPC(,I2,3H) =,E16.7,E25.4)
C                  FCLYNOMIALS OUTPUT
      IF( IBASE .EC. 0 ) WRITE(6,1C7C)
 1070 FCRMAT(4X32HBEST APPROX COEFS OF POWERS CF X )
C           GIVEN BASIS FUNCTICNS CUTPUT
      IF( IBASE .NE. 0 ) WRITE(6,1C75)
 1075 FCRMAT(4X,43HBEST APPROX COEFFS (FOR ORIGINAL FLNCTICNS))
      WRITE(6,1080) (J,COEF(J),J=1,MB)
 1080 FCRMAT(4X3HCF(,I2,3H) =, E16.7 )
  100 IF( MPRNT4 .NE. 1C1 ) WRITE(6,11C5) ERRMAX,ERRAVE
 1105 FCRMAT(3X,32H*** APPROXIMATION ACCEPTABLE *** /
     *7X11HMAX ERRCR = E12.4,3X11HAVE ERROR = E12.4 )
      IF( MPRNT4 .EQ. 101 ) WRITE(6,1110) ERRMAX,ERRAVE
 1110 FCRMAT(3X,49H*** UNABLE TO OBTAIN ACCEPTABLE APPROXIMATICN *** /
     *7X24HSTCPPEC WITH MAX ERROR = E12.4,3X11HAVE ERRCR = E12.4 )
      RETURN
      ENC
      SUBROUTINE PCWERS
C       THIS SUBRCUTINE COMPUTE THE CUEFFICIENTS CF THE FOWERS CF X
C       FRUM THE CRTHOCCNAL POLYNOMIAL CCEFFICIENTS PCCE. IT GENFRATES
C       THE MATRIX TR(I,J) THAT RELATES THESE TWO SETS CF COEFFICIE'TS
C-INPUT =   AC,AAA,EBB,CCC,MB,PCOE
C-CUTPUT=   TR,CCEF
      CCMMON/ SCUARA / ERRL2,ERRMAX,LOCMAX,ERRAVE,MB,MDEG,ERRSAV(3C),
     *  CCEF(30),PCOE(30),AO,AAA(30),BBB(3C),CCC(3C),TR(3C,30),LERR(5CO)
      TR(1,1) = AO
      TR(2,1) = EBB(1)*AO
      TR(2,2) = AAA(1)*AO
      CCEF(1) = (PCOE(1) + BBB(1)*PCOE(2))*AC
      CCEF(2) = AAA(1)*PCOE(2)*AO
      IF( ME .LT. 3 )                      RETLRN
      CC 145 J=3,MB
      J1 = J-1
      J2 = J-2
      TR(J,1) = EBB(J1)*TR(J1,1) - CCC(J1)*TR(J2,1)
      IF( J2 .LT. 2 )                      GO TC 144
      CC 143 I = 2,J2
  143 TR(J,I)= EEB(J1)*TR(J1,I)-CCC(J1)*TR(J2,I) + AAA(J1)*TR(J1,I-1)
  144 TR(J,J1)=EEB(J1)*TR(J1,J1) + AAA(J1)*TR(J1,J2)
      TR(J,J) = AAA(J1)*TR(J1,J1)
      CC 145 I = 1,J
  145 CCEF(I) = CCEF(I) + TR(J,I)*PCOE(J)
      RETURN
      ENC
      SUBROUTINE PAPPRX(MME,XX,U)
C-INPUT = MMF,IEASE,IREINF,MB,XX,LX,BIT,JSIZE,JORTHC,JLERR,MPRNT1
```

FIG. 3. (cont.)

```
C-CUTPUT= IWEHT,AO,AAA,EBB,CCC,PCOE,UERR,ERRL2,SIZE,TR,COEF
C
C         FAPPRX CCMPUTES ORTHCGCNAL FUNCTIONS (CNE PER CALL) FROM EITHER
C             1) THE PCLYNCMIALS, USING FORSYTHE RECURRENCE SCHEME, CR
C             2) AGIVEN SET OF FUNCTIONS DESIGNATED IN FUNCTION BASE,
C                USING GRAM SCHMIDT VIA THE TRANSFCRMATION MATRIX TR(I,J)
C             AT EACH CALL, THE COEFFICIENT OF BEST L2-APPROXIMATION, PCCE(MB)
C             IS CCMPUTEC ALCNC WITH THE ERROR FUNCTICN UERR AND THE ERRCR.
      REAL XX(1),U(1)
      LCGICAL IREINF
      EXTERNAL CEASE,PCL
      CCMMON/ SCUARA / ERRL2,ERRMAX,LOCMAX,ERRAVE,MB,MDEG,ERRSAV(3C),
     *   CCEF(30),PCOE(30),AO,AAA(30),BBB(30),CCC(3C),TR(3C,3C),LERR(5C0)
      CCMMON/ FDATAQ / LX,BIT,PRIOR,EPSMIN,SIZE,
     *   MAXEAS,MAXLNC,JUERR,JSIZE,JORTHO,TK(5CC),TTK(5CC)
      CCMMON/ SWTCHZ / IBASE,ITRENC,IWEGHT,MBPCL,ISUMSC,IWEHT,NP,IREINF,
     *       JPASS,MPRNT1,MPRNT2,MPRNT3,MPRNT4,NREINF
      MB = MMB
      IF( IBASE .EC. 1 )                     GO TC 5CC
C
C         FCRSYTHE SCHEME
C
C         SKIP GENERATICN CF THE ORTHOGCNAL FUNCTICNS DURING
C                    REFINEMENT MODE
      IF( IREINF )                        GO TC 3C
      IF( MB .EC. 1 )                     GO TC 1C
      MD = MB-1
      IWEHT = 2
      BBB(MC) = -TRAPC(POL,MC,MC,XX,U)
      IF( MB .EC. 2 )                     GO TC 2C
      CCC(MD) = TRAPC(PCL,MC,MC-1,XX,U)
C         **NCRMALIZATICN OF ORTHOGONAL POLYNOMIALS
      IWEHT = 1
      AAA(MC) = 1.0
      AN = SCRT(TRAPC(PCL,MB,MB,XX,U))
      AAA(MC) = 1.0/AN
      BBB(MC) = BBB(MC)/AN
      CCC(MD) = CCC(MC)/AN
                                          GO TC 3C
C         **FIRST CRTHO POLY
   10 AO = 1.0
      AO = 1./SCRT(TRAPC(PCL,1,1,XX,U))
                                          GO TC 3C
C         **SECCNC CRTHO POLY
   20 AAA(1) = 1.0
      IWEHT = 1
      AN = SCRT(TRAPC(PCL,2,2,XX,U))
      AAA(1) = 1.0/AN
      BBB(1) = BBB(1)/AN
      CCC(1) = 0.
C         **CCEFFICIENT CF BEST APPROX
   30 PCCE(MB) = TRAPC(POL,MB,JUERR,XX,U)
C         COMPUTE THE ERROR CURVE FROM THE ORTHCG. PCLYS.
      CC 70 I=1,LX
   70 UERR(I) = UERR(I) - POL(MB,XX(I))*PCOE(MB)*TTTT(XX(I))
   80 IF( IREINF )                          RETURN
      ERROR = TRAPC(POL,JSIZE,JSIZE,XX,U)
      IF( ISUMSC .NE. 1 ) ERROR = ERROR/(XX(LX)-XX(1))
      ERRL2 = SCRT(ERRCR)
      IF( JPASS .EC. JORTHO )               RETURN
      IF( MB .EC. 1 ) SIZE = SCRT(PCOE(1)**2 + ERRCR )
      ERRSAV(MB) = ERRL2
      RETURN
C
C         MCCIFIEC GRAM - SCHMIDT ORTHOGONALIZATICN
C
C         SKIP GENERATICN CF THE ORTHOGONAL FUNCTICNS DURING
C                    REFINEMENT MODE
  500 IF( IREINF )                          GO TC 55C
      AAA(MB) = TRAPC(CBASE,MB+JPASS,MB+JPASS,XX,U)
      IF( MB .EC. 1 )                       GO TO 525
      MD = MB-1
      CC 510 J = 1,MD
```

FIG. 3. (cont.)

```
C          COMPCNENT CF BASE(MB) IN THE ORTHOGONAL POLYS
           AAA(J)  = TRAPD(CBASE,J+JORTHO,MB+JPASS,XX,U)
           CC 510 K = 1,J
   510 TR(MB,K) = TR(MB,K) - AAA(J)*TR(J,K)
   525 XNCRM = AAA(MB)
           IF( MB .EC. 1 )                    GO TC 535
           CC 530 J=1,MC
   530 XNCRM = XNCRM - AAA(J)**2
C    XNCRM IS THE NORM CF THE NEWLY OBTAINED ORTHOG  FLNCIICN
C
C          **** CHECK LINEAR INDEPENCENCE   -  MACHINE CEPENDENT NLMBER = BIT
           IF( XNCRM .LE. AAA(MB)*BIT )       GO TC 58C
   535 XNCRM = SCRT(XNCRM)
           CC 540 I=1,MB
   540 TR(MB,I) = TR(MB,I)/XNORM
C          CCEF CF BEST APPROX( FOR ORTHOGONAL BASIS GENERATED )
   550 FCCE(MB) = TRAPD(CBASE,MB+JORTHO,JUERR,XX,U)
           CC 560 I=1,MB
   560 CCEF(I) = CCEF(I) + PCOE(MB)*TR(MB,I)
C                          COMPUTE ERROR CURVE
           CC 570 I=1,LX
   570 LERR(I) = UERR(I) - PCOE(MB)*QBASE(MB+JORTHC,XX(I))*TTTT(XX(I))
C                          RETURN VIA 80
                                          GO TO EC
   580 IF(MPRNT1 .EC. 101) WRITE(6,6CC) MB
   600 FCRMAT(2X6H****** 3HTHE I3,50H BASE FUNCTION IS LINEARLY CEFENDENT
     * CN THE CTHERS )
           FCCE(MB) = 0.
           CCEF(MP) = 0.
C          FREVENT ITERATIVE REFINEMENT FROM HAPPENING
           NREINF = -1
           RETURN
            END
           FUNCTICN TRAPD(FCT,I,J,XX,U)
C
C      TRAPD CCES TRAPEZCICAL RULE INTEGRATION FROM XX(1) TO XX(LX)
C      CF THE PRCCUCT OF TWC FUNCTIONS
C-INPUT = JUERR,JSIZE,XX,LX,ISUMSQ,UERR,TW,TTW,IWEHT
C-CUTPUT= TRAPD
C      NCTE  J = JUERR WHEN SECCND FUNCTION IS LERR
C            J = JSIZE WHEN BOTH FUNCTIONS ARE LERR
C
           REAL XX(1),U(1)
           CCMMON/ SCUARA / ERRL2,ERRMAX,LOCMAX,ERRAVE,MB,MDEG,ERRSAV(3C),
     *     CCEF(30),PCOE(30),AO,AAA(30),BBB(30),CCC(3C),TR(3C,30),LERR(5CC)
           CCMMON/ FCATAQ / LX,BIT,PRIOR,EPSMIN,SIZE,
     *         MAXEAS,MAXLNC,JUERR,JSIZE,JORTHO,TW(5CC),TTW(5CC)
           CCMMON/ SWTCHZ / IEASE,ITREND,IWEGHT,MBPCL,ISLMSC,IWEHT,NP,IREINF,
     *         JPASS,MPRNT1,MPRNT2,MPRNT3,MPRNT4,NREINF
           TRAPD = 0.
C      THIS SWITCH REPLACES THE TRAPEZOIDAL RULE BY THE RECTANGULAR RLLE
           IF(ISUMSC .EC. 1 )                 GO TC 3CC
           IF((J-JUERR)*(J-JSIZE))            10,6C,1C
   10 IF( IWEHT .EC. 2 )                      GO TC 3C
           CC 20 K = 1,LX
           XK = XX(K)
   20 TRAPD = TRAPC + FCT(I,XK)*FCT(J,XK)*TTW(K)
                                          RETURN
C
C          INTEGRATION WITH EXTRA X-FACTOR FOR FORSYTHE SCHEME
   30 CC 40 K = 1,LX
           XK = XX(K)
   40 TRAPD = TRAPC + FCT(I,XK)*FCT(J,XK)*TTW(K)*XK
                                          RETURN
C
   60 IF( J .EC. JSIZE )                      GO TC EC
C          INTEGRATE UERR TIMES ANOTHER FUNCTION
           CC 70 K = 1,LX
   70 TRAPD = TRAPC + FCT(I,XX(K))*UERR(K)*TW(K)
                                          RETLRN
C          INTEGRATE ERROR CURVE SQUARED
   80 CC 90 K = 1,LX
```

FIG. 3. (cont.)

```
   90 TRAPD = TRAPD + (UERR(K)*TW(K))**2/TTW(K)
                                         RETURN
C        SUM OF SQUARES INNER PRODUCT CODE = RECTANGULAR RULE
C        NO TREND OR WEIGHT FUNCTION ALLOWED HERE
C        THE CASES HERE ARE ORGANIZED AS ABOVE
  300 IF((J-JUERR)*(J-JSIZE))            31C,36C,31C
  310 IF( IWEHT .EC. 2 )                 GO TO 33C
      CC 320 K = 1,LX
      XK = XX(K)
  320 TRAPD = TRAPD + FCT(I,XK)*FCT(J,XK)
                                         RETURN
  330 CC 340 K = 1,LX
      XK = XX(K)
  340 TRAPD = TRAPD + FCT(I,XK)*FCT(J,XK)*XK
                                         RETURN
  360 IF( J .EC. JSIZE )                 GO TO 38C
      CC 370 K = 1,LX
  370 TRAPD = TRAPD + FCT(I,XX(K))*UERR(K)
                                         RETURN
  380 CC 390 K = 1,LX
  390 TRAPD = TRAPD + UERR(K)**2
                                         RETURN
      END
      FUNCTION FCL(II,X)
C-INPUT = II,X,AO,AAA,BBB,CCC,
C-CUTPUT= PCL
C        USES 3-TERM RECURRENCE FORMULA TO OBTAIN IT-TH ORTHOG POLY
C        P(K,X) = (AAA(K)*X BBB(K))*P(K-1,X) - CCC(K)*P(K-2,X)
C        FORWARD EVALUATION
      COMMON/ SQUARA / ERRL2,ERRMAX,LOCMAX,ERRAVE,MB,MDEG,ERRSAV(3C),
     *  COEF(30),PCOE(30),AO,AAA(30),BBB(30),CCC(3C),TR(3C,30),LERR(500)
      IF( II.LT.3)                       GO TO 1CC
      II=II-1
      V1 = AO
      V2 = (AAA(1)*X + BBB(1))*V1
      CC 10 I=2,II
      V3 = (AAA(I)*X + BBB(I))*V2 - CCC(I)*V1
      V1 = V2
   10 V2 = V3
      FCL = V3
                                         RETURN
  100 FCL = AO
      IF(II.EC.1)                        RETURN
      FCL = PCL*(AAA(1)*X + BBB(1))
      RETURN
      END
      FUNCTION PCLFIT(ZX)
C        THIS FUNCTION GIVES THE BEST LEAST SQUARES APPROXIMATION CBTAINED
C-INPUT = IBASE,ZX,MB,PCCE,AO,AAA,BBB,CCC,COEF
C-CUTPUT= PCLFIT
      COMMON/ SQUARA / ERRL2,ERRMAX,LOCMAX,ERRAVE,MB,MDEG,ERRSAV(3C),
     *  COEF(30),PCOE(30),AO,AAA(30),BBB(30),CCC(3C),TR(3C,30),LERR(500)
      COMMON/ SWTCHZ / IBASE,ITREND,IWEGHT,MBPOL,ISLMSQ,IWEHT,NP,IREINF,
     *  JPASS,MPRNT1,MPRNT2,MPRNT3,MPRNT4,NREINF
      IF( IBASE .EC. 1)                  GO TO 5CC
C        POLYNOMIAL EVALUATION FROM 3-TERM RECURRENCE RELATION
C        DESCENDING EVALUATION
      IF(MB.LE.2)                        GO TO 5C
      V1= 0.0
      V2= PCCE(MB)
      MD = MB - 1
      CC 40 J=1,MC
      JM= MB-J
      V3 = (AAA(JM)*ZX + BBB(JM))*V2 - CCC(JM+1)*V1 + PCOE(JM)
      V1=V2
   40 V2 = V3
                                         GO TO 7C
   50 V3 = PCCE(1)
      IF( MB .EC. 1 )                    GO TO 7C
   60 V3 = (AAA(1)*ZX + BBB(1))*PCOE(2) + V3
   70 FCLFIT = V3*TTTT(ZX)*AO
```

FIG. 3. (cont.)

```
      RETURN
C            EVALUATION USING GIVEN BASIS FUNCTICNS
  500 V3 = 0.
      CC 580 J=1,MP
  580 V3 = V3 + CCEF(J)*CBASE(J,ZX)
      FCLFIT = V3*TTTT(ZX)
      RETURN
      ENC
      FUNCTICN CEASE(II,X)
C  THIS FUNCTICN GIVES EITHER THE SPECIFIED BASIS (FOR II = 1 TC MAXBAS)
C  CR THE ASSCCIATED ORTHOCONAL BASIS (FOR II-JCRTHC = 1 TC MAXBAS)
C-INPUT = II,X,JCRTHO,TR
C-CUTPUT= QBASE
      CCMMON/ SCUARA / ERRL2,ERRMAX,LOCMAX,ERRAVE,MB,MDEG,ERRSAV(3C),
     *  CCEF(30),PCOE(30),AO,AAA(3C),BBB(3C),CCC(3C),TR(3C,30),LERR(5CO)
      CCMMON/ FDATAQ / LX,BIT,PRIOR,EPSMIN,SIZE,
     *       MAXEAS,MAXLNC,JUERR,JSIZE,JORTHO,TW(5CC),TTW(5CC)
      IF( II .GE. JORTHC )              GO TO 2CC
      CBASE = BASE(II,X)
                                RETURN
C  THE CRTHCC PCLYS OF THE BASE FUNCTIONS ARE INDEXED FROM JORTHC CN
  200 CBASE = 0.
      JJ = II - JORTHO
      CC 220 J= 1,JJ
  220 CBASE = CEASE + TR(JJ,J)*BASE(J,X)
      RETURN
      ENC
      FUNCTICN WWWW(I,XX,U)
C
C     THIS IS THE WEIGHT FUNCTION IN THE ACCLRACY
C-INPUT = I,XX,U,IWEGHT
C-CUTPUT= WWWW
C     IWEGHT IS SWITCH SO THAT IF
C           =1      USE CONSTANT WEIGHT FUNCTICN 1.
C           =2      USE RELATIVE ERROR WEIGHT FLNCTICN (1./L(>X))
C           =3      USE SPECIAL WEIGHT FUNCTION WRITTEN AT 40
C     **** NCTE THAT THE WEIGHTS ARE COMPUTED CNLY CNCE, THEN SAVEC
      REAL XX(1),U(1)
      CCMMON/ SCUARA / ERRL2,ERRMAX,LOCMAX,ERRAVE,MB,MDEG,ERRSAV(3C),
     *  COEF(30),PCOE(30),AO,AAA(3C),BBB(30),CCC(3C),TR(30,30),LERR(500)
      CCMMON/ SWTCHZ / IBASE,ITREND,IWEGHT,MBPCL,ISLMSQ,IWEHT,NP,IREINF,
     *           JPASS,MPRNT1,MPRNT2,MPRNT3,MPRNT4,NREINF
      GC TC(10,20,40), IWECHT
   10 WWWW = 1.
                                RETURN
   20 ABSU = ABS(U(I))
      WWWW = .2E+38
      IF( ABSU .NE. 0.0 ) WWWW = 1./ABSU
                                RETURN
   40 WWWW = WEICHT(XX(I))
   45 RETURN
      ENC
      FUNCTICN TTTT(X)
C-INPUT = X,ITRENC
C-CUTPUT= TTTT
C
C     THIS IS THE TRENC FUNCTION, IE THE PROBLEM IS TO
C     MINIMIZE NCRM CF (U(X)-T(X)*(APPROX))*W(X)
C       ITRENC = 1 SIGNALS SPECIFIED TREND, CTHERWISE 1. IS LSED
      CCMMON/ SWTCHZ / IBASE,ITREND,IWEGHT,MBPCL,ISLMSQ,IWEHT,NP,IREINF,
     *           JPASS,MPRNT1,MPRNT2,MPRNT3,MPRNT4,NREINF
      TTTT = 1.
      IF( ITRENC .NE. 1 )              RETURN
   15 TTTT = TRENC(X)
      RETURN
      ENC
```

Fig. 3. (cont.)

V. Example Program, Testing and Evaluation

A. A SAMPLE PROGRAM AND OUTPUT

We give a driver program (Fig. 4) which runs SQUARS through 8 illustrative examples. These are typical uses and they are not detailed further as the program and output is essentially self-explanatory.

```
      SUBROUTINE CRIVER
      REAL X(100), CATA(100)
      COMMON/ CUC/ KASSE
C     ****** CRIVER PROGRAM TO TEST SQLARS ******
      COMMON/ SQUARA / ERRL2,ERRMAX,LOCMAX,ERRAVE,MB,MDEG,ERRSAV(30),
     *   COEF(30),PCOE(30),AC,AAA(3C),BBB(3C),CCC(3C),TR(3C,30),UERR(500)
      KASSE = 2
C     *** GENERATE SIMPLE CATA AND APPROXIMATE IT BY A CLACRATIC ***
      WRITE(6,5)
    5 FCRMAT(1H1,10X22HEXAMPLE RUNS OF SQLARS//
     * 15X28HSIMPLE CUBIC POLYNOMIAL DATA )
C
      CC 10 J =1,21
      XJ   = .1*FLOAT(J-11)
      DATA(J) =   1.1  + 2.2*XJ - 4.4*XJ**2 + XJ**3
   10 X(J) = XJ
      ERR  = PCLCEG(X,DATA,20,2)
C
      WRITE(6,20) ERRMAX,LOCMAX,ERRAVE
   20 FCRMAT(5X,12HMAX ERROR = F9.5,9H AT POINT,I3,12H AVE ERRCR =,F9.6)
      WRITE(6,30) FIT1,FIT2
   30 FCRMAT(5X,35HVALUES OF BEST FIT AT C., .25 ARE = 2F12.8 )
C     *** ILLUSTRATE CUTPUT LEVELS ***
      CC 40 J = 1,4
      LV = J-2
      WRITE(6,35) LV
   35 FCRMAT(///14H CUTPUT LEVEL I2 )
   40 ERR = SCUARS(X,CATA,20,C,C,C,C,2,LV)
C
C     *** GENERATE NICE DATA - APPROXIMATE IT TO WITHIN .00001 ***
      WRITE(6,45)
   45 FCRMAT(///10X23HAPPROX SIN(X) TO .CCCC1)
      CC 50 J = 1,40
      X(J) = .1*FLOAT(J-1)
   50 DATA(J) = SIN(X(J))
      ERR = SCUARS(X,CATA,40,.CCCC1,C,C,C,-1,2)
      WRITE(6,60) MCEC,ERRL2,ERRMAX
   60 FCRMAT(/ 8H DEGREE ,I3,20H ACHIEVES L2-ERR OF F10.8,
     *12H MAX ERR OF F10.8 //)
C
C     *** GENERATE POCRLY CONDITIONED PROBLEM - SEE WAMPLER ***
C                  OUTPUT ASSUMES CDC65CC - 14 DECIMAL DIGITS
C                  FCR SHORTER WORC LENGTH PLT MDEG = 5 CR LFSS
      MCEG = 7
      MF = MCEC + 1
      CC 80 J = 1,21
      XJ = J-1
      V  = 0.
      CC 70 K = 1,MB
   70 V = V*XJ + 1.
      X(J) = XJ
   80 DATA(J) = V
      ERR = PCLEPS(X,CATA,21,.OC1)
      WRITE(6,90) ERRMAX,(J,COEF(J),J=1,MB)
   90 FCRMAT(///,5X9HMAX ERR = E12.4,14H WITH COEFS. =
     * (/I3,F23.14,I3,F23.14,I3,F23.14))
C
C     *** GENERATE HARC TO FIT CATA - ILLLSTRATE OTHER PCINTS ***
C     *** USES TRENC, EXPONENTIAL BASIS, RELATIVE ERROR       ***
      CC 100 J = 1,51
      XJ = .04*FLCAT(J-1)
```

FIG. 4. A driver program for SQUARS.

```
      DATA(J) = 1. + SQRT(ABS(XJ*(XJ+1.)*(XJ-2.)))
  100 X(J) = XJ
      WRITE(6,110)
  110 FORMAT(///5X38HTRY TO FIT TO WITHIN MAX ERROR OF .02 )
      ERR = SQUARS(X,DATA,51,.01 ,1,1,1,C,2)
      WRITE(6,90) ERRMAX,(J,COEF(J),J=1,MB)
      RETURN
      END
      FUNCTION BASE(I,T)
C     BASE GIVES 1, EXP(-X),EXP(X),EXP(-2X),EXP(2X),EXP(-3X),...
      SGN = 2*MOD(I,2) - 1
      BASE = EXP(SGN*FLOAT(I-1)*.5*T)
      RETURN
      END
      FUNCTION TREND(T)
      TREND = 1. + SQRT(ABS(T))
      RETURN
      END
```

```
                EXAMPLE RUNS OF SQUARS

               SIMPLE CUBIC POLYNOMIAL DATA
      *** APPROXIMATION ACCEPTABLE ***
         MAX ERROR = 3.3635E-01   AVE ERROR = 1.2444E-01
       MAX ERROR =   .33635 AT POINT 1 AVE ERROR =   .124440
       VALUES OF BEST FIT AT 0., .25 ARE = 0.00000000  0.00000000


   OUTPUT LEVEL -1


   OUTPUT LEVEL  0
      *** APPROXIMATION ACCEPTABLE ***
         MAX ERROR = 3.3635E-01   AVE ERROR = 1.2444E-01


   OUTPUT LEVEL  1
      *** LEAST SQUARES APPROXIMATION BY POLYNOMIALS OF DEGREE  2
      *** APPROXIMATION ACCEPTABLE ***
         MAX ERROR = 3.3635E-01   AVE ERROR = 1.2444E-01


   OUTPUT LEVEL  2
      *** LEAST SQUARES APPROXIMATION BY POLYNOMIALS OF DEGREE  2
      BEST APPROX COEF.(ORTHOG ONES)       LEAST SQUARES
          K                                  ERROR
       PC( 1) =  -5.4791594E-01            2.1559E+00
       PC( 2) =   2.4229225E+00            1.2483E+00
       PC( 3) =  -1.7107032E+00            1.3446E-01
      BEST APPROX COEFS OF POWERS OF X
       CF( 1) =   1.1272975E+00
       CF( 2) =   2.7409504E+00
       CF( 3) =  -4.5500000E+00
      *** APPROXIMATION ACCEPTABLE ***
         MAX ERROR = 3.3635E-01   AVE ERROR = 1.2444E-01


              APPROX SIN(X) TO .00001
      *** LEAST SQUARES APPROXIMATION BY POLYNOMIALS OF DEGREE  8
      BEST APPROX COEF.(ORTHOG ONES)       LEAST SQUARES
          K                                  ERROR
       PC( 1) =   8.7323137E-01            4.9059E-01
       PC( 2) =  -5.4940863E-01            4.0409E-01
       PC( 3) =  -7.8696234E-01            6.6995E-02
       PC( 4) =   1.1089786E-01            3.6536E-02
       PC( 5) =   7.1871005E-02            3.2210E-03
       PC( 6) =  -5.8521641E-03            1.2622E-03
       PC( 7) =  -2.4880472E-03            7.6150E-05
       PC( 8) =   1.4344916E-04            2.2950E-05
       PC( 9) =   4.5353964E-05            1.0486E-06
```

FIG. 4. (cont.)

```
BEST APPRCX CCEFS OF POWERS OF X
CF( 1) =   -1.9965213E-06
CF( 2) =    1.C0C0543E+00
CF( 3) =   -2.9243639E-04
CF( 4) =   -1.6606129E-01
CF( 5) =   -5.7018828E-04
CF( 6) =    8.5451919E-03
CF( 7) =    3.6479722E-05
CF( 8) =   -2.5881697E-04
CF( 9) =    2.0787558E-05
*** APFROXIMATICN ACCEPTABLE ***
    MAX ERRCR = 3.0625E-06    AVE ERROR = 9.4113E-C7

DEGREE  8 ACHIEVES L2-ERR OF .00000105 MAX ERR OF  .CCCCC306


*** APFROXIMATION ACCEPTABLE ***
    MAX ERRCR = 1.2206E-05    AVE ERROR = 4.2C9EE-C6


    MAX ERR = 1.2206E-05 WITH COEFS. =
1     1.0000000C000000   2    1.0C0CCCCCC(CCCC   3    .99999999999999

4     1.00000C0C000000   5    .99999999999999   6    1.0CC0CCC0C00000

7     .99999999999999   8    1.CC0CCCCCC(CCCC


TRY TC FIT TC WITHIN MAX ERROR OF .02
*** LEAST SCUARES APPROXIMATION WITH 1C BASIS FLNCTICNS
    TIMES THE FUNCTION TRENO(X)
    RELATIVE ERRCR CRITERION USEC
BEST APPRCX CCEF.(ORTHOC ONES)     LEAST SCLARES
    K                               ERROR
PC( 1) =    2.0095127E+00          2.8177E-C1
PC( 2) =    2.6516654E-01          2.1C32E-C1
PC( 3) =   -2.8815503E-01          5.2141E-C2
PC( 4) =    3.1822670E-02          4.7C36E-C2
PC( 5) =   -5.2770244E-02          2.8636E-C2
PC( 6) =    1.5423924E-02          2.647EE-C2
PC( 7) =   -2.8341395E-02          1.73C5E-C2
PC( 8) =    8.5001617E-03          1.6227E-C2
PC( 9) =   -1.7629785E-02          1.C389E-C2
PC(10) =    5.0390205E-03          9.7564E-C3
BEST APPRCX CCEFFS (FOR ORIGINAL FUNCTIONS)
CF( 1) =   -1.0213542E+03
CF( 2) =    1.6040680E+03
CF( 3) =    1.4439801E+02
CF( 4) =   -1.3703755E+03
CF( 5) =   -1.8295568E+01
CF( 6) =    1.C480620E+03
CF( 7) =    1.4035838E+00
CF( 8) =   -4.8488369E+02
CF( 9) =   -4.7311035E-02
CF(10) =    9.8042663E+01
*** APFROXIMATICN ACCEPTABLE ***
    MAX ERRCR = 3.7438E-02    AVE ERROR = 5.C692E-C3


    MAX ERR = 3.7438E-02 WITH COEFS. =
1    -1021.35422159605878   2    16C4.06796422716434   3    144.398C1371686090

4    -1370.37552953803242   5    -18.29556773987463   6    1C48.06196891286527

7     1.40358383027408   8    -484.883686C685C617   9    -.04731103526597

10    98.04266345009137
```

FIG. 4. (cont.)

B. TESTING AND EVALUATION

The algorithm SQUARS applies to a problem of relatively narrow scope and depends on a small number of arguments. Nevertheless it requires considerable effort to carry out even the obvious testing needed. For this purpose, nine subroutines were written to exercise and evaluate SQUARS in various ways. These are summarized in Table I. Two of these subroutines,

TABLE I

PROPERTIES OF SUBROUTINES USED TO EXERCISE SQUARS

Subroutine name	Number of calls on SQUARS	Brief Description
SQTEST	7	General variety of cases
SQREIN	16	Poorly conditioned problems
SQALL	96	Low degree—varies all of IB, IT, IW, LV
SQOLD	7	General variety of difficult cases
SQTIME	24	Selected arguments for timing data
SQMESS	28	Selected illegal nonsense arguments
SCALER	22	Badly scaled problems
DRIVER	8	Illustrative cases
SQKNOW	62	A spectrum of problems with known solutions

SQTEST and SQALL, receive the data as arguments and were run with several sets of data. Thus something over 250 test runs were made of SQUARS. The performance of SQUARS in these tests gives considerable confidence in its reliability, both from the point of view of protection from algorithm "blowup" and accuracy performance. Nevertheless, 250 is a very small number compared to the possible number of situations that can arise in SQUARS.

The primary tests used for accuracy are the problems proposed by Wampler (1969). He compares the performance of 27 different programs for the least-squares approximation problem. The accuracy obtained by SQUARS (whether using the three-term recurrence scheme or the modified Gram–Schmidt) is as good as any obtained in Wampler's tests (except the one that does exact rational arithmetic). SQUARS was changed so that it used ordinary Gram–Schmidt rather than the modified version, and this change had no effect on the accuracy obtained in these test problems.

The above tests yielded one result contrary to that reported by Wampler. The three-term recurrence scheme shows up to be intrinsically more accurate than Gram–Schmidt in the tests made with SQUARS.

REFERENCES

Businger, P., and Golub, G. (1965). Linear least squares solutions by Householder transformations. *Numer. Math.* **7,** 269–276.

Forsythe, G. E. (1957). Generation and use of orthogonal polynomials for data-fitting with a digital computer, *J. Soc. Indust. Appl. Math.* **5,** 74–88.

Golub, G., and Reinsch, C. (1970). Singular value decomposition and least squares solutions, *Numer. Math.* **14.**

Hanson, R., and Lawson, C. (1969). Extensions and applications of the Householder algorithm for solving linear least squares problems, *Math. Comput.* **23,** 787–812.

Hart, J. F. *et al.* (1968). "Computer Approximations." Wiley, New York.

Rice, J. R. (1964). "The Approximation of Functions," Vol. 1. Addison–Wesley, Reading, Massachusetts.

Rice, J. R. (1966). Experiments on Gram-Schmidt orthogonalization, *Math. Comput.* **20,** 325–328.

Rice, J. R. (1969). "The Approximation of Functions," Vol. 2. Addison–Wesley, Reading, Massachusetts.

Wampler, R. H. (1969). An evaluation of linear least squares computer programs, *J. Res. Nat. Bur. Standards,* **73B,** 59–87.

CHAPTER 9

DESUB: INTEGRATION OF A FIRST-ORDER SYSTEM OF ORDINARY DIFFERENTIAL EQUATIONS

P. A. Fox

NEWARK COLLEGE OF ENGINEERING

NEWARK, NEW JERSEY

I. Program Purpose and Use

A. PURPOSE

Integration of a system of first-order differential equations

$$dy/dx = \mathbf{f}(x, \mathbf{y})$$

subject to an initial condition

$$\mathbf{y}(x_0) = \mathbf{y}_0$$

B. USE

1. *Fortran call*

CALL DE(FUNC,N,YSTART,XSTART,XEND,H,EP,SE,DEØUT)

where

FUNC   Subroutine provided by the user to compute the right-hand sides, DY, of the differential equations, given the independent variable, X, and the vector of dependent variables, Y. DY and Y are vectors of dimension N. The routine DE calls FUNC(Y, X, DY).

N	Number of differential equations (N ≤ 20).
YSTART	Vector of dimension N containing initial values for the dependent variables. Not preserved by DE.
XSTART	Initial value of the independent variable.
XEND	Final value of the independent variable.
H	Initial step size ($|$ XEND–XSTART $|/2^{15} \leq |H| \leq |$ XEND–XSTART $|$) H can be negative.
EP	Error tolerance ($10^{-7} \leq EP \leq 10^{-2}$).
SE	Error routine for standard error treatment.
DEØUT	Print routine for standard output.

The external statement

EXTERNAL FUNC, SE, DEØUT

must be placed in the user's calling program.

The routine DE, in the interest of speed, provides output only at intervals in X automatically chosen to provide maximum spacing consistent with desired accuracy.

2. Printing at Specified Output Points

If the user wishes the values of the dependent variables at specified intermediate points $X_k = X_0 + k \cdot \Delta X, k = 0, 1, 2, \ldots$, in addition to the values at points automatically selected by the program the following call is used:

```
CALL DESP(SP,FUNC,N,YSTART,XSTART,XEND,H,EP,SE,DEØUT)
```

where SP is the increment, ΔX. Since the subroutine DESP is slower than DE, especially for small values of ΔX, this option should be used with discretion. The subroutine DEØUT, used with DESP, prints at both the specified points and the "natural" points automatically chosen by the program; the values at the specified points are indicated by an asterisk.

3. User-Supplied Output Routines

The user may supply his own output routine to be used in place of DEØUT. The routine must have the form

ØUT(Y,DY,N,X,SPTYPE)

where the arguments given to the user by the differential equation routine for output are:

Y   Array of current values of dependent variables.

DY   Array of current values of derivatives.

X   Current value of the independent variable.

SPTYPE   Logical variable with value .TRUE. if X is a specified output point; .FALSE. if X is the result of a natural step. (If the standard routine, DE, is used, SPTYPE can be treated as a dummy argument and need not be declared; otherwise it must be declared LOGICAL.)

The name of this user-supplied subroutine must replace the name DEØUT in the CALL and EXTERNAL statements.

4. *Double-Precision Version*

Double-precision versions of the Fortran programs are also given. The calls are

CALL DDE(FUNC,N,YSTART,XSTART,XEND,H,EP,DSE,DDEØUT)

or (for specified output points)

CALL DDESP(SP,FUNC,N,YSTART,XSTART,XEND,H,EP,DSE,
                                   DDEØUT)

together with the external statement

EXTERNAL FUNC, DSE, DDEØUT

The real variables YSTART, XSTART, XEND, H, EP, and SP must be declared DØUBLE PRECISIØN in the calling program. All user-supplied subroutines including FUNC must be in double precision. For alternate error tests (see below) the names DRE and DAE replace the names RE and AE. The limits on EP become $10^{-18} \leq EP \leq 10^{-2}$. The library provided print routine DDEØUT uses D25.16 format. The user may provide his own double-precision output routine to be used in place of DDEØUT. The arguments Y, DY, and X must be declared DØUBLE PRECISIØN.

5. *Error Information and Treatment*

An error return can occur on any one of the following six conditions:

(1) Failure to converge
(2) $N \leq 0$ or $N > 20$
(3) $EP < 10^{-7}$ or $EP > 10^{-2}$
(4) $H \cdot (XEND - XSTART) < 0$ (i.e., H has wrong sign)
(5) $H = 0$ or $XEND = XSTART$
(6) $|H| < |XEND - XSTART|/2^{15}$
    or $|H| > |XEND - XSTART|$.

If any of the above errors occur the subroutine, ERROR, included in the package, prints out the appropriate information.

The user may wish to replace the routine with his own subroutine ERROR; in this case he may access the error information by placing the following labeled CØMMØN statement in his calling program:

CØMMØN/INFØ/EX, ER, EH, NE, NERR

where

EX	Value, X_i, at which error occurred.
ER	Limiting error.
EH	Step size at time of error.
NE	Component with error ER.
NERR	Integer (1, 2, 3, 4, 5, or 6, as shown above, indicating type of error).

In double-precision versions the routine is named DERRØR. EX, ER, EH must be declared DØUBLE PRECISIØN, and the labeled CØMMØN statement to be used is

CØMMØN/DINFØ/EX, ER, EH, NE, NERR

II. Method

A. ALGORITHM

The algorithm used in the program to integrate a set of first-order ordinary differential equations is based on an extrapolation procedure. Extrapolation to zero interval size is a familiar device for improving a discrete approximation, $T(h)$, whose truncation error can be expressed in powers of the discretization interval h.

Let

$$T(h) = t_0 + t_1 h^{p_1} + t_2 h^{p_2} + \cdots$$

where t_0 is the exact answer and the rest of the sum represents the truncation error. In extrapolation, a sequence of h's, $h_0, h_1, h_2, \ldots$, tending to zero is used to compute successive approximations $T(h_0), T(h_1), T(h_2), \ldots$. The first two approximations can be used to eliminate the term of order h^{p_1}, the next one to eliminate the term of order h^{p_2}, etc. In other words, a polynomial in powers of h is fitted to the approximations, and extrapolated to zero interval size, $h = 0$.

Instead of a polynomial, a rational function of h may be used and extrapolated to zero, often with improved convergence. The program described in this chapter is based on the use of rational functions as developed by

Bulirsch and Stoer (1966). Rational extrapolation is applied, in their algorithm, to a modified midpoint integration rule.

Let

$$dy_i/dx = f_i(x, y_1, y_2, \ldots, y_N), \qquad i = 1, 2, \ldots, N$$

be the system of differential equations to be integrated with the initial conditions

$$y_i(0) = y_{i0}, \qquad i = 1, 2, \ldots, N$$

The algorithm, described below for a single equation,

$$dy/dx = f(x, y)$$

applies, for a system, to each equation in the system.

The modified midpoint rule at any point (see Fig. 1) is based on a "lower" value, y_l, of y, and a "middle" value, y_m, of y at an interval $h/2$ ahead of y_l. These values are used to compute an "upper" value, y_u, of y at an interval $h/2$ ahead of y_m. y_u is computed from

$$y_u = y_l + h \cdot f(x_m, y_m)$$

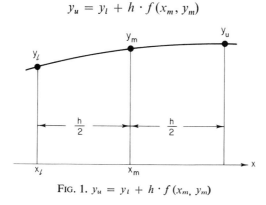

FIG. 1. $y_u = y_l + h \cdot f(x_m, y_m)$

At the next step y_m becomes y_l, y_u becomes y_m, and the process continues until the final y_m is the value at the end of the integration interval. A final correction sets the value of y at this point to an average

$$y = \tfrac{1}{2}[y_m + (y_l + (h/2) f(x_m, y_m))]$$

To start the process the algorithm uses

$$y_l = y_0, \qquad y_m = y_0 + (h/2) f(x_0, y_0)$$

Under suitable differentiability conditions the modified midpoint approximation to the solution $y(x)$ has the asymptotic expansion

$$T(h, x) = y(x) + t_1(x)h^2 + t_2(x)h^4 + \ldots$$

and an extrapolation to zero interval size can be applied. A rational extrapolation of order 6 is used in the programmed implementation; that is, at each integration step as many as six applications of the midpoint rule are computed for successively smaller h and extrapolated to $h = 0$ in attempting to achieve convergence.

A discussion of the implementation of the algorithm is available in Sect. VII.

B. Convergence Criteria

At each step of the integration, the extrapolation process is considered to have converged when each component y_i ($i = 1, 2, ..., n$) of the dependent variable vector has satisfied a convergence criterion specified by the user. The user may choose one of the three convergence criteria available to him. In testing for convergence, two successive extrapolated values (for each component) at the point in question are compared. Let the difference between the two for the jth component be δ_j, and let the error tolerance, given in the calling sequence, be EP. The three convergence error criteria provided can then be stated as:

1. Standard Error (SE)

Let $(y_j)_M$ be the largest absolute value attained so far in the integration by the dependent variable y_j. The convergence requirement is

$$| \delta_j/(y_j)_M | < \text{EP}, \qquad j = 1, 2, ..., n$$

If $(y_j)_M < \text{EP}$, it is replaced by EP.

2. Relative Error (RE)

Let y_j be the current approximation to the respective dependent variable. The convergence requirement is

$$| \delta_j/y_j | < \text{EP}, \qquad j = 1, 2, ..., n$$

If $| y_j | < \text{EP}$, it is replaced by EP in the test.

3. Absolute Error (AE)

The convergence requirement is

$$| \delta_j | < \text{EP}, \qquad j = 1, 2, ... n$$

To use the tests RE or AE in place of SE, replace the name SE in the calling sequence by the desired name.

III. History

The existence of expansions representing the truncation error in discretization processes in powers of a discretization parameter (or an interval size h) has been studied by Gragg (1963), Stetter (1965), and others. The parallel study of the convergence of extrapolative methods based on these expansions has been an increasingly important area of research. The improvement in convergence made possible through the use of rational extrapolation functions (in lieu of polynomial approximants) was investigated by Bulirsch and Stoer (1964).

In 1966, Bulirsch and Stoer applied the technique of rational extrapolation to the solution of ordinary differential equations and gave, in their paper, an Algol program implementing the technique. This program, in turn, was translated into Fortran by Clark (1966) of Argonne National Laboratories, and provided with a driver routine for input–output and control [Clark (1966a)]. Penelope Crane at Bell Telephone Laboratories in 1968 adapted the Argonne routine for use at the labs by writing a package containing a variety of error test options, printing options, and the option of calling the program from Algol programs through the local Fortran–Algol interface. This package, without the Algol option, is essentially the version presented in this chapter.

Tests on the method have been run by Clark (1968), and by Crane and Fox (1969). Although the technique is fairly new and experience with it is limited, the programs have performed very well on a variety of test problems studied in these two reports.

IV. Adaptation of the Program

The first part of this section discusses the adaptation of the program to other systems. This discussion is followed by information on how to change the current bounds set on N (the number of equations allowed in the system), on HMIN (the minimum value H is allowed to attain), on EMIN and EMAX (lower and upper bounds on EP), and how to change the order of the extrapolation method.

A. ADAPTATION OF THE PROGRAM TO OTHER SYSTEMS

In adapting the package to other systems the following comments may prove helpful. The subroutines whose names appear are discussed in more detail in Sect. VII, which deals with the programmed implementation.

1. Machine-Dependent Constants

There are five machine-dependent constants defined in the program and used by the routines XDE, DESUB, XDDE, and DDESUB. These are:

ZØT — Real numbers are considered to be equal if the difference between them is less than ZØT. The value of the constant ZØT should be chosen as $2^{(-B+6)}$ for a computer with a binary floating point fraction of length B.

DZØT — The double-precision version of ZØT.

ZØTUP — The constant used in the step halving decision in DESUB based on comparing ZØTUP $*$ DT with C. It is chosen as 2^{B-6}.

DZØTUP — The double-precision ZØTUP.

ICHØP — This integer constant with value 2^N is used to zero the rightmost N bits of the mantissa of the step-size H at each step in an attempt to reduce the buildup of round-off error. (For a 27-bit mantissa, the value of ICHØP used was 2^{15}. This leaves H with approximately four nonzero decimal digits at each step.)

2. Output Statements

There are four routines that use Fortran write statements. These are the single- and double-precision output routines DEØUT and DDEØUT, and the single- and double-precision error-commenting routines ERRØR and DERRØR. It should be noted that none of these routines are necessary in using the package. All are provided for user convenience and may be rewritten with any type of output statement.

B. Changing Basic Parameter Values

1. N, Maximum Number of Differential Equations Allowed

The subroutines XDE and DESUB (in the case of double precision these are XDDE and DDESUB) must be recompiled with all dimensions of 20 reset to the new value of NMAX. Then the labeled CØMMØN card

$$\text{CØMMØN/IPARAM/M, NMAX}$$

must be included in the calling program and the value of NMAX reset.

2. HMIN, the Minimum Value the Step Size Is Allowed to Attain

The current value of HMIN is set to

$$\text{HMIN} = \text{H/HDIV}$$

where H is the step size given in the calling sequence and HDIV $= 2^{10}$ $= 1024$. To change HDIV, place the labeled CØMMØN statement

CØMMØN/PARAM/ZØT, PØWER2, EMAX, EMIN, HDIV, ZØTUP, ICHØP

for single precision or

CØMMØN/DPARAM/DZØT, DP2, DEMAX, DEMIN, DHDIV, DZØTUP

for double precision in the program and reset the value of HDIV or DHDIV.

The same CØMMØN statements can be used to change the value, 2^{15}, used in testing that the initial H satisfies

$$| H | \geq | XEND - XSTART |/2^{15}$$

This quantity is stored in PØWER2 (or in double precision in DP2).

3. EMIN and EMAX, Lower and Upper Bounds on EP (DEMIN and DEMAX for the Double-Precision Version)

The current values of the parameters are:

$$EMIN = 10^{-7} \qquad DEMIN = 10^{-18}$$
$$EMAX = 10^{-2} \qquad DEMAX = 10^{-2}$$

To change these values place the labeled CØMMØN statement

CØMMØN/PARAM/ZØT, PØWER2, EMAX, EMIN, DHIV, ZØTUP, ICHØP

for single precision, or

CØMMØN/DPARAM/DZØT, DP2, DEMAX, DEMIN, DHDIV, DZØTUP

for double precision in the program and reset the values to be changed.

4. Changing the Order of the Extrapolation Procedure

The extrapolation procedure is currently set to sixth order. If an order of 4 or 5 is desired place the labeled CØMMØN statement

CØMMØN/IPARAM/M, NMAX

in the program and reset M to 4 or 5, respectively.

A change to an order higher than 6 requires more extensive knowledge of the programed implementation described in Sect. VII. In this case the user must not only reset M but must also recompile the routine DESUB and increase the second dimension of arrays DT, YG, HY, and SG, increase the dimension of the coefficient vector, D, and supply statements in the program to enter the additional values for the D vector. These additional values may be determined by noting that in each program path $D(J) = 4 * D(J - 2)$ for $J > 3$.

V. Testing and Results

The selection of the routine DESUB for inclusion in Bell Telephone Laboratories' Program Library I was preceded by a study of several programs for integrating differential equations. The results of that study are reported in some detail by Crane and Fox (1969a).

VI. Example Problems

A. REDUCTION OF A HIGH-ORDER SYSTEM OF DIFFERENTIAL EQUATIONS TO A FIRST-ORDER SYSTEM

Input equations to the routine DESUB should consist of a set of first-order differential equations. If the user wishes to integrate higher-order systems, they must be reduced to first-order systems. The example on the following page illustrates how the reduction may be accomplished.

```
C     TEST PROGRAM FOR CE SUBROUTINE
      CIMENSICN YSTART(2)
      CCMMON NCFNS
      EXTERNAL FEVAL,CECUT,SE
      NCFNS = 0
      N = 2
      YSTART(1) = 1.0
      YSTART(2) = -1.0
      XSTART = 0.0
      XEND = 4.0
      H = 1.0
      EPS = 1.0E-06
      CALL DE(FFVAL,N,YSTART,XSTART,XEND,H,EPS,SE,DEOLT)
      WRITE (6,97) NOFNS
97    FCRMAT(1H0,5X,36HTCTAL NC OF FUNCTION EVALUATIONS IS ,I6)
      STCP
      ENC

      SUBROUTINE FEVAL(Y,X,CY)
      CIMENSICN Y(2),CY(2)
      CCMMON NCFNS
      NCFNS = NCFNS + 1
      CY(1) = Y(2)
      CY(2) = Y(1)
      RETURN
      ENC
```

CE SCLUTICN FOR N = 2 EQUATIONS FROM XSTART = C.CCCCCE CO TC XEND = 0.40000E 01 WITH LOCAL ERROR TOLERANCE EP = 1.CCCCE-C6 AND INITIAL STEP SIZE H = 0.10000E 01. PRINTING OCCURS AT EACH NATURAL STEP IN TIME

THE OUTPUT COLUMNS ARE X, Y(1), Y(2),..., Y(N)

```
0.00000000E 00    0.1CCCCOCCE C1    -C.1CCCCCCCE C1
0.99023438E 00    0.37148976E CC    -0.37148976E CC
0.23085938E 01    0.99400C4C1E-C1   -C.994CC4C1E-C1
0.38320313E 01    0.216654C2E-C1    -C.216654C2F-C1
0.4CC0C000E 01    0.18315531E-C1    -C.18315531E-C1
```

TCTAL NC CF FUNCTICN EVALUATIONS IS   157

FIG. 2. First program and output.

Example **Given**

$$d^2u/dx^2 = f(u, du/dx, v, dv/dx, x)$$
$$d^2v/dx^2 = g(u, du/dx, v, dv/dx, x)$$

Set

$$dy_1/dx = y_2, \quad \text{where} \quad y_1 = u$$
$$dy_2/dx = (d^2y_1/dx^2 = d^2u/dx^2) = f(y_1, y_2, y_3, y_4, x)$$
$$dy_3/dx = y_4, \quad \text{where} \quad y_3 = v$$
$$dy_4/dx = (d^2y_3/dx^2 = d^2v/dx^2) = g(y_1, y_2, y_3, y_4, x)$$

Initial conditions are required for all the dependent variables $y_1, y_2, y_3, \ldots$.

```
C      TEST PRCGRAM FOR CCUBLE PRECISION VERSION
       CIMENSICN YSTART(2)
       CCUBLE PRECISION XSTART,XEND,F,EPS,YSTART
       CCMMON NCFNS
       EXTERNAL FEVAL,CCECUT,CSE
       NCFNS = 0
       SP = 1.0
       N = 2
       YSTART(1) = 1.0
       YSTART(2) = -1.0
       XSTART = 0.0
       XEND = 4.0
       H = 1.0
       EPS = 1.0E-06
       CALL DCESP(SP,FEVAL,N,YSTART,XSTART,XEND,F,EPS,DSE,DDEOLT)
       WRITE (6,97) NOFNS
97     FCRMAT (1H0,5X,36HTOTAL NO OF FUNCTION EVALUATIONS IS ,I6)
       STCP
       ENC

       SUBROUTINE FEVAL(Y,X,CY)
       CIMENSICN Y(2),CY(2)
       CCUBLE PRECISION Y,X,DY
       CCMMON NCFNS
       NCFNS = NCFNS + 1
       CY(1) = Y(2)
       CY(2) = Y(1)
       RETURN
       ENC
```

```
    CE SCLUTICN FOR N = 2 EQUATIONS FROM XSTART = C.CCCCCD CO TC XENC =
0.40000C 01 WITH LOCAL ERROR TOLERANCE EP = 1.CCCCCD-C6 AND INITIAL STEP
SIZE H = 0.1000CC 01.  PRINTING OCCURS AT EACH NATURAL STEP IN TIME AND AT
SPECIFIED PCINTS (XSTART+K*SP) FOR K=C,1,... AND SP = C.1CCCCD 01 (SPECIAL
PCINTS ARE ICENTIFIEC WITH *).

    THE CUTPUT CCLUMNS ARE X, Y(1), Y(2),..., Y(N)

0.000C000000000C00C C0    0.1C000CCOCOCOCCCCD C1   -C.1CCCOCCCCOCOOOOOC 01
0.999599999995999960 C0   0.3E787944109420370 CC  -0.3678794410942037C 00
0.100C00000000000000 01   0.3678794410942C370 CC  -0.3678794410942037C 00
0.200C00000000000000 01   0.13533528317578370 CC  -C.1353352831797837C 00
0.2333333333333332C 01    0.969719E7842C015ED-C1  -C.96971967842CO156C-01
0.300C000000000C0000 01    0.4978706824650032D-C1  -0.4978706824690032C-01
0.3888888888888883C 01    0.204680757C51403ED-C1  -0.2C4680757C514038C-01
0.400C0000000000000C 01    0.1831563880492525D-C1  -C.1831563E88C49252C-01

TCTAL NC CF FUNCTION EVALUATICNS IS    232
```

FIG. 3. Second program and output.

B. SINGLE-PRECISION EXAMPLE

$$dy_1/dx = y_2, \qquad dy_2/dx = y_1$$
$$y_1(0) = 1, \qquad y_2(0) = -1$$

The system is to be integrated from $x = 0$ to $x = 4$. The initial step size H is 1.0 and the error tolerance EPS is 10^{-6}.

Printing occurs at each natural step in time. (See Fig. 2, p. 486.)

C. DOUBLE-PRECISION EXAMPLE

The same example as above is done here except that printing occurs at points

$$\text{XSTART} + K * \text{SP} \qquad \text{for} \quad \text{SP} = 1$$
$$K = 0, 1, \dots$$

as well as at each natural step. (See Fig. 3, p. 487.)

VII. Organization of the Program

A. PROGRAM STRUCTURE

Figure 4 shows the program structure for the single-precision version. The double-precision version is similar. Standard usage paths are solid; optional paths are shown as dashed lines. The diagram shows the calling program, written by the user, which calls either DE or DESP (for the specified output point option).

DE and DESP sent control to XDE which monitors the entire integration, using the routine DESUB, which performs a single step of an integration, as a subroutine. After checking the initial input parameters, XDE sets up the integration, and after each integration step checks whether output is required and, if so, calls on the output routine (either DEØUT or the user supplied routine). If errors are found in the input parameters or if convergence cannot be achieved XDE returns to an error return in DE (or DESP) which in turn calls the subroutine ERRØR to print an error comment.

The integration routine DESUB calls on the routine ERR to provide the requested convergence test based on standard, relative, or absolute error.

The routine, supplied by the user, for evaluating the right-hand sides of the differential equations is used by both the single-step integrator, DESUB, and the routine XDE (for initial values and for output).

B. SUBROUTINES

The subroutines for both the single- and double-precision programs are listed below. Some routines, such as DESUB, appear in more than one set. The term D in a subroutine name indicates the double-precision package.

DDE	Double-precision version of DE.
DDEØUT	Double-precision version of DEØUT.
DDESP	Double-precision version of DESP.
DDESUB	Double-precision version of DESUB.
DE	Calls XDE after setting SP = 0 to indicate standard usage.
DEØUT	Output routine for recording results.
DERR	Double-precision version of ERR.
DERRØR	Double-precision version of ERRØR.
DESP	Calls XDE (when specified output point option used).
DESUB	Implements the algorithm to perform one integration step.
ERR	Checks for convergence using user-specified error criterion.
ERRØR	Routine for writing an error diagnostic in the case of error failure.
XDDE	Double-precision version of XDE.
XDE	Does housekeeping during integration: checks input parameters, checks for specified output points, steps H, calls DESUB, etc.

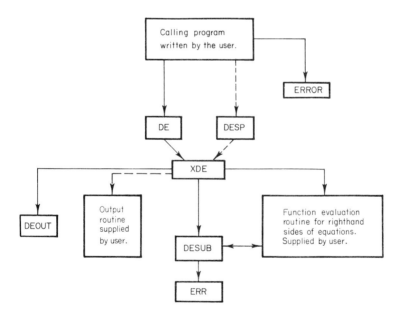

FIG. 4. The program structure.

C. PROGRAM VARIABLES

The variables used in the two major routines XDE and DESUB and in their double-precision counterparts XDDE and DDESUB are[1]:

---

[1] As much as possible the names have been assigned to correspond to the notation of the Algol program in Bulirsch and Stoer (1966).

A	Value of the independent variable at the end of the current integration step.
B	Temporary storage location.
BH	Logical variable; set .TRUE. when the current basic step size is halved (used to indicate whether previous values have been saved).
BØ	Logical variable; two-way switch used to determine which sequence of coefficients D to use in the current extrapolation step.
B1	Temporary storage location.
C	$C^{(km+1-k)}$ in the equations of Bulirsch and Stoer (1966).
D	An array of size JM + 1: $D(K) = (h_{m+1-k}/h_m)^2$.
DEMAX	The double-precision value of EMAX; currently 10^{-2}.
DEMIN	The double-precision counterpart of EMIN; currently 10^{-18}.
DHDIV	The double-precision value of HDIV; currently 1024.
DP2	The double-precision value of PØWER2; currently 2^{15}.
DT	A two-dimensional array of size NxJMAX: $DT(K) = \Delta T_k^{(m+1-k)}$
DY	An array of size N for the derivatives of the Y's.
DZ	An array of size N for the initial values of DY.
DZØT	The double-precision counterpart of ZØT; currently $2.17D-17$.
DZØTUP	The double-precision counterpart of ZØTUP; currently 3.6D16.
EMAX	The maximum value of EP allowed; currently 10^{-2}.
EMIN	The minimum value of EP allowed; currently 10^{-7}.
EH	Value of the step-size H for which DESUB failed to converge.
EPS	Local error tolerance.
EPSPRT	Labeled CØMMØN variable used to pass EPS to DEØUT.
ER	Limiting error in the case DESUB fails to converge.
EX	Value of X at which DESUB fails to get convergence.
FC	A factor used to adjust the step-size H for the next step to be taken, depending on the number of extrapolations needed for convergence in the current step.
FH	Relative difference between XP and X.
G	Current size of substep.
H	The step size of the step to be taken.
ICHØP	An integer constant, 2^k, used to insert zeros in the rightmost k bits of the step-size H; currently $k = 15$.
HDIV	Divisor used to compute HMIN; currently 1024.
HMAX	Largest step size allowed; defined as $XF-XI$.
HMIN	Smallest step size allowed; defined as $(XF-XI)/HDIV$.
HP	Size of step H taken by DESUB.
HIPRT	Labeled CØMMØN variable used to pass H to DEØUT.
HQ	Variable used to determine if current step taken by DESUB overstepped any specified output points.
J	Index for the order of the extrapolation being computed.
JH	Integer obtained by equivalencing an integer name to the step-size variable H.
JJ	Index for the array of values saved for use when the interval is halved; JJ gives the order of the extrapolation used to compute the value saved.
JM	The order of the rational function to be used; currently 6.
JMAX	If convergence does not occur after JMAX extrapolations the interval is halved.

JR The integers M, JR, and JS are denominators for three consecutive step sizes in the sequence $\{h_k\}$: H/M, H/JR, H/JS.

JS See JR.

KØNV Flag to indicate convergence; always set .FALSE. as long as the extrapolation order being computed is less than JM/2; set .TRUE. when error test satisfied.

KØNVF Second convergence flag; initially, .TRUE.; but set .FALSE. when the error test not satisfied for H $\leq$ HMIN.

KØUNT Counter used to compute next specified point.

M The routine XDE uses M as the order of the rational function to be used (see JM); the routine DESUB uses M in the current size of substep G = H/M (see JR).

N The number of differential equations.

NE Number of the component in which convergence fails to occur.

NERR Error code number.

NMAX Maximum number of differential equations; currently 20.

NP Flag to indicate whether or not specified output points are requested; NP = 1 if the user called DESP; NP = 0 if the user called DE.

NPPRT Labeled CØMMØN variable used to pass NP to DEØUT.

PØWER2 (XEND − XSTART)/PØWER2 used as lower bound for initial stepsize H; currently 2^{15}.

R An array of size N containing the values of the absolute errors in each component.

S An array of size N containing either (1) the largest value of each Y computed since the start of the integration for the *standard error test*, (2) the largest value of each Y computed during the current step being taken for *relative* error test, or (3) the value 1 for the *absolute error test*.

SA An array of size N containing initial values of S.

SG A two-dimensional array of size Nx(JMAX − 2) containing values of S saved in case the interval must be halved.

SP Increment, ΔX, for printing at specified points.

SPPRT Labeled CØMMØN variable used to pass SP to DEØUT.

STYPE Logical variable; a call to XØUTX has the value .TRUE. if X is a specified output point; .FALSE. if X is the result of a natural step.

TA $= T_L^{(1)} = \sum_{k=1}^{L} \Delta T_k^{(L+1-k)}$ = latest (Lth) extrapolated approximation in Bulirsch and Stoer (1966).

TITLE Logical variable; a call to DEØUT with value .TRUE. indicates title is to be printed.

TTL XF−XI, length of interval over which integration is to be performed.

U $\Delta T_{k-1}^{(m+1-k)}$ in Bulirsch and Stoer (1966).

X The independent variable.

XF The final value of the independent variable.

XFPRT Labeled CØMMØN variable used to pass XF to DEØUT.

XI The initial value of the independent variable.

XIPRT Labeled CØMMØN variable used to pass XI to DEØUT.

XP Value of X for next specified output point.

XPMX Directed distance from next specified output point to current position of X.

XR   Retrieves previous value of X if specified points have been stepped over by DESUB (used in REDIF call).

XT   Saves value of X before call to DESUB in case specified points are stepped over.

Y   The dependent variable array of size N.

YA   An array of size N containing initial values of Y.

YG   A two-dimensional array of size Nx(JMAX − 2) containing values of YL saved in case the interval must be halved.

YH   A two-dimensional array of size Nx(JMAX − 2) containing values of YM saved in case the interval must be halved.

YL   An array of size N containing the lower point in the modified mid-point rule.

YM   An array of size N containing the middle point in the modified midpoint rule.

ZØT   Real numbers are assumed to be equal if the relative difference between them is less than ZØT; currently 4.76837158E − 07.

ZØTUP   A constant used in step halving decision based on comparing ZØTUP * DT with C; currently 2097152.

VIII. DESUB

```
      SUBROUTINE DE(XFX,N,Y,XI,XF,HI,EPS,ERR,XOUTX)
C
C     PURPOSE        INTEGRATION OF A SYSTEM OF FIRST ORDER DIFFERENTIAL
C                    EQUATIONS (INITIAL VALUE PROBLEM).
C     METHOD         RATIONAL EXTRAPOLATION APPLIED TO THE MODIFIED
C                    MIDPOINT METHOD.
C     SOURCE         BULIRSCH, ROLAND AND JOSEF STOER, 'NUMERICAL
C                    TREATMENT OF ORDINARY DIFFERENTIAL EQUATIONS BY
C                    EXTRAPOLATION METHODS,' NUMERISCHE MATHEMATIK 8,
C                    PP. 1-13 (1966).  THIS PAPER CONTAINS AN ALGOL
C                    PROGRAM WHICH WAS IMPLEMENTED IN FORTRAN BY
C                    NANCY W. CLARK AT ARGONNE NATIONAL LABORATORIES
C                    AS ANL9250 - DIFSUB.
C     ADAPTED BY     P. C. CROCKETT, P. A. FOX, AND C. B. HERGENHAN
C     DATE           AUGUST 13, 1968
C
C     USING DE CAUSES VALUES OF DEPENDENT VARIABLES TO BE PRINTED AT
C     POINTS AUTOMATICALLY SELECTED BY THE PROGRAM.
C
      EXTERNAL XFX,ERR,XOUTX
      COMMON /DESPCM/ NP,KOUNT
      NP = 0
      SP = 0.0
      CALL XDE(SP,XFX,N,Y,XI,XF,HI,EPS,ERR,XOUTX)
      RETURN
      END

      SUBROUTINE DESP(SP,XFX,N,Y,XI,XF,HI,EPS,ERR,XOUTX)
C
C     USING DESP CAUSES VALUES OF DEPENDENT VARIABLES TO BE PRINTED AT
C     SPECIFIED POINTS IN ADDITION TO THOSE AUTOMATICALLY SELECTED.
C
      EXTERNAL XFX,ERR,XOUTX
      COMMON /DESPCM/ NP,KOUNT
      NP = 1
      KOUNT = 0
      IF(SP*(XF-XI)) 2,4,10
2     SP=SIGN(SP,XF-XI)
      GO TO 10
4     IF(SP .NE. 0.0) GO TO 10
      NP = 0
10    CALL XDE(SP,XFX,N,Y,XI,XF,HI,EPS,ERR,XOUTX)
      RETURN
      END

      SUBROUTINE XDE(SP,XFX,N,Y,XI,XF,HI,EPS,ERR,XOUTX)
C
C            DOES HOUSEKEEPING DURING INTEGRATION, CHECKS INPUT
C            PARAMETERS, CHECKS FOR SPECIFIED OUTPUT POINTS,
C            STEPS H, ETC.
C
      DIMENSION Y(N),DY(20),S(20),R(20),YR(20)
      COMMON /DESPCM/ NP,KOUNT
      COMMON /PARAM/ ZOT,POWER2,EMAX,EMIN,HDIV,ZOTUP,ICHCP
      COMMON /IPARAM/ M,NMAX
      COMMON /INFO/ EX,ER,EH,NE,NERR
      COMMON /OTPUT/ SPPRT,HIPRT,XIPRT,XFPRT,EPSPRT,NPPRT,TITLE
      LOGICAL STYPE,KCNVF,TITLE
      EQUIVALENCE (JH,H)
      EXTERNAL XFX,ERR
C******** INITIALIZE CONSTANTS **************************************
      ZOT=4.76837158E-07
      POWER2 = 32768.0
      EMAX = 1.E-02
      EMIN=1.E-07
      HDIV = 1024.
      ZOTUP=2097152.
      ICHCP = 32768
      M = 6
      NMAX = 20
```

Fig. 5.

```
C******** CHECK PARAMETERS AND PERFORM INITIALIZATION              ********
      TITLE = .TRUE.
      STYPE=.TRUE.
      NPPRT=NP
      SPPRT=SP
      HIPRT=HI
      XIPRT=XI
      XFPRT=XF
      EPSPRT=EPS
      IF ((N.LE.0).OR.(N.GT.NMAX)) GOTO 84
      IF ((EPS.LT.EMIN).OR.(EPS.GT.EMAX)) GOTO 85
      TTL=XF-XI
      H=HI
      IF (TTL*H) 86,87,12
   12 IF (((H/TTL)*POWER2.LT.1.).OR.((H/TTL).GT.1.)) GOTO 88
      DO 14 I=1,N
      S(I)= ABS(Y(I))
   14 CONTINUE
      KONVF=.TRUE.
      HMIN=H/HDIV
      HMAX=TTL
      HP=0.
      XP=XI
      X=XI
C******** BEGIN SOLUTION OF DIFFERENTIAL EQUATIONS                 ********
   20 H=H/3.
C******** NOTE ************************************************************
C    THE FOLLOWING STATEMENT IS MEANT TO ZERO THE LEAST SIGNIFICANT
C    BITS IN THE MANTISSA OF H.  IT WILL WORK ONLY IF THE FORTRAN
C    INTEGER DIVIDE AND MULTIPLY OPERATIONS ARE CARRIED OUT WITH
C    HARDWARE INTEGER ARITHMETIC COMMANDS (E. G., CDC 6000 SERIES
C    USERS MUST USE AN ALTERNATE METHOD).
      JH=JH/ICHOP*ICHOP
      H=3.*H
      IF ((NP.EQ.0).AND.(.NOT.STYPE)) GOTO 50
      XPMX=XP-X
      FH=XPMX/H
      IF (FH.GT.ZOT) GOTO 50
C******** GET TO SPECIFIED VALUE OF X AND CALL XOUTX               ********
   30 IF (ABS(FH).GT.ZOT) GOTO 34
      DO 32 I=1,N
      YR(I)=Y(I)
   32 CONTINUE
      HG=HP
      XR=X
      GOTO 36
   34 HG=XPMX+HP
      HR=HQ
      XR=XT
      CALL REDIF (N,XP,YR,DY,HR,HQ,EPS,M,S,R,KONVF,XFX,ERR)
      HG=XR-XT
   36 CALL XFX (YR,XR,DY)
      STYPE=.TRUE.
      CALL XOUTX(YR,DY,N,XR,STYPE)
      STYPE=.FALSE.
      IF (KONVF) GOTO 40
      GOTO 82
C******** STEP TO NEXT SPECIFIED OUTPUT VALUE                      ********
   40 IF ((XF-XR)/TTL.LE.0.) GOTO 70
      KOUNT = KOUNT+1
      XP = XI+FLOAT(KOUNT)*SP
      IF ((XP-XF)/H.GT.0.) XP=XF
      GOTO 20
C******** OUTPUT RESULTS AT A NATURAL STEP                         ********
   50 IF ((ABS((X-XR)/H)).LE.ZOT) GOTO 60
      CALL XFX (Y,X,DY)
      CALL XOUTX(Y,DY,N,X,STYPE)
C******** PERFORM A STEP IN SOLUTION OF DIFFERENTIAL EQUATIONS     ********
   60 IF ((XF-X)/TTL.LE.0.) GOTO 70
      IF (ABS(H).LT.HMIN) H=SIGN(HMIN,H)
      IF (ABS(H).GT.HMAX) H=SIGN(HMAX,H)
      IF ((XF-X-H)/TTL.LT.0.) H=XF-X
```

FIG. 5. (cont.)

```
        XT=X
        CALL DESUB (N,X,Y,DY,H,HMIN,EPS,M,S,R,KONVF,XFX,ERR)
        HP=X-XT
        IF (KONVF) GOTO 20
        GCTC 80
C******** STANDARD RETURN ON COMPLETION OF SOLUTION          ********
 70     RETURN
C******** SET UP ERROR RETURN AND DIAGNOSTICS                ********
 80     NERR=1
        EP=0.
        CC 81 I=1,N
        IF (ER*S(I).GE.R(I)) GOTO 81
        ER=R(I)/S(I)
        NE=I
 81     CCNTINUE
        EH=HP
        EX=X
        CALL XFX (Y,X,CY)
        CALL XOUTX(Y,CY,N,X,STYPE)
        GCTC 92
 82     NERR=1
        ER=0.
        CC 83 I=1,N
        IF (ER*S(I).GE.R(I)) COTO 83
        ER=R(I)/S(I)
        NE=I
 83     CCNTINUE
        EH=HQ
        EX=XR
        GCTC 92
 84     NERR=2
        GCTC 90
 85     NERR=3
        GCTC 90
 86     NERR=4
        CCTC 90
 87     NERR=5
        GCTC 90
 88     NERR=6
 90     CALL XFX (Y,XI,CY)
        CALL XOUTX(Y,CY,N,XI,STYPE)

 92     CALL ERROR
        RETURN
        END

        SUBROUTINE CESUB(N,X,Y,DY,H,HMIN,EPS,JM,S,R,KONVF,FCT,ERR)
C
C******** PERFCRM INITIALIZATION TO CALL DERESB **********************
C
        CCMMON /CERECM/ JMAX,YA,SA,DZ
        CIMENSICN Y(N),S(N),YA(20),SA(20),DZ(20)
        EXTERNAL FCT,ERR
C********FOR AN EXTRAPOLATION OF ORDER JM,  JM+1 LNEXTRAPCLATED********
C          APPROXIMATIONS ARE REQUIRED. THREE MORE ARE ALLCWED IN
C          ATTEMPTING TO ACHIEVE CONVERGENCE
        JMAX=JM+4
C********SAVE THE INITIAL VALUES FOR THE DEPENDENT VARIABLES AND*******
C          THE ERROR TEST VECTOR FOR THE STEP
        CC 100 I=1,N
        YA(I)=Y(I)
        SA(I)=S(I)
 100    CONTINUE
C********USE THE FUNCTION ROUTINE TO OBTAIN THE INITIAL SLCPES*********
C                   DZ = DY/DX
        CALL FCT(Y,X,CZ)
C******** PERFORM AN INTEGRATION STEP *******************************
        CALL DERESB(N,X,Y,DY,H,HMIN,EPS,JM,S,R,KONVF,FCT,ERR)
        RETURN
        END

        SUBROUTINE RECIF(N,X,Y,DY,H,HMIN,EPS,JM,S,R,KONVF,FCT,ERR)
```

FIG. 5. (cont.)

```
      C
      C******** PERFORM INITIALIZATION TO CALL DERESB ********************
      C
            CCMMON /DERECM/ JMAX,YA,SA,DZ
            DIMENSION Y(N),YA(20),SA(20),DZ(20)
            EXTERNAL FCT,ERR
            DC 300 I=1,N
              Y(I)=YA(I)
      300   CONTINUE
      C******** PERFORM AN INTEGRATION STEP ***************************
            CALL DERESB(N,X,Y,DY,H,HMIN,EPS,JM,S,R,KONVF,FCT,ERR)
            RETURN
            END

            SUBROUTINE DERESB(N,X,Y,DY,H,HMIN,EPS,JM,S,R,KONVF,FCT,ERR)
      C
      C          IMPLEMENTS THE ALGORITHM AND PERFORMS ONE INTEGRATION STEP
      C
            CCMMON /DERECM/ JMAX,YA,SA,DZ
            DIMENSION Y(N),CY(N),S(N),R(N),YA(20),YL(20),YM(20),DZ(20),SA(20),
           1          C(7),CT(20,8),YG(20,8),YH(20,8),SG(20,8)
            CCMMON /PARAM/ ZOT,P2,EMAX,EMIN,HDIV,ZOTLP,ICHOP
            LCGICAL KCNVF,KONV,BO,BH
      C*********THE LOGICAL VARIABLE,BH, DETERMINES WHETHER THE**********
      C          STEPSIZE HAS BEEN HALVED. INITIALLY FALSE.
      C          LATER BH IS FALSE IF THE STEPSIZE IS CUT BY A FACTCR NOT 2
      10    BH=.FALSE.
      C*********PRESET THE CONVERGENCE SUCCESS FLAG TRUE*************
            KCNVF=.TRUE.
      C*********ADVANCE THE INDEPENDENT VARIABLE BY THE STEPSIZE,H*********
      20    A=H+X
      C*********SET THE SWITCH BO FOR THE FIRST SET OF CCEFFICIENTS,C*********
            BC=.FALSE.
      C******** INITIALIZE THE H SEQUENCE-- H/M,H/JR,H/JS ****************
            M=1
            JR=2
            JS=3
      C*********JJ IS THE INDEX FOR THE ARRAY OF VALUES SAVED IN*********
      C          CASE THE INTERVAL MUST BE HALVED
            JJ=0
      C****************************************************************
      C                     INTEGRATION STEP
      C                 MIDPOINT + EXTRAPOLATION
      C****************************************************************
            DC 200 J=1,JMAX
      C******** SET THE VALUES OF THE EXTRAPOLATION COEFFICIENTS TC **********
      C          THEIR CORRECT VALUES FOR THIS EXTRAPOLATION STEP
            IF (.NOT.BO) GOTO 201
            C(2)=16./9.
            C(4)=64./9.
            C(6)=256./9.
            GOTO 202
      201   C(2)=9./4.
            C(4)=9.
            C(6)=36.
      C*********IF THE ORDER OF THE EXTRAPOLATION STEP BEING CCMPUTED IS******
      C          LESS THAN JM/2 , SET KONV FALSE
      202   KCNV=.TRUE.
            IF (J.LE.(JM/2)) KONV=.FALSE.
            IF (J.LE.(JM+1)) GOTO 203
      C*********RESTRICT THE ORDER OF THE EXTRAPOLATION TO JM*************
      C          ADJUST THE EXTRAPOLATION COEFFICIENT
            L=JM+1
            C(L)=4.*C(L-2)
      C*********DISCOURAGE THE STEP-INCREASING FACTOR,FC, BY A FACTCR CF******
      C          SQRT(2) SINCE CONVERGENCE WAS NCT OBTAINED IN JM
      C          EXTRAPOLATIONS
            FC=.7071068*FC
            GOTO 204
      C*********THE NUMBER, J, OF EXTRAPOLATIONS HAS NOT EXCEEDED JM**********
      C          FIND C(J) = (H DIVIDED BY H/M)**2
      C          ADJUST THE FACTOR,FC, USED TO ADJUST THE STEPSIZE FCR
      C          THE NEXT STEP TO BE TAKEN
```

FIG. 5. (cont.)

```
203    L=J
       C(L)=FLOAT(M*M)
       FC=1.0+FLOAT(JM+1-J)/6.0
C*******************************************************************
C          MODIFIED MIDPOINT RULE USED TO FIND FIRST
C          VALUE FOR THIS EXTRAPOLATION STEP
C*******************************************************************
204    M=M+M
       G=H/FLOAT(M)
       B=C+G
C*********IF THE STEPSIZE HAS NOT BEEN HALVED OR IF THE ORDER OF THE****
C          EXTRAPOLATION STEP EXCEEDS THAT FOR WHICH PREVIOUSLY
C          COMPUTED VALUES WERE SAVED, THEY MUST BE COMPUTED
       IF ((.NOT.BH).OR.(J.GE.(JMAX-1))) GOTO 205
C*********OTHERWISE THE VALUES HAVE BEEN SAVED AND CAN BE RESTORED******
       DO 210 I=1,N
       YM(I)=YH(I,J)
       YL(I)=YG(I,J)
       S(I)=SG(I,J)
210    CONTINUE
       GOTO 206
C*********COMPUTE STARTING VALUES FOR THE MODIFIED MIDPOINT RULE********
205    DO 220 I=1,N
       YL(I)=YA(I)
       YM(I)=YA(I)+G*DZ(I)
       S(I)=SA(I)
220    CONTINUE
       KH=M/2
       XU=X
C*********THE MEMBER OF THE H SEQUENCE BEING USED BY THE MIDPOINT*******
C          INTEGRATION RULE IS H/M.  COMPUTE TO THE END OF THE STEP
C          CONTINUOUSLY UPDATING THE VECTOR,S,CONTAINING THE LARGEST-
C          VALUE-SO-FAR FOR EACH DEPENDENT VARIABLE
       DO 230 K=2,M
       XU=XU+G
       CALL FCT (YM,XU,DY)
       DO 231 I=1,N
       U=YL(I)+B*DY(I)
       YL(I)=YM(I)
       YM(I)=U
       U=ABS(U)
       IF (U.GT.S(I)) S(I)=U
231    CONTINUE
C*********IN CASE THE INTERVAL MUST BE HALVED NEXT TIME, SAVE THE*******
C          VALUES AT HALFWAY ALONG (KH=M/2) THE STEP UNLESS K=3
       IF ((K.NE.KH).OR.(K.EQ.3)) GOTO 230
       JJ=1+JJ
       DO 232 I=1,N
       YH(I,JJ)=YM(I)
       YG(I,JJ)=YL(I)
       SG(I,JJ)=S(I)
232    CONTINUE
230    CONTINUE
206    CALL FCT (YM,A,DY)
       DO 240 I=1,N
C*********V IS USED TO SAVE THE VALUE OBTAINED BY THE MIDPOINT RULE*****
C          USING THE PREVIOUS MEMBER OF THE H SEQUENCE
C          (THE FIRST TIME THROUGH THIS VALUE IS MEANINGLESS, BUT IT
C          IS NOT USED SINCE L IS LESS THAN 2)
       V=DT(I,1)
C*********COMPUTE THE FINAL VALUE OBTAINED FOR THIS MEMBER OF THE*******
C          H SEQUENCE BY THE MODIFIED MIDPOINT RULE
       DT(I,1)=(YM(I)+YL(I)+G*DY(I))*.5
       C=DT(I,1)
       TA=C
C*********AT LEAST TWO VALUES ARE NEEDED TO START EXTRAPOLATION********
       IF (L.LT.2) GOTO 242
C*********IF THE VALUE JUST COMPUTED BY THE MIDPOINT RULE SHOWS A*******
C          LARGE JUMP FROM THE PREVIOUS, HALVE THE INTERVAL
       IF((ABS(V)*ZOTUP.LT.ABS(C)).AND.(H.NE.HMIN).AND.(J.GT.JM/2+1))
     1                    GO TO 30
C*********PERFORM THE L STEPS FOR THE CURRENT LTH ORDER ***************
```

FIG. 5. (cont.)

498 P. A. FOX

```
C          EXTRAPOLATION STEP.  IF THE DENOMINATOR CF THE RATICNAL
C          FUNCTION GOES TO ZERO AT ANY STEP, SET DT AT THAT STEP
C          TO ITS VALUE JUST BEFORE
           DC 241 K=2,L
           B1=D(K)*V
           B=B1-C
           U=V
           IF (B.EQ.0.) GOTO 243

           P=(C-V)/B
           U=C*P
           C=B1*P
  243      V=DT(I,K)
           DT(I,K)=U
           TA=U+TA
  241      CCNTINUE
C**********USE THE ERROR ROUTINE FOR EACH DEPENDENT VARIABLE TC CHECK****
C          WHETHER CONVERGENCE HAS BEEN ACHIEVED
  242      CALL ERR(TA,Y(I),S(I),R(I),EPS,KCNV)
  240      CCNTINUE
           IF (KCNV) COTO 40
C**********RESET THE EXTRAPOLATION CCEFFICIENTS**************************
           C(3)=4.
           C(5)=16.
C**********FLIP THE BO SWITCH FOR THE NEXT SET CF CCEFFICIENTS***********
           BO=(.NOT.BO)
C**********TAKE THE NEXT MEMBER OF THE H SEQUENCE***********************
           M=JR
           JR=JS
           JS=M+M
C**********AND CC BACK FOR THE NEXT EXTRAPOLATICN**********************
  200      CCNTINUE
C**********IF, AFTER ALL THE EXTRAPOLATIONS ALLCWED, CCNVERGENCE HAS*****
C          NOT BEEN ACHIEVED, ATTEMPT TO HALVE H SO THAT THE SAVED
C          VALUES CAN BE USED (SET BH TRUE FOR THIS PURPOSE)
C          IF HALVING H MAKES IT LESS THAN HMIN, SET H=HMIN
C          IN THIS CASE THE SAVED VALUES CANNOT BE USED
C          IF H HAD ALREADY BEEN AT HMIN, CONVERGENCE CANNCT BE
C          ACHIEVED FOR THIS HMIN AND THIS EPS CRITERION. SET KCNV FALSE
           BH=(.NOT.BH)
  30       IF (ABS(H).LE.HMIN) GOTO 50
           H=H/2.
           IF (ABS(H).CE.HMIN) COTO 20
           H=SIGN(HMIN,H)
           GCTO 10
  50       KCNVF=.FALSE.
C**********WHETHER OR NOT CONVERGENCE HAS BEEN ACHIEVED***************
C          SET A NEW SUGGESTED STEPSIZE FOR THE NEXT STEP
C          ASSIGN THE END OF STEP VALUE TO THE INDEPENDENT VARIABLE
  40       H=FC*H
           X=A
           RETURN
           END

           SUBROUTINE DEOUT(Y,DY,N,X,STYPE)
C
C          OUTPUT ROUTINE FOR RECORDING RESULTS
C
           DIMENSION Y(N),DY(N)
           LOGICAL STYPE,TITLE
           COMMON /OTPUT/ SP,H,XI,XF,EPS,NP,TITLE
C*********IF TITLE =.TRUE. WRITE HEADINGS AND SET TITLE = .FALSE**********
           IF (.NOT.TITLE) GOTO 10
           TITLE=.FALSE.
           WRITE (6,89) N,XI,XF,EPS,H
           IF (NP.EC.1) WRITE (6,88) SP
           WRITE (6,87)
C********DETERMINE WHETHER OUTPUT POINT IS SPECIFIED POINT**************
C          OR NATURAL STEP AND OUTPUT IT
  10       IF ((NP.EC.1).AND.STYPE) GOTO 20

           WRITE (6,86) X, (Y(I),I=1,N)
           RETURN
  20       WRITE (6,85) X,(Y(I),I=1,N)
```

Fɪɢ. 5. (cont.)

```
        RETURN
89    FCRMAT (1H1,10X,19HDE SOLUTION FOR N =,I2,24H EQUATICNS FRCM XSTAR
     1T =,F12.5,10H TO XEND =,E12.5/8X,31HWITH LOCAL ERRCR TCLERANCE EP
     2=,1PE12.5,26H AND INITIAL STEP SIZE H =,CPE12.5, 1H./8X,44HPRINTIN
     3G OCCURS AT EACH NATURAL STEP IN TIME)
88    FCRMAT (1H+,52X,37HAND AT SPECIFIED POINTS (XSTART+K*SP)/8X,22HFCR
     1 K=0,1,... AND SP =,E12.5,42H (SPECIFIED POINTS ARE IDENTIFIED WIT
     2H *).)
87    FCRMAT (1H0,14X,47HTHE OUTPUT COLUMNS ARE   X, Y(1), Y(2),..., Y(N)
     1/)
86    FCRMAT (1H ,10X,6(E15.8,5X)/(31X,5(E15.8,5X)))
85    FCRMAT (1H ,4X,1H*,5X,6(E15.8,5X)/(31X,5(E15.8,5X)))
        END


        SUBROUTINE ERROR
C
C           WRITES AN ERROR DIAGNOSTIC IN THE CASE OF ERROR FAILURE
C
        COMMON /INFO/EX,ER,EH,NE,NERR
        GCTO (10,20,30,40,50,60),NERR
10    WRITE (6,91) EX,EH,ER,NE
91    FCRMAT (5H0****,5X,35HNO CONVERGENCE IN ABOVE STEP TC X =,E12.5, 9
     1H WITH H =,E12.5, 1H,,5X, 4H****,/1CX,22HTHE LIMITING ERRCR IS ,
     2E12.5,13H IN EQUATION ,I2//)
        RETURN
20    WRITE (6,92)
92    FCRMAT (5H0****,5X,19HN.LT.C .OR. N.GT.2C,5X,4H****)
        RETURN
30    WRITE (6,93)
93    FCRMAT (5H0****,5X,29HEP.LT.1.E-C7 .OR. EP.GT.1.E-2,5X,4H****)
        RETURN
40    WRITE (6,94)
94    FCRMAT (5H0****,5X,22HH*(XEND-XSTART) .LT. C,5X,4H****)
        RETURN
50    WRITE (6,95)
95    FCRMAT (5H0****,5X,21HH=O. .OR. XEND=XSTART,5X,4H****)
        RETURN
60    WRITE (6,96)
96    FCRMAT (5H0****,5X,48HH.LT.(XEND-XSTART)/2**15 .CR. H.GT.(XEND-XST
     1ART),5X,4H****)
        RETURN
        END


        SUBROUTINE SE(TA,Y,S,R,EPS,KONV)
C
C           CHECKS FOR CONVERGENCE LSING STANDARD ERROR TEST
C
        U=ABS(TA)
        IF(U.GT.S) S=U
        CALL ERR(TA,Y,S,R,EPS,KONV)
        RETURN
        END


        SUBROUTINE RE(TA,Y,S,R,EPS,KONV)
C
C           CHECKS FOR CONVERGENCE LSING RELATIVE ERROR TEST
C
        S=ABS(Y)
        CALL ERR(TA,Y,S,R,EPS,KONV)
        RETURN
        END


        SUBROUTINE AE(TA,Y,S,R,EPS,KONV)
C
C           CHECKS FOR CONVERGENCE LSING ABSOLUTE ERROR TEST
C
        S=1.0
        CALL ERR(TA,Y,S,R,EPS,KONV)
        RETURN
        END


        SUBROUTINE ERR(TA,Y,S,R,EPS,KONV)
C
```

FIG. 5. (cont.)

```
C              PERFORMS CONVERGENCE TEST
C
       LOGICAL KONV
       R=ABS(Y-TA)
       Y=TA
       IF(S.LT.EPS) S=EPS
       IF(R.GT.EPS*S) KONV=.FALSE.
       RETURN
       END
       SUBROUTINE DDE(XFX,N,Y,XI,XF,HI,EPS,DERR,XOUTX)
C
C   DOUBLE PRECISION VERSION OF DE
C
C   USING DDE CAUSES VALUES OF DEPENDENT VARIABLES TO BE PRINTED AT
C   POINTS AUTOMATICALLY SELECTED BY THE PROGRAM.
C
       EXTERNAL XFX,DERR,XOUTX
       DOUBLE PRECISION SP
       COMMON /DDESPC/ NP, KOUNT
       NP = 0
       SP = 0.0D0
       CALL XDDE(SP,XFX,N,Y,XI,XF,HI,EPS,DERR,XOUTX)
       RETURN
       END

       SUBROUTINE DDESP(SP,XFX,N,Y,XI,XF,HI,EPS,DERR,XOUTX)
C
C   DOUBLE PRECISION VERSION OF DESP
C
C   USING DDESP CAUSES VALUES OF DEPENDENT VARIABLES TO BE PRINTED AT
C   SPECIFIED POINTS IN ADDITION TO THOSE AUTOMATICALLY SELECTED.
C
       EXTERNAL XFX,DERR,XOUTX
       DOUBLE PRECISION SP,XF,XI
       COMMON /DDESPC/ NP, KOUNT
       NP = 1
       KOUNT = 0
       IF(SP*(XF-XI)) 2,4,10
2      SP = DSIGN(SP,XF-XI)
       GO TO 10
4      IF(SP .NE. 0.0D0) GO TO 10
       NP = 0
10     CALL XDDE(SP,XFX,N,Y,XI,XF,HI,EPS,ERR,XOUTX)
       RETURN
       END

       SUBROUTINE XDDE(SP,XFX,N,Y,XI,XF,HI,EPS,DERR,XOUTX)
C
C          DOES DOUBLE PRECISION HOUSEKEEPING DURING INTEGRATION,
C          CHECKS INPUT PARAMETERS, CHECKS FOR SPECIFIED OUTPUT
C          POINTS, STEPS H, ETC.
C
       DIMENSION Y(N),DY(20),S(20),R(20),YR(20)
       DOUBLE PRECISION Y,DY,YR,XI,XF,XIPRT,XFPRT,TTL,X,XP,XPMX,XR,XT,S,
      1R,DZOT,DP2,DEMIN,DEMAX,EX,ER,EH,DHDIV,DZOTLP,HI,HIPRT,HMIN,HMAX,HP
      2,HC,HR,HF,FF,EPS,EPSPRT,SP,H,SPPRT
       COMMON /DDESPC/ NP, KOUNT
       COMMON /IPARAM/N,NMAX
       COMMON /DPARAM/DZOT,DP2,DEMAX,DEMIN,DHDIV,DZOTLP
       COMMON /DINFO/ FX,ER,EH,NF,NERR
       COMMON /DOTPUT/ SPPRT,HIPRT,XIPRT,XFPRT,EPSPRT,NPPRT,TITLE
       LOGICAL STYPE,KONVF,TITLE
       EXTERNAL XFX,DERR
C******** INITIALIZE CONSTANTS *********************************************
       DZOT = 2.77D-17
       DP2 = 32768.0D0
       DEMAX = 1.0D-02
       DEMIN = 1.0D-18
       DHDIV = 1024.0D0
```

FIG. 5. (cont.)

```
      DZOTUP = 3.6D16
      M = 6
      NMAX = 20
C******** CHECK PARAMETERS AND PERFORM INITIALIZATION           ********
      TITLE=.TRUE.
      STYPE=.TRUE.
      NPPRT=NP
      SPPRT=SP
      HIPRT=HI
      XIPRT=XI
      XFPRT=XF
      EPSPRT=EPS
      IF ((N.LE.0).OR.(N.GT.NMAX)) GOTO 84
      IF ((EPS.LT.DEMIN).OR.(EPS.GT.DEMAX)) GOTO 85
      TTL=XF-XI
      H=HI
      IF (TTL*H) 86,87,12
   12 IF (((H/TTL)* DP2  .LT.1.).OR.((H/TTL).GT.1.)) GOTO 88
      DO 14 I=1,N
      S(I)=DABS(Y(I))
   14 CONTINUE
      KCNVF=.TRUE.
      HMIN=H/DHDIV
      HMAX=TTL
      HP=0.
      XP=XI
      X=XI
C******** BEGIN SOLUTION OF DIFFERENTIAL EQUATIONS               ********
C******** BEGIN SOLUTION OF DIFFERENTIAL EQUATIONS               ********
   20 H=H/3.
      H=3.*H
      IF ((NP.EQ.0).AND.(.NOT.STYPE)) GOTO 50
      XPMX=XP-X
      FH=XPMX/H
      IF (FH.GT.DZOT) GOTO 50
C******** GET TO SPECIFIED VALUE OF X AND CALL XOUTX             ********
   30 IF (DABS(FH).GT.DZOT) GOTO 34
      DO 32 I=1,N
      YR(I)=Y(I)
   32 CONTINUE
      HQ=HP
      XR=X
      GOTO 36
   34 HQ=XPMX+HP
      HR=HQ
      XR=XT
      CALL DRECIF (N,XR,YR,DY,HR,HQ,EPS,M,S,R,KCNVF,XFX,DERR)
      HQ=XR-XT
   36 CALL XFX (YR,XR,DY)
      STYPE=.TRUE.
      CALL XOUTX(YR,DY,N,XR,STYPE)
      STYPE=.FALSE.
      IF (KONVF) GOTO 40
   38 IF (KONVF) GOTO 70
      GOTO 82
C******** STEP TO NEXT SPECIFIED OUTPUT VALUE                    ********
   40 IF ((XF-XR)/TTL.LE.0.DC) GOTO 70
      KCUNT = KOUNT+1
      XP = XI+FLOAT(KCUNT)*SP
      IF ((XP-XF)/H.GT.0.DC) XP=XF
      GOTO 20
C******** OUTPUT RESULTS AT A NATURAL STEP                       ********
   50 IF ((DABS((X-XR)/H)).LE.DZOT) GOTO 60
      CALL XFX (Y,X,DY)
      CALL XOUTX(Y,DY,N,X,STYPE)
C******** PERFORM A STEP IN SOLUTION OF DIFFERENTIAL EQUATIONS   ********
   60 IF ((XF-X)/TTL.LE.0.DC) GOTO 70
      IF (DABS(H).LT.HMIN) H=DSIGN(HMIN,H)
      IF (DABS(H).GT.HMAX) H=DSIGN(HMAX,H)
      IF ((XF-X-H)/TTL.LT.0.DC) H=XF-X
      XT=X
      CALL DCESUB (N,X,Y,DY,H,HMIN,EPS,M,S,R,KCNVF,XFX,DERR)
      HP=X-XT
```

FIG. 5. (cont.)

```
      IF (KONVF) GOTO 20
      GCTC 80
C******** STANDARD RETURN ON COMPLETICN OF SOLLTION          ********
 70   RETURN
C******** SET UP ERROR RETURN AND DIAGNOSTICS                ********
 80   NERR=1
      ER=0.DO
      DC 81 I=1,N
      IF (ER*S(I).GE.R(I)) GOTO 81
      ER=R(I)/S(I)
      NE=I
 81   CCNTINUE
      EF=FP
      EX=X
      CALL XFX (Y,X,CY)
      CALL XOUTX(Y,CY,N,X,STYPE)
      GCTC 92
 82   NERR=1
      ER=0.DO
      DC 83 I=1,N
      IF (ER*S(I).GE.R(I)) GOTO 83
      ER=R(I)/S(I)
      NF=I
 83   CCNTINUE
      EF=HQ
      EX=XR
      GCTC 92
 84   NERR=2
      GCTC 90
 85   NERR=3
      GCTC 90
 86   NERR=4
      GCTC 90
 87   NERR=5
      GCTG 90
 88   NERR=6
 90   CALL XFX (Y,XI,CY)
      CALL XOUTX(Y,CY,N,XI,STYPE)
 92   CALL DERROR
      RETURN
      END

      SUBROUTINE CDESUB(N,X,Y,DY,H,HMIN,EPS,JM,S,R,KCNVF,FCT,DERR)
C
C******** PERFORM INITIALIZATION TO CALL CDERSB ************************
C
      COMMON /CDERCM/ JMAX,YA,SA,DZ
      DIMENSION Y(N),S(N),YA(20),SA(2C),DZ(2C)
      CCUBLE PRECISION Y,S,YA,SA,DZ
      EXTERNAL FCT,CERR
C***********************************************************************
C*********FOR AN EXTRAPOLATION OF ORDER JM,  JM+1 LNEXTRAPCLATED********
C          APPROXIMATIONS ARE REQUIRED. THREE MORE ARE ALLCWED IN
C          ATTEMPTING TO ACHIEVE CONVERGENCE
      JMAX = JM + 4
C*********SAVE THE INITIAL VALUES FOR THE DEPENDENT VARIABLES AND*******
C          THE ERROR TEST VECTOR FOR THE STEP
      DC 100 I = 1,N
      YA(I) = Y(I)
      SA(I) = S(I)
100   CCNTINUE
C*********USE THE FUNCTION ROUTINE TO OBTAIN THE INITIAL SLCPES*********
C          DZ = DY/DX
      CALL FCT(Y,X,CZ)
C******** PERFORM AN INTEGRATION STEP *********************************
      CALL DCERSB(N,X,Y,DY,H,HMIN,EPS,JM,S,R,KONVF,FCT,DERR)
      RETURN
      END

      SUBROUTINE CREDIF(N,X,Y,DY,H,HMIN,EPS,JM,S,R,KCNVF,FCT,CERR)
C
C******** PERFORM INITIALIZATION TO CALL CDERSB ************************
C
```

FIG. 5. (cont.)

```
       COMMON /CDERCM/ JMAX,YA,SA,DZ
       DIMENSION Y(N),YA(20),SA(20),DZ(20)
       DOUBLE PRECISION Y,YA,SA,DZ
       EXTERNAL FCT,DERR
       DO 300 I = 1, N
       Y(I) = YA(I)
300    CONTINUE
C******** PERFORM AN INTEGRATION STEP ********************************
       CALL DDERSB(N,X,Y,DY,H,HMIN,EPS,JM,S,R,KONVF,FCT,DERR)
       RETURN
       END

       SUBROUTINE DDERSB(N,X,Y,DY,H,HMIN,EPS,JM,S,R,KONVF,FCT,DERR)
C
C             IMPLEMENTS THE ALGORITHM IN DOUBLE PRECISION
C             AND PERFORMS ONE INTEGRATION STEP
C
       COMMON /CDERCM/ JMAX,YA,SA,DZ
       COMMON /CPARAM/DZOT,DP2,DEMAX,DEMIN,DHDIV,DZCTLP
       DIMENSION Y(N),DY(N),S(N),R(N),YA(20),YL(20),YM(20),DZ(20),
      1SA(20),C(7),DT(20,7),YG(20,8),YH(20,8),SG(20,8)
       DOUBLE PRECISION Y,DY,S,R,YA,YL,YM,DZ,SA,D,DT,YG,YH,SG,X,H,HMIN,A,
      1EPS,FC,G,B,B1,XU,U,V,C,TA,DZOT,DP2,DEMIN,DEMAX,DHDIV,DZCTLP
       LOGICAL KONVF,KONV,BO,BH
C*********THE LOGICAL VARIABLE,BH, DETERMINES WHETHER THE***************
C          STEPSIZE HAS BEEN HALVED. INITIALLY FALSE.
C          LATER BH IS FALSE IF THE STEPSIZE IS CUT BY A FACTOR NOT 2
10     BH=.FALSE.
C*********PRESET THE CONVERGENCE SUCCESS FLAG TRUE*********************
       KONVF=.TRUE.
C*********ADVANCE THE INDEPENDENT VARIABLE BY THE STEPSIZE,H***********
20     A=H+X
C*********SET THE SWITCH BO FOR THE FIRST SET OF COEFFICIENTS,C********
       BO=.FALSE.
C*********INITIALIZE THE H SEQUENCE C  H/M,H/JR,H/JS *****************
       M=1
       JR=2
       JS=3
C*********JJ IS THE INDEX FOR THE ARRAY OF VALUES SAVED IN************
C          CASE THE INTERVAL MUST BE HALVED
       JJ=0
C*******************************************************************
C                         INTEGRATION STEP
C                   MIDPOINT + EXTRAPOLATION
C*******************************************************************
       DO 200 J=1,JMAX
C*********SET THE VALUES OF THE EXTRAPOLATION COEFFICIENTS TO**********
C          THEIR CORRECT VALUES FOR THIS EXTRAPOLATION STEP
       IF (.NOT.BO) GOTO 201
       C(2)=16.D0/9.D0
       C(4)=64.D0/9.D0
       C(6)=256.D0/9.D0
       GOTO 202
201    C(2)=9.D0/4.D0
       C(4)=9.D0
       C(6)=36.D0
C*********IF THE ORDER OF THE EXTRAPOLATION STEP BEING COMPUTED IS******
C          LESS THAN JM/2 , SET KONV FALSE
202    KONV=.TRUE.
       IF (J.LE.(JM/2)) KONV=.FALSE.
       IF (J.LE.(JM+1)) GOTO 203
C*********RESTRICT THE ORDER OF THE EXTRAPOLATION TO JM****************
C          ADJUST THE EXTRAPOLATION COEFFICIENT
       L=JM+1
       C(L)=4.D0*C(L-2)
C*********DISCOURAGE THE STEP-INCREASING FACTOR,FC, BY A FACTOR OF******
C          SQRT(2) SINCE CONVERGENCE WAS NOT OBTAINED IN JM
C          EXTRAPOLATIONS
       FC=.70710680D0*FC
       GOTO 204
C*********THE NUMBER, J, OF EXTRAPOLATIONS HAS NOT EXCEEDED JM*********
C          FIND C(J) = (H DIVIDED BY H/M)**2
C          ADJUST THE FACTOR,FC, USED TO ADJUST THE STEPSIZE FOR
```

FIG. 5. (cont.)

504   P. A. FOX

```
C            THE NEXT STEP TO BE TAKEN
  203   L=J
        D(L)=FLOAT(M*M)
        FC=1.D0+FLOAT(JM+1-J)/6.D0
C***********************************************************************
C            MODIFIED MIDPOINT RULE USED TO FIND FIRST
C            VALUE FOR THIS EXTRAPOLATION STEP
C***********************************************************************
  204   M=M+M
        G=H/FLOAT(M)
        B=G+G
C*********IF THE STEPSIZE HAS NOT BEEN HALVED OR IF THE ORDER OF THE****
C            EXTRAPOLATION STEP EXCEEDS THAT FOR WHICH PREVIOUSLY
C            COMPUTED VALUES WERE SAVED, THEY MUST BE COMPUTED
        IF ((.NOT.BH).OR.(J.GE.(JMAX-1))) GOTO 205
C*********OTHERWISE THE VALUES HAVE BEEN SAVED AND CAN BE RESTORED******
        DO 210 I=1,N
        YM(I)=YH(I,J)
        YL(I)=YG(I,J)
        S(I)=SG(I,J)
  210   CONTINUE
        GOTO 206
C*********COMPUTE STARTING VALUES FOR THE MODIFIED MIDPOINT RULE********
  205   DO 220 I=1,N
        YL(I)=YA(I)
        YM(I)=YA(I)+G*DZ(I)
        S(I)=SA(I)
  220   CONTINUE
        KH=M/2
        XU=X
C*********THE MEMBER OF THE H SEQUENCE BEING USED BY THE MIDPOINT*******
C            INTEGRATION RULE IS H/M. COMPUTE TO THE END OF THE STEP
C            CONTINUOUSLY UPDATING THE VECTOR,S,CONTAINING THE LARGEST-
C            VALUE-SO-FAR FOR EACH DEPENDENT VARIABLE
        DO 230 K=2,M
        XU=XU+G
        CALL FCT (YM,XU,DY)
        DO 231 I=1,N
        U=YL(I)+B*DY(I)
        YL(I)=YM(I)
        YM(I)=U
        U=DABS(U)
        IF (U.GT.S(I)) S(I)=U
  231   CONTINUE
C*********IN CASE THE INTERVAL MUST BE HALVED NEXT TIME. SAVE THE*******
C            VALUES AT HALFWAY ALONG (KH=M/2) THE STEP UNLESS K=3
        IF ((K.NE.KH).OR.(K.EQ.3)) GOTO 230
        JJ=1+JJ
        DO 232 I=1,N
        YH(I,JJ)=YM(I)
        YG(I,JJ)=YL(I)
        SG(I,JJ)=S(I)
  232   CONTINUE
  230   CONTINUE
  206   CALL FCT (YM,A,DY)
        DO 240 I=1,N
C*********V IS USED TO SAVE THE VALUE OBTAINED BY THE MIDPOINT RULE*****
C            USING THE PREVIOUS MEMBER OF THE H SEQUENCE
C            (THE FIRST TIME THROUGH THIS VALUE IS MEANINGLESS, BUT IT
C            IS NOT USED SINCE L IS LESS THAN 2)
        V=DT(I,1)
C*********COMPUTE THE FINAL VALUE OBTAINED FOR THIS MEMBER OF THE*******
C            H SEQUENCE BY THE MODIFIED MIDPOINT RULE
        DT(I,1)=(YM(I)+YL(I)+G*DY(I))*.5
        C=DT(I,1)
        TA=C
C*********AT LEAST TWO VALUES ARE NEEDED TO START EXTRAPOLATION********
        IF (L.LT.2) GOTO 242
C*********IF THE VALUE JUST COMPUTED BY THE MIDPOINT RULE SHOWS A*******
C            LARGE JUMP FROM THE PREVIOUS, HALVE THE INTERVAL
        IF((DABS(V)*DZOTUP.LT.DABS(C)).AND.(H.NE.HMIN).AND.(J.GT.JM/2+1))
       1                            GO TO 3C
```

FIG. 5. (cont.)

```
C**********PERFORM THE L STEPS FOR THE CURRENT LTH ORDE *****************
C          EXTRAPOLATION STEP.  IF THE DENOMINATOR OF THE RATIONAL
C          FUNCTION GOES TO ZERO AT ANY STEP, SET DT AT THAT STEP
C          TO ITS VALUE JUST BEFORE
           DO 241 K=2,L
           B1=C(K)*V
           B=B1-C
           U=V
           IF (B.EQ.0.DC) GOTO 243
           B=(C-V)/B
           U=C*B
           C=B1*B
  243      V=DT(I,K)
           DT(I,K)=U
           TA=U+TA
  241      CONTINUE
C**********USE THE ERROR ROUTINE FOR EACH DEPENDENT VARIABLE TO CHECK****
C          WHETHER CONVERGENCE HAS BEEN ACHIEVED
  242      CALL DERR(TA,Y(I),S(I),R(I),EPS,KONV)
  240      CONTINUE
           IF (KONV) GOTO 40
C**********RESET THE EXTRAPOLATION COEFFICIENTS*************************
           C(3)=4.DO
           C(5)=16.DO
C**********FLIP THE BO SWITCH FOR THE NEXT SET OF COEFFICIENTS**********
           BO=(.NOT.BO)
C**********TAKE THE NEXT MEMBER OF THE H SEQUENCE*********************
           M=JR
           JR=JS
           JS=M+M
C**********AND GO BACK FOR THE NEXT EXTRAPOLATION********************
  200      CONTINUE
C**********IF, AFTER ALL THE EXTRAPOLATIONS ALLOWED, CONVERGENCE HAS*****
C          NOT BEEN ACHIEVED, ATTEMPT TO HALVE H SO THAT THE SAVED
C          VALUES CAN BE USED (SET BH TRUE FOR THIS PURPOSE)
C          IF HALVING H MAKES IT LESS THAN HMIN, SET H=HMIN
C          IN THIS CASE THE SAVED VALUES CANNOT BE USED
C          IF H HAD ALREADY BEEN AT HMIN, CONVERGENCE CANNOT BE
C          ACHIEVED FOR THIS HMIN AND THIS EPS CRITERION. SET KONV FALSE
           BH=(.NOT.BH)
  30       IF (DABS(H).LE.HMIN) GOTO 50
           H=H/2.DO
           IF (DABS(H).GE.HMIN) GOTO 20
           H=DSIGN(HMIN,H)
           GOTO 10
  50       KONVF=.FALSE.
C**********WHETHER OR NOT CONVERGENCE HAS BEEN ACHIEVED*****************
C          SET A NEW SUGGESTED STEPSIZE FOR THE NEXT STEP
C          ASSIGN THE END OF STEP VALUE TO THE INDEPENDENT VARIABLE
  40       H=FC*H
           X=A
           RETURN
           END

           SUBROUTINE CDEOUT(Y,DY,N,X,STYPE)
C
C          OUTPUT ROUTINE FOR DOUBLE PRECISION RESULTS
C
           DIMENSION Y(N),DY(N)
           DOUBLE PRECISION Y,DY,X,SP,H,XI,XF,EPS
           LOGICAL STYPE,TITLE
           COMMON /DOTPUT/ SP,H,XI,XF,EPS,NP,TITLE
C********IF TITLE =.TRUE. WRITE HEADINGS AND SET TITLE = .FALSE**********
           IF (.NOT.TITLE) GOTO 10
           TITLE=.FALSE.
           WRITE (6,89) N,XI,XF,EPS,H
           IF (NP.EC.1) WRITE (6,88) SP
           WRITE (6,87)
C******DETERMINE WHETHER OUTPUT POINT IS SPECIFIED POINT***************
C          OR NATURAL STEP AND OUTPUT IT
  10       IF ((NP.EC.1).AND.STYPE) GOTO 20
           WRITE (6,86) X, (Y(I),I=1,N)
```

Fig. 5. (cont.)

```
           RETURN
    20     WRITE (6,85) X,(Y(I),I=1,N)
           RETURN
    89     FCRMAT (1H1,10X,19HDE SOLUTION FOR N =,I2,24H EQUATICNS FRCM XSTAR
          1T =,D12.5,10H TC XEND =,D12.5/8X,31HWITH LOCAL ERRCR TCLERANCE EP
          2=,1PD12.5,26H AND INITIAL STEP SIZE H =,OPD12.5, 1H./8X,44HPRINTIN
          3G OCCURS AT EACH NATURAL STEP IN TIME)
    88     FCRMAT (1H+,52X,37HAND AT SPECIFIED POINTS (XSTART+K*SP)/8X,22HFCR
          1 K=0,1,... AND SP =,D12.5,42H (SPECIFIED POINTS ARE IDENTIFIED WIT
          2H *).)
    87     FCRMAT (1H0,14X,47HTHE OUTPUT COLUMNS ARE   X, Y(1), Y(2),..., Y(N)
          1/)
    86     FCRMAT (1H ,10X,4(D25.16,5X)/(41X,3(D25.16,5X)))
    85     FCRMAT (1H ,4X,1H*,5X,4(D25.16,5X)/(41X,3(D25.16,5X)))
           END

           SUBROUTINE CERRCR
    C
    C          WRITES AN ERROR CIAGNOSTIC IN THE CASE OF ERRCR FAILURE
    C          IN COUBLE PRECISION
    C
           DCUBLE PRECISION EX,EH,ER
           CCMMON /INFO/EX,ER,EH,NE,NERR
           GCTC (10,20,30,40,50,60),NERR
    10     WRITE (6,91) EX,EH,ER,NE
    91     FCRMAT (5H0****,5X,35HNO CONVERGENCE IN ABOVE STEP TC X =,D12.5, 9
          1H WITH H =,D12.5, 1H,,5X,  4H****,/1CX,22HTHE LIMITING ERRCR IS ,
          2D12.5,13H IN EQUATION ,I2//)
           RETURN
    20     WRITE (6,92)
    92     FCRMAT (5H0****,5X,19HN.LT.C .OR. N.GT.2C,5X,4H****)
           RETURN
    30     WRITE (6,93)
    93     FCRMAT (5H0****,5X,29HEP.LT.1.D-18 .OR. EP.GT.1.D-2,5X,4H****)
           RETURN
    40     WRITE (6,94)
    94     FCRMAT (5H0****,5X,22HH*(XEND-XSTART) .LT. C,5X,4H****)
           RETURN
    50     WRITE (6,95)
    95     FCRMAT (5H0****,5X,21HH=0. .OR. XEND=XSTART,5X,4H****)
           RETURN
    60     WRITE (6,96)
    96     FCRMAT (5H0****,5X,48HH.LT.(XEND-XSTART)/2**15 .CR. H.GT.(XEND-XST
          1ART),5X,4H****)
           RETURN
           END

           SUBROUTINE CSE(TA,Y,S,R,EPS,KCNV)
    C
    C          CHECKS FOR DOUBLE PRECISION CCNVERGENCE USING
    C          STANDARD ERROR TEST
    C
           DCUBLE PRECISION TA,S,U
           U = DABS(TA)
           IF(U .GT. S) S = U
           CALL DERR(TA,Y,S,R,EPS,KONV)
           RETURN
           END

           SUBROUTINE CRE(TA,Y,S,R,EPS,KCNV)
    C
    C          CHECKS FOR COUBLE PRECISION CCNVERGENCE USING
    C          RELATIVE ERROR TEST
    C
           DCUBLE PRECISION Y,S
           S = DABS(Y)
           CALL DERR(TA,Y,S,R,EPS,KONV)
           RETURN
           END
```

FIG. 5. (cont.)

```
SUBROUTINE CAE(TA,Y,S,R,EPS,KONV)
C
C            CHECKS FOR DOUBLE PRECISION CONVERGENCE USING
C            ABSOLUTE ERROR TEST
C
      DOUBLE PRECISION S
      S = 1.000
      CALL CERR(TA,Y,S,R,EPS,KONV)
      RETURN
      END

      SUBROUTINE CERR(TA,Y,S,R,EPS,KONV)
C
C            PERFORMS DOUBLE PRECISION CONVERGENCE TEST
C
      DOUBLE PRECISION TA,Y,S,R,EPS
      LOGICAL KONV
      R = DABS(Y-TA)
      Y = TA
      IF(S .LT. EPS) S = EPS
      IF(R .GT. EPS*S) KONV = .FALSE.
      RETURN
      END
```

FIG. 5. (cont.)

REFERENCES

Bulirsch, R., and Stoer, J. (1964). Fehlerabschätzungen und Extrapolation mit rationalen Funktionen bei Verfahren vom Richardson-Typus. *Numer. Math.* **6,** 413–427.

Bulirsch, R., and Stoer, J. (1966). Numerical Treatment of Ordinary Differential Equations by Extrapolation Methods. *Numer. Math.* **8,** 1–13.

Clark, N. W. (1966). ANL D250-DIFSUB (3600 Fortran Program). Argonne National Laboratories.

Clark, N. W. (1966a). ANL D251-DIFSYS (3600 Fortran Program for use with ANL D250). Argonne National Laboratories.

Clark, N. W. (1968). A Study of Some Numerical Methods for the Integration of Systems of First-Order Ordinary Differential Equations. Rep. ANL-7428, Argonne National Laboratories, Argonne, Illinois.

Crane, P. C., and Fox, P. (1969). DESUB—Integration of a First Order System of Ordinary Differential Equations, Vol. 2, Issue 1. Numerical Mathematics Program Library Project, Computing Science Research Center, Bell Telephone Laboratories, Inc., Murray Hill, New Jersey.

Crane, P. C., and Fox, P. (1969a). A Comparative Study of Computer Programs for Integrating Differential Equations, Vol. 2, Issue 2. Numerical Mathematics Program Library Project, Computing Science Research Center, Bell Telephone Laboratories, Inc., Murray Hill, New Jersey.

Gragg, W. (1963). Repeated extrapolation to the limit in the numerical solution of ordinary differential equations. Thesis UCLA.

Stetter, H. J. (1965). Asymptotic expansions for the error of discretization algorithms for non-linear functional equations. *Numer. Math.* **7,** 18–31.

INDEX